NONLINEAR STOCHASTIC CONTROL SYSTEMS

NONLINEAR STOCHASTIC CONTROL SYSTEMS

Edited by

A. T. Fuller

Assistant Director of Research
Engineering Department
Cambridge University

TAYLOR AND FRANCIS LTD
LONDON
BARNES & NOBLE INC
NEW YORK

1970

First published 1970 *by Taylor & Francis Ltd., London and Barnes & Noble Inc., New York.*

Printed photo-offset and bound in Great Britain by Warren & Son Ltd. (a subsidiary of Taylor & Francis Ltd.), The Wykeham Press, Winchester, Hampshire.

Taylor & Francis ISBN 0 85066 034 3
Barnes & Noble SBN 389 03991 8

A. T. Fuller graduated at Cambridge University in 1945. He then spent nine years in the Civil Service and in Industry, working on research into control and feedback problems. In 1954 he returned to Cambridge University, joining Professor J. F. Coales' control engineering group. He obtained Ph.D. in 1960. At present he is Assistant Director of Research in the Engineering Department at Cambridge, and is a Fellow of Churchill College.

Dr. Fuller is the author of more than thirty papers relating to control problems, and has specialized in the study of nonlinear control systems. He is a member of the editorial board of the *International Journal of Control.*

Preface

Control theory has advanced apace in recent years. Yet there remain many unsolved problems, so much so that practicians often complain about the inapplicability of contemporary theories. Thus control systems are usually subject to several complicating factors, such as stochastic disturbances, noisy measurements, nonlinearities, distributed parameters, plant variability and interaction with other systems. To make progress, theoreticians have often ignored all except one of these factors, and so have ended up with unrealistic designs.

For example, well-developed theory is available for the optimal design of linear controllers for plants described by linear constant coefficient ordinary differential equations with stochastic disturbances. Again, theory is available for optimizing nonlinear controllers for low-order deterministic plants subject to control signal saturation. Either theory often seems over-idealized. What is wanted is a technique for dealing with systems which are subject to both stochastic disturbances and nonlinearities, especially control signal saturation.

The papers collected in the present book represent an attack on this problem. They have been selected from the *International Journal of Control* and its predecessor the *Journal of Electronics and Control*, and were originally published from 1960 onwards. Many excellent pioneering papers on nonlinear stochastic systems have appeared in these two journals, and the task of selection has not been easy. However, an attempt has been made to narrow the field and bring together those papers which form a coherent group.

Part I gives an account of two general tools for analysing nonlinear stochastic systems, namely the Fokker-Planck equation and the theory of functionals.

Part II treats the optimization of deterministic nonlinear (saturating) systems with quadratic performance indices. There are three reasons for treating the deterministic rather than the stochastic problem here. First the deterministic optimization theory gives a simple introduction to the more difficult stochastic optimization theory of Part III. Secondly, the deterministic optimal controllers approximate corresponding stochastic optimal controllers in regions of state space remote from the origin. Thirdly, the available books on deterministic optimal nonlinear theory give little attention to the case of quadratic performance indices—a case which is especially relevant when the transition to stochastic problems is considered.

Part III deals with the optimization of stochastic nonlinear (saturating) systems.

Part IV considers sub-optimal controllers for nonlinear deterministic and stochastic systems.

The papers are written in straightforward expository style. The level of rigour is such that the book should be assimilable by a broad spectrum of engineers and mathematicians.

A. T. FULLER

Engineering Department
Cambridge University
November 1968

Contents

Part I.

Analysis of nonlinear stochastic systems

INTRODUCTION TO PART I

The papers in Part I deal with methods for the analysis of given nonlinear stochastic systems. Paper 1 gives an introduction to the Fokker-Planck equation, derives a number of solutions for the steady-state probability density in state space, and relates these solutions to the Maxwell-Boltzmann distribution. Paper 2 discusses the theory of functionals. Paper 3 applies functionals to the calculation of the autocorrelation and spectrum of the output of nonlinear system excited by Gaussian white noise.

Paper 1

Reprinted from INTERNATIONAL JOURNAL OF CONTROL, Vol. 9, No. 6, pp. 603–655, June, 1969.

Paper 2

Reprinted from JOURNAL OF ELECTRONICS AND CONTROL, Vol. 15, No. 6, pp. 567–615, December, 1963.

Paper 3

Reprinted from JOURNAL OF ELECTRONICS AND CONTROL, Vol. 16, No. 1, pp. 107–113, January, 1964.

Analysis of nonlinear stochastic systems by means of the Fokker–Planck equation†

By A. T. Fuller
Engineering Department, Cambridge University

[Received August 15, 1968]

Abstract

Nonlinear systems disturbed by Gaussian white noises (or by signals obtained from Gaussian white noises) can sometimes be analysed by setting up and solving the Fokker–Planck equation for the probability density in state space. In the present paper a simplified derivation of the Fokker–Planck equation is given. The uniqueness of the steady-state solution is discussed. Steady-state solutions are obtained for certain classes of system. These solutions correspond to or slightly generalize the Maxwell-Boltzmann distribution which is well known in classical statistical mechanics.

Part I. Derivation of the Fokker–Planck equation

1. Introduction

In practice control systems and other dynamic systems involve non-linearities such as saturation and are also subject to disturbances which vary randomly with time. The mathematical analysis of such systems is difficult. However, progress has been made in the case when the stochastic disturbances are Gaussian white noises or signals obtained from Gaussian white noises. Barrett (1958, 1960) and Chuang and Kazda (1959) showed how to set up for such control systems the Fokker–Planck partial differential equation which describes the evolution of the probability density in the state space. They also obtained some exact solutions for the steady-state probability density, for some simple control systems of first and second order. It is rather surprising that exact results can be obtained for nonlinear stochastic systems of second order and even (as we shall see) higher order. After all, the exact analysis of the corresponding deterministic nonlinear systems, i.e. with the stochastic disturbances removed, is not an easy problem.

In the present paper an exposition of the application of the Fokker–Planck equation to nonlinear systems will be given, with emphasis on steady-state solutions. It will also be shown that the known steady-state solutions are closely related to the Maxwell–Boltzmann distribution which belongs to classical statistical mechanics.

To bring out the relation of our results with the Maxwell–Boltzmann distribution we shall begin with an examination of the historical development of the Fokker–Planck equation. We shall then derive the Fokker–Planck equation by a straightforward method.

Various questions of rigour will be ignored. The aim is to emphasize the main structure of the argument rather than to explore the boundaries of the region of its validity.

† Communicated by the Author.

Recently the Fokker–Planck equation has been used in the study of linear systems with stochastic coefficients, and in the study of systems with white noise disturbances multiplied by state-dependent functions. This work will not be reviewed, as much of it is outside the scope of the present paper.

The Fokker–Planck equation can also be applied to the study of first-passage times; however, this application will not be treated in the present paper.

For expositions of the Fokker–Planck equation with emphasis on communications engineering problems see Middleton (1960) and Stratonovich (1963).

2. The Maxwell–Boltzmann distribution

The development of the Fokker–Planck equation, and of the theory of stochastic processes in general, began in the nineteenth century when physicists were trying to show that heat in a medium is essentially a random motion of the constituent molecules. Maxwell (1860) considered a gas as consisting of small spherical rigid molecules distributed randomly but with uniform average density in a container. In Maxwell's model the molecules are supposed to have random velocities and to collide elastically with one another and with the walls of the container. Suppose that the position of a molecule is represented by Cartesian coordinates (x_1, x_2, x_3) and its velocity by (u_1, u_2, u_3), so that

$$\left\{\begin{aligned} \frac{dx_1}{dt} &= u_1 \\ \frac{dx_2}{dt} &= u_2 \\ \frac{dx_3}{dt} &= u_3 \end{aligned}\right. \tag{2.1}$$

where t is time. Maxwell asked what is the steady-state probability $p(u_1, u_2, u_3)\delta u_1 \delta u_2 \delta u_3$ that the velocity components lie in small elements between u_1 and $u_1+\delta u_1$, u_2 and $u_2+\delta u_2$, and u_3 and $u_3+\delta u_3$. After ingenious but heuristic arguments he concluded that the probability density is given by

$$p(u_1, u_2, u_3) = C \exp\left[-A(u_1^2+u_2^2+u_3^2)\right] \tag{2.2}$$

where A is a constant determined by the mass of a molecule and the temperature of the gas, and C is a normalizing constant chosen to make the total probability unity

$$C = \left\{\int_{-\infty}^{\infty}\int_{-\infty}^{\infty}\int_{-\infty}^{\infty} \exp\left[-A(u_1^2+u_2^2+u_3^2)\right] du_1\, du_2\, du_3\right\}^{-1}. \tag{2.3}$$

Boltzmann, in several papers beginning in 1868, generalized Maxwell's result as follows. Suppose the molecules not only collide with one another and the container but are additionally subject to a conservative field of force. A molecule then has potential energy $G(x_1, x_2, x_3)$ as well as kinetic energy, so that its total energy is

$$H(x_1, x_2, x_3, u_1, u_2, u_3) = G(x_1, x_2, x_3) + \tfrac{1}{2}m(u_1^2+u_2^2+u_3^2) \tag{2.4}$$

where m is the mass of the molecule. Boltzmann's result was that the steady-state probability density in $(x_1, x_2, x_3, u_1, u_2, u_3)$ state space is

$$p(x_1, x_2, x_3, u_1, u_2, u_3) = C' \exp\left[-\frac{2A}{m} H(x_1, x_2, x_3, u_1, u_2, u_3)\right] \tag{2.5}$$

where C' is a constant. Boltzmann's arguments, like Maxwell's, were ingenious but heuristic.

A gas whose molecules have probability density (2.2) is said to have a *Maxwell distribution* of velocity, and a gas whose molecules have probability density (2.5) is said to have a *Maxwell–Boltzmann distribution* of state.

Boltzmann also gave arguments indicating that a gas with an arbitrary initial probability density tends to approach probability density (2.5) as time increases. However his methods are regarded as controversial (see Moyal 1949).

Maxwell and Boltzmann subsequently elaborated and refined their theories (see their collected works and also Boltzmann's book (1896)). However, certain difficulties were encountered. The gas model treated was basically a conservative autonomous system, with randomness only entering in the initial positions and velocities of the molecules. As pointed out by Loschmidt (1876), in such a system, if at a certain instant the signs of all the velocities are reversed, the system retraces its previous motion. An ensemble of such systems then tends to diverge from rather than approach the Maxwell–Boltzmann distribution. This phenomenon, known as the *reversibility paradox*, contradicts Boltzmann's assertion that the Maxwell–Boltzmann distribution must be approached from any initial conditions. Again, Zermelo (1896) pointed out that since a conservative system must return arbitrarily closely to any given initial state (provided it remains in the finite part of the state space), the gas model must revert, arbitrarily closely to any initial spatial distribution of temperature, however irregular. This *recurrence paradox* contradicted the empirical law of thermodynamics that heat does not flow from a cold region to a hot region.

Nevertheless the theories of Maxwell and Boltzmann were successful in accounting for many of the properties of gases. In particular it follows from (2.4) and (2.5) that the mean kinetic energy for each degree of freedom is the same, i.e.

$$\tfrac{1}{2}m\overline{u_1^2} = \tfrac{1}{2}m\overline{u_2^2} = \tfrac{1}{2}m\overline{u_3^2} \tag{2.6}$$

This property, which generalizes to the case when there are several different kinds of molecules in the gas, is known as the *equipartition of energy*. With the mean kinetic energy interpreted as temperature, equipartition of energy implies that gases in contact reach a common temperature—a result in agreement with experiment. Consequently there was a tendency for physicists to accept the validity of the Maxwell–Boltzmann distribution. Gibbs (1902) in effect took the Maxwell–Boltzmann distribution as a primary postulate in the theory of gases, contenting himself with only a sketchy discussion of how this distribution was arrived at.

The writings of Maxwell, Boltzmann and Gibbs were not always easy to follow, partly because these authors did not always clearly distinguish between various kinds of averages, i.e. between time-average for a single gas molecule, time-average for a single gas model, ensemble-average for an ensemble of gas models, and average over the constituent molecules of a single gas model. For an early critical account of their theories see Ehrenfest and Ehrenfest† (1912). For a recent introduction to statistical mechanics see ter Haar (1966).

3. The early development of the Fokker–Planck equation

Around the end of the nineteenth century a new trend appeared in the treatment of the random processes of physics. Instead of considering random motion as due to collisions among objects having a random distribution of initial positions and velocities, some writers began to adopt more direct mathematical models of random disturbances. These models involved random walks, i.e. discrete-time systems with well-specified random disturbances entering continually. Although these models were no longer autonomous conservative systems (as in the gas models of Maxwell and Boltzmann) they were amenable to simple mathematical analysis.

Rayleigh (1880, 1894, 1899) was in effect the first to treat a random walk for a problem in physics‡, although his formulation of the problem was rather disguised. His version of the problem was to find the statistical distribution of a sum of n sinusoidal motions all having the same period and amplitude, but having phases distributed at random. This problem is mathematically equivalent to the random walk problem first formulated explicitly by Pearson (1905) in the following words: "A man starts from a point 0 and walks l yards in a straight line; he then turns through any angle whatever and walks another l yards in a second straight line. He repeats this process n times. I require the probability that after these n stretches he is at a distance between r and $r+dr$ from his starting point 0." For $n \to \infty$ and various cases of random walk, Rayleigh (1894, 1899) obtained a partial differential equation for the probability density of the displacement. Moreover, Rayleigh (1891) applied similar techniques to the theory of gases, and arrived at a partial differential equation for the probability density of the velocity of the molecules. This was the first example in physics of what was later called the Fokker–Planck equation.

Bachelier (1900) made a mathematical model of the French stock exchange, and obtained a simple case of the Fokker–Planck equation. Later Bachelier (1910, 1912) studied the related problem of the gambler's ruin, which is in effect a type of random walk problem. He showed that when the sequence of bets placed by the gambler is large, it is simpler to formulate a continuous model of the process. By this technique he was led to a moderately general version of the Fokker–Planck equation.

This work of Rayleigh and Bachelier seems to have been largely unnoticed by contemporary physicists, who subsequently proceeded to develop the Fokker–Planck equation for themselves.

---

† Note however that according to Brush (1964) the Ehrenfests misrepresented some of the opinions and terminology of Maxwell and Boltzmann, especially in connection with what the Ehrenfests called the ergodic hypothesis.

‡ Mathematicians had treated the problem of the gambler's ruin, which is a kind of random walk problem, in the early days of the development of probability theory, Related problems occur in the mixing of balls drawn from and replaced in two urns, and in this context Laplace (1812) obtained a partial differential equation similar to the Fokker–Planck equation (see Hostinsky, 1931, 1933, 1937).

Einstein (1905) brought together the Maxwell–Boltzmann theory and the random walk method in a paper on Brownian motion. This, the continual irregular motion of microscopic particles immersed in a fluid, had long been conjectured as due to the impacts on the particles of the surrounding molecules of the fluid†. Einstein's analysis showed that on this basis the mean-square displacement of an ensemble of such particles should increase linearly with time. This result was soon verified experimentally, thus giving considerable support to the molecular theory of matter and the kinetic theory of heat.

When a particle in a frictionless fluid receives an impulse due to impact with a molecule, the result is a change in the *velocity* of the particle. If, on the other hand, the fluid is highly viscous, the change of velocity is soon dissipated, and the net result of an impact is a change in the *displacement* of the particle. Thus Einstein assumed that collisions produced random jumps in the position of the particle, i.e. the particle performed a kind of random walk. Taking the random jumps to be small, he obtained a partial differential equation for the probability density of the displacement in one dimension—the simplest case of the Fokker–Planck equation. He obtained the solution of this equation. By invoking thermodynamic relations based on the Maxwell–Boltzmann theory he was able to express the constants in his solution in terms of the temperature and viscosity of the fluid.

The theory of Brownian motion was subsequently generalized in various directions by many physicists, notably Smoluchowski, Langevin, Fokker, Planck, Ornstein, Burger, Fürth and Kramers. For references see Uhlenbeck and Ornstein (1930), Ming Chen Wang and Uhlenbeck (1945), and other papers in the collection edited by Wax (1954). The generalizations included the treatment of (i) particles in three dimensions, (ii) particles in a field of force, and (iii) particles in non-highly-viscous fluids, i.e. when the inertia of a particle is no longer negligible, so that the state coordinates of a particle include both velocities and positions. Partial differential equations were obtained which were versions of what was subsequently called the Fokker–Planck equation.

Langevin (1908) introduced an idea which although at first sight appears trivial is helpful in clarifying and classifying various types of Brownian motion. (In this respect it is rather like the introduction of block diagrams into control theory.) He simply wrote down the ordinary differential equations of the motion of the particle, interpreting the random disturbances as an additional forcing function. Thus for the case of a free particle (when there is no field of force), Langevin's equation for the x_1 coordinate is

$$m\frac{d^2x_1}{dt^2} = -k\frac{dx_1}{dt} + \xi(t) \tag{3.1}$$

Here the term $-k(dx_1/dt)$ represents damping due to the viscosity of the fluid, and $\xi(t)$ represents the impulsive forces due to the fluid molecules. The impulses are of infinitesimal magnitude and occur at infinite rate, so that $\xi(t)$ is a type of 'white noise' (a term subsequently used in the discussion of random electrical signals). Langevin's equation is formal in nature, i.e. strictly speaking the limits implicit in the definition of the derivatives should be taken

† See the historical notes by Fürth in the collection of papers by Einstein on Brownian motion (1926).

before the limits implicit in the definition of the forcing function. However, this mathematical difficulty does not affect the usefulness of Langevin's equation as a classifying device.

Fokker (1913, 1914, 1918) treated a first-order system in which the intensity of the white noise was dependent on the system state†. Planck (1915, 1917 a, b) applied Fokker's equation to problems in quantum physics, and also generalized Fokker's equation considerably, treating an nth-order system with state-dependent white noise.

4. The later development of the Fokker–Planck equation

The theory of the Fokker–Planck equation was made considerably more general and abstract by Kolmogorov (1931). Following Bachelier he assumed the process to be continuous with respect to time. He also assumed the process to be a Markov process, i.e. one for which the future probability density conditional on the present and past, is actually independent of the past. On introducing further assumptions Kolmogorov was able to show that the probability density of the process obeys a partial differential equation of the Fokker–Planck type. He also gave another partial differential equation which the future probability density, conditional on the present state, obeys with respect to the present state. The latter equation is called Kolmogorov's first equation, and the Fokker–Planck type equation is called Kolmogorov's second equation. Sometimes the Fokker–Planck equation is called the Fokker–Planck–Kolmogorov equation. Kolmogorov's two equations are adjoints of one another. In a subsequent paper, Kolmogorov (1933) extended his treatment to vector processes, and also discussed the uniqueness of the solution of the Fokker–Planck equation.

Andronov *et al.* (1933) applied Kolmogorov's second equation, i.e. the Fokker–Planck equation, to the study of general dynamic systems, subject to random disturbances, and obtained solutions for the probability density in simple cases.

Useful papers reviewing and consolidating the theory of Brownian motion were given by Chandrasekhar (1943), and Ming Chen Wang and Uhlenbeck (1945).

During 1953–60 Stratonovich and his colleagues applied the Fokker–Planck equation to problems in electronic engineering (see the translated collection of papers by Kuznetsov *et al.* 1965). Stratonovich's book (1963) contains a useful exposition of applied Fokker–Planck theory, and includes non-steady-state solutions.

Chuang and Kazda (1959) applied the Fokker–Planck equation to the study of nonlinear control systems; however they introduced certain approximations which tended to obscure the argument. Barrett (1958, 1960) gave a more complete account of the application of the Fokker–Planck equation to control systems. Rather special control problems were studied by means of the Fokker–Planck equation by Bolshakov (1959), Tikhonov

† Fokker considered the motion of an electron in a field of random radiation. In his 1914 paper an integrated version of the Fokker–Planck equation rather than the Fokker–Planck equation itself was given, and proofs were omitted. More details were given in the 1918 paper, which had been written in 1914 but was not then published, doubtless owing to the war.

(1959, 1960), Blachman (1960), and Ruina and Valkenburg (1960). Further work on control applications has been done by Benedict† (1959), Fuller (1959, 1961), Sawaragi *et al.* (1960, 1961 a, b), Khazen (1961 a, b, c), Pugachev (1961), Pervosvanski (1962) and Butchart (1965, 1967). Merklinger (1963 a, b) and Buhr (1966) have studied the numerical solution of the Fokker–Planck equation.

The Fokker–Planck equation has also been used in the study of nonlinear vibrations. Work along these lines has been done by Ariaratnam (1960, 1962), Lyon (1960, 1961, 1963 a, b), Smith (1962), Caughey and Dienes (1961), Crandall (1961, 1962 a, b, 1963 a, b) and Caughey (1963 a, b).

Many papers have been written on problems of rigour associated with the Fokker–Planck equation. No attempt will be made here to review such papers as they are outside the scope of the present treatment. See Bharucha-Reid (1960) and Dynkin (1965) for references.

Several of the solutions obtained by the above writers seem to have been found piece-meal by trial and error. Barrett (1959) and Lyon (1963 a) suggested that such solutions are related to the Maxwell–Boltzmann distribution. One aim of the present paper is to confirm this suggestion, and to derive further solutions related to the Maxwell–Boltzmann distribution. Another aim is to give an elementary derivation (essentially Bachelier's and Einstein's) of the Fokker–Planck equation which is more straightforward than the usual abstract derivation by Kolmogorov's method.

For a wider survey of the historical development of the theory of stochastic processes see Barrett (1961).

5. Markov processes

The stochastic processes we shall be dealing with have the property of being Markov processes. This means that the probability density of the process at any future instant, conditional on past and present values of the process, is actually independent of the past values. Such processes were studied by Markov in several papers during 1907–1924‡.

More specifically, let us consider any scalar random process $x(t)$ and a set of instants $t_1, t_0, t_{-1}, t_{-2}, \ldots$ such that

$$\ldots t_{-3} < t_{-2} < t_{-1} < t_0 < t_1 \tag{5.1}$$

Let us call the t's with negative subscripts past instants, t_0 the present, and t_1 a future instant. Those members of the ensemble which have given past and present values

$$x(t_0) = x_0, \;\; x(t_{-1}) = x_{-1}, \;\; x(t_{-2}) = x_{-2}, \;\ldots \tag{5.2}$$

have values with a certain probability density function at the future instant t_1, namely the conditional probability density function

$$p(x_1, t_1 | x_0, t_0 ; x_{-1}, t_{-1} ; x_{-2}, t_{-2} ; \ldots) \tag{5.3}$$

† Benedict treated second-order relay control systems. However, his solutions of the Fokker–Planck equation are inappropriate since they are discontinuous at the switching curve.

‡ See Hostinsky (1931) for references. See Markov (1910, 1912) for papers in French and German.

What characterizes a Markov process is that the function (5.3) is actually independent of the past arguments $x_{-1}, t_{-1}, x_{-2}, t_{-2}, \ldots$ so that

$$p(x_1, t_1|x_0, t_0; x_{-1}, t_{-1}; x_{-2}, t_{-2}; \ldots) = p(x_1, t_1|x_0, t_0) \tag{5.4}$$

If, more generally, $x(t)$ is a vector Markov process with n components, (5.4) still represents its defining property. In this case each argument $x_1, x_0, x_{-1}, \ldots$ in (5.4) is a vector with n components.

A Markov process satisfies an equation known as the Chapman–Kolmogorov equation† (Chapman 1928, Kolmogorov 1931). This equation is basic in the derivation of the Fokker–Planck equation, and is itself easily derived as follows.

Let us consider any stochastic process $x(t)$, not necessarily a Markov process, and for simplicity let us suppose initially that x is scalar. Let us consider the ensemble at a sequence of three instants t_1, t_2, t_3 satisfying

$$t_1 < t_2 < t_3 \tag{5.5}$$

Those members of the ensemble which have the fixed value x_1 at time t_1

$$x(t_1) = x_1 \tag{5.6}$$

constitute a sub-ensemble. A fraction $p(x_2, t_2|x_1, t_1)\delta x_2$ of the sub-ensemble have values in an element‡ δx_2 at x_2 at time t_2. Of these, a fraction $p(x_3, t_3|x_2, t_2; x_1, t_1)\delta x_3$ have values in an element δx_3 at x_3 at time t_3. Thus the fraction

$$p(x_3, t_3|x_2, t_2; x_1, t_1)p(x_2, t_2|x_1, t_1)\delta x_3 \delta x_2 \tag{5.7}$$

of the sub-ensemble have values both in an element δx_2 at x_2 at time t_2 and in an element δx_3 at x_3 at time t_3. Hence the fraction of the sub-ensemble which have values in an element δx_3 at x_3 at time t_3, which is simply $p(x_3, t_3|x_1, t_1)\delta x_3$, is obtained by adding the contributions (5.7) of all the elements δx_2, i.e.

$$p(x_3, t_3|x_1, t_1) = \int_{-\infty}^{\infty} p(x_3, t_3|x_2, t_2; x_1, t_1)p(x_2, t_2|x_1, t_1)\, dx_2 \tag{5.8}$$

Formula (5.8) holds for any process. Let us now introduce the restriction that $x(t)$ is a Markov process. Then

$$p(x_3, t_3|x_2, t_2; x_1, t_1) = p(x_3, t_3|x_2, t_2) \tag{5.9}$$

so that (5.8) simplifies to

$$p(x_3, t_3|x_1, t_1) = \int_{-\infty}^{\infty} p(x_3, t_3|x_2, t_2)p(x_2, t_2|x_1, t_1)\, dx_2 \tag{5.10}$$

---

† Special cases of the Chapman–Kolmogorov equation were already used by Einstein (1905), Bachelier (1912) and Smoluchowski (1913). The equation is sometimes called the Smoluchowski equation.

‡ Unsophisticated language like 'in an element δx_2' is deliberately used in this paper because of its intuitive appeal. Such language can be avoided, at the cost of greater abstractness, by working with probability distribution functions rather than probability density functions.

This is the required Chapman–Kolmogorov equation†. It still holds when $x(t)$ is a vector process, since the above derivation still holds. For example if x is a two-dimensional vector (ρ, σ), (5.10) means that

$$p(\rho_3, \sigma_3, t_3|\rho_1, \sigma_1, t_1) = \int_{-\infty}^{\infty}\int_{-\infty}^{\infty} p(\rho_3, \sigma_3, t_3|\rho_2, \sigma_2, t_2)p(\rho_2, \sigma_2, t_2|\rho_1, \sigma_1, t_1)\, d\rho_2\, d\sigma_2 \tag{5.11}$$

6. Derivation of the Fokker–Planck equation in the simplest case

We shall first derive the Fokker–Planck equation in the simplest case, namely for a scalar process $x(t)$ obtained by integrating a white noise $\xi(t)$. This is essentially the case treated by Einstein (1905), and the method we shall use is essentially Bachelier's and his. What is meant by white noise here will appear from various postulated properties in the ensuing development.

The Langevin equation of the process is

$$\frac{dx}{dt} = \xi(t) \tag{6.1}$$

Formal integration of (6.1) yields

$$x(t) = x(0) + \int_0^t \xi(\theta)\, d\theta \tag{6.2}$$

Since $\xi(t)$ consists of infinitesimal impulses occurring at infinite rate, (6.1) must be interpreted with care (see § 3). Mathematicians often prefer to avoid this difficulty by taking (6.2) rather than (6.1) as the defining equation of the process. The integral on the right is then interpreted as a sum of increments

$$\int_0^t \xi(\theta)\, d\theta = \sum_i \int_{\tau_i}^{\tau_{i+1}} \xi(\theta)\, d\theta = \sum_i \Delta_i \tag{6.3}$$

and the statement that $\xi(t)$ is a white noise is interpreted as meaning that each increment Δ_i is statistically independent of $x(t)$ for $t < \tau_i$. $x(t)$ is said to be a process with 'independent increments', and this property is postulated to hold no matter how the time interval 0–t is divided into segments by the instants τ_i.

Since the increments of $x(t)$ are statistically independent of its past $x(t)$ is a Markov process. Hence the Chapman–Kolmogorov eqn. (5.10) applies. To simplify the notation let us omit the arguments x_1 and t_1 from (5.10). (It will turn out that these arguments do not play an essential role in the derivation of the Fokker–Planck equation.) (5.10) then becomes

$$p(x_3, t_3) = \int_{-\infty}^{\infty} p(x_3, t_3|x_2, t_2)p(x_2, t_2)\, dx_2 \tag{6.4}$$

Let us further write

$$t_2 = t \tag{6.5}$$

$$t_3 = t + \delta t \tag{6.6}$$

† Kolmogorov (1931) took this equation as a starting point in the theory of Markov processes; however it is more natural to derive it as above. Smoluchowski (1913) gave (5.10) in the special case when $p(x_j, t_j|x_i, t_i)$ depends only on x_i, x_j and the difference $t_j - t_i$.

so that (6.4) becomes

$$p(x_3, t+\delta t) = \int_{-\infty}^{\infty} p(x_3, t+\delta t | x_2, t) p(x_2, t)\, dx_2 \tag{6.7}$$

Let us write the change in x which occurs during the interval δt as z

$$z = x_3 - x_2 \tag{6.8}$$

Then x_3 determines z, if we consider x_2 as fixed. Hence the probability that $x(t+\delta t)$ is in an element δx_3 at x_3 (given that $x(t) = x_2$) is the same thing as the probability that the change during the interval δt is in an element δz at z (given that $x(t) = x_2$). If we designate the latter probability $q(z, \delta t | x_2, t)\delta z$, we have

$$p(x_3, t+\delta t | x_2, t)\delta x_3 = q(z, \delta t | x_2, t)\delta z \tag{6.9}$$

In (6.8) x_2 is fixed, so that

$$\delta z = \delta x_3 \tag{6.10}$$

Hence (6.9) implies that

$$p(x_3, t+\delta t | x_2, t) = q(z, \delta t | x_2, t) \tag{6.11}$$

We call $q(z, \delta t | x_2, t)$ the transition probability density.

Let us confine attention to the simplest case in which the transition probability density q is actually independent of the value x_2 from which the transition is made, and independent of the time t at which it is made. Then (6.11) simplifies to

$$p(x_3, t+\delta t | x_2, t) = q(z, \delta t) \tag{6.12}$$

From (6.8) and (6.12), (6.7) may be written

$$p(x_3, t+\delta t) = \int_{-\infty}^{\infty} q(z, \delta t) p(x_3 - z, t)\, dz \tag{6.13}$$

We may now drop the subscript from x_3 and write (6.13) as

$$p(x, t+\delta t) = \int_{-\infty}^{\infty} q(z, \delta t) p(x - z, t)\, dz \tag{6.14}$$

Result (6.14) is intuitively obvious; it states that the probability that the system will be in a differential element at x at time $t+\delta t$ is equal to the probability that it changes from a differential element at $x-z$ at time t, multiplied by the transition probability of the change z, summed with respect to all change values z. However, we have derived (6.14) at length to clarify how the Markov property enters.

We now postulate that when δt is small the probability that a large change z occurs is small, so that the transition probability density $q(z, \delta t)$ has appreciable magnitude only when z is near zero. We postulate that only this small range of values of z contributes appreciably to the integral in (6.14), and that over this small range we may represent the term $p(x-z, t)$ by the first few terms of a Taylor series

$$p(x-z, t) = p(x, t) - z\frac{\partial p(x,t)}{\partial x} + \frac{z^2}{2!}\frac{\partial^2 p(x,t)}{\partial x^2} - \ldots \tag{6.15}$$

Similarly we represent the left side of (6.14) by the first few terms of a Taylor series

$$p(x, t+\delta t) = p(x, t) + \delta t \frac{\partial p(x, t)}{\partial t} + \frac{(\delta t)^2}{2!} \frac{\partial^2 p(x, t)}{\partial t^2} + \dots \tag{6.16}$$

If the terms in (6.16) of order of magnitude $(\delta t)^2$ are neglected (6.14) then becomes

$$\begin{aligned} p(x, t) + \delta t \frac{\partial p(x, t)}{\partial t} &= p(x, t) \int_{-\infty}^{\infty} q(z, \delta t)\, dz - \frac{\partial p(x, t)}{\partial x} \int_{-\infty}^{\infty} z q(z, \delta t)\, dz \\ &+ \frac{1}{2!} \frac{\partial^2 p(x, t)}{\partial x^2} \int_{-\infty}^{\infty} z^2 q(z, \delta t)\, dz - \frac{1}{3!} \frac{\partial^3 p(x, t)}{\partial x^3} \int_{-\infty}^{\infty} z^3 q(z, \delta t)\, dz + \dots \end{aligned} \tag{6.17}$$

The first integral on the right of (6.17) is the integral of a probability density and is therefore unity. Hence the first term on the right cancels the first term on the left.

The second integral on the right of (6.17) is the mean of the transition probability density $q(z, \delta t)$. Let us restrict attention to the simplest case in which the mean value of the change z is zero. Then the second term on the right of (6.17) vanishes.

The third integral on the right of (6.17) represents the mean-square value of the change z, and we write this as $\overline{z^2}(\delta t)$

$$\overline{z^2}(\delta t) = \int_{-\infty}^{\infty} z^2 q(z, \delta t)\, dz \tag{6.18}$$

The fourth term on the right of (6.17) has as a factor the third moment $\overline{z^3}(\delta t)$. Since the density function $q(z, t)$ is negligible except for small values of $|z|$, we postulate that $\overline{z^3}(\delta t)$ is negligible compared with $\overline{z^2}(\delta t)$. Thus in (6.17) we neglect the fourth term, and similarly we neglect the subsequent terms.

Then (6.17) simplifies to

$$\frac{\partial p(x, t)}{\partial t} \delta t = \tfrac{1}{2} \overline{z^2}(\delta t) \frac{\partial^2 p(x, t)}{\partial x^2} \tag{6.19}$$

When $\delta t \to 0$, we postulate that the ratio $\overline{z^2}(\delta t)/\delta t$ approaches a constant value b (which is a parameter of the white noise)

$$\lim_{\delta t \to 0} \frac{\overline{z^2}(\delta t)}{\delta t} = b \tag{6.20}$$

Postulate (6.20) is a defining property of the white noise. That it is a natural postulate may be seen if we reflect that z is the sum of independent increments taken over sub-intervals of the interval δt. The variance of the sum of a set of independent variables is well known to be the sum of the separate variances; hence if δt is increased by the addition of a sub-interval we expect the variance of z to increase *linearly*.

From (6.19) and (6.20) we obtain finally

$$\frac{\partial p}{\partial t} = \frac{b}{2}\frac{\partial^2 p}{\partial x^2} \tag{6.21}$$

which is the required Fokker–Planck equation for the conditional probability density $p(x, t|x_1, t_1)$ for system (6.1).

7. Discussion of the Fokker–Planck equation in the simplest case

The above derivation of the Fokker–Planck equation lacks mathematical rigour at a number of points, but is relatively straightforward. The usual, more roundabout, approach is to derive Kolmogorov's first equation and then obtain from it Kolmogorov's second equation (Kolmogorov 1931). Kolmogorov's approach has the advantage that the postulates on which the Fokker–Planck equation is based can be framed more compactly. On the other hand, the postulates used above have more intuitive appeal.

To convince oneself that the postulates made above about the white noise are reasonable, i.e. that stochastic processes exist which satisfy (6.21), it is helpful to consider limiting cases of random walks when the step length is made small. This is the approach expounded by Chandrasekhar (1943), and his paper should be consulted by readers who find the above argument lacking in concreteness.

Equation (6.21) is the same as the well-known equation for the diffusion of heat in a one-dimensional homogeneous medium. The heat equation was derived and solved by Fourier (1822, chap. 9); he found that a particular solution is

$$p(x, t) = [2\pi b(t - t_1)]^{-1/2} \exp\left[-\frac{(x - x_1)^2}{2b(t - t_1)}\right] \tag{7.1}$$

This is the required solution for $p(x, t|x_1, t_1)$, since it yields for $t = t_1$

$$p(x, t_1) = \delta(x - x_1) \tag{7.2}$$

where the right side is a Dirac delta function, i.e. at $t = t_1$ all members of the ensemble have $x = x_1$, in agreement with (5.6).

In mathematical literature the term Brownian motion is sometimes reserved for a process having density function (7.1). This process (which is integrated white noise) is taken as a starting point in certain theories of stochastic processes, since it is easier to discuss rigorously than white noise. This approach is due to Wiener. For example, in his book *Extrapolation, Interpolation and Smoothing of Stationary Time Series* (1949) he writes of the response of a system to Brownian motion input, rather than to white noise input. A process satisfying (7.1) is also called a Wiener process.

Expression (7.1) implies that $x(t)$ is a Gaussian process†, i.e. its joint probability distributions at any set of instants are Gaussian. For this reason it is appropriate to call the white noise which generates $x(t)$ a Gaussian white

† The concept of a Gaussian process goes back at least to Khintchine (1934), although he did not call it that. The term Gaussian process was perhaps introduced by Doob (1944). Note that the Gaussian distribution was actually discovered by de Moivre (see Pearson 1924), so the Gaussian process ought really to be called the de Moivre process.

noise. More generally any white noise which on being smoothed by a linear system generates a Gaussian process is called a Gaussian white noise; this terminology† distinguishes it from other white noises, e.g. Poisson white noise (see Fuller 1960 c).

8. Derivation of the Fokker–Planck equation in the general case

Let us consider the general case when x is an N-dimensional vector $(x_1, x_2, \ldots, x_N)$. (Here subscripts are used to distinguish between the different components of x, and no longer to distinguish between values of x at different times as in §§ 5 and 6.) We assume that $x(t)$ satisfies the generalized Langevin equation

$$\frac{dx_i}{dt} = f_i(x_1, x_2, \ldots, x_N, t) + \xi_i(t) \quad (i = 1, 2, \ldots, N) \tag{8.1}$$

where the $\xi_i(t)$ are white noises, and the f_i are nonlinear or linear functions of the arguments shown. When the white noise forcing functions are absent, (8.1) represents the differential equations of a dynamic system written in normal form, and x is the state vector of the system; we retain the same terminology in the stochastic case. (See, e.g., Fuller 1960 a, b, c for an introduction to state space or phase space methods.)

Usually when treating nonlinear functions in this paper we shall not stop to specify which class of functions they must be limited to. However, we may mention that in (8.1) the f's have to be differentiable for the subsequent arguments to be formally valid. By suitable generalization we can allow the f's to be piece-wise differentiable —thus, e.g., relay control systems can be dealt with.

Formal integration of (8.1) yields

$$x_i(t) - x_i(0) = \int_0^t f_i(x_1(\theta), x_2(\theta), \ldots, x_N(\theta), \theta)\, d\theta + \int_0^t \xi_i(\theta)\, d\theta \; (i = 1, 2, \ldots, N) \tag{8.2}$$

As in § 6, the integral of white noise appearing in (8.2) is interpreted as a sum of increments

$$\int_0^t \xi_i(\theta)\, d\theta = \sum_j \int_{\tau_j}^{\tau_{j+1}} \xi_i(\theta)\, d\theta = \sum_j \Delta_{ij} \tag{8.3}$$

In (8.3) Δ_{ij} is assumed to be statistically independent of any value of $x(t)$ occurring before $t = \tau_j$. (However, the increment vector $(\Delta_{1j}, \Delta_{2j}, \ldots, \Delta_{Nj})$ has components which are not necessarily statistically independent of one another.)

It follows from (8.2) that any future value of x depends only on the present value of x and on the future increments Δ_{ij}. Since the latter are statistically independent of the past, $x(t)$ is a Markov process. Hence the Chapman–Kolmogorov equation applies, and we can write it in the form, analogous to (6.7)

$$p(x, t + \delta t) = \int_{-\infty}^{\infty} p(x, t + \delta t | x', t) p(x', t)\, dx' \tag{8.4}$$

† There is a tendency in the literature to imply that a white noise is Gaussian if its joint probability distributions at any set of instants are Gaussian (e.g. Ming Chen Wang and Uhlenbech 1945). This definition is unsatisfactory since the distributions do not exist except in a limiting sense.

Let the change which occurs during the interval δt be the vector z (compare (6.8)):

$$z = x - x' \tag{8.5}$$

Then the term $p(x, t+\delta t|x', t)$ appearing in (8.4) can be written

$$p(x, t+\delta t|x', t) = q(z, \delta t|x', t) \tag{8.6}$$

where q is the transition probability density (compare (6.11)). Unlike our treatment in § 6, we shall not assume that $q(z, \delta t|x', t)$ is actually independent of the conditional arguments x' and t. From (8.4), (8.5) and (8.6) the Chapman–Kolmogorov equation can be written

$$p(x, t+\delta t) = \int_{-\infty}^{\infty} q(z, \delta t|x-z, t)\, p(x-z, t)\, dz \tag{8.7}$$

which is a generalization of (6.14). In (8.7) x and z are vectors, and p, q, t and δt are scalars. dz is the volume element $dz_1 dz_2 \ldots dz_N$.

As in § 6, we postulate that when δt is small, the probability of a large change z is small, so that only a small range of values of z around $z=0$ contributes appreciably to the integral in (8.7). Thus we postulate that we can approximate $p(x-z, t)$ by the first few terms of a Taylor series with respect to its first argument, and similarly approximate $q(z, \delta t|x-z, t)$ by a Taylor series with respect to its third argument. Rather than expand p and q separately, it is neater to treat the integrand in (8.7) as a single function r

$$r(x-z, t, z, \delta t) = q(z, \delta t|x-z, t)p(x-z, t) \tag{8.8}$$

and expand r in a Taylor series with respect to its first argument. For this purpose we shall write r as a function of its first argument $x-z$ only, leaving the other arguments implicit. Also we shall write the components of the vector argument $x-z$ explicitly. Thus

$$qp = r(x_1 - z_1, x_2 - z_2, \ldots, x_N - z_N) \tag{8.9}$$

The formal Taylor series expansion of r with respect to its first argument can be written symbolically as

$$r(x_1 - z_1, x_2 - z_2, \ldots, x_N - z_N) = \mathrm{e}^{-z_1 \frac{\partial}{\partial x_1}} r(x_1, x_2 - z_2, \ldots, x_N - z_N) \tag{8.10}$$

in which the exponential term means the exponential series

$$\mathrm{e}^{-z_1 \frac{\partial}{\partial x_1}} = 1 - z_1 \frac{\partial}{\partial x_1} + \frac{1}{2!}\left(z_1 \frac{\partial}{\partial x_1}\right)^2 - \ldots \tag{8.11}$$

Expanding the right side of (8.10) successively with respect to $z_2, z_3, \ldots, z_N$, we get

$$\begin{aligned} r(x_1 - z_1, x_2 - z_2, \ldots, x_N - z_N) &\\ = \mathrm{e}^{-z_N \frac{\partial}{\partial x_N}} \ldots \mathrm{e}^{-z_2 \frac{\partial}{\partial x_2}} \mathrm{e}^{-z_1 \frac{\partial}{\partial x_1}} r(x_1, x_2, \ldots, x_N) & \qquad (8.12) \\ = \mathrm{e}^{-\left(z_1 \frac{\partial}{\partial x_1} + z_2 \frac{\partial}{\partial x_2} + \ldots + z_N \frac{\partial}{\partial x_N}\right)} r(x_1, x_2, \ldots, x_N) & \qquad (8.13) \end{aligned}$$

For this symbolic method of representing Taylor's series, which goes back to Lagrange (1772), see, e.g., Pipes (1946).

Thus (8.13) implies

$$\begin{aligned} r(x_1-z_1, x_2-z_2, \ldots, x_N-z_N) &= r(x_1, \ldots, x_N) - \left(z_1\frac{\partial}{\partial x_1}+\ldots+z_N\frac{\partial}{\partial x_N}\right) r(x_1, \ldots, x_N) \\ &\quad + \frac{1}{2!}\left(z_1\frac{\partial}{\partial x_1}+\ldots+z_N\frac{\partial}{\partial x_N}\right)^2 r(x_1, \ldots, x_N) - \ldots \end{aligned} \tag{8.14}$$

$$\begin{aligned} &= r(x_1, \ldots, x_N) - \sum_{i=1}^{N} z_i \frac{\partial}{\partial x_i} r(x_1, \ldots, x_N) \\ &\quad + \frac{1}{2!}\sum_{i=1}^{N}\sum_{j=1}^{N} z_i z_j \frac{\partial^2}{\partial x_i \partial x_j} r(x_1, \ldots, x_N) \\ &\quad - \frac{1}{3!}\sum_{i=1}^{N}\sum_{j=1}^{N}\sum_{k=1}^{N} z_i z_j z_k \frac{\partial^3}{\partial x_i \partial x_j \partial x_k} r(x_1, \ldots, x_N) \\ &\quad + \ldots . \end{aligned} \tag{8.15}$$

From (8.9) and (8.15), (8.7) may be written as

$$\begin{aligned} p(x, t+\delta t) &= \int_{-\infty}^{\infty}\cdots\int_{-\infty}^{\infty} r(x_1, \ldots, x_N)\, dz_1 \ldots dz_N \\ &\quad - \sum_{i=1}^{N}\int_{-\infty}^{\infty}\cdots\int_{-\infty}^{\infty} z_i \frac{\partial r(x_1, \ldots, x_N)}{\partial x_i}\, dz_1 \ldots dz_N \\ &\quad + \frac{1}{2!}\sum_{i=1}^{N}\sum_{j=1}^{N}\int_{-\infty}^{\infty}\cdots\int_{-\infty}^{\infty} z_i z_j \frac{\partial^2 r(x_1, \ldots, x_N)}{\partial x_i \partial x_j}\, dz_1 \ldots dz_N \\ &\quad - \ldots \end{aligned} \tag{8.16}$$

Assuming that the order of differentiations and integrations may be interchanged, we write (8.16) as

$$\begin{aligned} p(x, t+\delta t) &= \int_{-\infty}^{\infty}\cdots\int_{-\infty}^{\infty} r(x_1, \ldots, x_N)\, dz_1 \ldots dz_N \\ &\quad - \sum_{i}\frac{\partial}{\partial x_i}\int_{-\infty}^{\infty}\cdots\int_{-\infty}^{\infty} z_i r(x_1, \ldots, x_N)\, dz_1 \ldots dz_N \\ &\quad + \frac{1}{2!}\sum_{i,j}\frac{\partial^2}{\partial x_i \partial x_j}\int_{-\infty}^{\infty}\cdots\int_{-\infty}^{\infty} z_i z_j r(x_1, \ldots, x_N)\, dz_1 \ldots dz_N \\ &\quad - \ldots \end{aligned} \tag{8.17}$$

From (8.8) and (8.9)

$$r(x_1, \ldots, x_N) = r(x, t, z, \delta t) \tag{8.18}$$

$$= q(z, \delta t \,|\, x, t)\, p(x, t) \tag{8.19}$$

Substituting (8.19) in (8.17) and taking the factor $p(x, t)$ outside the integrals we find

$$
\begin{aligned}
&p(x_1, \ldots, x_N, t+\delta t) \\
&= \int_{-\infty}^{\infty} \cdots \int_{-\infty}^{\infty} q(z_1, \ldots, z_N, \delta t | x_1, \ldots, x_N, t)\, dz_1 \ldots dz_N p(x_1, \ldots, x_N, t) \\
&- \sum_i \frac{\partial}{\partial x_i}\left[\int_{-\infty}^{\infty} \cdots \int_{-\infty}^{\infty} z_i q(z_1, \ldots, z_N, \delta t | x_1, \ldots, x_N, t)\, dz_1 \ldots dz_N \right. \\
&\qquad\qquad \left. p(x_1, \ldots, x_N, t)\right] \\
&+ \frac{1}{2!}\sum_{i,j} \frac{\partial^2}{\partial x_i \partial x_j}\left[\int_{-\infty}^{\infty} \cdots \int_{-\infty}^{\infty} z_i z_j q(z_1, \ldots, z_N, \delta t | x_1, \ldots, x_N, t) \right. \\
&\qquad\qquad \left. dz_1 \ldots dz_N p(x_1, \ldots, x_N, t)\right] \\
&- \ldots
\end{aligned} \tag{8.20}
$$

The first multiple integral on the right of (8.20) is the integral of a probability density and is thus unity. The next multiple integral is the first moment $\bar{z}_i$ and since this depends on the arguments $\delta t, x, t$ of q we write it as

$$\bar{z}_i(\delta t, x_1, \ldots, x_N, t)$$

Similarly the next multiple integral is the second moment $\overline{z_i z_j}(\delta t, x_1, \ldots, x_N, t)$, the next will be the third moment $\overline{z_i z_j z_k}(\delta t, x_1, \ldots, x_N, t)$, and so on.

Since the probability of a large value z_i is small, we postulate that the third, fourth, ... moments are negligible compared with the first and second moments. We also expand the left side of (8.20) in a Taylor series with respect to t, and neglect terms of order of magnitude $(\delta t)^2$. Then (8.20) simplifies to

$$
\begin{aligned}
\frac{\partial p(x_1, \ldots, x_N, t)}{\partial t}\delta t = &- \sum_i \frac{\partial}{\partial x_i}[\bar{z}_i(\delta t, x_1, \ldots, x_N, t) p(x_1, \ldots, x_N, t)] \\
&+ \tfrac{1}{2}\sum_{i,j} \frac{\partial^2}{\partial x_i \partial x_j}[\overline{z_i z_j}(\delta t, x_1, \ldots, x_N, t) p(x_1, \ldots, x_N, t)]
\end{aligned} \tag{8.21}
$$

When $\delta t \to 0$ we postulate that the ratio $\bar{z}_i/\delta t$ approaches a limiting value a_i, and that the ratio $\overline{z_i z_j}/\delta t$ approaches a limiting value b_{ij}. Thus

$$\lim_{\delta t \to 0} \frac{\bar{z}_i(\delta t, x_1, \ldots, x_N, t)}{\delta t} = a_i(x_1, \ldots, x_N, t) \tag{8.22}$$

and

$$\lim_{\delta t \to 0} \frac{\overline{z_i z_j}(\delta t, x_1, \ldots, x_N, t)}{\delta t} = b_{ij}(x_1, \ldots, x_N, t) \tag{8.23}$$

Then (8.21) yields

$$
\begin{aligned}
\frac{\partial p(x_1, \ldots, x_N, t)}{\partial t} = &- \sum_{i=1}^{N} \frac{\partial}{\partial x_i}[a_i(x_1, \ldots, x_N, t) p(x_1, \ldots, x_N, t)] \\
&+ \tfrac{1}{2}\sum_{i=1}^{N}\sum_{j=1}^{N} \frac{\partial^2}{\partial x_i \partial x_j}[b_{ij}(x_1, \ldots, x_N, t) p(x_1, \ldots, x_N, t)]
\end{aligned} \tag{8.24}
$$

This is Kolmogorov's version of the Fokker–Planck equation.

It remains for us to express the coefficients a_i and b_{ij}, which we shall call *moment rates*, in terms of the Langevin system (8.1). Kolmogorov (1931) did not concern himself with this problem; it was in effect treated by Andronov *et al.* (1933). For system (8.1) we have (compare (8.2))

$$x_i(t+\delta t)-x_i(t)=\int_t^{t+\delta t} f_i(x_1(\theta),\ldots,x_N(\theta),\theta)\,d\theta+\int_t^{t+\delta t}\xi_i(\theta)\,d\theta \quad (i=1,2,\ldots,N) \tag{8.25}$$

Here the left side is the change z_i. Since $x(t)$ is (by definition) a continuous solution of eqns. (8.1), the first integrand in (8.25) is nearly a constant when δt is small. Hence we approximate the first integral by $f_i(x_1(t),\ldots,x_N(t),t)\delta t$. The second integral in (8.25) is an increment Δ_i which in general depends on δt, $x(t)$ and t. Hence (8.25) yields

$$z_i=f_i(x_1,\ldots,x_N,t)\delta t+\Delta_i(\delta t,x_1,\ldots,x_N,t) \tag{8.26}$$

The mean of (8.26) taken over the ensemble of increments $\Delta_i(\delta t,x_1,\ldots,x_N,t)$ is

$$\overline{z_i}(\delta t,x_1,\ldots,x_N,t)=f_i(x_1,\ldots,x_N,t)\delta t+\overline{\Delta_i}(\delta t,x_1,\ldots,x_N,t) \tag{8.27}$$

We now postulate that the mean of the white noise increment $\overline{\Delta_i}$ is zero:

$$\overline{\Delta_i}(\delta t,x_1,\ldots,x_N,t)=0 \quad (i=1,2,\ldots,N) \tag{8.28}$$

(This is a defining property of the white noise.) Then (8.27) yields

$$\overline{z_i}(\delta t,x_1,\ldots,x_N,t)=f_i(x_1,\ldots,x_N,t)\delta t \tag{8.29}$$

From (8.22) and (8.29) the required expression for the first moment rate is

$$a_i(x_1,\ldots,x_N,t)=f_i(x_1,\ldots,x_N,t) \quad (i=1,\ldots,N) \tag{8.30}$$

Let us now calculate the second moment rate b_{ij}. From (8.26)

$$z_iz_j=f_if_j(\delta t)^2+f_i\Delta_j\delta t+f_j\Delta_i\delta t+\Delta_i\Delta_j \tag{8.31}$$

where for simplicity we have not written the arguments of the f's and Δ's explicitly. The mean of (8.31) with respect to the ensemble of the Δ's is

$$\overline{z_iz_j}=f_if_j(\delta t)^2+f_i\overline{\Delta_j}\delta t+f_j\overline{\Delta_i}\delta t+\overline{\Delta_i\Delta_j} \tag{8.32}$$

Here the first term on the right is of order of magnitude $(\delta t)^2$ and will be neglected since δt is small. The second and third terms are zero in view of (8.28). Hence (8.32) simplifies to

$$\overline{z_iz_j}=\overline{\Delta_i\Delta_j} \tag{8.33}$$

From (8.23) and (8.33)

$$b_{ij}(x_1,\ldots,x_N,t)=\lim_{\delta t\to 0}\frac{\overline{\Delta_i\Delta_j}(\delta t,x_1,\ldots,x_N,t)}{\delta t} \quad (i=1,2,\ldots,N;\ j=1,2,\ldots,N) \tag{8.34}$$

Thus the second moment rates are properties of the white noise vector ξ, being the limiting properties of the increments as indicated on the right of (8.34). We take these rates as defining properties of the white noise vector, and retain the symbols b_{ij} for them.

From (8.24) and (8.30), the Fokker–Planck equation for system (8.1) is

$$\frac{\partial p(x_1,\ldots,x_N,t)}{\partial t} = -\sum_{i=1}^{N}\frac{\partial}{\partial x_i}[f_i(x_1,\ldots,x_N,t)p(x_1,\ldots,x_N,t)]$$
$$+\tfrac{1}{2}\sum_{i=1}^{N}\sum_{j=1}^{N}\frac{\partial^2}{\partial x_i \partial x_j}[b_{ij}(x_1,\ldots,x_N,t)p(x_1,\ldots,x_N,t)] \qquad (8.35)$$

where the b_{ij} are the second moment rates of the white noise vector ξ.

9. Initial conditions

We have derived the Fokker–Planck equation for the conditional probability density which may be written $p(x_1,\ldots,x_N,t|x_1^*,\ldots,x_N^*,t^*)$. This is the fraction of an ensemble in a differential element at x at time t, given that all members of the ensemble were at x^* at an earlier time t^*, divided by the volume of the differential element. However, the Fokker–Planck eqn. (8.35) applies equally to the more general conditional probability density

$$p(x_1,\ldots,x_N,t|s(x_1^*,\ldots,x_N^*),t^*);$$

viz. the fraction of an ensemble in a differential element at x at time t, given that the ensemble was distributed with probability density $s(x_1^*,\ldots,x_N^*)$ at an earlier time t^*, divided by the volume of the differential element. This result follows from the fact that in the derivation of the Fokker–Planck equation from the Chapman–Kolmogorov equation

$$p(x,t|x^*,t^*) = \int_{-\infty}^{\infty} p(x,t|x',t')p(x',t'|x^*,t^*)\,dx' \qquad (9.1)$$

the conditional arguments x^*, t^* do not enter materially, and can be replaced by conditional arguments $s(x^*), t^*$.

To verify that the Chapman–Kolmogorov equation still holds when the conditional arguments are replaced in this way, note that the fraction of the ensemble in a differential element $\delta x_1^*\delta x_2^*\ldots\delta x_N^*$ at $(x_1^*,\ldots,x_N^*)$ at time t^* is $s(x_1^*,\ldots,x_N^*)\delta x_1^*\ldots\delta x_N^*$. This sub-ensemble has at time t a density

$$p(x_1,\ldots,x_N,t|x_1^*,\ldots,x_N^*,t^*)s(x_1^*,\ldots,x_N^*)\delta x_1^*\ldots\delta x_N^* \qquad (9.2)$$

Hence at t the probability density of the total ensemble, which we have expressed as $p(x_1,\ldots,x_N,t|s(x_1^*,\ldots,x_N^*),t^*)$, is also expressed by

$$\int_{-\infty}^{\infty}\cdots\int_{-\infty}^{\infty} p(x_1,\ldots,x_N,t|x_1^*,\ldots,x_N^*,t^*)s(x_1^*,\ldots,x_N^*)\,dx_1^*\ldots dx_N^* \qquad (9.3)$$

Therefore if the Chapman–Kolmogorov eqn. (9.1) is multiplied by $s(x^*)$, integrated with respect to x^*, and the order of integration is interchanged on the right side, it becomes

$$p(x,t|s(x^*),t^*) = \int_{-\infty}^{\infty} p(x,t|x',t')p(x',t'|s(x^*),t^*)\,dx' \qquad (9.4)$$

Alternatively one can derive the Fokker–Planck equation for $p(x,t|s(x^*),t^*)$ directly from the Fokker–Planck equation for $p(x,t|x^*,t^*)$, by multiplying the latter equation by $s(x^*)$, integrating with respect to x^*, and interchanging the order of derivative and integral operations.

Boundary conditions for the Fokker–Planck equation can be obtained as follows. When $t=t^*$, p equals the given initial probability density at t^*. Thus when in the Fokker–Planck equation $p(x,t)$ is an abbreviation for $p(x,t|s(x^*),t^*)$, the appropriate boundary condition is

$$p(x,t^*)=s(x^*) \tag{9.5}$$

On the other hand, when $p(x,t)$ is an abbreviation for $p(x,t|x^*,t^*)$, the appropriate boundary condition is

$$p(x,t^*)=\delta(x-x^*) \tag{9.6}$$

Here $\delta(x-x^*)$ is a Dirac delta function with vector argument. This is equivalent to a product of Dirac delta functions with scalar arguments

$$\delta(x-x^*)=\delta(x_1-x_1{}^*)\delta(x_2-x_2{}^*)\dots\delta(x_N-x_N{}^*) \tag{9.7}$$

10. Stationarity

In many applications both the f_i and b_{ij} are time-independent, so that the Fokker–Planck equation becomes

$$\frac{\partial p}{\partial t}=-\sum_i\frac{\partial}{\partial x_i}(f_i(x)p)+\tfrac{1}{2}\sum_{i,j}\frac{\partial^2(b_{ij}(x)p)}{\partial x_i\partial x_j} \tag{10.1}$$

In such cases we are often interested in the steady-state probability density (if any) which p approaches when t becomes large. In the steady state, $\partial p/\partial t$ is zero, i.e. the probability density is stationary, and (10.1) simplifies to the stationary Fokker–Planck equation:

$$-\sum_i\frac{\partial}{\partial x_i}(f_i(x)p)+\tfrac{1}{2}\sum_{i,j}\frac{\partial^2(b_{ij}(x)p)}{\partial x_i\partial x_j}=0 \tag{10.2}$$

or, written in full

$$-\sum_{i=1}^{N}\frac{\partial}{\partial x_i}[f_i(x_1,\dots,x_N)p(x_1,\dots,x_N)]$$
$$+\tfrac{1}{2}\sum_{i=1}^{N}\sum_{j=1}^{N}\frac{\partial^2}{\partial x_i\partial x_j}[b_{ij}(x_1,\dots,x_N)p(x_1,\dots,x_N)]=0 \tag{10.3}$$

A solution $p(x_1,\dots,x_N)$, if it is to represent a probability density function, must also satisfy

$$p(x_1,\dots,x_N)\geqslant 0 \tag{10.4}$$

and

$$\int_{-\infty}^{\infty}\dots\int_{-\infty}^{\infty}p(x_1,\dots,x_N)\,dx_1\dots dx_N=1 \tag{10.5}$$

We call any solution of (10.3) which also satisfies (10.4) and (10.5) a *stationary solution* of the Fokker–Planck equation.

Boundary conditions for the stationary Fokker–Planck equation (10.3) can usually be assigned as follows. We assume that p is zero at infinity, i.e.

$$p(x_1,\dots,x_N)\to 0\quad(x_1{}^2+x_2{}^2+\dots+x_N{}^2\to\infty) \tag{10.6}$$

and, as a consequence,

$$\frac{\partial p}{\partial x_i}(x_1,\ldots,x_N)\to 0 \quad (x_1^2+x_2^2+\ldots+x_N^2\to\infty;\quad i=1,2,\ldots,N) \tag{10.7}$$

For if (10.6) is not satisfied, (10.5) is not satisfied, unless $p(x_1,\ldots,x_N)$ is of very special form†.

11. Stability

It is often plausible that a system possesses a stationary distribution which it approaches from any given initial distribution. For example, in a control system the white noise tends to disperse the ensemble over the state space, whereas the controlling action tends to concentrate the ensemble in a desirable region of the state space. In a well-designed system we expect these two tendencies eventually to balance, with the ensemble distributed in a stationary condition and most of it in the desirable region.

Suppose an ensemble is in a stationary distribution and at a certain instant the distribution is perturbed slightly. If as time increases the ensemble tends uniformly to its original stationary distribution, let us say that the stationary distribution is asymptotically stable.

11.1. *Uniqueness of the stationary distribution*

In general eqns. (10.3), (10.4) and (10.5) might have several stationary solutions, and we should then have the difficult question of deciding which of these are asymptotically stable, and which of the latter are relevant to the problem in hand. Fortunately however it can be shown that *if there exists an asymptotically stable stationary solution, it is the only stationary solution.*

The proof is simple and proceeds by contradiction. Let $p_1(x)$ be an asymptotically stable stationary solution, and suppose there exists another stationary solution $p_2(x)$. Consider the probability density function $p_3(x)$ defined by

$$p_3(x)=(1-\lambda)p_1(x)+\lambda p_2(x) \tag{11.1}$$

where λ is a parameter satisfying

$$0\leqslant\lambda\leqslant 1 \tag{11.2}$$

Then $p_3(x)$ is also a stationary solution, in view of the linearity of differential eqn. (10.3). But when λ is small, $p_3(x)$ is a slightly perturbed version of $p_1(x)$, and its ensemble density must therefore approach $p_1(x)$ as time increases. Hence $p_3(x)$ is not stationary. This contradiction shows that our initial supposition of the existence of $p_2(x)$ is invalid.

Therefore, if the system possesses an asymptotically stable stationary solution, and if we can find, by any means whatever, a solution of (10.3), (10.4), (10.5), the two solutions are identical.

A somewhat stronger result can be stated. If the system possesses an asymptotically stable stationary solution then this solution is asymptotically stable in-the-large; i.e. an ensemble with any initial distribution approaches

† e.g. if the ensemble is distributed in a small region at infinity (say near the x_1-axis), (10.6) does not hold. However, this situation does not seem to arise in typical physical problems.

this stationary solution as time increases. A proof is as follows. Consider the following special perturbation of the stationary solution $p_1(x)$. A small fraction λ of members of the ensemble are taken from the stationary ensemble in such a way that the remaining sub-ensemble α is distributed with density

$$p_\alpha = (1-\lambda)p_1(x) \tag{11.3}$$

The removed sub-ensemble β is redistributed at time t_0 with arbitrary density $\lambda p_\beta(x, t_0)$. As time increases the sub-ensemble α remains stationary since p_α satisfies (10.3). On the other hand, the sub-ensemble β evolves with density $\lambda p_\beta(x, t)$. But since the perturbation is small, the complete perturbed distribution satisfies

$$p_\alpha(x) + \lambda p_\beta(x, t) \rightarrow p_1(x) \quad (t \rightarrow \infty) \tag{11.4}$$

From (11.3) and (11.4)

$$p_\beta(x, t) \rightarrow p_1(x) \quad (t \rightarrow \infty) \tag{11.5}$$

(11.5) states that an ensemble with arbitrary initial probability density $p_\beta(x, t_0)$ approaches the stationary density $p_1(x)$ as time increases, which is the desired result.

11.2. *Existence of a stable distribution*

Note that if a system possesses more than one stationary distribution, it possesses an infinite number of stationary distributions. This follows from the fact that a linear combination of stationary distributions is also a stationary distribution. In view of the above remarks none of this infinity of stationary distributions is asymptotically stable. An example of such a system is given in § 15.

Note also that there are systems which do not possess any stationary distribution. For example the ensemble may disperse to infinity as time increases, i.e. for any finite x

$$p(x, t) \rightarrow 0 \quad (t \rightarrow \infty) \tag{11.6}$$

(Let us agree not to count

$$p(x) = 0 \tag{11.7}$$

as a stationary solution since, strictly speaking, it does not satisfy (10.5).) A case in point is Brownian motion, for which the density $p(x, t)$ is given by (7.1).

However, in most cases of practical interest, it seems that a system possesses just one stationary distribution, which is moreover stable in-the-large. To justify this statement would require more rigour than is used in the present paper. Nevertheless the following remarks may be helpful.

Let us consider a *Markov chain*, i.e. a discrete-time Markov process, with a discretized state space. (The random walks mentioned in Part I are cases of Markov chains.) For a Markov chain with a finite number of states, it is easy to show (see, e.g., Kolmogorov 1931) that there is a stationary distribution which is asymptotically stable in-the-large, provided there is a positive (i.e. non-zero) probability of transition from every state to every other state.

Kolmogorov (1931) showed that this result generalizes to the case of a continuous Markov process provided there is a positive probability of transition from every point to every region of positive volume, in a finite time. Strictly speaking, however, Kolmogorov's result only applies if the volume of the total state space involved is finite (since for an infinite state space the probability of transition from the origin to a region of positive volume at infinity is zero). Thus Kolmogorov's result applies if the ensemble is restricted to a finite region of state space by means of 'reflecting barriers'.

Suppose we gradually move the reflecting barriers out to infinity. It is plausible that the stable stationary distribution will gradually approach a limiting distribution, and that this limiting distribution will be either the trivial distribution (11.7) or a stable stationary distribution for the system without barriers.

Thus we can heuristically assume the existence of a unique stationary distribution which is moreover asymptotically stable if (i) no part of the system is completely isolated from the effects of the white noise (ii) the system has restoring forces which prevent the ensemble from dispersing to infinity.

Therefore, whenever we find a stationary solution of the Fokker–Planck equation and these circumstances (i) and (ii) seem to apply, we shall call it *the steady-state solution.*

For more rigorous discussions of the existence of stationary solutions and related properties, see Kolmogorov (1933), Khasminski (1960), an argument due to Gray given in the paper of Caughey (1963), Wonham (1966) and Kushner (1968).

12. White noise properties

The moment rates b_{ij} of the white noise cannot be assigned completely arbitrarily. In fact, from (8.34), they must satisfy

$$b_{ii} \geqslant 0 \quad (i = 1, 2, \dots, N) \tag{12.1}$$

and

$$b_{ij} = b_{ji} \quad (i, j = 1, 2, \dots, N) \tag{12.2}$$

More generally, consider the following inequality

$$\overline{(V_1\Delta_1 + V_2\Delta_2 + \dots + V_N\Delta_N)^2} \geqslant 0 \tag{12.3}$$

where the V's are arbitrary parameters and the Δ's are the white noise increments (see (8.26)). Dividing (12.3) by δt and letting $\delta t \to 0$ we obtain

$$\sum_{i,\,j} V_i V_j b_{ij} \geqslant 0 \tag{12.4}$$

i.e.

$$V'[b_{ij}]V \geqslant 0 \tag{12.5}$$

where V is the vector with arbitrary components $V_1, V_2, \dots, V_N$ and $[b_{ij}]$ is the matrix of the b's. (12.5) shows that $[b_{ij}]$ must be positive semi-definite (Kolmogorov 1933). Hence the following inequalities (criteria of Sylvester 1852) must hold:

$$b_{11} \geqslant 0, \begin{vmatrix} b_{11} & b_{12} \\ b_{21} & b_{22} \end{vmatrix} \geqslant 0 \begin{vmatrix} b_{11} & b_{12} & b_{13} \\ b_{21} & b_{22} & b_{23} \\ b_{31} & b_{32} & b_{33} \end{vmatrix} \geqslant 0, \dots |b_{ij}| \geqslant 0 \tag{12.6}$$

From (8.35) we require that the $b_{ij}(x_1, \ldots, x_N, t)$ possess second partial derivatives with respect to the x's.

It can be shown that the b_{ij} represent the spectral densities and cross-spectral densities of the white noises ξ_i (see, e.g., Fuller 1963).

13. Probability flow

Consider a compressible fluid flowing in three dimensions. Suppose the density at point (x_1, x_2, x_3) and time t is $W(x_1, x_2, x_3, t)$ and suppose the velocity components there are $u_1(x_1, x_2, x_3, t)$, $u_2(x_1, x_2, x_3, t)$ and $u_3(x_1, x_2, x_3, t)$. Then as is well known the density obeys the partial differential equation (Euler 1755)

$$\frac{\partial W}{\partial t} = -\frac{\partial}{\partial x_1}(u_1 W) - \frac{\partial}{\partial x_2}(u_2 W) - \frac{\partial}{\partial x_3}(u_3 W) \tag{13.1}$$

Now $u_i W$ is the rate of mass flow per unit area in the x_i direction. Let us call the vector with components $(u_1 W, u_2 W, u_3 W)$ the mass flow vector J

$$J = (J_1, J_2, J_3) = (u_1 W, u_2 W, u_3 W) \tag{13.2}$$

Then (13.1) is

$$\frac{\partial W}{\partial t} = -\frac{\partial J_1}{\partial x_1} - \frac{\partial J_2}{\partial x_2} - \frac{\partial J_3}{\partial x_3} \tag{13.3}$$

Comparison of (13.3) with the Fokker–Planck eqn. (8.35) shows that p is analogous to W and

$$f_i p - \tfrac{1}{2}\sum_{j=1}^{N} \frac{\partial}{\partial x_j}(b_{ij} p) \tag{13.4}$$

is analogous to J_i. Thus we say that the ensemble 'flows in the state space', and its probability flow vector J has components

$$J_i = f_i p - \tfrac{1}{2}\sum_{j=1}^{N} \frac{\partial}{\partial x_j}(b_{ij} p) \tag{13.5}$$

The Fokker–Planck equation can then be written

$$\frac{\partial p}{\partial t} = -\sum_{i=1}^{N} \frac{\partial J_i}{\partial x_i} \tag{13.6}$$

The analogy with fluid flow goes back to Fokker (1918).

It is sometimes helpful to picture the ensemble as flowing in this way. However, it has to be remembered that individual members of the ensemble do not usually follow regular paths or stream lines. A sub-ensemble initially in a small element of the state space tends to disperse in various directions as time increases, and only on average is there a tendency to move in the direction of the flow vector.

Part II. Stationary solutions of the Fokker–Planck equation

14. Conservative systems

Statistical mechanics was originally developed for conservative systems (see § 2). It will be helpful if we also begin with conservative systems. In this section a few facts about such systems will be recalled.

Consider a body with mass m moving in a straight line in a conservative field of force. If the displacement of the body is x and the potential energy is $G(x)$ then the force on the body is $-\partial G/\partial x$ so that the equation of motion is

$$m\frac{d^2x}{dt^2} = -\frac{\partial G(x)}{\partial x} \tag{14.1}$$

Thus if y represents the momentum, the phase plane coordinates x and y satisfy

$$\left.\begin{aligned} \frac{dx}{dt} &= \frac{y}{m} \\ \frac{dy}{dt} &= -\frac{\partial G(x)}{\partial x} \end{aligned}\right\} \tag{14.2}$$

If the total energy (potential plus kinetic) is designated $H(x, y)$ we have

$$H(x,y) = G(x) + \frac{1}{2m}y^2 \tag{14.3}$$

Thus (14.2) can be written

$$\left.\begin{aligned} \frac{dx}{dt} &= \frac{\partial H(x,y)}{\partial y} \\ \frac{dy}{dt} &= -\frac{\partial H(x,y)}{\partial x} \end{aligned}\right\} \tag{14.4}$$

More generally, for a mechanical system with n degrees of freedom and energy $H(x_1, \ldots, x_n, y_1, \ldots, y_n)$, the equations of motion can be written

$$\left.\begin{aligned} \frac{dx_i}{dt} &= \frac{\partial H}{\partial y_i} \\ \frac{dy_i}{dt} &= -\frac{\partial H}{\partial x_i} \end{aligned}\right\} \quad (i = 1, 2, \ldots, n) \tag{14.5}$$

Here the x_i are the generalized coordinates ('positions') and the y_i are called the momenta (mass times 'velocity'). System (14.5) is a special case of (8.1) in the noise-free case ($\xi = 0$) with N even and

$$\left.\begin{aligned} n &= \tfrac{1}{2}N \\ y_i &= x_{1/2N+i} \end{aligned}\right\} \quad (i = 1, 2, \ldots, \tfrac{1}{2}N) \tag{14.6}$$

Equations (14.5) were derived by Hamilton (1835) and are called Hamiltonian equations, or canonical equations (Jacobi 1837). However, they had already been touched on by Lagrange† (1809, 1811). They were discussed extensively by Jacobi (1842). H is called the Hamiltonian.

The same terminology is retained for any dynamic system (i.e. not necessarily a mechanical system) whose equations of motion can be put in form (14.5).

† Hamiltonian equations evolved from the efforts of astronomers to calculate the perturbations of planets from their otherwise elliptical orbits; these perturbations arising from the mutual interaction of the planets. See Cayley (1857) for the history.

Such a system is called conservative since the Hamiltonian remains constant during the motion of the system. This property is easily proved, as follows (Hamilton 1835)

$$\frac{d}{dt}H(x_1,\ldots,x_n,y_1,\ldots,y_n)=\frac{\partial H}{\partial x_1}\frac{dx_1}{dt}+\ldots+\frac{\partial H}{\partial x_n}\frac{dx_n}{dt}$$
$$+\frac{\partial H}{\partial y_1}\frac{dy_1}{dt}+\ldots+\frac{\partial H}{\partial y_n}\frac{dy_n}{dt} \tag{14.7}$$

Consider the terms involving x_1 and y_1 on the right of (14.7). Their contribution is

$$\frac{\partial H}{\partial x_1}\frac{dx_1}{dt}+\frac{\partial H}{\partial y_1}\frac{dy_1}{dt} \tag{14.8}$$

or, in view of (14.5)

$$\frac{\partial H}{\partial x_1}\frac{\partial H}{\partial y_1}+\frac{\partial H}{\partial y_1}\left(-\frac{\partial H}{\partial x_1}\right)=0 \tag{14.9}$$

Similarly all the terms on the right of (14.7) cancel, so that

$$\frac{dH}{dt}=0 \tag{14.10}$$

i.e.

$$H(t)=\text{const.} \tag{14.11}$$

Control engineers are to some extent acquainted with Hamiltonian equations, thanks to Pontryagin's maximum principle (Pontryagin *et al.* 1961). For further discussion of Hamiltonian mechanics see, e.g., Whittaker (1937) or ter Haar (1961).

15. Ensembles of conservative systems

Our aim is to find solutions of the stationary Fokker–Planck equation for systems of gradually increasing complexity. We begin with the case when the white noise disturbances are absent.

The b_{ij} are then all zero, so that the Fokker–Planck eqn. (8.35) reduces to

$$\frac{\partial}{\partial t}p(x_1,\ldots,x_N,t)=-\sum_{i=1}^{N}\frac{\partial}{\partial x_i}[f_i(x_1,\ldots,x_N,t)p(x_1,\ldots,x_N,t)] \tag{15.1}$$

Equation (15.1) describes how the probability density of an ensemble of identical deterministic systems evolves, given that the ensemble was initially distributed in state space with probability density $s(x_1^*,\ldots,x_N^*)$ at time t^*.

Suppose the members of the ensemble are conservative systems, each satisfying (14.5). Then (15.1) becomes

$$\frac{\partial}{\partial t}p(x_1,\ldots,x_n,y_1,\ldots,y_n,t)$$
$$=-\sum_{i=1}^{n}\frac{\partial}{\partial x_i}\left[p(x_1,\ldots,x_n,y_1,\ldots,y_n,t)\frac{\partial}{\partial y_i}H(x_1,\ldots,x_n,y_1,\ldots,y_n)\right]$$
$$+\sum_{i=1}^{n}\frac{\partial}{\partial y_i}\left[p(x_1,\ldots,x_n,y_1,\ldots,y_n,t)\frac{\partial}{\partial x_i}H(x_1,\ldots,x_n,y_1,\ldots,y_n)\right] \tag{15.2}$$

On differentiating the products on the right of (15.2) we see that the terms $p(\partial^2 H/\partial x_i \partial y_i)$ all cancel. Hence (15.2) simplifies to

$$\frac{\partial p}{\partial t} = -\sum_{i=1}^{n} \frac{\partial p}{\partial x_i}\frac{\partial H}{\partial y_i} + \sum_{i=1}^{n} \frac{\partial p}{\partial y_i}\frac{\partial H}{\partial x_i} \tag{15.3}$$

Let us seek a stationary solution of (15.3), i.e. a solution of

$$\sum_{i=1}^{n} \left(\frac{\partial p}{\partial x_i}\frac{\partial H}{\partial y_i} - \frac{\partial p}{\partial y_i}\frac{\partial H}{\partial x_i}\right) = 0 \tag{15.4}$$

This problem, treated by Boltzmann (1871), is well known in statistical mechanics and can be solved as follows. Consider a special ensemble, all of which is initially distributed with uniform density between two neighbouring surfaces in the phase space. Suppose the surfaces are constant energy surfaces, one being given by

$$H(x_1, \ldots, x_n, y_1, \ldots, y_n) = I \tag{15.5}$$

and the other by

$$H(x_1, \ldots, x_n, y_1, \ldots, y_n) = I + \delta I \tag{15.6}$$

We call such an ensemble a shell ensemble†. Then the ensemble will stay between these two surfaces, since for each member of the ensemble $H(t)$ is constant (see (14.11)). Moreover a theorem of Liouville (1838) states that (for conservative systems) the motion of the ensemble is similar to that of an incompressible fluid‡; i.e. any set of representative points is displaced under the phase motion in such a way that the phase volume occupied by the set remains unchanged. It follows that the shell ensemble not only remains between the two surfaces, but also retains its initial uniform density. Hence the shell ensemble is a stationary ensemble.

Now consider an ensemble which is built of non-overlapping shell sub-ensembles. This ensemble will also be stationary. Each shell has uniform density, but the density can vary from shell to shell. Consequently the ensemble has a probability density which can be a general function of energy H.

The above argument suggests that for conservative systems any probability density which is simply a function of H is stationary. We can immediately verify this result from (15.4). Thus substituting

$$p(x_1, \ldots, x_n, y_1, \ldots, y_n) = p(H(x_1, \ldots, x_n, y_1, \ldots, y_n)) \tag{15.7}$$

in (15.4) we find the left side becomes

$$\sum_{i=1}^{n} \left(\frac{dp}{dH}\frac{\partial H}{\partial x_i}\frac{\partial H}{\partial y_i} - \frac{dp}{dH}\frac{\partial H}{\partial y_i}\frac{\partial H}{\partial x_i}\right) \tag{15.8}$$

which is zero, i.e. (15.4) is satisfied identically by (15.7).

---

† This term was used by Ehrenfest and Ehrenfest (1912). Shell ensembles are in effect what were called ergoden by Boltzmann (1884) and microcanonical ensembles by Gibbs (1902).

‡ Liouville did not actually state the theorem in this picturesque fashion. It was Boltzmann (1868, 1871 a) who interpreted one of Liouville's equations as a condition for incompressible flow.

Thus the noise-free conservative system has an infinity of stationary probability densities, all of the form $p(H)$. From § 11 none of the stationary distributions is asymptotically stable.

16. Solution of the Fokker–Planck equation for a second-order system

Let us generalize the simple conservative system (14.1) by adding a damping term $k(dx/dt)$ and a (scalar) white noise forcing function $\xi(t)$ with second moment rate b. k and b are positive constants. The system then becomes the (non-linear, non-conservative, stochastic) system

$$m\frac{d^2x}{dt^2}+k\frac{dx}{dt}+\frac{\partial G(x)}{\partial x}=\xi(t) \tag{16.1}$$

i.e. with $y=m(dx/dt)$, the system

$$\left.\begin{aligned}\frac{dx}{dt}&=\frac{y}{m}\\ \frac{dy}{dt}&=-\frac{\partial G(x)}{\partial x}-\frac{k}{m}y+\xi\end{aligned}\right\} \tag{16.2}$$

Retaining the symbol H for the energy (potential plus kinetic), so that

$$H=G(x)+\frac{1}{2m}y^2 \tag{16.3}$$

we see that (16.2) is a special case of the system

$$\left.\begin{aligned}\frac{dx}{dt}&=\frac{\partial H}{\partial y}\\ \frac{dy}{dt}&=-\frac{\partial H}{\partial x}-k\frac{\partial H}{\partial y}+\xi\end{aligned}\right\} \tag{16.4}$$

Let us seek the steady-state probability density for the general system (16.4).

The stationary Fokker–Planck eqn. (10.2) for system (16.4) is

$$\left[-\frac{\partial}{\partial x}\left(\frac{\partial H}{\partial y}p\right)+\frac{\partial}{\partial y}\left(\frac{\partial H}{\partial x}p\right)\right]+\frac{\partial}{\partial y}\left(k\frac{\partial H}{\partial y}p\right)+\tfrac{1}{2}b\frac{\partial^2 p}{\partial y^2}=0 \tag{16.5}$$

The terms in the square brackets are the terms which would remain if the damping and white noise were removed from the system, i.e. if the system were conservative. From § 15 these terms cancel if p is any function of the energy H. Therefore if we can find a particular such function $p(H)$ which makes the remaining two terms in (16.5) cancel, we shall have found a solution of (16.5). With this end in view, we substitute

$$p(x,y)=p(H(x,y)) \tag{16.6}$$

in (16.5), which becomes

$$\frac{\partial}{\partial y}\left(k\frac{\partial H}{\partial y}p(H)\right)+\tfrac{1}{2}b\frac{\partial^2}{\partial y^2}p(H)=0 \tag{16.7}$$

Integration with respect to y yields

$$k\frac{\partial H}{\partial y}p(H)+\frac{b}{2}\frac{\partial}{\partial y}p(H)=L(x) \tag{16.8}$$

where $L(x)$ is an arbitrary function.

In view of (10.6) and (10.7) we assume that the left side of (16.8) vanishes when $y \to \infty$. Hence

$$L(x)=0 \tag{16.9}$$

and (16.8) simplifies to

$$k\frac{\partial H}{\partial y}p(H)+\frac{b}{2}\frac{dp}{dH}\frac{\partial H}{\partial y}=0 \tag{16.10}$$

We assume that $\partial H/\partial y$ is not identically zero. (Otherwise H is independent of y and hence so is p. Hence (10.5) implies that p is identically zero.) Then (16.10) yields

$$\frac{dp}{dH}+\frac{2k}{b}p=0 \tag{16.11}$$

The general solution of (16.11) is

$$p=C\exp\left[-\frac{2k}{b}H\right] \tag{16.12}$$

Here C is a constant, which is to be chosen to normalize the solution.

Our result is that for the second-order system

$$\left.\begin{aligned}\frac{dx}{dt}&=\frac{\partial H}{\partial y}\\ \frac{dy}{dt}&=-\frac{\partial H}{\partial x}-k\frac{\partial H}{\partial y}+\xi\end{aligned}\right\} \tag{16.13}$$

with

$$b=\text{const.},\quad k=\text{const.} \tag{16.14}$$

the steady-state probability density is

$$p(x,y)=C\exp\left[-2\frac{k}{b}H(x,y)\right] \tag{16.15}$$

Example

If we write the nonlinear term in (16.1) as $g(x)$ we have

$$g(x)=\frac{\partial G(x)}{\partial x} \tag{16.16}$$

Then from (16.3)

$$H=\int_0^x g(\theta)\,d\theta+g(0)+\frac{1}{2m}y^2 \tag{16.17}$$

Thus for the system

$$m\frac{d^2x}{dt^2}+k\frac{dx}{dt}+g(x)=\xi \quad (b=\text{const.}) \tag{16.18}$$

result (16.15) yields for the steady-state probability density

$$p(x,y)=C\exp\left[-\frac{k}{b}\left(2\int_0^x g(\theta)\,d\theta+\frac{1}{m}y^2\right)\right] \tag{16.19}$$

where $y=m(dx/dt)$ and C is a normalizing constant†. Solution (16.19) is well known (e.g. Chuang and Kazda 1959, Barrett 1960, Ariaratnam 1960).

Example

Consider a control system with a double-integrator plant satisfying

$$M\frac{d^2X}{dt^2}=U(t) \tag{16.20}$$

where X is the system output and U is the plant input. U is obtained from a nonlinear amplifier with characteristic

$$U=F(V) \tag{16.21}$$

The command signal $S(t)$ is Brownian motion (in the restricted sense of § 7), thus

$$\frac{dS}{dt}=\xi \tag{16.22}$$

The system error $E(t)$ is

$$E=S-X \tag{16.23}$$

The control signal $V(t)$ is a linear combination of system error and output rate

$$V=E-k\frac{dX}{dt} \tag{16.24}$$

With the output rate designated as Y

$$\frac{dX}{dt}=Y \tag{16.25}$$

the system equations are

$$\left.\begin{aligned}\frac{dY}{dt}&=\frac{1}{M}F(V)\\ \frac{dV}{dt}&=-Y-\frac{k}{M}F(V)+\xi\end{aligned}\right\} \tag{16.26}$$

This is a case of system (16.13), with $x=Y, y=V$ and

$$H=\tfrac{1}{2}Y^2+\frac{1}{M}\int_0^V F(\theta)\,d\theta \tag{16.27}$$

Hence result (16.15) shows that the steady-state probability density is

$$p=C'\exp\left[-\frac{k}{b}\left(Y^2+\frac{2}{M}\int_0^V F(\theta)\,d\theta\right)\right] \tag{16.28}$$

† C represents different values in (16.5) and (16.19).

i.e. in view of (16.24) and (16.25)

$$p(Y, E) = C \exp\left[-\frac{k}{b}\left(Y^2 + \frac{2}{M}\int_0^{E-kY} F(\theta)\,d\theta\right)\right] \tag{16.29}$$

In particular, for a relay control system, when

$$F(V) = \operatorname{sgn} V \tag{16.30}$$

result (16.29) becomes

$$p(Y, E) = \frac{k^{3/2}}{\pi^{1/2} b^{3/2} M} \exp\left[-\frac{k}{b}\left(Y^2 + \frac{2}{M}|E - kY|\right)\right] \tag{16.31}$$

Result (16.31) was originally obtained by trial and error by the writer and was quoted in the paper of his colleague Barrett (1960).

17. Solution of the Fokker–Planck equation for a higher-order system

Starting with the general conservative system (14.5), let us make it non-conservative and stochastic by adding to the last equation a damping term $-k(\partial H/\partial y_n)$ and a scalar white noise ξ with second moment rate b. k and b are positive constants. The system then becomes

$$\left.\begin{aligned} \frac{dx_i}{dt} &= \frac{\partial H}{\partial y_i} \quad (i = 1, 2, \ldots, n) \\ \frac{dy_i}{dt} &= -\frac{\partial H}{\partial x_i} \quad (i = 1, 2, \ldots, n-1) \\ \frac{dy_n}{dt} &= -\frac{\partial H}{\partial x_n} - k\frac{\partial H}{\partial y_n} + \xi \end{aligned}\right\} \tag{17.1}$$

which is a higher-order generalization of system (16.4). The stationary Fokker–Planck equation is

$$\left[\sum_{i=1}^{n}\left\{-\frac{\partial}{\partial x_i}\left(\frac{\partial H}{\partial y_i}p\right) + \frac{\partial}{\partial y_i}\left(\frac{\partial H}{\partial x_i}p\right)\right\}\right] + \frac{\partial}{\partial y_n}\left(k\frac{\partial H}{\partial y_n}p\right) + \frac{b}{2}\frac{\partial^2 p}{\partial y_n^2} = 0 \tag{17.2}$$

The terms in the square brackets are the terms that would remain if the damping and white noise were absent, i.e. terms which correspond to a conservative system. As in § 16 these terms cancel if p is a function of energy H. Substituting

$$p = p(H) \tag{17.3}$$

in (17.2) we obtain

$$\frac{\partial}{\partial y_n}\left(k\frac{\partial H}{\partial y_n}p\right) + \frac{b}{2}\frac{\partial^2 p}{\partial y_n^2} = 0 \tag{17.4}$$

This is the same as (16.7) with y_n written for y. Hence the solution is obtained in the same way as for (16.7) and is

$$p(x_1, \ldots, x_n, y_1, \ldots, y_n) = C \exp\left[-\frac{2k}{b}H(x_1, \ldots, x_n, y_1 \ldots, y_n)\right] \tag{17.5}$$

where C is a normalizing constant.

Solution (17.5) is the probability density function for a Maxwell–Boltzmann distribution (compare (2.5)). This result is not surprising in view of the connection of the Maxwell–Boltzmann distribution with the historical development of the Fokker–Planck equation (see §§ 2 and 3). Thus (17.1) is analogous to a Langevin equation for a particle with n degrees of freedom immersed in a gas.

Example

Consider a vibration isolating suspension with two degrees of freedom satisfying the equations (Ariaratnam 1960)

$$\left.\begin{aligned} m_1 \frac{d^2x_1}{dt^2} + \frac{\partial G(x_1, x_2)}{\partial x_1} &= 0 \\ m_2 \frac{d^2x_2}{dt^2} + k\frac{dx_2}{dt} + \frac{\partial G(x_1, x_2)}{\partial x_2} &= \xi \quad (b = \text{const.}) \end{aligned}\right\} \tag{17.6}$$

where $G(x_1, x_2)$ represents potential energy of springs. In normal form the system is

$$\left.\begin{aligned} \frac{dx_1}{dt} &= \frac{y_1}{m_1} \\ \frac{dx_2}{dt} &= \frac{y_2}{m_2} \\ \frac{dy_1}{dt} &= -\frac{\partial G}{\partial x_1} \\ \frac{dy_2}{dt} &= -\frac{\partial G}{\partial x_2} - \frac{k}{m_2} y_2 + \xi \end{aligned}\right\} \tag{17.7}$$

Here

$$H = G(x_1, x_2) + \frac{1}{2m_1} y_1{}^2 + \frac{1}{2m_2} y_2{}^2 \tag{17.8}$$

From (17.5) the steady-state probability density is

$$p(x_1, x_2, y_1, y_2) = C \exp\left[-\frac{k}{b}\left\{ 2G(x_1, x_2) + \frac{y_1{}^2}{m_1} + \frac{y_2{}^2}{m_2} \right\}\right] \tag{17.9}$$

which is the result given by Ariaratnam (1960).

18. Systems with uncorrelated white noises

In §§ 16 and 17 the case of a scalar white noise was treated. We now generalize to a vector white noise, treating the system

$$\left.\begin{aligned} \frac{dx_i}{dt} &= \frac{\partial H}{\partial y_i} && (i = 1, 2, \ldots, n) \\ \frac{dy_i}{dt} &= -\frac{\partial H}{\partial x_i} - k_i \frac{\partial H}{\partial y_i} + \xi_i && (i = 1, 2, \ldots, n) \end{aligned}\right\} \tag{18.1}$$

This represents a conservative system to which has been added damping terms $k_i(\partial H/\partial y_i)$ and white noise terms ξ_i. Let the k_i and the b_{ij} be constants, and to begin with let the ξ_i have no cross-correlation, so that

$$b_{ij}=0 \quad (i \neq j) \tag{18.2}$$

and we can write

$$b_{ii}=b_i \quad (i=1,2,\ldots,n) \tag{18.3}$$

The stationary Fokker–Planck equation is then

$$\sum_{i=1}^{n}\left[-\frac{\partial}{\partial x_i}\left(\frac{\partial H}{\partial y_i}p\right)+\frac{\partial}{\partial y_i}\left(\frac{\partial H}{\partial x_i}p\right)\right]+\sum_{i=1}^{n}\left[\frac{\partial}{\partial y_i}\left(k_i\frac{\partial H}{\partial y_i}p\right)+\frac{b_i}{2}\frac{\partial^2 p}{\partial y_i^2}\right]=0 \tag{18.4}$$

As before we seek a solution of the form

$$p=p(H) \tag{18.5}$$

so that (18.4) reduces to

$$\sum_{i=1}^{n}\left[\frac{\partial}{\partial y_i}\left(k_i\frac{\partial H}{\partial y_i}p(H)\right)+\frac{b_i}{2}\frac{\partial^2 p(H)}{\partial y_i^2}\right]=0 \tag{18.6}$$

Let us write (18.6) as

$$\sum_{i=1}^{n} A_i=0 \tag{18.7}$$

and seek a solution of (18.6) which makes each term A_i separately zero. A_1 equated to zero is an equation similar to (18.4), and its solution, analogous to (17.5), is

$$p=C_1\exp\left[-\frac{2k_1}{b_1}H\right] \tag{18.8}$$

Equating all the A_i to zero, we obtain

$$p=C_i\exp\left[-\frac{2k_i}{b_i}H\right] \quad (i=1,2,\ldots,n) \tag{18.9}$$

For (18.9) to be self-consistent we must have

$$\frac{2k_1}{b_1}=\frac{2k_2}{b_2}=\ldots=\frac{2k_n}{b_n}\equiv l, \quad \text{say} \tag{18.10}$$

Thus, if the damping coefficients k_i satisfy (18.10), the steady-state solution for system (18.1) is

$$p(x_1,\ldots,x_n,y_1,\ldots,y_n)=C\exp\left[-lH(x_1,\ldots,x_n,y_1,\ldots,y_n)\right] \tag{18.11}$$

It should be noted however that the restriction (18.10) on the damping coefficients k_i implies that the range of systems for which (18.11) is applicable is quite limited.

In statistical mechanics the result, that each degree of freedom is associated with the same mean energy (equipartition of energy) is analogous to requirement (18.10). Thus it is not surprising that with (18.11) we have again arrived at a Maxwell–Boltzmann distribution.

Result (18.11) includes a corrected version of an assertion of Chuang and Kazda (1959). They treated a higher-order system but did not specify the conditions for the validity of their solution. Their system, in our notation†, is

$$\left.\begin{aligned}\frac{dx_i}{dt} &= y_i \\ \frac{dy_i}{dt} &= -\sum_{j=1}^{n} g_{ij}(x_j) - k_i y_i + \xi_i\end{aligned}\right\} \quad (i=1,2,\ldots,n) \tag{18.12}$$

They imply that if "$g_{ij}(x_j)$ and $g_{ji}(x_j)$ are of the same functional form" the probability density function 'usually' has the form

$$p = C \exp\left[\sum_{i=1}^{n}\left\{\sum_{j=1}^{n} d_i \int_0^{x_i} g_{ij}(x_j)\,dx_i + e_i y_i^2\right\}\right] \tag{18.13}$$

Actually, however (18.13) usually fails to satisfy the Fokker–Planck equation, as one may check for the case $n=2$. Investigation shows that, to be valid, (18.13) should have d_i replaced by d_{ij}, the cross-coupling terms should be linear

$$g_{ij}(\theta) = g_{ji}(\theta) = c_{ij}\theta \quad (i \neq j) \tag{18.14}$$

and (18.10) should hold. But in this case system (18.12) is a special case of system (18.1) with

$$H = \sum_i \left\{\sum_{j\neq i} \tfrac{1}{2} c_{ij} x_i x_j + \int_0^{x_i} g_{ii}(\theta)\,d\theta + \tfrac{1}{2} y_i^2\right\} \tag{18.15}$$

so that our result (18.11) applies.

Example

Consider the following generalization of the vibration isolating suspension (17.6)

$$\left.\begin{aligned}m_1\frac{d^2x_1}{dt^2} + k_1\frac{dx_1}{dt} + \frac{\partial G(x_1,x_2)}{\partial x_1} &= \xi_1 \\ m_2\frac{d^2x_2}{dt^2} + k_2\frac{dx_2}{dt} + \frac{\partial G(x_1,x_2)}{\partial x_2} &= \xi_2\end{aligned}\right\} \tag{18.16}$$

where the k's and b's are constants and satisfy

$$\left.\begin{aligned} b_{12} = b_{21} &= 0 \\ \frac{k_1}{b_{11}} = \frac{k_2}{b_{22}} &= \frac{k}{b}\end{aligned}\right\} \tag{18.17}$$

In normal form the system is

$$\left.\begin{aligned}\frac{dx_1}{dt} &= \frac{y_1}{m_1} \\ \frac{dx_2}{dt} &= \frac{y_2}{m_2} \\ \frac{dy_1}{dt} &= -\frac{\partial G}{\partial x_1} - \frac{k_1}{m_1} y_1 + \xi_1 \\ \frac{dy_2}{dt} &= -\frac{\partial G}{\partial x_2} - \frac{k_2}{m_2} y_2 + \xi_2\end{aligned}\right\} \tag{18.18}$$

† Their $K_i(x_i)$ is included in our $g_{ii}(x_i)$.

so that

$$H = G(x_1, x_2) + \frac{1}{2m_1} y_1^2 + \frac{1}{2m_2} y_2^2 \tag{18.19}$$

From (18.11) the steady-state probability density is

$$p(x_1, x_2, y_1, y_2) = C \exp\left[-\frac{k}{b}\left\{ 2G(x_1, x_2) + \frac{1}{m_1} y_1^2 + \frac{1}{m_2} y_2^2 \right\}\right] \tag{18.20}$$

which is a result obtained by Ariaratnam (1960). (18.20) is exactly the same as (17.9).

Caughey (1963 b) has treated the generalization of this example to the case when there are n degrees of freedom.

Example

Caughey (1960) and Ariaratnam (1962) have considered the response to random excitation of a light elastic string loaded at equal intervals by a number of equal masses. If the string is stretched between two fixed supports and carries n particles each of mass m, the equations of motion, with nonlinear effects taken into account, are

$$m\frac{d^2x_i}{dt^2} + k\frac{dx_i}{dt} - \left\{ \omega + \sum_{j=1}^{n+1} (x_j - x_{j-1})^2 \right\}(x_{i-1} - 2x_i + x_{i+1}) = \xi_i \; (i = 1, 2, \ldots, n) \tag{18.21}$$

with boundary conditions

$$x_0(t) = x_{n+1}(t) = 0 \tag{18.22}$$

The white noise satisfies

$$\left.\begin{aligned} b_{ij} &= 0 && (i \neq j) \\ b_{ii} &= b = \text{const.} && (i = 1, 2, \ldots, n) \end{aligned}\right\} \tag{18.23}$$

The potential energy in the stretched string is

$$G(x_1, \ldots, x_n) = \tfrac{1}{4}\left\{ \omega + \sum_{j=1}^{n+1} (x_j - x_{j-1})^2 \right\}^2 \tag{18.24}$$

Thus system (18.21) may be written

$$\left.\begin{aligned} \frac{dx_i}{dt} &= \frac{y_i}{m} \\ \frac{dy_i}{dt} &= -\frac{\partial G}{\partial x_i} - \frac{k}{m} y_i + \xi_i \end{aligned}\right\} (i = 1, 2, \ldots, n) \tag{18.25}$$

This is a case of system (18.1) with

$$H = G + \frac{1}{2m}\sum_{j=1}^{n} y_j^2 \tag{18.26}$$

Hence, from (18.11), the steady-state probability density is

$$p(x_1,\ldots,x_n,y_1,\ldots,y_n)=C\exp\left[-\frac{k}{2b}\left\{\omega+\sum_{j=1}^{n+1}(x_j-x_{j-1})^2\right\}^2-\frac{k}{bm}\sum_{j=1}^{n}y_j^2\right] \tag{18.27}$$

Ariaratnam (1962) applied the Fokker–Planck equation to this example, but his results were expressed somewhat differently. He used a linear transformation of coordinates, working with what would be the modes of the system if it were linear. This approach enabled him to compare his results with those of Caughey (1960) who had investigated the system by means of an approximate linearization technique. A related example treated by Caughey (1963 a) is a case of system (18.1), and can similarly be solved directly rather than by the linear transformation technique used by Caughey.

19. Systems with correlated white noises

Let us generalize the results of the last section to the case when the ξ_i are cross-correlated, i.e. we drop restriction (18.2). At the same time we add 'cross-damping' terms $k_{ij}(\partial H/\partial y_j)$ to the system, so that it becomes

$$\left.\begin{aligned}\frac{dx_i}{dt}&=\frac{\partial H}{\partial y_i}\\ \frac{dy_i}{dt}&=-\frac{\partial H}{\partial x_i}-\sum_{j=1}^{n}k_{ij}\frac{\partial H}{\partial y_j}+\xi_i\end{aligned}\right\}\quad(i=1,2,\ldots,n) \tag{19.1}$$

The stationary Fokker–Planck equation is now

$$\sum_{i=1}^{n}\left[-\frac{\partial}{\partial x_i}\left(\frac{\partial H}{\partial y_i}p\right)+\frac{\partial}{\partial y_i}\left(\frac{\partial H}{\partial x_i}p\right)\right]+\sum_{i=1}^{n}\sum_{j=1}^{n}\left[\frac{\partial}{\partial y_i}\left(k_{ij}\frac{\partial H}{\partial y_j}p\right)+\tfrac{1}{2}b_{ij}\frac{\partial^2 p}{\partial y_i\partial y_j}\right]=0 \tag{19.2}$$

As before we seek a solution

$$p=p(H) \tag{19.3}$$

(19.2) then becomes

$$\sum_{i=1}^{n}\sum_{j=1}^{n}\left[\frac{\partial}{\partial y_i}\left(k_{ij}\frac{\partial H}{\partial y_j}p(H)\right)+\tfrac{1}{2}b_{ij}\frac{\partial^2 p(H)}{\partial y_i\partial y_j}\right]=0 \tag{19.4}$$

i.e.

$$\sum_{i=1}^{n}\sum_{j=1}^{n}A_{ij}=0 \tag{19.5}$$

We seek a solution which makes each term A_{ij} separately zero, i.e.

$$\frac{\partial}{\partial y_i}\left[k_{ij}\frac{\partial H}{\partial y_j}p(H)+\tfrac{1}{2}b_{ij}\frac{\partial p(H)}{\partial y_j}\right]=0\quad(i,j=1,2,\ldots,n) \tag{19.6}$$

(19.6) is analogous to (17.4), and its solution is therefore

$$p=C_{ij}\exp\left[-2\frac{k_{ij}}{b_{ij}}H\right]\quad(i,j=1,2,\ldots,n) \tag{19.7}$$

For eqns. (19.7) to be consistent with one another we must impose the restriction that all the ratios k_{ij}/b_{ij} are equal

$$\frac{k_{ij}}{b_{ij}} = \tfrac{1}{2}l \quad \text{say} \quad (i, j = 1, 2, \ldots, n) \tag{19.8}$$

Thus if the damping coefficients k_{ij} satisfy (19.8), the steady-state probability density for system (19.1) is

$$p(x_1, \ldots, x_n, y_1, \ldots, y_n) = C \exp\left[-lH(x_1, \ldots, x_n, y_1, \ldots, y_n)\right] \tag{19.9}$$

Here again restriction (19.8) limits considerably the applicability of the result.

Note that if $l > 0$ it follows from (19.9) and (10.6) that H must approach $+\infty$ when the state point approaches infinity. It also follows from (19.8) and (12.6) that the matrix of damping coefficients $[k_{ij}]$ must be positive semi-definite.

20. Generalized damping for a second-order system

So far we have restricted attention to the case when the damping terms added to the conservative system have the form $k_{ij}(\partial H/\partial y_j)$ where the k_{ij} are constants. Let us now investigate whether the damping terms can be generalized in such a way that we still obtain a probability density of the form

$$p = p(H) \tag{20.1}$$

For simplicity we begin with the second-order system

$$\left.\begin{aligned} \frac{dx}{dt} &= \frac{\partial H(x,y)}{\partial y} \\ \frac{dy}{dt} &= -\frac{\partial H(x,y)}{\partial x} - K(x,y) + \xi \end{aligned}\right\} \tag{20.2}$$

where $K(x, y)$ is the 'damping' term whose form is to be chosen. The stationary Fokker–Planck equation is

$$-\frac{\partial}{\partial x}\left(\frac{\partial H}{\partial y} p\right) + \frac{\partial}{\partial y}\left(\frac{\partial H}{\partial x} p\right) + \frac{\partial}{\partial y}(Kp) + \tfrac{1}{2}b\frac{\partial^2 p}{\partial y^2} = 0 \tag{20.3}$$

When p is of the form $p(H)$, (20.3) reduces to

$$\frac{\partial}{\partial y}(Kp) + \tfrac{1}{2}b\frac{\partial^2 p}{\partial y^2} = 0 \tag{20.4}$$

Integration with respect to y yields

$$K(x,y)p(H(x,y)) + \tfrac{1}{2}b\frac{\partial p(H(x,y))}{\partial y} = L(x) \tag{20.5}$$

We assume that p and $\partial p/\partial y$ approach zero at infinity so fast that the left side of (20.5) approaches zero, so that $L(x)$ is zero. Then (20.5) yields

$$K(x,y) = -\tfrac{1}{2}b\frac{\dfrac{\partial p(H)}{\partial y}}{p(H)} \tag{20.6}$$

$$= -\tfrac{1}{2}b \frac{\partial}{\partial y} \log p(H) \tag{20.7}$$

The right side of (20.7) is the derivative with respect to y of a function of H. Hence so must be the left side, i.e. K must be chosen to be of the form

$$K(x,y) = \frac{\partial}{\partial y}(E(H)) = \frac{\partial E(H)}{\partial H}\frac{\partial H}{\partial y} \tag{20.8}$$

i.e.

$$K(x,y) = D(H(x,y))\frac{\partial H(x,y)}{\partial y} \tag{20.9}$$

where D and E are related by

$$E(H) - E(0) = \int_0^H D(\theta)\,d\theta \tag{20.10}$$

From (20.7) and (20.8)

$$\frac{\partial}{\partial y}\log p(H) = \frac{\partial}{\partial y}\left(-\frac{2}{b}E(H)\right) \tag{20.11}$$

integration of which gives

$$p = C' \exp\left[-\frac{2}{b}E(H)\right] \tag{20.12}$$

i.e. in view of (20.10),

$$p = C \exp\left[-\frac{2}{b}\int_0^H D(\theta)\,d\theta\right] \tag{20.13}$$

Our result is that for the system

$$\left.\begin{aligned} \frac{dx}{dt} &= \frac{\partial H}{\partial y} \\ \frac{dy}{dt} &= -\frac{\partial H}{\partial x} - D(H)\frac{\partial H}{\partial y} + \xi \end{aligned}\right\} \tag{20.14}$$

where $D(H)$ is an arbitrary function of H and b is a constant, the steady-state probability density is

$$p(x,y) = C \exp\left[-\frac{2}{b}\int_0^{H(x,y)} D(\theta)\,d\theta\right] \tag{20.15}$$

Example

When

$$H = \tfrac{1}{2}(x^2 + y^2) \tag{20.16}$$

and

$$D(H) = 2H \tag{20.17}$$

the system is

$$\left.\begin{aligned} \frac{dx}{dt} &= y \\ \frac{dy}{dt} &= -x - x^2 y - y^3 + \xi \end{aligned}\right\} \tag{20.18}$$

and its steady-state probability density is

$$p(x, y) = C \exp\left[-\frac{1}{2b}(x^2+y^2)^2 \right] \tag{20.19}$$

This density function is very flat near the origin, since the damping there is small. It is also very small far from the origin, because the damping there is very large.

Example

When

$$\left.\begin{aligned} H &= \tfrac{1}{2}(x^2+y^2) \\ D(H) &= \frac{1}{2H} \\ b &= 1 \end{aligned}\right\} \tag{20.20}$$

the system is

$$\left.\begin{aligned} \frac{dx}{dt} &= y \\ \frac{dy}{dt} &= -x - \frac{y}{x^2+y^2} + \xi \end{aligned}\right\} \tag{20.21}$$

and its steady-state density is formally

$$p(x, y) = C\frac{1}{x^2+y^2} \tag{20.22}$$

Evaluation of C would show $C=0$. Presumably this implies that (20.22) is a Dirac delta function

$$p(x, y) = \delta(x, y) = \delta(x)\delta(y) \tag{20.23}$$

Result (20.23) is to be expected since the damping is infinite at the origin.

21. Generalized damping for higher-order systems

We can extend the results of § 20 to higher-order systems in the same way that §§ 17, 18 and 19 extend to higher-order systems the results of § 16.

Thus, replacing the damping term in (17.1) by the more general term $D(H)(\partial H/\partial y)$, we obtain the system

$$\left.\begin{aligned} \frac{dx_i}{dt} &= \frac{\partial H}{\partial y_i} \quad (i=1, 2, \ldots, n) \\ \frac{dy_i}{dt} &= -\frac{\partial H}{\partial x_i} \quad (i=1, 2, \ldots, n-1) \\ \frac{dy_n}{dt} &= -\frac{\partial H}{\partial x_n} - D(H)\frac{\partial H}{\partial y} + \xi \end{aligned}\right\} \tag{21.1}$$

with

$$b_{nn} = b = \text{const.} \tag{21.2}$$

The steady-state probability density, analogous to (20.15), is then

$$p(x_1, \ldots, x_n, y_1, \ldots, y_n) = C \exp\left[-\frac{2}{b} \int_0^{H(x_1, \ldots, x_n, y_1, \ldots, y_n)} D(\theta)\, d\theta \right] \tag{21.3}$$

Similarly modifying system (19.1), we obtain the system

$$\left.\begin{aligned} \frac{dx_i}{dt} &= \frac{\partial H}{\partial y_i} \\ \frac{dy_i}{dt} &= -\frac{\partial H}{\partial x_i} - \sum_{j=1}^{n} k_{ij} D(H) \frac{\partial H}{\partial y_j} + \xi_i \end{aligned}\right\} \quad (i = 1, 2, \ldots, n) \tag{21.4}$$

with

$$\left.\begin{aligned} k_{ij} &= \text{const.} \\ b_{ij} &= \text{const.} \end{aligned}\right\} \tag{21.5}$$

With the (severe) restriction

$$\frac{k_{ij}}{b_{ij}} = \tfrac{1}{2} l \quad (i, j = 1, 2, \ldots, n) \tag{21.6}$$

the steady-state probability density for system (21.4) is

$$p(x_1, \ldots, x_n, y_1, \ldots, y_n) = C \exp\left[-l \int_0^{H(x_1, \ldots, x_n, y_1, \ldots, y_n)} D(\theta)\, d\theta \right] \tag{21.7}$$

22. State-dependent white noises

Let us revert to system (20.2)

$$\left.\begin{aligned} \frac{dx}{dt} &= \frac{\partial H(x, y)}{\partial y} \\ \frac{dy}{dt} &= -\frac{\partial}{\partial x} H(x, y) - K(x, y) + \xi \end{aligned}\right\} \tag{22.1}$$

but with b no longer a constant, so that

$$b = b(x, y) \tag{22.2}$$

We ask what must be the form of the 'damping' term $K(x, y)$ in order that the steady-state probability density be of the form

$$p = p(H(x, y)) \tag{22.3}$$

The calculations are similar to those of § 20. The stationary Fokker–Planck equation is

$$-\frac{\partial}{\partial x}\left(\frac{\partial H}{\partial y} p\right) + \frac{\partial}{\partial y}\left(\frac{\partial H}{\partial x} p\right) + \frac{\partial}{\partial y}(Kp) + \tfrac{1}{2} \frac{\partial^2}{\partial y^2}(bp) = 0 \tag{22.4}$$

or, in view of (22.3),

$$\frac{\partial}{\partial y}(Kp) + \tfrac{1}{2} \frac{\partial^2}{\partial y^2}(bp) = 0 \tag{22.5}$$

Hence

$$Kp+\tfrac{1}{2}\frac{\partial b}{\partial y}p+\tfrac{1}{2}b\frac{\partial p}{\partial y}=0 \tag{22.6}$$

i.e.

$$\frac{K}{b}+\tfrac{1}{2}\frac{\partial}{\partial y}\log b=-\tfrac{1}{2}\frac{\partial}{\partial y}\log p(H) \tag{22.7}$$

Hence the left side of (22.7) must be the derivative with respect to y of a function of H, say $E(H)$

$$\frac{K}{b}+\tfrac{1}{2}\frac{\partial}{\partial y}\log b=\frac{\partial}{\partial y}E(H)=\frac{dE}{dH}\frac{\partial H}{\partial y} \tag{22.8}$$

i.e.

$$K(x,y)=-\tfrac{1}{2}\frac{\partial b(x,y)}{\partial y}+b(x,y)D(H(x,y))\frac{\partial H(x,y)}{\partial y} \tag{22.9}$$

where D satisfies

$$E(H)-E(0)=\int_0^H D(\theta)\,d\theta \tag{22.10}$$

From (22.7) and (22.8)

$$\frac{\partial}{\partial y}\log p(H)=-2\frac{\partial}{\partial y}E(H) \tag{22.11}$$

i.e.

$$p=C'\exp\left[-2E(H)\right] \tag{22.12}$$

or, from (22.10)

$$p=C\exp\left[-2\int_0^H D(\theta)\,d\theta\right] \tag{22.13}$$

Our result is that for the system

$$\left.\begin{aligned}\frac{dx}{dt}&=\frac{\partial H}{\partial y}\\ \frac{dy}{dt}&=-\frac{\partial H}{\partial x}+\tfrac{1}{2}\frac{\partial b}{\partial y}-bD(H)\frac{\partial H}{\partial y}+\xi\end{aligned}\right\} \tag{22.14}$$

where b is a function of x and y, and D is a function of $H(x,y)$, the steady-state probability density is

$$p(x,y)=C\exp\left[-2\int_0^{H(x,y)} D(\theta)\,d\theta\right] \tag{22.15}$$

This result can be generalized to higher-order systems by the technique of previous sections. We readily find that for the system

$$\left.\begin{aligned}\frac{dx_i}{dt}&=\frac{\partial H}{\partial y_i}\\ \frac{dy_i}{dt}&=-\frac{\partial H}{\partial x_i}+\sum_{j=1}^{n}\left[-\tfrac{1}{2}\frac{\partial b_{ij}}{\partial y_i}+b_{ij}D(H)\frac{\partial H}{\partial y_i}\right]+\xi_i\end{aligned}\right\}\ (i=1,2,\ldots,n) \tag{22.16}$$

where the white noises are state-dependent

$$b_{ij} = b_{ij}(x_1, \ldots, x_n, y_1, \ldots, y_n) \tag{22.17}$$

the steady-state probability density is

$$p(x_1, \ldots, x_n, y_1, \ldots, y_n) = C \exp\left[-2\int_0^{H(x_1,\ldots,x_n,y_1,\ldots,y_n)} D(\theta)\, d\theta\right] \tag{22.18}$$

Example

For

$$\left.\begin{aligned} H &= \tfrac{1}{2}(x^2+y^2) \\ D(H) &= 1 \\ b &= y^2 \end{aligned}\right\} \tag{22.19}$$

the system is

$$\left.\begin{aligned} \frac{dx}{dt} &= y \\ \frac{dy}{dt} &= -x + y - y^3 + \xi \end{aligned}\right\} \tag{22.20}$$

and its steady-state probability density is

$$p(x, y) = C \exp\left[-(x^2+y^2)\right] \tag{22.21}$$

This corresponds to a Gaussian distribution, although the system is nonlinear.

23. Second-order systems with small inertia

Consider a mass m moving in one dimension and satisfying the equation

$$m\frac{d^2x}{dt^2} + \frac{dx}{dt} + \frac{\partial G(x)}{\partial x} = \xi \quad (b = \text{const.}) \tag{23.1}$$

In normal form the equations of motion are, with y representing momentum,

$$\left.\begin{aligned} \frac{dx}{dt} &= \frac{y}{m} \\ \frac{dy}{dt} &= -\frac{\partial G(x)}{\partial x} - \frac{1}{m}y + \xi \end{aligned}\right\} \tag{23.2}$$

for which

$$H = G(x) + \frac{1}{2m}y^2 \tag{23.3}$$

Hence for system (23.2) result (16.12) is

$$p(x, y) = C' \exp\left[-\frac{2}{b}G(x) - \frac{y^2}{bm}\right] \tag{23.4}$$

The marginal probability density for the x-coordinate alone, namely

$$p(x) = \int_{-\infty}^{\infty} p(x, y)\, dy \tag{23.5}$$

is therefore

$$p(x)=C\exp\left[-\frac{2}{b}G(x)\right] \tag{23.6}$$

Now let $m\to 0$. (23.1) becomes the first-order system

$$\frac{dx}{dt}=-\frac{\partial G(x)}{\partial x}+\xi \quad (b=\text{const.}) \tag{23.7}$$

(23.6) remains valid and is thus the stationary probability density for system (23.7).

Example

For the system

$$\frac{dx}{dt}=f(x)+\xi \quad (b=\text{const.}) \tag{23.8}$$

the steady-state probability density is

$$p(x)=C\exp\left[\frac{2}{b}\int_0^x f(\theta)\,d\theta\right] \tag{23.9}$$

This result, which goes back to Smoluchowski (1915), can also be obtained directly by setting up and solving the Fokker–Planck equation for system (23.8) (see, e.g., Andronov *et al.* (1933), Barrett (1960)). Our approach shows that (23.9) corresponds to a special case of the Maxwell–Boltzmann distribution.

24. Higher-order systems with small inertia

The preceding section can be readily generalized to higher-order systems. Thus the system

$$m\frac{d^2x_i}{dt}+\frac{dx_i}{dt}+\frac{\partial G(x_1,\ldots,x_n)}{\partial x_i}=\xi_i \quad (i=1,2,\ldots,n) \tag{24.1}$$

can be written

$$\left.\begin{aligned}\frac{dx_i}{dt}&=\frac{y_i}{m}\\ \frac{dy_i}{dt}&=-\frac{\partial G}{\partial x_i}-\frac{1}{m}y_i+\xi_i\end{aligned}\right\}\quad (i=1,2,\ldots,n) \tag{24.2}$$

for which

$$H=G(x_1,\ldots,x_n)+\frac{1}{2m}(y_1{}^2+y_2{}^2+\ldots+y_n{}^2) \tag{24.3}$$

Let all the white noises ξ_i $(i=1,2,\ldots,n)$ have the same moment rate b, with zero cross-correlation

$$\left.\begin{aligned}b_{ii}&=b=\text{const.} \quad &(i=1,2,\ldots,n)\\ b_{ij}&=0 &(i\neq j)\end{aligned}\right\} \tag{24.4}$$

Then result (18.11) for system (24.2) is

$$p(x_1,\ldots,x_n,y_1,\ldots,y_n)=C'\exp\left[-\frac{2}{b}G(x_1,\ldots,x_n)-\frac{1}{bm}(y_1{}^2+y_2{}^2+\ldots+y_n{}^2)\right] \tag{24.5}$$

The marginal probability density (in x space) is therefore

$$p(x_1, \ldots, x_n) = C \exp\left[-\frac{2}{b} G(x_1, \ldots, x_n)\right] \tag{24.6}$$

Let $m \to 0$. (24.1) becomes the system

$$\frac{dx_i}{dt} = -\frac{\partial G(x_1, \ldots, x_n)}{\partial x_i} + \xi_i \quad (i = 1, 2, \ldots, n) \tag{24.7}$$

Our result is that for system (24.7) with white noises satisfying (24.4) the stationary probability density is

$$p(x_1, \ldots, x_n) = C \exp\left[-\frac{2}{b} G(x_1, \ldots, x_n)\right] \tag{24.8}$$

Result (24.8) can be reached by alternative routes (Barrett 1958, Stratonovich 1963). Once again our approach has emphasized the connection with the Maxwell–Boltzmann distribution.

Note that for system (24.7) with (24.4) the flow vector components (13.5) are

$$J_i = -\frac{\partial G}{\partial x_i} p - \tfrac{1}{2} b \frac{\partial p}{\partial x_i} \quad (i = 1, 2, \ldots, n) \tag{24.9}$$

From (24.8) and (24.9)

$$J_i = 0 \quad (i = 1, 2, \ldots, n) \tag{24.10}$$

i.e. in the steady state the flow vector is zero everywhere. This means that everywhere the diffusive effect of the white noise cancels the flow due to the restoring forces of the system. Such a special situation seems to arise only rarely in practice, so that result (24.8) is of rather academic interest.

25. Transformation of coordinates

If we have say a second-order system with state coordinates x and y and know its steady-state probability density $p(x, y)$, we can obtain the probability density $P(X, Y)$ for the new system which results on making the substitution

$$\left.\begin{aligned} x &= U(X, Y) \\ y &= V(X, Y) \end{aligned}\right\} \tag{25.1}$$

where U and V are given functions. Such a transformation of coordinates constitutes basically only a trivial generalization of the original system. However, it does sometimes lead to results which might be difficult to discover otherwise. Transformation of coordinates was discussed by Fokker (1918) and Kolmogorov (1931).

As a simple example consider the system

$$\left.\begin{aligned} \frac{dx}{dt} &= y \\ \frac{dy}{dt} &= -g(x) - ky + \xi \quad (b = \text{const.}) \end{aligned}\right\} \tag{25.2}$$

and the transformation

$$\left.\begin{aligned} x &= U(X, Y) = U(X) \\ y &= V(X, Y) = Y \end{aligned}\right\} \tag{25.3}$$

The new system is

$$\left.\begin{aligned} \frac{dU}{dX}\cdot\frac{dX}{dt} &= Y \\ \frac{dY}{dt} &= -g(U(X)) - kY + \xi \end{aligned}\right\} \tag{25.4}$$

Writing

$$\left(\frac{dU}{dX}\right)^{-1} = R(X) \tag{25.5}$$

and

$$g(U(X)) = Q(X) \tag{25.6}$$

we see that system (25.4) is

$$\left.\begin{aligned} \frac{dX}{dt} &= YR(X) \\ \frac{dY}{dt} &= -Q(X) - kY + \xi \end{aligned}\right\} \tag{25.7}$$

Comparison with (25.2) shows that (25.7) has a more general form.

The differential element $\delta X\,\delta Y$ is, from (25.3),

$$\delta X\delta Y = \delta x\left(\frac{dU}{dX}\right)^{-1}.\,\delta y = R(X)\delta x\delta y \tag{25.8}$$

The members of the ensemble in this element are the same as the members in the element $\delta x\delta y$, i.e.

$$P(X, Y)\delta X\delta Y = p(x, y)\delta x\delta y \tag{25.9}$$

From (25.8) and (25.9)

$$P(X, Y) = \frac{1}{R(X)}p(x, y) \tag{25.10}$$

We know that for system (25.2) the steady-state probability density is (see (16.16))

$$p(x, y) = C'\exp\left[-\frac{k}{b}\left(2\int_0^x g(\theta)\,d\theta + y^2\right)\right] \tag{25.11}$$

$$= C'\exp\left[-\frac{k}{b}\left(2\int_0^{U(X)} g(\theta)\,d\theta + Y^2\right)\right] \tag{25.12}$$

Writing

$$\theta = U(\phi) \tag{25.13}$$

we have, in view of (25.6) and (25.5),

$$\int_0^{U(X)} g(\theta)\,d\theta = \int_\gamma^X \frac{Q(\phi)}{R(\phi)}\,d\phi \tag{25.14}$$

where γ is given by

$$U(\gamma) = 0 \tag{25.15}$$

From (25.10), (25.12) and (25.14)

$$P(X, Y) = \frac{C}{R(X)} \exp\left[-\frac{k}{b}\left(2\int_0^X \frac{Q(\phi)}{R(\phi)}\,d\phi + Y^2\right)\right] \tag{25.16}$$

This result for system (25.7) generalizes result (25.11) for system (25.2).

26. State-multiplied white noises

If we attempt a more general transformation than that of the previous section, we find that in the new equations the white noise is multiplied by a function of the state coordinates. There arises the question of interpretation of such a white noise. To discuss this question let us restrict attention to the simple first-order system

$$\frac{dx}{dt} = f(x) + \xi \quad (b = \text{const.}) \tag{26.1}$$

With the transformation

$$x = U(X) \tag{26.2}$$

the system is

$$\frac{dU}{dX}\frac{dX}{dt} = f(U(X)) + \xi \tag{26.3}$$

or, with $Q(X)$ and $R(X)$ defined as

$$Q(X) = f(U(X))\left(\frac{dU}{dX}\right)^{-1} \tag{26.4}$$

and

$$R(X) = \left(\frac{dU}{dX}\right)^{-1} \tag{26.5}$$

the system is

$$\frac{dX}{dt} = Q(X) + R(X)\xi \quad (b = \text{const.}) \tag{26.6}$$

In (26.6) the white noise ξ is multiplied by a function R of state X. We say that (26.6) has *state-multiplied* white noise (a term which is to be distinguished from *state-dependent* white noise). The difficulty is that the interpretation of state-multiplied white noise is not immediately obvious.

The correct interpretation can be found by examining the Fokker–Planck equation for the steady-state probability density $P(X)$ of the variable X. To derive this equation we note that the stationary Fokker–Planck equation for system (26.1) is

$$-\frac{\partial}{\partial x}(f(x)p(x)) + \frac{b}{2}\frac{\partial^2 p(x)}{\partial x^2} = 0 \tag{26.7}$$

and that (compare (25.10))

$$p(x) = R(X)P(X) \tag{26.8}$$

Substituting (26.8) in (26.7) and using (26.4) and (26.5), we find

$$-\frac{\partial}{\partial x}(Q(X)P(X)) + \frac{b}{2}\frac{\partial^2}{\partial x^2}(R(X)P(X)) = 0 \tag{26.9}$$

i.e.

$$-\frac{\partial}{\partial X}(Q(X)P(X))\frac{dX}{dx}+\frac{b}{2}\frac{\partial}{\partial x}\left[\frac{\partial}{\partial X}(R(X)P(X))\frac{dX}{dx}\right]=0 \tag{26.10}$$

or

$$\frac{\partial}{\partial X}(QP)=\frac{b}{2}\frac{\partial}{\partial X}\left[\frac{\partial}{\partial X}(RP)\frac{dX}{dx}\right] \tag{26.11}$$

$$=\frac{b}{2}\frac{\partial}{\partial X}\left[\frac{\partial}{\partial X}(RP)R\right] \tag{26.12}$$

$$=\frac{b}{2}\frac{\partial}{\partial X}\left[\frac{\partial}{\partial X}(R^2P)-RP\frac{\partial R}{\partial X}\right] \tag{26.13}$$

i.e.

$$-\frac{\partial}{\partial X}(QP)-\frac{\partial}{\partial X}\left(\frac{b}{2}R\frac{\partial R}{\partial X}P\right)+\frac{\partial^2}{\partial X^2}\left(\frac{b}{2}R^2P\right)=0 \tag{26.14}$$

This is the required steady-state Fokker–Planck equation for $P(X)$.

Examination of (26.14) shows that the white noise term $R(X)\xi$ in (26.6) gives not only a second moment rate $(b/2)R^2$, but also a contribution

$$\tfrac{1}{2}bR\frac{\partial R}{\partial X} \tag{26.15}$$

to the first moment rate. Thus the state-multiplied white noise $R(X)\xi$ does not have zero mean, in general, This is confirmed by a direct calculation of the first moment rate of $R(X)\xi$ in the Appendix.

The contribution (26.15) to the first moment rate is the same as would be produced by an extra restoring term (26.15) on the right side of the system equation. Thus system (26.6) is equivalent to the system

$$\frac{dX}{dt}=Q(X)+\tfrac{1}{2}\frac{\partial R}{\partial X}Rb+\Xi \tag{26.16}$$

where $\Xi(t)$ is a white noise with zero mean and a second moment rate B given by

$$B(X)=bR^2(X) \tag{26.17}$$

We can now find the steady-state solution $P(X)$ for system (26.16) from that for system (26.1) which is (see (23.9))

$$p(x)=C'\exp\left[\frac{2}{b}\int_0^x f(\theta)\,d\theta\right] \tag{26.18}$$

From (26.8) and (26.18)

$$P(X)=\frac{C'}{R(X)}\exp\left[\frac{2}{b}\int_0^{U(X)} f(\theta)\,d\theta\right] \tag{26.19}$$

Writing

$$\theta=U(\phi) \tag{26.20}$$

we have in view of (26.4) and (26.5)

$$\int_0^{U(X)} f(\theta)\,d\theta=\int_\gamma^X \frac{Q(\phi)}{R^2(\phi)}\,d\phi \tag{26.21}$$

and thus

$$P(X)=\frac{C''}{R(X)}\exp\left[\frac{2}{b}\int_0^X \frac{Q(\phi)}{R^2(\phi)}\,d\phi\right] \tag{26.22}$$

If we write

$$F(X)=Q(X)+\tfrac{1}{2}\frac{\partial R}{\partial X}Rb \tag{26.23}$$

system (26.16) is

$$\frac{dX}{dt}=F(X)+\Xi \quad (B=B(X)) \tag{26.24}$$

and its steady-state solution is, from (26.22) and (26.17),

$$P(X)=\frac{C'''}{\sqrt{[B(X)]}}\exp\left[2\int_0^X\left\{\frac{F(\phi)}{B(\phi)}-\tfrac{1}{2}\frac{\partial R(\phi)}{\partial\phi}\frac{1}{R(\phi)}\right\}d\phi\right] \tag{26.25}$$

$$=\frac{C''''}{\sqrt{[B(X)]}}\exp\left[2\int_0^X\frac{F(\phi)}{B(\phi)}\,d\phi-\log R(x)\right] \tag{26.26}$$

$$=\frac{C}{B(X)}\exp\left[2\int_0^X\frac{F(\phi)}{B(\phi)}\,d\phi\right] \tag{26.27}$$

Result (26.27) can also be obtained directly by integrating the **stationary** Fokker–Planck equation for system (26.23) (Andronov *et al.* 1933, Kuznetsov *et al.* 1954, Barrett 1960). Our approach shows that (26.27) is closely related to the Maxwell–Boltzmann distribution.

27. Systems with positional disturbances

In this paper we have usually considered the white noises and damping terms as analogous to forces, i.e. these terms have been introduced in the momenta equations (y equations) and not in the position equations (x equations). Thus in the system

$$\left.\begin{aligned}\frac{dx}{dt}&=\frac{\partial H}{\partial y}\\ \frac{dy}{dt}&=-\frac{\partial H}{\partial x}-k\frac{\partial H}{\partial y}+\xi \quad (b=\text{const.})\end{aligned}\right\} \tag{27.1}$$

the damping and white noise enter into the second equation rather than the first. However, this restriction is not necessary; we could instead treat the system

$$\left.\begin{aligned}\frac{dx}{dt}&=\frac{\partial U}{\partial y}-k\frac{\partial H}{\partial x}+\xi \quad (b=\text{const.})\\ \frac{dy}{dt}&=-\frac{\partial H}{\partial x}\end{aligned}\right\} \tag{27.2}$$

which has white noise and damping added to the first equation. The argument of § 16 is not affected materially by this change, so that the steady-state probability density for system (27.2) is

$$p=C\exp\left[-\frac{2k}{b}H\right] \tag{27.3}$$

the same as for system (27.1).

Similarly we can add white noise and damping terms to all the equations of a conservative system, and obtain results analogous to those in previous sections. Thus for example the appropriate generalization of section 20 yields the following result. For the system

$$\left.\begin{aligned} \frac{dx}{dt} &= \frac{\partial H}{\partial y} - k_1 D(H)\frac{\partial H}{\partial x} + \xi_1 \\ \frac{dy}{dt} &= -\frac{\partial H}{\partial x} - k_2 D(H)\frac{\partial H}{\partial y} + \xi_2 \end{aligned}\right\} \tag{27.4}$$

where

$$\left.\begin{aligned} & b_{12} = b_{21} = 0, \quad b_{11} = \text{const.}, \quad b_{22} = \text{const.} \\ & \frac{k_1}{b_{11}} = \frac{k_2}{b_{22}} = \tfrac{1}{2}l = \text{const.} \end{aligned}\right\} \tag{27.5}$$

the steady-state probability density is

$$p = C \exp\left[-l \int_0^{H(x,y)} D(\theta)\, d\theta\right] \tag{27.6}$$

Example

When

$$\left.\begin{aligned} H &= \tfrac{1}{2}(x^2 + y^2) \\ D(H) &= 2H - 1 \\ k_1 &= k_2 = 1 \\ b_{11} &= b_{22} = b = \text{const.} \\ b_{12} &= b_{21} = 0 \end{aligned}\right\} \tag{27.7}$$

the system (27.4) is

$$\left.\begin{aligned} \frac{dx}{dt} &= y + \{1 - (x^2 + y^2)\}x + \xi_1 \\ \frac{dy}{dt} &= -x + \{1 - (x^2 + y^2)\}y + \xi_2 \end{aligned}\right\} \tag{27.8}$$

From (27.6) the steady-state probability density is

$$p = C \exp\left[\frac{1}{b}\{(x^2 + y^2) - \tfrac{1}{2}(x^2 + y^2)^2\}\right] \tag{27.9}$$

which is equivalent† to a result given by Andronov *et al.* (1933).

28. Conclusions

The steady-state solutions we have found for the Fokker–Planck equation all correspond to cases of the Maxwell–Boltzmann distribution, or to slight generalizations of it. This result is not surprising in view of the historical beginnings of the Fokker–Planck equation in statistical mechanics.

---

† Their system is equivalent to our system (27.8) with the symbols x and y interchanged.

Our technique involves restricting attention to systems for which the terms in the stationary Fokker–Planck equation cancel in pairs. By this means the Fokker–Planck equation is reduced to an ordinary differential equation for which explicit solutions can be found. Roughly speaking we restrict attention to systems analogous to conservative systems to which have been added damping terms and white noise terms in certain proportions. This restriction limits considerably the applicability of our solutions. Nevertheless certain nonlinear stochastic control systems and vibration systems can be analysed explicitly.

It is difficult to see how closed-form solutions could be found for realistic† systems of greater complexity and moderate generality. One might introduce transformations aimed at reducing the Fokker–Planck equation to a partial differential equation with fewer terms. But the problems involved in making this reduction and in solving the resulting partial differential equation seem formidable. Chandrasekhar (1943) has proposed a somewhat similar approach to the solution of the non-steady-state Fokker–Planck equation. However, his method requires the explicit solution of a nonlinear non-conservative system of ordinary differential equations and is therefore unfeasible in general.

For further discussion of solutions of the non-steady-state Fokker–Planck equation, see Caughey and Dienes (1961), who solve a simple case, and Stratonovich (1963), who uses eigenfunction expansions.

Appendix

First moment rate of state-multiplied white noise

In this Appendix the first moment rate of state-multiplied white noise is calculated. The aim is to verify result (26.15). The system is described by the first-order equation

$$\frac{dX}{dt} = Q(X) + R(X)\xi \quad (b = \text{const.}) \tag{29.1}$$

(see (26.6)).

We interpret (29.1) as

$$X(t+\delta t) - X(t) = \int_t^{t+\delta t} Q(X(\theta))\,d\theta + \int_t^{t+\delta t} R(X(\theta))\xi(\theta)\,d\theta \tag{29.2}$$

i.e.

$$Z = Z_1 + Z_2 \tag{29.3}$$

where Z is the change in X during δt, and

$$Z_1 = \int_t^{t+\delta t} Q(X(\theta))\,d\theta \tag{29.4}$$

$$Z_2 = \int_t^{t+\delta t} R(X(\theta))\xi(\theta)\,d\theta \tag{29.5}$$

† One can construct artificial systems by starting with any expression for p, substituting this in the Fokker–Planck equation, and solving the latter for one of the f's, the other f's being assigned arbitrarily.

Since $X(t)$ is continuous (by definition), in (29.5) $R(X(\theta))$ varies little and almost linearly during the interval

$$t<\theta<t+\delta t \tag{29.6}$$

It is reasonable to assume therefore that in (29.5) we can replace $R(X(\theta))$ by its average value over the interval (29.6), i.e. by the value $R(X(\theta_1))$ where θ_1 is the centre of the range (29.6). Then

$$Z_2 = R(X(\theta_1)) \int_t^{t+\delta t} \xi(\theta)\, d\theta \tag{29.7}$$

$$= R(X(\theta_1))\Delta(\delta t, t) \tag{29.8}$$

where Δ is the increment of the white noise. Since

$$\theta_1 = t + \tfrac{1}{2}\delta t \tag{29.9}$$

and $X(t)$ varies approximately linearly during the interval δt,

$$X(\theta_1) = X(t) + \tfrac{1}{2}[X(t+\delta t) - X(t)] \tag{29.10}$$

$$= X(t) + \tfrac{1}{2}Z \tag{29.11}$$

Hence

$$R(X(\theta_1)) = R(X(t) + \tfrac{1}{2}Z) \tag{29.12}$$

$$= R(X(t)) + \tfrac{1}{2}\frac{\partial R(X)}{\partial X}Z \tag{29.13}$$

approximately, since the change Z is small. From (29.3), (29.8) and (29.13)

$$Z_2 = \left[R(X(t)) + \tfrac{1}{2}\frac{\partial R(X)}{\partial X}(Z_1 + Z_2)\right]\Delta(\delta t, t) \tag{29.14}$$

Let us calculate Z_2 to within terms of order of magnitude Δ^2 by means of formula (29.14). To do this we can substitute for the term Z_2 on the right of (29.14) an approximation to within first order terms. From (29.8) this approximation is

$$Z_2 \simeq R(X(t))\Delta \tag{29.15}$$

Substituting (29.15) in the right side of (29.14) we obtain

$$Z_2 = \left[R(X(t)) + \tfrac{1}{2}\frac{\partial R(X)}{\partial X}Z_1\right]\Delta + \tfrac{1}{2}\frac{\partial R(X)}{\partial X}R(X)\Delta^2 \tag{29.16}$$

Now from (29.4) Z_1 is of order of magnitude δt, i.e. of order of magnitude Δ^2. Hence the term involving Z_1 in (29.16) is negligible, and (29.16) reduces to

$$Z_2 = R(X(t))\Delta + \tfrac{1}{2}\frac{\partial R(X)}{\partial X}R(X)\Delta^2 \tag{29.17}$$

Averaging with respect to Δ, we obtain

$$\overline{Z_2} = \tfrac{1}{2}\frac{\partial R(X)}{\partial X}R(X)\overline{\Delta^2} \tag{29.18}$$

or, since

$$\overline{\Delta^2} = b\delta t \tag{29.19}$$

$$\overline{Z_2} = \tfrac{1}{2}\frac{\partial R(X)}{\partial X}R(X)b\delta t \tag{29.20}$$

Thus the first moment rate of the state-multiplied white noise $R(X)\xi$ is

$$\frac{\overline{Z_2}}{\delta t} = \tfrac{1}{2}\frac{\partial R}{\partial X}Rb \tag{29.21}$$

and is not, in general, zero. This confirms result (26.15).

In interpreting state-multiplied white noise in this appendix we have in effect used the calculus of Stratonovich (1964). We could have used instead the calculus of Ito (1944, 1946, 1951) in which state-multiplied white noises are interpreted as having zero first moment rates, and whenever a transformation is made, extra terms are (somewhat arbitrarily) added to the right sides of the system equations. Stratonovich's calculus seems more natural in the present context. For an introduction to Ito's and Stratonovich's methods, see Wonham (1965).

References

Andronov, A. A., Vitt, A. A., and Pontryagin, L. S., 1933, *Zh. exp. teor. Fiz.*, **3**, (3), 165: German translation *Phys. Z. Sowjet*, **6** (1934), 1.

Ariaratnam, S. T., 1960, *J. mech. engng Sci.*, **2**, 195; 1962, *J. appl. Mech.*, **29**, 483.

Bachelier, L., 1900, *Ann. Sci. de l'École Norm. Sup.*, **17**, 21; 1910, *Ibid.*, **27**, 339; 1912, *Calcul des probabilités* (Paris: Gauthier-Villars), Chaps 16–18.

Barrett, J. F., 1958, Ph.D. Thesis, Cambridge University; 1959, Private note; 1960, *Proc. IFAC Congress, Moscow*, **2**, 724 (London: Butterworths, 1961); 1961, *Lecture notes on random processes* (Southampton University).

Benedict, T. R., 1959, *I.R.E. Trans. autom. Control*, **4**, 232.

Bharucha-Reid, A. T., 1960, *Elements of the theory of Markov processes and their application* (New York: McGraw-Hill), Chap. 3.

Blachman, N. M., 1960, *Proc. IFAC Congress, Moscow*, **2**, 770 (London: Butterworths, 1961).

Bolshakov, I. A., 1959, *Avt. i Telemekh.*, **20**, 1611.

Boltzmann, L., 1868, *Sitz. d. k. Akad. Wiss. Wien*, **58**, 517; 1871 a, *Ibid.*, **63**, 397; 1871 b, *Ibid.*, **63**, 679; 1884, *Ibid.*, **90**, 231; 1896–8, *Vorlesungen über Gastheorie* (Leipzig: Barth); translation: *Lectures on gas theory* (Berkeley: University of California Press, 1964); 1909, *Wiss. Abhandlungen* (Leipzig: Barth).

Brooks, R. E., 1961, Bibliography of articles on the development and application of the Fokker–Planck equation to problems in system engineering. (Unpublished note.)

Brush, S. G., 1964, Translator's remarks in Boltzmann's *Lectures on gas theory*, pp. 10–12, 297.

Buhr, R. J. A., 1966, Ph.D. Thesis, Cambridge University.

Butchart, R. L., 1965, *Int. J. Control*, **1**, 201; 1967, Ph.D. Thesis, Southampton University.

Caughey, T. K., 1960, *J. appl. Mech.*, **27**, 575; 1963 a, *Ibid.*, **30**, 634; 1963 b, *J. acoust. Soc. Am.*, **35**, 1683.

Caughey, T. K., and Dienes, J. K., 1961, *J. appl. Phys.*, **32**, 2476.

Cayley, A., 1857, *Rep. Brit. Ass. Advance. Sci.*, 1: *Math. papers*, **3**, 156 (Cambridge University Press, 1890).

Chandrasekhar, S., 1943, *Rev. mod. Phys.*, **15**, 1.

Chapman, S., 1928, *Proc. R. Soc.* A, **119**, 34.

CHUANG, K., and KAZDA, L. F., 1959, *Applic. and Industry* (*Trans. A.I.E.E.*), p. 100.
CRANDALL, S. H., 1961, *Proc. Int. Symp. Nonlin. Osc., Kiev*; 1962 a, *Second Conf. Nonlin. Osc., Warsaw*; 1962 b, *J. appl. Mech.*, **29,** 477; 1963 a, *Random vibration,* edited by S. H. Crandall, **2,** 85 (M.I.T. Press); 1963 b, *J. acoust. Soc. Am.*, **35,** 1700.
DOOB, J. L., 1944, *Ann. Math. Statistics,* **15,** 229.
DYNKIN, E. B., 1965, *Markov Processes* (Berlin : Springer).
EHRENFEST, P., and EHRENFEST, T., 1912, *Enc. math. Wiss.*, **IV 2,** II: English translation *The conceptual foundations of the statistical approach in mechanics* (New York: Cornell University Press, 1959).
EINSTEIN, A., 1905, *Ann. d. Phys.*, **17,** 549; *Investigations on the theory of Brownian movement* (London: Methuen, 1926; New York: Dover, 1956).
EULER, L., 1755, *Histoire de l'Academie Royale des Sciences et Belles Lettres de Berlin,* **11,** 274.
FOKKER, A., 1913, *Over Brownsche bewegingen in het stralingsveld* (Dissertation, Leiden University); 1914, *Annln. Phys,* **43,** 810; 1918, *Archives Néerlandaises Sci. Exactes Nat. III A,* **4,** 379.
FOURIER, J. B. J., 1822, *Théorie Analytique de la Chaleur* (Paris: Didot): English translation (Cambridge University Press, 1878; New York: Dover, 1955), Chap. 9.
FULLER, A. T., 1959, Ph.D. Thesis, Cambridge University; 1960 a, *J. Electron. Control,* **8,** 381; 1960 b, *Ibid.*, **8,** 465; 1960 c, *Ibid.*, **9,** 65; 1961, *Ibid.*, **10,** 157; 1963, *Ibid.*, **14,** 669.
GIBBS, J. W., 1902, *Elementary Principles in Statistical Mechanics* (Yale University Press; New York: Dover, 1960).
HAMILTON, W. R., 1835, *Phil. Trans. R. Soc.*, Part I, 95; *Math. papers,* **2,** 162 (Cambridge University Press, 1940).
HOSTINSKY, B., 1931, *Mémorial Sci. Math.*, **52,** 1; 1933, *Ann. Inst. H. Poincaré,* **3,** 1; 1937, *Ibid.*, **7,** 69.
ITO, K., 1944, *Proc. Imp. Acad. Tokyo,* **20,** 519; 1946, *Proc. Japan Acad.*, Nos.1–4, 32; 1951, *Mem. Am. math. Soc.*, **4,** 51.
JACOBI, C. G. J., 1837, *C. r. hebd. Séanc. Acad. Sci. Paris,* **5,** 61; *Gesamm. Werke,* **4,** 129 (Berlin: Reimer, 1886); 1842, *Vorlesungen über Dynamik* (Lectures given at Königsberg University; Berlin: Reimer, 1866): *Gesamm. Werke, Supplementband.*
KHASMINSKI, R. Z., 1960, *Teor. Ver. Prim.*, **5,** 196 (translation : *Th. Prob. Appl.*, **5,** 179).
KHAZEN, E. M., 1961 a, *Teor. Ver. Prim.*, **6,** 130 (translation: *Prob. Theory Appl.*, **6,** 117); 1961 b, *Ibid.*, **6,** 234 (translation: **6,** 214); 1961 c, *Izv. Akad. Nauk SSSR, OTN, Energet. Avtomat.*, No. 3, 58.
KHINTCHINE, A., 1934, *Math. Ann.*, **109,** 604.
KOLMOGOROV, A. N., 1931, *Math. Ann.*, **104,** 415; 1933, *Ibid.*, **108,** 149; 1938, *Usp. mat. Nauk,* **5.**
KRAMERS, H. A., 1940, *Physica,* **7,** 284.
KUSHNER, H. J., 1968, *Technical Report 68-3, Centre for Dynamical Systems* (Brown University).
KUZNETSOV, P. I., STRATONOVICH, R. L., and TIKHONOV, V. I., 1953, *Avt. i Telemekh.*, **14,** 375; 1954 a, *Ibid.*, **15,** 200; 1954 b, *Dokl. Akad. Nauk SSSR,* **97,** 639; 1955, *Zh. exp. teor. Fiz.*, **28,** 509; 1965, *Non-linear Transformations of Stochastic Processes* (Oxford: Pergamon).
LAGRANGE, J. L., 1772, *Nouv. Mém. Acad. Roy. Sci. Belles-Lettres Berlin,* 185; *Oeuvres,* **3,** 441 (Paris: Gauthier-Villars, 1869); 1809, *Mémoires de l'Institut de France,* 343; *Oeuvres,* **6,** 807; 1811, *Méchanique Analytique,* 2nd edition, (Paris: Courcier), 336.
LANGEVIN, P., 1908, *C. r. hebd. Séanc. Acad. Sci. Paris,* **146,** 530.
LAPLACE, P. S., 1812, *Théorie analytique des probabilités,* 275 (Paris: Courcier): *Oeuvres,* **7,** 280 (Paris: Gauthier-Villars, 1886).
LIOUVILLE, J., 1838, *J. Math. pures appl.*, **3,** 342.

LOSCHMIDT, J., 1876, *Sbr. Akad. Wiss. Wien.*, **73**, 139; 1877, *Ibid.*, **75**, 67.
LYON, R. H., 1960, *J. acoust. Soc. Am.*, **32**, 716; 1961, *Ibid.*, **33**, 1395; 1963 a, *Jl appl. Mech.*, **30**, 636; 1963 b, *J. acoust. Soc. Am.*, **35**, 1712.
MARKOV, A. A., 1910, *Acta Math.*, **33**, 87; 1912, *Wahrscheinlichkeitsrechnung* (Leipzig: Teubner), Appendix II.
MAXWELL, J. C., 1860, *Phil. Mag.*, **19**, 19; **20**, 21, 33; 1890, *Scientific papers* (Cambridge University Press; New York: Dover, 1952).
MERKLINGER, K. J., 1963 a, *Proc. IFAC Congress, Basle, Theory*, 81 (London: Butterworths, 1964); 1963 b, Ph.D. Thesis, Cambridge University.
MIDDLETON, D., 1960, *Introduction to Statistical Communication Theory* (New York: McGraw-Hill), Chap. 10.
MING CHEN WANG, and UHLENBECK, G. E., 1945, *Rev. mod. Phys.*, **17**, 323.
MOYAL, J. E., 1949, *Jl R. statist. Soc.* B, **11**, 150, 278.
PEARSON, K., 1905, *Nature, Lond.*, **77**, 294; 1924, *Biometrika*, **16**, 402.
PERVOSVANSKI, A. A., 1962, *Random processes in nonlinear control systems* (Moscow: Fizmatgiz. English translation, New York: Academic Press, 1965).
PIPES, L. A., 1946, *Applied Mathematics for Engineers and Physicists* (New York: McGraw-Hill), p. 272.
PLANCK, M., 1915, *Sbr. preuss. Akad. Wiss.*, p. 512; 1917 a, *Ibid.*, p. 324; 1917 b, *Annln Phys.*, **52**, 491, **53**, 241.
PONTRYAGIN, L. S., BOLTYANSKI, V. G., GAMKRELIDZE, R. V., and MISHCHENKO, E. F., 1961, *Mathematical Theory of Optimal Processes* (Moscow: Fizmatgiz); translation (New York: Wiley, 1962).
PUGACHEV, V. S., 1961, *Izv. Akad. Nauk SSSR, OTN, Energ. Avtomat.*, No. 3, 46.
RAYLEIGH, LORD, 1880, *Phil. Mag.*, **10**, 73; *Scientific papers*, **1**, 491 (Cambridge University Press, 1899); 1891, *Ibid.*, **32**, 424: *Scientific papers*, **3**, 473; 1894, *Theory of Sound*, 2nd edition (London: Macmillan), **1**, 35; 1899, *Phil. Mag.*, **47**, 246; *Scientific papers*, **4**, 370.
RUINA, J. P., and VAN VALKENBURG, M. E., 1960, *Proc. IFAC Congress, Moscow*, **2**, 810 (London: Butterworths, 1961).
SAWARAGI, Y., and SUNAHARA, Y., 1960, *Tech. Rep. Engng Res. Inst. Kyoto*, **10**, 43.
SAWARAGI, Y., SUNAHARA, Y., and NAKAMIZO, T., 1961, *Tech. Rep. Engng Res. Inst. Kyoto*, **11**, 1.
SAWARAGI, Y., SUNAHARA, Y., and SOEDA, T., 1961 a, *Tech. Rep. Engng Res. Inst. Kyoto*, **11**, 19; 1961 b, *Ibid.*, **11**, 37.
SMITH, P. W., 1962, *J. acoust. Soc. Am.*, **34**, 827.
SMOLUCHOWSKI, M., 1906, *Annln. Phys.*, **21**, 756; 1915, *Ibid.*, **48**, 1103; 1916, *Phys. Z.*, **17**, 557, 585; 1923, *Ostwald's Klassiker der Exacten Wiss.*, No. 207.
STRATONOVICH, R. L., 1958, *Radiotekh. Élektron.*, **3**, 497; 1963, *Topics in the Theory of Random Noise* (translation, New York: Gordon and Breach), **1**, Chap. 4; 1964, *Vestnik Mosk. Univ., Mat. Mekh.*, No. 1, 3: translation, *S.I.A.M. Jl Control*, **4** (1966) 362.
SYLVESTER, J. J., 1852, *Phil. Mag.*, **4**, 138; *Math. papers*, **1**, 378 (Cambridge University Press, 1904).
TER HAAR, D., 1961, *Elements of Hamiltonian Mechanics* (Amsterdam: North-Holland Publishing Co.); 1966, *Elements of Thermostatistics*, 2nd edition (New York: Holt, Rhinehart & Winston).
TIKHONOV, V. I., 1959, *Avt. Telemekh.*, **20**, 1188; 1960, *Ibid.*, **21**, 301.
UHLENBECK, G. E., and ORNSTEIN, L. S., 1930, *Phys. Rev.*, **36**, 823.
WAX, N., 1954, *Selected Papers on Noise and Stochastic Processes* (New York: Dover).
WHITTAKER, E. T., 1937, *Analytical dynamics*, 4th edition (Cambridge University Press).
WIENER, N., 1949, *Extrapolation, interpolation, and smoothing of stationary time series* (Mass. Inst. Tech.: Technology Press).
WISHNER, R., 1960, Ph.D. Thesis, University of Illinois.
WONHAM, W. M., 1965, Advances in nonlinear filtering (Lecture notes, M.I.T.); 1966, *J. Diff. Eqns*, **2**, 195; **2**, 365.
ZERMELO, E., 1896, *Annln Phys.*, **57**, 485; **59**, 793.

The Use of Functionals in the Analysis of Non-linear Physical Systems†

By J. F. Barrett
Department of Mechanical Engineering, University of Birmingham

[Received February 4, 1963]

Abstract

This report is an attempt to develop a method of analysis applicable equally to linear or non-linear systems. The main method discussed is the expansion of the input–output relation in a functional power series—an idea first due to Volterra for general functional relationships and to Wiener in its application to non-linear communication problems. The report attempts to present a systematic development of this idea. The last part of the report discusses analogous expansions in a series of terms orthogonal with respect to input statistics.

§ 1. Statistical Design

1.1. *The Black Box*

The widely used 'black box' method of description of engineering systems shows that, for many purposes, the actual nature of an engineering system is unimportant and only the way it responds to certain input signals is of interest. Denote input signal symbolically by s and response by r. Then mathematically, the point of interest is the functional relation $r = F(s)$ between r and s.

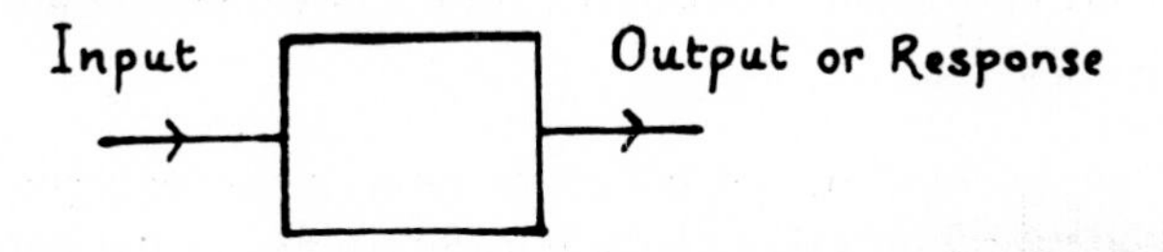

Example: If (i) the input and response can be represented as real or complex valued functions of time, e.g. they are fluctuating voltages, (ii) the system is linear and time-invariant (see below for the precise definition of these terms), then the functional relation F may be conveniently described by a transfer function, i.e. the ratio of the Laplace

† Communicated by A. T. Fuller. The present paper was first written in April 1955 as a Cambridge University Engineering Laboratory report and was duplicated and distributed in 1956 by the Ministry of Supply as S.A.U. report 1/57. The text is unchanged apart from the correction of errors, etc.

transforms of the response and input. At present no restriction is made to this class of systems and so the general F-notation must be retained.

1.2. *Performance Criteria*

In order to characterize the response of a system to some input s as a good or bad one, some method of comparison must be made between the actual response r and ideal response r_i. The nearer r is to r_i, in some sense, the better the response r. r_i depends on s, the input signal, and so it is possible to write $r_i = F_i(s)$, where F_i describes the ideal system. The measure of success of the system in dealing with signal s is then obtained by comparison of r with r_i ($F(s)$ with $F_i(s)$).

Examples:

1.2.1. *Duplicators*

In an important class of systems, sometimes called *duplicators* (Trimmer 1950), the system is required to reproduce the input signal exactly. Systems for transmitting speech and music come into this class as do also a large number of servo systems.

Here, with suitable choice of units, etc., the ideal response to a signal s is s itself: $r_i = s$. The success of the system in dealing with s is thus measured by the closeness in some sense of r to s. The method adopted to measure this closeness is largely arbitrary and can often be chosen for convenience. Thus, for a simple (angle) position control system, if $\theta_1(t)$ and $\theta_2(t)$ are input and response angles at time t, possible numerical measures of the amount of discrepancy between θ_2 and θ_1 in an interval of time, $0 \leqslant t \leqslant T$ are:

$$\max_{0 \leqslant t \leqslant T} |\theta_2(t) - \theta_1(t)|, \quad \frac{1}{T}\int_0^T |\theta_2(t) - \theta_1(t)|\, dt$$

or

$$\frac{1}{T}\int_0^T (\theta_2(t) - \theta_1(t))^2\, dt.$$

All these are equivalent in the sense that if any one is small, so must the others be†.

1.2.2. *Filters*

In a system for filtering out noise, the input signal consists of a wanted signal and certain unwanted disturbances—'noise'. So, for example, if the input signal is in the form of a varying voltage,

$$V(t) = V_1(t) + V_2(t),$$

where $V(t)$, $V_1(t)$ and $V_2(t)$ are the input, wanted signal and noise voltages at time t. Here the input signal s is the record of $V(t)$ over the time interval of observation.

The ideal response r_i of the system to the signal $V(t)$ is a voltage $V_1(t)$ at time t or more generally $V(t - t_1)$ allowing for a delay of t_1. Let the actual response at time t be $V_3(t)$.

† θ_1 and θ_2 are assumed continuous functions of time.

As a measure of success of the system when operating on the particular signal $V(t)$ for the given time interval, any of the expressions of the preceding example may be used with $\theta_2(t)$ replaced by $V_3(t)$, $\theta_1(t)$ by $V_1(t)$ and the time interval $(0, T)$ by the interval of observation.

More complicated filtering problems arise, e.g. the received signal may be of the form $V(t) = a(t) \sin(\omega t + n_1(t)) + n_2(t)$, where $n_1(t)$ and $n_2(t)$ are now unwanted disturbances. Here the wanted signal $a(t)$ is modulated. The ideal system for simultaneous demodulation and filtering out noise would be one giving $a(t)$ from $V(t)$ at any time t (with possibly a delay).

1.2.3. *Predictors*

Here again, the received signal is a function of time. If the value of x at time t_1 ahead is required, the ideal response at time t is $x(t+t_1)$.

In the above examples, signals were in each case represented by time varying quantities and it was thus possible to give a numerical value to the discrepancy between two signals. This is the most common case. A signal may usually be adequately described by the time variation of some physical quantity. Thus a piece of music can be adequately described for present purposes by the voltage variations produced in a suitably adjusted microphone, or more basically by the pressure variations in the air near the microphone.

It will consequently be assumed that signals and responses are represented by the time variation of physical magnitudes and that a numerical measure of discrepancy between two responses is given. If the discrepancy between r and r_i is $\delta(r, r_i)$ the function will always be assumed to have the property:

$$\delta(r, r_i) \geqslant 0 \text{ for all } r \text{ and } \delta(r, r_i) = 0 \text{ if } r = r_i.$$

This condition is satisfied with the measures given above (§ 1.2.1). A function having this property has been called a *distance function* by Shannon and Weaver (1949).

The inaccuracy of the system in responding to a signal s is thus measured by $\delta(r, r_i) = \delta(F(s), F_i(s))$.

1.3. *Optimal Systems*

In general, the problem is to make the system optimal with a given distance function. In other words an F must be found which minimizes the express $\delta(r, r_i) = \delta(F(s), F_i(s))$. A trivial solution is $F = F_i$ but in general it is not possible to achieve this because of the limitation of means in realizing F. Viewed mathematically, the problem is one of minimization under certain constraints which exclude the possibility $F = F_i$.

This is merely the optimization of the response to a single input and will usually not be what is required. Any signal transmission system will be required to operate satisfactorily over a whole range of inputs. A message which is known before it is sent conveys no information and so a communication system designed to receive a single message is redundant. The signal which is actually the input on any one working run of the system must be one of a range of possible input signals. This range of possible input signals will be called the *signal ensemble.* When a signal transmission system is being designed, it must be designed to deal efficiently with every possible signal that might occur, i.e. with all signals of the signal ensemble. In this case it is necessary to make some compromise between the relative efficiencies for different signals (relative to some distance function). A system S_1 may have a better response than S_2 on some signals but a worse one on others. How is a choice to be made between the two systems ?

The solution proposed by Wiener in 1942 (see Wiener 1949) is based on the relative frequencies of occurrence of the individual signals of the signal ensemble. In generalized form, the method is to choose F so as to minimize the average value $\overline{\delta(F(s), F_i(s))}$ of $\delta(F(s), F_i(s))$ the average value being over the relative frequencies of occurrence of the signals s†. A system designed on this principle will consequently have, on the average, a response at least as good, or better than, every other system of the range of systems considered.

This value of F will depend on δ, F_i, the input statistics and the range of variation allowed in F. The general mathematical problem in statistical design is to give an explicit expression for F in terms of these. The following theory is centred round the problem of finding this F when signals are functions of time (time-series) and δ is the mean square value.

The chief drawbacks of statistically designed systems should be mentioned. The two most important of these are probably:

(i) a system designed according to this method will only give a better performance than others on the average. On occasions its performance might be very poor;

(ii) a statistically designed system may be sensitive to changes in input statistics. If these change, the system will no longer be statistically optimal. Consequently it might appear that a system designed on the statistical principle is too highly specialized to justify the trouble of constructing it.

(i) seems quite valid in certain cases. The problem is often not to design a system which will be good on the average but one which will never be bad. If this is the case, the statistics of the signal ensemble do not have to be known and F will be chosen so as to minimize $\max \delta(f(s), F_i(s))$, the maximum being taken over all s of the signal ensemble. (ii) should be met by designing a system which adjusts itself, or can be adjusted, to changes of input statistics.

§ 2. Operators and Functionals

In the relation $r = F(s)$, if s and r are functions of time, as they will always be assumed to be from now on, F is a function with argument and value

† Convenient though not precise. $\overline{\delta(F(s), F_1(s))}$ depends on $\delta(F, F_1)$ and signal statistics, not s.

which are also functions. To avoid confusion of language, F will be termed an *operator*. An operator is defined by a rule for deriving one function from another, familiar examples being derivation and indefinite integration. Another way of expressing the relationship between s and r in the case where both are functions involves the idea of a *functional*. A functional is a function whose argument is a function and whose value is a number. Definite integration is an example: every definite integral will define a functional the argument being the integrand and the value being the value of the definite integral.

Given a general 'black box', the relation between the input signal s and the response r, both of which are time varying quantities x and y with values $x(t)$ and $y(t)$ at time t, may be described mathematically in two ways—both of which are really equivalent. In the first method, attention is given to the relation between the whole response as represented by a record or graph of y over the period of working of the system and the whole input signal as represented by a record or graph of x over the period of working. Here attention is directed to the functional relation between the two functions of time x and y. This is the *operational description*†. In the second method, the dependence of the response at a particular time t on the previous input is considered. This is a relation between a function and a number and is thus a *functional*. In this case the relation can be written $y(t)=f(t;s)$ or alternatively, $y(t)=f(t;\ x(t'), t' \leqslant t)$ since the only part of the signal s that y is dependent on at time t, is the part which occurs before time t. Generally speaking, the functional description is more suitable for concrete problems and more attention will be paid to it here. Before stating the most important results on functionals for the present theory, a few remarks are necessary on the representation of signals.

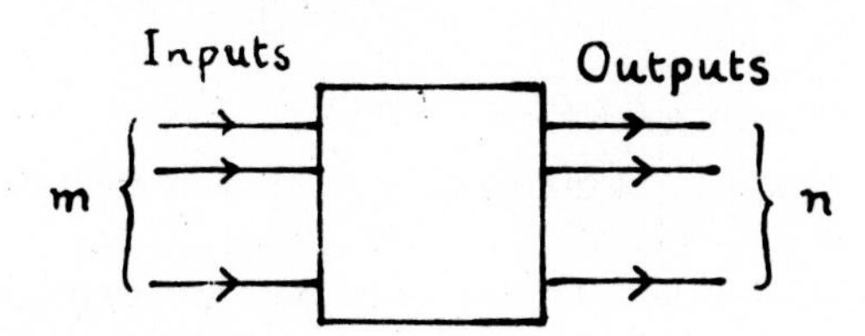

The input signal and response as represented by the time variation of physical quantities x and y will usually be real valued functions of time in the appropriate range of time. More generally they can be complex or vector quantities or both. The advantages of complex representation of electrical quantities is well known (see Gabor 1946 for the theory). Multi-channel transmission is a case where the signal is conveniently described by a vector: if there are n input signals to a system by the time variation of $x_1, x_2, \ldots, x_n$ the input may be described by the single vector quantity

† This usage of the term 'operational' should not be confused with that frequent in physics. 'Operational' here means 'described by a mathematical operator'.

$\mathbf{x}=(x_1, x_2, \ldots, x_n)$. The response may also be a vector. For instance, the response of the system may be recorded on n dials with readings $y_1, y_2, \ldots, y_n$. In what follows, functions for signals will usually be real valued functions of time, the extension to other cases being more or less straightforward. The range of time considered will be from a fixed time denoted by t_0, the time of switching on, to 'now', usually represented by the time t. t_0 may be at $-\infty$ in the case where only steady-state operation is considered.

With these preliminaries we return to the consideration of functionals. The most important result of this section is the power series expansion of a non-linear functional. This will be preceded by a characterization of two important classes of functionals, time-invariant and linear functionals respectively.

2.1. *Time-invariance*

In many systems the form of response is independent of the particular time at which the input signal is received. Electrical filters with lumped L, R, C elements with constant values come into this class. Such systems and the functionals (or operators) which describe them will be called *time-invariant*.

Let $f(t;s)$ be the response of the system at time t to a signal s. Let s' represent the same signal a time τ later (earlier if τ is negative). Then if f represents a time invariant functional the equation

$$f(t;\, s)=f(t+\tau;\, s')$$

must hold for all times t. Another way of expressing the same property is that in the notation $f(t;\, x(t'),\, t'\leqslant t)$ for the response at time t, the first t need not be written and the response is only dependent on the actual values of x before t and not on t itself. This may be deduced from the above equation as follows:

$$f(t;\, x(t'),\, t'\leqslant t)=f(t+\tau;\, x(t''),\, t''\leqslant t+\tau)$$

from the above equation. Putting $t'''=t''-\tau$, the right-hand side of the equation becomes $f(t+\tau;\, x(t'''),\, t'''\leqslant t)$. This is the same as $f(t+\tau;\, x(t'),\, t'\leqslant t)$ since the primed t's are only dummy variables and can be replaced by any other variables without any change of meaning. Thus $f(t;\, x(t'), t'\leqslant t)=f(t+\tau;\, x(t'), t'\leqslant t)$ for any τ and so the first t is redundant: its value has no influence on the value of the functional f.

Time-invariant operators are defined in a similar way.

2.2. *Linearity*

Many systems, or more precisely the operators or functionals describing them, show the *additive property* or *principle of superposition* with respect to input signals, i.e. if input signals $x(t)$, $x'(t)$ give output signals $y(t)$, $y'(t)$, then an input signal $x(t)+x'(t)$ gives an output signal $y(t)+y'(t)$. (Here 'an input signal $x(t)$', etc., means a signal which is represented by a physical quantity denoted by x which has a value $x(t)$ at any time t, etc.)

It may be shown† that the additive property is equivalent to the *homogeneity property* with respect to input signals, i.e. that if a signal $x(t)$ gives output signal $y(t)$, then an input signal: constant $\times x(t)$, say $kx(t)$, gives an output signal $ky(t)$. This is equivalent to saying that, apart from a change of scale—in fact the same change of scale for both input and output—the performance of the system is independent of amplitude and there is no distortion due to changes of amplitude of the input‡.

A system (or the operator, functional describing it) will be called *linear* if it satisfies these conditions. If it is not linear in this sense, it will be called *non-linear*§.

2.3. *Examples of Functionals*

From now on the behaviour of systems will be discussed in terms of functionals rather than operators. (Any functional may be thought of as defining an operator.) The value of the functional at time t will be denoted by $y(t)$.

2.3.1. *The identity*

This is defined by the equation:

$$y(t)=f(t;\ x(t)',\ t'\leqslant t)=x(t).$$

It is obviously time-invariant and linear. The device described by this functional is the ideal duplicator: output always equals input.

2.3.2. *Time translation*

The defining equation is:

$$y(t)=f(t;\ x(t'),\ t'\leqslant t)=x(t+\tau).$$

Again this is obviously time-invariant and linear. The device described by this functional is the ideal predictor if $\tau>0$ and a pure delay if $\tau<0$. If $\tau=0$ it reduces to the identity functional.

2.3.3. *Differentiation*

$$y(t)=f(t;\ x(t'),\ t'\leqslant t)=\dot{x}(t).$$

Once more, time-invariant and linear. In this example and in example 1, the value of the functional for time t depends only on the values of $x(t)$ in the immediate neighbourhood of the time t. A functional with this property is called *instantaneous*. So is the device it describes.

2.3.4. *Integral and convolution functionals*

f is a linear integral functional if it is of the form:

$$y(t)=\int_{t_0}^{t} g(t,t')x(t')\,dt.$$

$g(t,t')$ is called the *kernel function* of the integral functional, (kernel for short).

---

† Subject to certain mild continuity conditions.

‡ This behaviour is not solely characteristic of linear systems and is shared by those described by homogeneous operators (functionals) of any degree.

§ This definition of a non-linear system is by no means universal among engineers, e.g. see James *et al.* 1947, p. 29. A linear filter in the sense of this book is described here as a stable, time-invariant linear filter.

It is convenient to have the integral over the range $(-\infty, \infty)$. This is accomplished by defining $x(t)=0$ for $t<t_0$ if t_0 is finite and also defining $g(t,t)=0$ for $t<t'$. The integral is then:

$$\int_{-\infty}^{\infty} g(t,t')x(t)\,dt'.$$

Having g vanish for $t<t'$ ensures that only the past values of x at any time influence the value of the functional at that time. A kernel of this kind is said to be of *Volterra type*.

If $g(t,t')$ is a function only of the difference $t-t'$, the functional is time-invariant. This may be seen by a change of variable. Suppose $g(t,t')=h(t-t')$. Put $t-t'=t''$. Then

$$\begin{aligned}\int_{-\infty}^{\infty} g(t,t')x(t')\,dt' &= \int_{-\infty}^{\infty} h(t-t')x(t')\,dt' \\ &= \int_{-\infty}^{\infty} h(t'')x(t-t'')\,dt'' \\ &= \int_{-\infty}^{\infty} h(t'')x(t')\,dt''\end{aligned}$$

Thus f depends only on the values of $x(t')$ for $t'\leqslant t$ (=values of $x(t-t'')$ for $t''\geqslant 0$). Such a functional will be called a *convolution integral* (convolution for short).

By introduction of the Dirac δ-function and its derivatives†, the functionals in examples 2.3.1, 2.3.2 and 2.3.3 may be written as convolutions. They are respectively:

$$\int_{-\infty}^{\infty} \delta(t-t')x(t')\,dt', \qquad \int_{-\infty}^{\infty} \delta(t+\tau_0-t')x(t')\,dt'$$

and

$$\int_{-\infty}^{\infty} \delta(t-t')x(t')\,dt.$$

2.3.5. *Modulation*

Defined by

$$f(t;\, x(t'), t'\leqslant t) = M(t)x(t).$$

It is linear, instantaneous, but not time-invariant if the modulating function $M(t)$ varies with time.

2.3.6. *Square law devices*

$$f(t;\, x(t'),\, t'\leqslant t) = kx^2(t),$$

where k is a constant. This is non-linear, instantaneous and time-invariant.

2.3.7. *Regular homogeneous functional of degree n*

$$f(t;\, x(t'),\, t'\leqslant t) =$$

$$\int_{t_0}^{t}\int_{t_0}^{t}\cdots\int_{t_0}^{t} g(t;\, t_1, t_2, \ldots, t_n)x(t_1)x(t_2)\ldots x(t_n)\,dt_1\,dt_2\ldots dt_n.$$

† The use of these may be justified by the theory of distributions of L. Schwartz (1950, 1951). See, e.g., Lafleur and Namias (1954).

g is again called the kernel function (kernel). It is of Volterra type, i.e. $g(t; t_1, t_2, \ldots, t_n) = 0$ if any $t_0 > t_i$, $i = 1, 2, 3, \ldots, n$. If the kernel is a function only of the time differences $t - t_i$, $i = 1, 2, 3, \ldots, n$, the functional is time-invariant. This follows from a change of variables as in the linear case.

The functional is said to be homogeneous of degree n because changing x to kx at all times multiplies the value of f by k^n: $n = 1$ gives the linear integral functional.

By using the Dirac δ-function again, the square law device may be described by a regular homogeneous functional of degree 2:

$$kx^2(t) = k\int \delta(t-t')x(t')\,dt' \quad \int \delta(t-t'')x(t'')\,dt''$$

$$= \int\int k\delta(t-t')\;\delta(t-t'')x(t')x(t'')\,dt'\,dt''.$$

Thus the kernel is $k\delta(t-t')\,\delta(t-t'')$. Note that the kernel depends only on the differences $t-t'$, $t-t''$, so verifying time-invariance.

In general the nth order device:

$$f(t;\; x(t'),\, t' \leqslant t) = kx^n(t)$$

may be expressed as a regular homogeneous functional with the kernel $k\delta(t-t_1)\;\delta(t-t_2)\;\delta(t-t_3)\ldots\delta(t-t_n)$.

2.3.8. *Functional power series*

These are defined by expressions of the form:

$$\Sigma\int\int\cdots\int g(t; t_1, t_2, \ldots, t_n)x(t_1)x(t_2)\ldots x(t_n)\,dt_1\,dt_2\ldots dt_n,$$

where the summation is over an infinity of terms. If the summation is over only a finite number of terms, the expression is called a *functional polynomial*. For the general expression to have a meaning, the series will have to be convergent.

The general instantaneous device with a characteristic given by a function $f(\;)$ may be written in this form if f has a power series expansion:

$$f(x) = \sum_{n=0}^{\infty} k_n x^n.$$

Each term may now be expressed in integral form as noted in the last example.

Note: The same symbol g can be used for all the kernels. There is no risk of confusion because each is dependent on a different number of variables.

2.3.9. *Linear functionals*

Before going on to general functionals, the main results about the structure of linear functionals will be briefly summarized. It was seen

C

from the preceding examples that many linear functionals may be expressed in integral form:

$$\int g(t, t')x(t')\, dt.'$$

If the kernel is of Volterra type, viz. $g(t, t') = 0$ if $t' > t$, the functional is dependent only on the past of x at any time. If $g(t, t')$ is a function of the time difference $t - t'$, the functional is time-invariant.

The main result is that these properties are quite general if delta functions and their derivatives are used in the kernel.

If delta functions are avoided, any linear functional is expressible in the form:

$$\int g(t, t')x(t')\, dt' + \sum_{m,n} a_{m,n} x^{(m)}(t - t_n),$$

where the summation is taken over a number, possibly infinite, of values of m and n.

If the functional acts only on past values of the input x at any time, g is of Volterra type and all the t_n are non-positive: $t_n \leqslant 0$ for all n.

If the functional is time-invariant, $g(t, t')$ depends only on the time difference $t - t'$.

The results may be expressed more concisely with δ-functions. The functional may be put in integral form with the kernel:

$$k(t, t') = g(t, t') + \sum_{m,n} a_{m,n} \delta^{(m)}(t' - t - t_n).$$

The customary derivation of the main result is as follows†: The function $x(t)$ may be thought of as the sum of an infinite number of impulses —one for each time t. The magnitude of the impulse corresponding to time t is proportional to $x(t)$.

This decomposition of x may be written:

$$x(t) = \sum x(t')\delta(t - t').$$

It is seen that this corresponds to the formula:

$$x(t) = \int \delta(t - t')x(t')\, dt'.$$

By an extension of the linearity property of f,

$$f(t;\, x(t'), t' \leqslant t) = \int x(t')g(t, t')\, dt',$$

where $g(t, t'') = f(t;\, \delta(t - t'')t'' \leqslant t)$, i.e. $g(t, t')$ is the effect at time t due to a unit impulsive input at time t'. This is the required integral form.

Since $f(t;\, x(t'), t' \leqslant t)$ is dependent only on the past, $g(t, t') = 0$ if $t' > t$ by definition.

---

† A proof on these lines may be given using distribution theory (see footnote to page 62). For continuous linear functionals, the result follows from an extension of the theorem of F. Riesz. See Bourbaki (1952), Book VI, p. 57 et seq.

If the system is time-invariant, a unit impulse applied at time $t'+\tau$ will give the same output at time $t+\tau$ as a unit impulse applied at time t' will give at time t: this follows immediately from the definition of time-invariance. In terms of the impulse response function, this statement is expressed by the equation: $g(t+\tau, t'+\tau) = g(t, t')$ for all values of t, t' and τ. Putting $\tau = -t'$ gives $g(t, t') = g(t-t', 0)$, for all values of t and t', and so $g(t, t')$ depends only on the difference $t-t'$.

2.3.10. *Stability*

When dealing with input signals $x(t)$ which are non-zero over an infinite range of time, it is necessary to ensure that the infinite integral

$$\int_{-\infty}^{\infty} g(t, t')x(t')\, dt'$$

exists at the lower limit. (The upper limit gives no trouble as the integral is only apparently an infinite one at $+\infty$ when the functional depends on the past values only.)

A sufficient condition for this is that

$$\int_{-\infty}^{\infty} |g(t, t')|\, dt' < \text{some constant},$$

say M, for all times t. For then

$$\left|\int_{\infty}^{\infty} g(t, t')x(t')\, dt'\right| \leqslant \int_{-\infty}^{\infty} |g(t, t')x(t')|\, dt'$$

$$\leqslant \max_{-\infty < t' < \infty} |x(t')| \int_{-\infty}^{\infty} |g(t, t')|\, dt'$$

$$< \max_{-\infty < t' < \infty} |x(t')|\, M.$$

Thus, if x is bounded, so is the infinite integral.

Alternatively we may state that every bounded ensemble of inputs will give a bounded ensemble of outputs. This is the definition of stability proposed by Zadeh (1952). For time-invariant linear systems, it may be shown equivalent to all the usual definitions (James *et al.* 1947). It has the advantage of applying equally well to non-linear systems whereas other definitions fail.

§ 3. General Functionals

We wish to find an explicit expression for a general functional which will exhibit its structure in much the same way as the integral form does in the case of a linear functional. This is accomplished by the power series expansion which seems to be valid for a wide class of functionals†. The basic functionals turn out to be the regular homogeneous functionals already mentioned (p. 62). They share with the linear functionals the homogeneous (non-distortion) property, but not that of additivity.

† The power series expansion for a functional is due to Volterra. Accounts of the theory may be found in V. Volterra's book and in those of Lévy (1951) and Hille (1948) (abstract treatment).

The situation will be seen to be quite analogous to that which occurs with functions of many variables. If $f(x_1, x_2, \ldots, x_n)$ is a real valued function of n real variables $x_1, x_2, \ldots, x_n$, we may say that f is additive with respect to these variables if

$$f(x_1+x_1', x_2+x_2', \ldots, x_n+x_n') = f(x_1, x_2, \ldots, x_n) + f(x_1', x_2', \ldots, x_n')$$

and is homogeneous of degree 1 if

$$f(kx_1, kx_2 \ldots, kx_n) = kf(x_1, x_2, \ldots, x_n) \quad (k \text{ any constant}).$$

If f is both additive and homogeneous of degree 1, it must be a linear sum of x's with constant coefficients.

For

$$f(x_1, x_2, \ldots, x_n) = f(x_1, 0, \ldots, 0) + f(0, x_2, \ldots, x_n)$$

by additivity,

$$= \sum_{i=1}^{n} f(0, 0, \ldots, x_i, \ldots, 0)$$

by repeated application of the additivity property:

$$= \sum_{i=1}^{n} x_i f(0, 0, \ldots, 1, \ldots, 0)$$

by the homogeneity property,

$$= \sum_{i=1}^{n} k_i x_i, \text{ where } k_i = f(0, 0, \ldots, 1, \ldots, 0)$$

which is the required result.

Subject to conditions on good behaviour† of the function f, it is easily shown that the properties of additivity and homogeneity of degree 1 are equivalent. Thus any sufficiently regular function satisfying either condition is a linear expression:

$$\sum_{i=1}^{n} k_i x_i.$$

This is the stage so far in the classification of functionals. The equivalent may be seen by regarding a functional as a function of a continuous infinity of variables $x_{t_1}, x_{t_2}, \ldots$, where x_t is $x(t)$, the value of x at time t. In this case, in the expression $\sum k_i x_i$, summation must be made over a continuous infinity of values of i, viz. all the values of t in the appropriate range. This implies an integration: $\int k(t)x(t)\,dt$, giving again the integral form for a linear functional.

The analogy between functionals and functions of many variables may be pushed further. It is known that if f is sufficiently regular (analytic) near the values $x_1 = x_2 = \ldots = x_n = 0$ it has a power series expansion:

$$f(x_1, x_2, \ldots, x_n) = k_0 + \sum_{i=1}^{n} k_i x_i + \sum_{i=1}^{n}\sum_{j=1}^{n} k_{ij} x_i x_j + \sum_{i=1}^{n}\sum_{j=1}^{n}\sum_{k=1}^{n} k_{ijk} x_i x_j x_k + \ldots,$$

which is convergent for sufficiently small values of the x's, i.e. there is some number $r > 0$ with the property that the expansion is convergent if each of

† It is sufficient for f to be differentiable.

the x's is less than r in absolute value. The greatest such number r may be called the *modulus of convergence* of the series. r may, of course, be infinite.

To carry over this expansion to functionals, the previous procedure may be tentatively adopted, i.e. the method of replacing summation over a finite set of suffices by summation (integration) over a continuous infinity of values of the variable t:

$$\sum_{i=1}^{n} k_i x_i \quad \text{is replaced by} \quad \int k(t)x(t)\,dt,$$

$$\sum_{i=1}^{n}\sum_{j=1}^{n} k_{ij}x_i x_j \quad \text{is replaced by} \quad \int\int k(t_1, t_2)x(t_1)x(t_2)\,dt_1\,dt_2,$$

etc.

Thus the functional expansion might be expected to take the form:

$$f(x(t')\,t') = k_0 + \int k(t)x(t)\,dt + \int\int k(t_1, t_2)x(t_1)x(t_2)\,dt_1\,dt_2$$
$$+ \int\int\int k(t_1, t_2, t_3)x(t_1)x(t_2)x(t_3)\,dt_1\,dt_2\,dt_3$$
$$+ \dots .$$

The modulus of convergence of the series is defined to be the greatest $r > 0$ with the property that the expansion is convergent if $|x(t)| < r$ (for the range of t considered). If it is possible to find some modulus of convergence of the above series, f will be called *analytic* at $x = 0$. It will be noticed that if f is a function analytic at $x = 0$, its power series expansion as a functional may be obtained from its power series expansion as a function (cf. example 2.3.8, page 63).

Not all functions of n variables can be expanded in a power series. Thus with $n = 1$, the function defined by $f(x) = 0$, $x < 0$ and $f(x) = mx$, $x > 0$ $(m \neq 0)$, which has a simple discontinuity in the gradient at $x = 0$, has no power series expansion about the origin. It is, however, continuous, and so according to the theorem of Weierstrass can be approximated by polynomials to any required degree of accuracy in the neighbourhood of the origin. Weierstrass' theorem extends to continuous functionals. Any continuous functional may be approximated by polynomials, i.e. functional polynomials†. This approximation theorem should suffice in the majority of cases of engineering systems which do not respond critically to certain changes in input (e.g. as a flip-flop would). *From now on functionals will generally be assumed analytic.* It should be borne in mind that according to the above approximation theorem any continuous functional will be arbitrarily near to an analytic functional (in fact a polynomial).

† Volterra (1930), page 20. Continuity is defined as follows : $(f(x_1(t')) - f(x_2(t'))$ is to be small whenever $x_1(t') - x_2(t')$ is small (for all times t').

The power series expansion may be shown unique if all the kernels are completely symmetrical in the variables, i.e.

$$k(t_1, t_2, \ldots, t_n) = k(t_2, t_3, \ldots, t_n, t_1), \text{ etc.},$$

for any rearrangement of the variables t. For example, for $n=3$ we would require that

$$k(t_1, t_2, t_3) = k(t_3, t_1, t_2) = k(t_2, t_3, t_1) = k(t_3, t_2, t_1)$$
$$= k(t_1, t_3, t_2) = k(t_2, t_1, t_3)$$

identically. Any unsymmetrical kernel may always be changed to a symmetrical kernel, i.e. if $k(t_1, t_2, \ldots, t_n)$ is not symmetrical, there is another kernel $k^*(t_1, t_2, \ldots, t_n)$ such that

$$\int\int \cdots \int k(t_1, t_2, \ldots, t_n) x(t_1) x(t_2) \ldots x(t_n)\, dt_1\, dt_2 \ldots dt_n$$
$$= \int\int \cdots \int k^*(t_1, t_2, \ldots, t_n) x(t_1) x(t_2) \ldots x(t_n)\, dt_1\, dt_2 \ldots dt_n$$

identically. It is merely necessary to put

$$k^*(t_1, t_2, \ldots, t_n) = \frac{1}{n} \sum k(t_{i1}, t_{i2}, \ldots, t_{in}),$$

where the sum is over all permutations of the suffices. Thus, for $n=3$ $k^*(t_1, t_2, t_3) = \frac{1}{6}(k(t_1, t_2, t_3) + k(t_3, t_2, t_1) + \ldots)$. Clearly, if k is symmetrical, $k^* = k$.

Return now to the functional relation between the output signal $y(t)$ at any time say, and the input signal $x(t)$:

$$y(t) = f(t\,;\, x(t'), t' \leqslant t).$$

Expand f in a power series:

$$f(t\,;\, x(t'), t' \leqslant t) = \sum_n \int\int \cdots \int k(t_1, t_2, \ldots, t_n)\, x(t_1) x(t_2) \ldots x(t_n)\, dt_1\, dt_2 \ldots dt_n,$$

where the summation and integration are over the appropriate ranges.

The symmetric kernels in this expansion are uniquely determined by the functional f. Since the value of the left-hand side depends on t in the time-variant case, so do the kernels. This dependence should be explicit in the notation. Thus denoting $k(t_1, t_2, \ldots, t_n)$ corresponding to time t by $k(t\,;\, t_1, t_2, \ldots, t_n)$ we have:

$$f(t\,;\, x(t'), t' \leqslant t) = \sum \int\int \cdots \int k(t\,;\, t_1, t_2, \ldots, t_n) x(t_1) x(t_2) \ldots x(t_n)\, dt_1\, dt_2 \ldots dt_n.$$

Because f acts only on the past values of x, viz. the values $x(t')$, $t' < t$ at any time t, the kernels will have the property: $k(t: t_1, t_2, \ldots, t_n) = 0$ if any of the $t_1, t_2, \ldots, t_n$ is greater than t. The kernel with this property will be said to be of Volterra type.

If f is time-invariant, the kernels depend only on time differences $t-t_1$, $t-t_2$, etc. This may be shown as follows: if

$$y(t)=\sum\int\int\ldots\int k(t;\, t_1, t_2, \ldots, t_n)x(t_1)x(t_2)\ldots x(t_n)\, dt_1\, dt_2 \ldots dt_n, \quad \text{(A)}$$

and the functional on the right-hand side is time-invariant, then

$$y(t+\tau)=\sum\int\int\ldots\int k(t;\, t_1, t_2, \ldots, t_n)x(t_1+\tau)x(t_2+\tau)\ldots x(t_n+\tau)\, dt_1\, dt_2 \ldots dt_n \quad \text{(B)}$$

since the right-hand side represents the output corresponding to an input $x(t'+\tau)$ at time t'; i.e. to $x(t')$ a time τ earlier. This by time-invariance is $y(t+\tau)$, i.e. $y(t)$ a time τ earlier.

Changing the variables in (A),

$$y(t+\tau)=\sum\int\int\ldots\int k(t+\tau;\, t_1+\tau, \ldots)x(t_1+\tau)\ldots dt_1 \ldots; \quad \text{(C)}$$

$\therefore$ by uniqueness of the expansion (symmetric kernels),

$$k(t;\, t_1, t_2, \ldots, t_n)=k(t+\tau;\, t_1+\tau, \ldots t_n+\tau) \text{ identically.}$$

Put $\tau=-t$:

$$k(t;\, t_1, t_2, \ldots, t_n)=k(0, t_1-t, t_2-t, \ldots t_n-t).$$

Hence putting

$$k(t;\, t_1, t_2, \ldots, t_n)=h(t-t_1, t-t_2, \ldots, t-t_n),$$

we have the general expression:

$$y(t)=\sum\int\int\ldots\int h(t-t_1, t-t_2, \ldots, t-t_n)x(t_1)x(t_2)\, x(t_n)\, dt_1\, dt_2 \ldots dt_n$$

in the time-invariant case which, of course, reduces to the ordinary convolution formula

$$y(t)=\int h(t-t_1)x(t_1)\, dt_1$$

for a linear time-invariant f.

Stability and convergence

Suppose that the functional power series

$$\sum\int\int\ldots\int f(t;\, t_1, t_2, \ldots, t_n)x(t_1)x(t_2)\ldots x(t_n)\, dt_1\, dt_2 \ldots dt_n$$

has modulus of convergence r and suppose that $\sum\limits_n a_n r^n$ is convergent, where

$$a_n=\int\int\ldots\int |f(t;\, t_1, t_2, \ldots, t_n)|\, dt_1\, dt_2 \ldots dt_n.$$

Put $\sum\limits_n a_n r^n = A$.

Then if

$$y(t)=\sum_n \underbrace{\int\int\ldots\int}_{n} f(t;\, t_1, t_2, \ldots, t_n)x(t_1)x(t_2)\ldots x(t_n)\, dt_1\, dt_2 \ldots dt_n$$

and if $|x(t)| < r$ for all t, then $|y(t)| < A$ for all t, i.e. the system described by the functional is stable (in the sense of Zadeh) for the regime $|x| < r$.

This connection between the stability of a system under continuous input and the modulus of convergence of the power series expansion of its functional makes the study of the last of some importance.

Analysis with functionals

It has been seen that in certain ways a functional is rather like an ordinary function of real variables. In fact a function of real variables can be considered as a special type of functional. As the literature shows, quite a large number of the properties of functions extend to functionals and in particular it is possible to define derivatives and show the validity of MacLaurin and Taylor expansions in some cases. Further discussion and references are to be found in the works of Volterra, Lévy and Hille.

Determination of functional expansions from differential equations

The relation between input and output of engineering systems is frequently given in the form of a differential equation or more generally, by a differential-difference equation. The connection between this form of relation and the relations of the type considered here will be briefly discussed.

A general differential equation relating input x and output y will be of the form:

$$f(x, \dot{x}, \ldots, x^{(m)}; y, \dot{y}, \ldots, y^{(n)}; t) = 0, \qquad (1)$$

where dots denote time-derivation†.

An example of a differential-difference equation would be

$$\dot{y}(t) + ay(t) = bx(t) + c\dot{x}(t)x(t-\tau),$$

where a, b, c and τ are constants.

If we only consider signals $x(t)$ and $y(t)$ which are zero up to a certain time, say t_0 (which will represent the switching-on time of the system), the equation will in general determine the value of y at any instant uniquely in terms of past values of x; i.e. $y(t)$ will be a functional depending only on past values of x. The explicit expression may be given if f is a linear sum of derivatives with constant coefficients: if

$$f = a_0x + a_1\dot{x} + \ldots + a_mx^{(m)} + b_0y + b_1\dot{y} + \ldots + b_ny^{(n)}$$

then the equation $f = 0$ may be solved by Laplace transformation to give:

$$y(t) = \int h(t-t_1)x(t_1)\,dt_1, \qquad (2)$$

where the Laplace transform of $h(t)$ is given by:

$$L_p(h) = \int_{-\infty}^{\infty} h(t)\exp(-pt)\,dt$$

$$= -\frac{a_0 + a_1p + \ldots + a_mp^m}{b_0 + b_1p + \ldots + b_np^n}. \qquad (3)$$

† Millar gives some discussion on relations of this form in his paper. Existence theorems for equations of this sort are given in L. M. Graves *Theory of Functions of Real Variables* (McGraw-Hill), 1946, Chap. IX.

The method of Laplace transformation will extend to the treatment of linear differential-difference equations with constant coefficients (Bellman 1954) but not to any case of differential or differential-difference equation which involves t explicitly. However, it may easily be verified that the functional defined by such a differential equation must be linear and so by the general theory of linear functionals, the relation between x and y will be of the form:

$$y(t)=\int_{-\infty}^{\infty} k(t,t_1)x(t_1)\,dt_1,$$

where $k(t, t_1)$ is the output at time t due to a unit impulse input at time t_1. Thus in the linear case, the derivation of the functional expression from the differential or differential-difference equation is, at least in principle, determinable.

For a general function f, it is not at all clear yet when the functional relation between x and y is analytic. It seems that if $x=y=0$ is a solution of eqn. (1) and that f is sufficiently regular for small values of x, y and their derivatives, then y is a continuous functional of x.

It can actually be shown in certain cases that if f is analytic in all the derivatives $x^{(r)}$, $y^{(s)}$, $r=0,1,\ldots,m$; $s=1,2,\ldots,n$, then $y(t)$ is an analytic functional of values of $x(t_1)$, $t_1 \leqslant t$ if $\max x(t)$ is sufficiently small and that the power series expansion of this functional may be obtained by:

(*a*) expanding f as a multiple power series in the derivatives $x^{(r)}$, $y^{(s)}$, $r=0,1,\ldots,m$; $s=0,1,\ldots,n$;

(*b*) writing $D^r x$ for $x^{(r)}$, $r=0,1,\ldots,m$; $D^s y$ for $y^{(s)}$, $s=0,1,\ldots,n$;

(*c*) formally solving the resulting double power series in x and y for y;

(*d*) interpreting the operators in the coefficients suitably.

Example:

$$\ddot{y}+a\dot{y}+by+ey^3=x(t) \tag{4}$$

(Duffing's equation with arbitrary forcing).

Step (*a*) is not necessary as the function f is a simple polynomial expression. Introduce the operator $D=d/dt$ and write the equation as:

$$Ly+ey^3=x, \tag{5}$$

where $L=D^2+aD+b$. Now invert this relation in the neighbourhood of $x=y=0$, treating L as an ordinary number to obtain a power series expansion:

$$y=c_1x+c_2x^2+\ldots. \tag{6}$$

The coefficients c may be obtained by substituting this power series for y in the left-hand side of eqn. (5) and equating coefficients of powers of x on left- and right-hand sides.

C1

We get

$$L(c_1x+c_2x^2+\ldots)+e(c_1x+c_2x^2+\ldots)^3=x;$$

i.e.

$$Lc_1x+Lc_2x^2+\ldots+ec_1{}^3x^3+3ec_1{}^2c_2x^4+\ldots=x.$$

$$\therefore\ Lc_1=1,$$
$$Lc_2=0,$$
$$Lc_3+ec_1{}^3=0.$$
$$\ldots\ldots\ldots\ldots$$

$$\therefore\ c_2=c_4=c_6=\ldots=0 \quad\text{and}\quad c_1=L^{-1}, c_3=-L^{-1}eL^{-3},\ldots.$$

The meaning of L^{-1} is fairly clear. For the equation $Ly=x$ has solution $y=L^{-1}x$.

Comparing this with the solution

$$y(t)=\int_{-\infty}^{\infty} h(t-t_1)x(t_1)\,dt_1$$

of $\ddot{y}+a\dot{y}+by=x$, it is seen that $L^{-1}x$ is to be interpreted as the linear functional operation defined by

$$\int_{-\infty}^{\infty} h(t-t_1)x(t_1)\,dt_1.$$

Here of course $h(t)$ is given by its Laplace transform:

$$L_p(h)=\frac{1}{p^2+ap+b}.$$

The first two non-zero terms of the solution are thus:

$$y(t)=\int_{-\infty}^{\infty} h(t-t_1)x(t_1)\,dt_1-e\int_{-\infty}^{\infty} h(t-t_1)\left(\int_{-\infty}^{\infty} h(t_1-t_2)x(t_2)\,dt_2\right)^3 dt_1,$$

corresponding to $y=L^{-1}x-eL^{-1}(L^{-1}x)^3$. This may be re-written:

$$y(t)=\int_{-\infty}^{\infty} h(t-t_1)x(t_1)\,dt_1-e\int_{-\infty}^{\infty} dt_1\iiint_{-\infty}^{\infty} h(t-t_1)h(t_1-t_2)$$
$$\times h(t_1-t_3)h(t_1-t_4)x(t_2)x(t_3)x(t_4)\,dt_2\,dt_3\,dt_4$$

$$=\int_{-\infty}^{\infty} h(t-t_1)x(t_1)\,dt_1-e\iiint_{-\infty}^{\infty} h(t-t_1,t-t_2,t-t_3)x(t_1)x(t_2)x(t_3)\,dt_1\,dt_2\,dt_3$$

on writing

$$h(t-t_2,t-t_3,t-t_4)=\int_{-\infty}^{\infty} h(t-t_1)h(t_1-t_2)h(t_1-t_3)h(t_1-t_4)\,dt_1$$

and re-naming the variables t_2, t_3, t_4 as t_1, t_2, t_3.

This series expansion may be obtained rather more conveniently in this case by writing the equation:

$$Ly+ey^3=x \quad\text{as}\quad y+eL^{-1}y^3=L^{-1}x=\xi,$$

say, and then inverting the relation

$$y+eL^{-1}y^3=\xi,$$

for instance, by the method of successive approximations if e is small. In the case where $b>0$, $a^2>4b$ (linear part over-damped) the functional power series may quite easily be shown convergent (i) for all $x(t)$ when

$e \geqslant 0$, (ii) for all $x(t)$ satisfying

$$\max_{-\infty<t<\infty} |x(t)| < \frac{\sqrt{2b}}{3}\sqrt{\frac{b}{3|e|}}$$

when $e \leqslant 0$.

Examples of the Use of Functional Expansions

1. *The static characteristic of a device*

Suppose that a device is described by the analytical functional relation:

$$y(t) = \sum_{n=0}^{\infty} \int\int \cdots \int k_n(t;\, t_1, t_2, \ldots, t_n) x(t_1) x(t_2) \ldots x(t_n)\, dt_1\, dt_2 \ldots dt_n$$

between input ($x(t)$ at time t) and output ($y(t)$ at time t). The 'static characteristic' relating x and y is obtained by fixing the input $x(t)$ at a constant value say X, observing the resulting constant value of $y(t)$, say Y, and plotting the relation between X and Y. Put

$$\left.\begin{matrix} x(t) = X \\ y(t) = Y \end{matrix}\right\} \text{for all time } t.$$

Then

$$\begin{aligned} Y &= \sum_n \int\int \cdots \int k_n(t;\, t_1, t_2, \ldots, t_n) X\,.\,X \ldots X\, dt_1\, dt_2 \ldots dt_n \\ &= \sum_n X^n \int\int \cdots \int k_n(t;\, t_1, t_2, \ldots, t_n)\, dt_1\, dt_2 \ldots dt_n \\ &= \sum_n a_n X^n, \end{aligned}$$

where

$$a_n = \int\int \cdots \int k_n(t;\, t_1, t_2, \ldots, t_n)\, dt_1\, dt_2 \ldots dt_n.$$

Thus the characteristic is given in power series form. If the device is time-invariant, it is easily verified by the change of variables $t - t_i = \tau_i$; $i = 1, 2, \ldots, n$, that a_n is independent of time.

2. *Response to a sinusoidal input*

Put $x(t) = A \cos \omega t$.
Then

$$\begin{aligned} y(t) &= \sum_{n=0}^{\infty} \int\int \cdots \int k_n(t;\, t_1, t_2, \ldots, t_n) x(t_1) x(t_2) \ldots x(t_n)\, dt_1\, dt_2 \ldots dt_n \\ &= \sum \int\int \cdots \int k_n(t;\, t_1, t_2, \ldots, t_n) \cos \omega t_1 \cos \omega t_2 \ldots \cos \omega t_n\, dt_1\, dt_2 \ldots dt_n . A^n \end{aligned}$$

The integral

$$\int\int \cdots \int k_n(t;\, t_1, t_2, \ldots, t_n) \cos \omega t_1 \cos \omega t_2 \ldots \cos \omega t_n\, dt_1\, dt_2 \ldots dt_n$$

may be evaluated by putting

$$\cos \omega t_r = \tfrac{1}{2}[\exp(i\omega t_r) + \exp(-i\omega t_r)],\ r = 1, 2, \ldots n,$$

and introducing the n-dimensional Fourier transform (or Laplace transform if k_n is of Volterra type).

Consider, e.g., the case $n=3$ and when the kernel is a function only of time differences.

Suppose

$$k_3(t\,;\,t_1,t_2,t_3)=h(t-t_1,t-t_2,t-t_3).$$

The 3-dimensional Fourier transform of h, $H(\omega_1,\omega_2,\omega_3)$ is defined by

$$H(\omega_1,\omega_2,\omega_3)=\iiint_{-\infty}^{\infty} h(\tau_1,\tau_2,\tau_3)\exp\left[-i(\omega_1\tau_1+\omega_2\tau_2+\omega_3\tau_3)\right]d\tau_1\,d\tau_2\,d\tau_3,$$

so that
$$\iiint k_3(t\,;\,t_1,t_2,t_3)\cos\omega t_1\cos\omega t_2\cos\omega t_3\,dt_1\,dt_2\,dt_3$$

$$=\frac{1}{2^3}\iiint h(t-t_1,t-t_2,t-t_3)[\exp(i\omega t_1)+\exp(-i\omega t_1)][\exp(i\omega t_2)$$

$$+\exp(-i\omega t_2)]\,[\exp(i\omega t_3)+\exp(-i\omega t_3)]\,dt_1\,dt_2\,dt_3$$

$$=\frac{1}{2^3}\iiint h(t-t_1,t-t_2,t-t_3)[\textstyle\sum\exp(\pm i\omega t_1\pm i\omega t_2\pm i\omega t_3)]dt_1\,dt_2\,dt_3,$$

where the summation is over all sign combinations:

$$=\frac{1}{2^3}\,[\exp(-3i\omega t)H(\omega,\omega,\omega)+\exp(-i\omega t)H(-\omega,\omega,\omega)$$

$+6$ more terms corresponding to all possible sign changes].

These eight terms may be grouped in four pairs corresponding to conjugate complex quantities and then can be replaced by sines and cosines of ωt and harmonics; e.g.

$$\tfrac{1}{2}[\exp(-i\omega t)H(-\omega,\omega,\omega)+\exp(i\omega t)H(\omega,-\omega,-\omega)$$

$$=\mathscr{R}H(\omega,-\omega,-\omega)\cos\omega t-\mathscr{I}H(\omega,-\omega,-\omega)\sin\omega t,$$

where $\mathscr{R}$ and $\mathscr{I}$ denote real and imaginary parts.

3. *Response to a random input*

Using an analytical function expansion it is possible to find the value of any output statistic in terms of the input statistics. This method has been used by Wiener (1930, 1942, 1949) and Ikehara (1951) to find the response of a non-linear device to noise.

For example, consider the auto-covariance of the output of a time-invariant non-linear device described by an odd analytical functional. Let

$$y(t)=\int k_1(t-t_1)x(t_1)\,dt_1+\iiint k_3(t-t_1,t-t_2,t-t_3)x(t_1)x(t_2)x(t_3)\,dt_1\,dt_2\,dt_3$$

$$+\dots \qquad \qquad . \quad . \quad . \quad \text{(A)}$$

Then, after a few changes of variable under the integral signs, we get:

$$y(t)y(t+\tau)=\iint k_1(t-t_1)k_1(t+\tau-t_2)x(t_1)x(t_2)\,dt_1\,dt_2$$

$$+\iiiint k_1(t-t_1)k_3(t+\tau-t_2,t+\tau-t_3,t+\tau-t_4)x(t_1)x(t_2)x(t_3)x(t_4)dt_1\,dt_2\,dt_3\,dt_4$$

$$+\iiiint k_1(t+\tau-t_1)k_3(t-t_2,t-t_3,t-t_4)x(t_1)x(t_2)x(t_3)x(t_4)\,dt_1\,dt_2\,dt_3\,dt_4$$

$$+\dots \qquad \qquad . \quad . \quad . \quad \text{(B)}$$

The required relation follows by taking statistical averages of both sides:

$$\overline{y(t)y(t+\tau)} = \int\int k_1(t-t_1)k_1(t+\tau-t_2)\overline{x(t_1)x(t_2)}\,dt_1\,dt_2$$
$$+\int\int\int\int k_1(t-t_1)k_3(t+\tau-t_2, t+\tau-t_3, t+\tau-t_4)$$
$$\overline{x(t_1)x(t_2)x(t_3)x(t_4)}\,dt_1\,dt_2\,dt_3\,dt_4$$
$$+\int\int\int\int k_1(t+\tau-t_1)k_3(t-t_2, t-t_3, t-t_4),$$
$$\overline{x(t_1)x(t_2)x(t_3)x(t_4)}\,dt_1\,dt_2\,dt_3\,dt_4$$
$$+\ldots \qquad \ldots \quad \text{(C)}$$

If the input is a stationary stochastic process, moments such as $\overline{x(t_1)x(t_2)x(t_3)x(t_4)}$ will depend only on time differences t_1-t_2, t_1-t_3, etc. and the integrals may be re-written by using a Fourier transformation in a way which may be more useful for practical calculation, as follows:

By the n-dimensional form of Parseval's theorem (Bochner and Chandrasekharan 1949):

$$\int\int\ldots\int f_1(t_1, t_2, \ldots, t_n)f_2(t_1, t_2, \ldots, t_n)\,dt_1\,dt_2\ldots dt_n$$
$$= \frac{1}{(2\pi)^n}\int\int\ldots\int F_1(\omega_1, \omega_2, \ldots, \omega_n)F_2(\omega_1, \omega_2, \ldots, \omega_n)\,d\omega_1\,d\omega_2\ldots d\omega_n, \quad \text{(D)}$$

where

$$\begin{cases} F_j(\omega_1, \omega_2, \ldots, \omega_n) = \int\int\ldots\int f_j(t_1, t_2, \ldots, t_n)\exp\left(-i\sum_{r=1}^{n}\omega_r t_r\right)dt_1\,dt_2\ldots dt_n, \\ f_j(t_1, t_2, \ldots, t_n) = \frac{1}{(2\pi)^n}\int\int\ldots\int F_j(\omega_1, \omega_2, \ldots, \omega_n)\exp\left(i\sum_{r=1}^{n}\omega_r t_r\right) d\omega_1\,d\omega_2\ldots d\omega_n, \end{cases}$$
$$j=1,2$$

are inverse Fourier transforms.

In the first term on the right-hand side, of eqn. (C), $n=2$,

$\therefore$ put $f_1(t_1, t_2) = k_1(t-t_1)k_1(t+\tau-t_2)$, $f_2(t_1, t_2) = \overline{x(t_1)x(t_2)}$

in the Parseval formula:

$$F_1(\omega_1, \omega_2) = \int\int f_1(t_1, t_2)\exp\left[-i(\omega_1 t_1 + \omega_2 t_2)\right]dt_1\,dt_2$$
$$= \int\int k_1(t-t_1)k_1(t+\tau-t_2)\exp\left[-i(\omega_1 t_1+\omega_2 t_2)\right]dt_1\,dt_2$$
$$= \left(\int k_1(t-t_1)\exp(-i\omega_1 t_1)\,dt_1\right)\left(\int k_2(t+\tau-t_2)\exp(-i\omega_2 t_2)\,dt_2\right)$$
$$= \left(K_1{}^*(\omega_1)\exp(-i\omega_1 t)\right)\left(K_1{}^*(\omega_2)\exp\left[-i\omega_2(t+\tau)\right]\right)$$
$$= \quad K_1{}^*(\omega_1)K_1{}^*(\omega_2)\exp\left[-i(\omega_1+\omega_2)t - i\omega_2\tau\right].$$

Here $K_1(\omega) = \int k_1(t)\exp(-i\omega t)\,dt$, the Fourier transform of $k_1(t)$ and the star denotes complex conjugate:

$$F_2(\omega_1, \omega_2) = \int\int \overline{x(t_1)x(t_2)} \exp[-i(\omega_1 t_1 + \omega_2 t_2)]\, dt_1\, dt_2$$

$$= \int\int \phi(\tau) \exp[-i(\omega_1 + \omega_2)t_2 - i\tau\omega_1)]\, d\tau\, dt_2,$$

where $\tau = t_1 - t_2$ and $\phi(\tau) = \overline{x(t)x(t+\tau)}$ is the auto-covariance function. It is dependent only on τ if $x(t)$ is a stationary process.

Now $\int_{-\infty}^{\infty} \exp(-i\omega t)\, dt = 2\pi\delta(\omega)$, where $\delta(\omega)$ is the Dirac δ-function.

$$\therefore\ F_2(\omega_1, \omega_2) = \left(\int \phi(\tau) \exp(-i\tau\omega_1)\, d\tau\right)\left(\int \exp[-i(\omega_1 + \omega_2)t_2)]\, dt_2\right)$$

$$= \Phi(\omega_1) 2\pi\delta(\omega_1 + \omega_2),$$

where $\Phi(\omega)$ is the spectral density function since $\Phi(\omega) = \int \phi(\tau) \exp(-i\tau\omega)\, d\tau$ by the Wiener Khinchin theorem.

$$\therefore \int\int f_1(t_1, t_2) f_2(t_1, t_2)\, dt_1\, dt_2 = \frac{1}{(2\pi)^2} \int\int K_1^*(\omega_1) K_1^*(\omega_2)$$

$$\times \exp[-i(\omega_1 + \omega_2)t - i\tau\omega_2] 2\pi\Phi(\omega_1)\, \delta(\omega_1 + \omega_2)\, d\omega_1\, d\omega_2$$

$$= \frac{1}{2\pi} \int K_1^*(-\omega_2) K_1^*(\omega_2) \exp(-i\tau\omega_2) \Phi(\omega_2)\, d\omega_2$$

$$= \frac{1}{2\pi} \int |K_1(\omega)|^2 \Phi(\omega) \exp(-i\tau\omega)\, d\omega \ldots \qquad \text{(E)}$$

on changing from ω_2 to ω, because $K_1(-\omega) = K_1^*(\omega)$. This formula is well known from linear theory: it is the only term on the right-hand side of (C) if the functional is linear.

Consider the second term on the right-hand side of (C) which will only occur when the functional is non-linear.

In the Parseval formula, put $n = 4$ and

$$f_1(t_1, t_2, t_3, t_4) = k_1(t - t_1) k_3(t + \tau - t_2, t + \tau - t_3, t + \tau - t_4),$$

$$f_2(t_1, t_2, t_3, t_4) = \overline{x(t_1)x(t_2)x(t_3)x(t_4)},$$

$$F_1(\omega_1, \omega_2, \omega_3, \omega_4) = \int\int\int\int k_1(t - t_1) k_3(t + \tau - t_2, t + \tau - t_3, t + \tau - t_4)$$

$$\exp[-i(\omega_1 t_1 + \omega_2 t_2 + \omega_3 t_3 + \omega_4 t_4)]\, dt_1\, dt_2\, dt_3\, dt_4$$

$$= \left(\int k_1(t - t_1) \exp(-i\omega_1 t_1)\, dt_1\right)\left(\int\int\int k_3(t + \tau - t_2, t + \tau - t_3,\right.$$

$$\left. t + \tau - t_4) \exp[-i(\omega_2 t_2 + \omega_3 t_3 + \omega_4 t_4)]\, dt_2\, dt_3\, dt_4\right)$$

$$= (K_1^*(\omega_1) \exp(-i\omega_1 t))(K_3^*(\omega_2, \omega_3, \omega_4) \exp[-i(t + \tau)$$

$$(\omega_2 + \omega_3 + \omega_4)])$$

$$= K_1^*(\omega_1) K_3^*(\omega_2, \omega_3, \omega_4) \exp[-it(\omega_1 + \omega_2 + \omega_3 + \omega_4)]$$

$$\exp[-i\tau(\omega_2 + \omega_3 + \omega_4)], \qquad \text{(F)}$$

where

$$K_3(\omega_1, \omega_2, \omega_3) = \iiint k_3(t_1, t_2, t_3) \exp[-i(\omega_1 t_1 + \omega_2 t_2 + \omega_3 t_3)]\, dt_1\, dt_2\, dt_3$$

is the Fourier transform of $k_3(t_1, t_2, t_3)$:

$$F_2(\omega_1, \omega_2, \omega_3, \omega_4) = \iiiint \overline{x(t_1)x(t_2)x(t_3)x(t_4)} \exp[-i(\omega_1 t_1 + \ldots + \omega_4 t_4)]\, dt_1\, dt_2\, dt_3\, dt_4$$

$$= \iiiint \overline{x(t_1)x(t_1+\tau_2)x(t_1+\tau_3)x(t_1+\tau_4)} \exp[-i(\omega_1+\omega_2+\omega_3+\omega_4)t_1] \exp[-i(\omega_2\tau_2+\omega_3\tau_3+\omega_4\tau_4)]\, dt_1\, d\tau_2\, d\tau_3\, d\tau_4$$

(on putting $\tau_j = t_j - t_1$, $j = 2, 3, 4$.)

$$= \int \exp[-i(\omega_1+\omega_2+\omega_3+\omega_4)t_1]\, dt_1 \iiint \overline{x(t_1)x(t_1+\tau_2)x(t_1+\tau_3)x(t_1+\tau_4)} \exp[-i(\omega_2\tau_2+\omega_3\tau_3+\omega_4\tau_4)]\, d\tau_2\, d\tau_3\, d\tau_4$$

$$= 2\pi\delta(\omega_1+\omega_2+\omega_3+\omega_4)\Phi(\omega_2, \omega_3, \omega_4),$$

where

$$\Phi(\omega_2, \omega_3, \omega_4) = \iiint \overline{x(t_1)x(t_1+\tau_2)x(t_1+\tau_3)x(t_1+\tau_4)} \exp[-i(\omega_2\tau_2+\omega_3\tau_3+\omega_4\tau_4)]\, d\tau_2\, d\tau_3\, d\tau_4 \quad . \; . \; . \quad (G)$$

is a generalized spectral density function.

∴ by Parseval's formula and eqns. (F) and (G):

$$\iiiint k_1(t-t_1)k_3(t+\tau-t_2, t+\tau-t_3, t+\tau-t_4)\,\overline{x(t_1)x(t_2)x(t_3)x(t_4)}\, dt_1\, dt_2\, dt_3\, dt_4$$

$$= \left(\frac{1}{(2\pi)^4}\right) \iiiint K_1^*(\omega_1)K_3^*(\omega_2, \omega_3, \omega_4) \exp[-it(\omega_1+\omega_2+\omega_3+\omega_4)] \times \exp[-i\tau(\omega_2+\omega_3+\omega_4)]\,.\,2\pi\delta(\omega_1+\omega_2+\omega_3+\omega_4) \times \Phi(\omega_2, \omega_3, \omega_4)\, d\omega_1\, d\omega_2\, d\omega_3\, d\omega_4$$

$$= \left(\frac{1}{(2\pi)^3}\right) \iiiint_{\omega_1+\omega_2+\omega_3+\omega_4=0.} K_1^*(\omega_1)K_3^*(\omega_2, \omega_3, \omega_4,)\Phi(\omega_2, \omega_3, \omega_4) \exp[i\tau(\omega_2+\omega_3+\omega_4)]\, d\omega_1\, d\omega_2\, d\omega_3\, d\omega_4$$

This is thus the second term of the expansion (C). Higher terms may be transformed in the same way. Note that the third term is obtained from the second by changing τ to $-\tau$.

3.1. *The Effect of a Non-linear Element on a Signal Disturbed by Noise*

As a further application of functional expansions to statistical calculations, the interaction terms between signal and noise due to passage through a non-linear device will be found.

If the input to a linear element consists of a signal perturbed by noise, it follows from the principle of superposition, that the output signal will be the sum of filtered signal and filtered noise. However, in the case where the element is not linear, this will not be the case and further terms will arise due to interaction between signal and noise.

Consider first the linear case. Let $x(t)$ be the input at time t which is

$$x(t) = s(t) + n(t), \qquad (1)$$

the sum of a signal component $s(t)$ and noise component, $n(t)$. Suppose that the input is related to the output $y(t)$ at time t by the equation

$$y(t) = \int k(t, t_1) x(t_1)\, dt_1. \qquad (2)$$

Then by substitution, it is seen that:

$$y(t) = s'(t) + n'(t),$$
$$s'(t) = \int k(t, t_1) s(t_1)\, dt_1 \qquad (3)$$

where

$$n'(t) = \int k(t, t_1) n(t_1)\, dt_1, \qquad (4)$$

which may be called filtered signal and filtered noise respectively.

The mean square deviation of the output from its true value due to the input disturbance $n(t)$ is:

$$\overline{[n'(t)]^2} = \overline{\left[\int k(t, t_1) n(t_1)\, dt_1\right]^2}$$
$$= \overline{\left[\int\int k(t, t_1) k(t, t_2) n(t_1) n(t_2)\, dt_1\, dt_2\right]}$$
$$= \int\int k(t, t_1) k(t, t_2) \overline{n(t_1) n(t_2)}\, dt_1\, dt_2.$$

It is seen to depend only on the value of the input disturbance. This will not hold in the non-linear case.

In the non-linear case the output will be assumed to be related to the input by a functional equation:

$$y(t) = f(t; x(t'), t' \leqslant t),$$

which, if analytic, may be expanded in a convergent power series. As the general case would lead to tedious computation, the special case will be considered where the functional is odd and almost linear so that it may be adequately represented by the linear and cubic terms of the expansion, viz.

$$y(t) = \int k_1(t, t_1) x(t_1)\, dt_1 + \int\int\int k_3(t; t_1, t_2, t_3) x(t_1) x(t_2) x(t_3)\, dt_1\, dt_2\, dt_3.$$

Suppose, as before, that

$$x(t) = s(t) + n(t).$$

Then

$$y(t) = \int k_1(t, t_1)s(t_1)\,dt_1 + \int k_1(t, t_1)n(t_1)\,dt_1$$

linear part filtered signal + filtered noise

$$+ \int\int\int k_3(t; t_1, t_2, t_3)s(t_1)s(t_2)s(t_3)\,dt_1\,dt_2\,dt_3$$

non-linear part of filtered signal

$$+ 3\int\int\int k_3(t; t_1, t_2, t_3)s(t_1)s(t_2)n(t_3)\,dt_1\,dt_2\,dt_3$$

interaction terms

$$+ 3\int\int\int k_3(t; t_1, t_2, t_3)s(t_1)n(t_2)n(t_3)\,dt_1\,dt_2\,dt_3$$

$$+ \int\int\int k_3(t; t_1, t_2, t_3,)n(t_1)n(t_2)n(t_3)\,dt_1\,dt_2\,dt_3$$

non-linear part of filtered noise.

The output is seen to include a number of interaction terms due to the simultaneous presence of signal and noise. These remain, even if signal and noise are statistically independent.

The mean square deviation of the output from the value it would have in the absence of noise may be calculated as before. It is:

$$\overline{\left[y(t) - \int k_1(t, t_1)s(t_1)\,dt_1 - \int\int\int k_3(t; t_1, t_2, t_3)s(t_1)s(t_2)s(t_3)\,dt_1 dt_2\,dt_3\right]^2}$$

$$= \overline{\left[\int k_1(t, t_1)n(t_1)\,dt_1 + 3\int\int\int k_3(t; t_1, t_2, t_3)s(t_1)s(t_2)n(t_3)\,dt_1\,dt_2\,dt_3 + \ldots\right]^2},$$

and is seen to depend on the value of s as well as n. The right-hand side on expansion gives 10 terms involving multi-dimensional integrals of the higher cross-moments of s and n; e.g. one term is:

$$6\int\int\int\int k_1(t, t_1)k_3(t; t_2, t_3, t_4)\overline{n(t_1)s(t_2)s(t_3)n(t_4)}\,dt_1\,dt_2\,dt_3\,dt_4.$$

To evaluate these integrals, the values of the kernels and the joint statistics of signal and noise need to be known.

3.2. *Chain Product of Functionals*

Since most signal transmission systems consist of sequences of devices in series, it is of interest to know the rule of combination of the corresponding functionals.

With the notation of the diagram, the basic relations are:

$$y(t) = f(t; x(t'), t' \leqslant t), \quad \ldots\ldots\ldots \quad \text{(A)}$$

$$z(t) = g(t; y(t'), t' \leqslant t). \quad \ldots\ldots\ldots \quad \text{(B)}$$

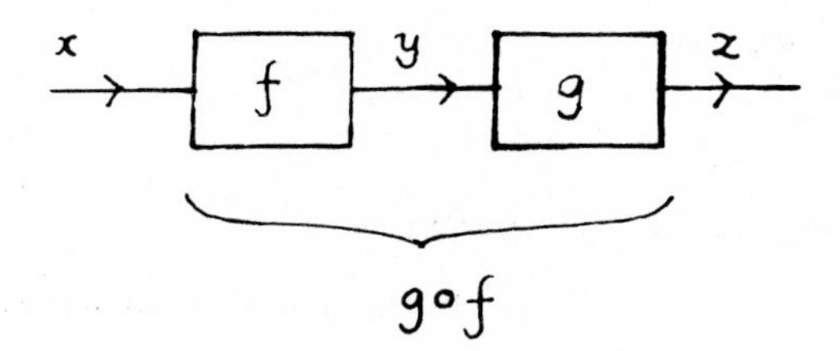

These two relations determine a relation of the same type between z and x, say:

$$z(t)=h(t;x(t'),\ t'\leqslant t). \qquad \text{(C)}$$

The functional h will be called the chain product of f and g written $g^\circ f$.

With three functionals f_1, f_2 and f_3, it is clear by definition that $f_3{}^\circ(f_2{}^\circ f_1)=(f_3{}^\circ f_2)^\circ f_1$ and so these expressions may be written $f_3{}^\circ f_2{}^\circ f_1$ without ambiguity.

It is not true in general that $g^\circ f=f^\circ g$. If this is so, the functionals will be said to commute. The physical meaning of commuting functionals may be been in the case when the signals corresponding to the variables x and y are of the same nature, e.g. both voltages of roughly the same strength. If f and g commute then it does not matter whether the operation corresponding to f or that corresponding to g is applied first to the signal.

3.2.1. *f and g linear*

Denote the kernels of the functionals by $f(t,t_1)$ and $g(t,t_1)$ respectively. Then eqns. (A) and (B) are:

$$y(t)=\int f(t,t_1)x(t_1)\,dt_1, \qquad \text{(A')}$$

$$z(t)=\int g(t,t_1)y(t_1)\,dt_1. \qquad \text{(B')}$$

The result of eliminating y is a linear relation of the same form which may be obtained by straightforward substitution:

$$\begin{aligned}
z(t)&=\int g(t,t_1)y(t_1)\,dt_1\\
&=\int g(t,t_1)\left(\int f(t_1,t_2)x(t_2)\,dt_2\right)dt_1\\
&=\int\!\!\int g(t,t_1)f(t_1,t_2)x(t_2)\,dt_1\,dt_2\\
&=\int\left(\int g(t,t_1)f(t_1,t_2)\,dt_1\right)x(t_2)\,dt_2\\
&=\int h(t,t_2)x(t_2)\,dt_2,
\end{aligned}$$

where $h(t,t_2)=\int g(t,t_1)f(t_1,t_2)\,dt_1$. This is the kernel of $g^\circ f$. The kernel of $f^\circ g$ will be $\int f(t,t_1)g(t_1,t_2)\,dt_1$. These are not in general the same and so the functionals will not, in general, commute (Volterra 1930). However if f and g are time-invariant, the functionals will always commute. This is seen immediately, since in this case the Fourier transforms are related by a simple algebraic identity. Alternatively, it may be seen by a straightforward change of variable: f and g depend only on time differences in the time-invariant case. Write $f(t, t_1) = f_1(t-t_1)$, $g(t, t_1) = g_1(t-t_1)$. Then from above:

$$h(t, t_2) = \int g(t, t_1) f(t_1, t_2)\, dt_1$$

$$= \int g_1(t-t_1) f_1(t_1-t_2)\, dt_1$$

$$= \int g_1(t_1'-t_2) f_1(t-t_1')\, dt_1'$$

(after the change of variable $t_1' = t - t_1 + t_2$)

$$= \int f(t, t_1') g(t_1', t_2)\, dt_1'$$

as required.

3.2.2. *f and g analytic*

Let the expansions of f and g be:

$$y(t) = \sum_{n=0}^{\infty} \int\int \ldots \int f(t;\, t_1, t_2, \ldots, t_n) x(t_1) x(t_2) \ldots x(t_n)\, dt_1\, dt_2 \ldots dt_n \quad (\mathrm{A}'')$$

$$z(t) = \sum_{n=0}^{\infty} \int\int \ldots \int g(t;\, t_1, t_2, \ldots, t_n) y(t_1) y(t_2) \ldots y(t_n)\, dt_1\, dt_2 \ldots dt_n, \quad (\mathrm{B}'')$$

where the letters f and g have been used also for the kernels corresponding to the functionals f and g. The relation between z and x may be obtained by direct substitution of the values of y from the first equation in the second equation. As in the corresponding case with ordinary power-series (Goursat 1904) it is seen that if both series have non-zero moduli of convergence the resultant has a non-zero modulus of convergence. The result tends to be algebraically complicated. To reduce the length of the equations, an abbreviated notation will be introduced.

Equations (A″) and (B″) will be written, in this abbreviated notation:

$$\left.\begin{aligned} y_t &= \sum f_{t;\, t_1 t_2 \ldots t_n} x_{t_1} x_{t_2} \ldots x_{t_n}, \\ z_t &= \sum g_{t;\, t_1, t_2, \ldots, t_n} y_{t_1} y_{t_2} \ldots y_{t_n}. \end{aligned}\right\} \quad \ldots\ldots \quad (\mathrm{C}'')$$

The convention is that functional dependence is indicated by the suffices and any suffix which occurs just twice in the same term is to be integrated over the appropriate range of time†. Note that any repeated suffix in the same term may be replaced by any other pair of repeated suffices which are not already present. For example,

$$g_{t;\, t_1, t_2} x_{t_1} x_{t_2} = g_{t;\, \alpha, t_2} x_\alpha x_{t_2},$$

† This is a generalization of the summation convention of the tensor calculus (the quantities are actually tensors).

both being contracted forms of the integral:

$$\iint g(t;\, t_1, t_2)x(t_1)x(t_2)\, dt_1\, dt_2.$$

Changing the variable t_1 to α merely means that α is used as a variable of integration instead of t_1.

If attention is restricted to the case where $f_t = g_t = 0$ (this means that for both devices, zero input implies zero output), the first few terms of the expansion are given by the equations:

$$z_t = h_{t\alpha}x_\alpha + h_{t\alpha_1\alpha_2}x_{\alpha_1}x_{\alpha_2} + \dots,$$

where

$$\left.\begin{aligned} h_{t\alpha} &= g_{t\beta} f_{\beta\alpha}, \\ h_{t\alpha_1\alpha_2} &= g_{t\beta_1\beta_2} f_{\beta_1\alpha_1} f_{\beta_2\alpha_2} + g_{t\beta} f_{\beta\alpha_1\alpha_2}, \\ h_{t\alpha_1\alpha_2\alpha_3} &= g_{t\beta_1\beta_2\beta_3} f_{\beta_1\alpha_1} f_{\beta_2\alpha_2} f_{\beta_3\alpha_3} + g_{t\beta_1\beta_2} f_{\beta_1\alpha_1} f_{\beta_2\alpha_2\alpha_3} + g_{t\beta_1\beta_2} f_{\beta_1\alpha_1\alpha_2} f_{\beta_2\alpha_3} + g_{t\beta} f_{\beta\alpha_1\alpha_2\alpha_3}. \end{aligned}\right\} \quad \text{(D)}$$

Greek letters are used here to avoid too many suffices on the t's. As remarked above, it does not matter what symbols are used. Thus the third kernel is, in full notation:

$$\begin{aligned} h(t;\, \alpha_1, \alpha_2, \alpha_3) = &\iiint g(t;\, \beta_1, \beta_2, \beta_3) f(\beta_1, \alpha_1) f(\beta_2, \alpha_2) f(\beta_3, \alpha_3)\, d\beta_1\, d\beta_2\, d\beta_3 \\ &+ \iint g(t;\, \beta_1, \beta_2) f(\beta_1, \alpha_1) f(\beta_2;\, \alpha_2, \alpha_3)\, d\beta_1\, d\beta_2 \\ &+ \iint g(t;\, \beta_1, \beta_2) f(\beta_1;\, \alpha_1, \alpha_2) f(\beta_2, \alpha_3)\, d\beta_1\, d\beta_2 \\ &+ \int g(t, \beta) f(\beta;\, \alpha_1, \alpha_2, \alpha_3)\, d\beta. \quad \dots \quad \text{(E)} \end{aligned}$$

Time-invariant case: Use of Fourier transform:

In the time-invariant case, all these integrals will take the form of generalized convolutions, the linear parts of the expressions naturally corresponding to the result found above for the linear case. By taking Fourier transforms of both sides of eqns. (D) it is possible to arrive at algebraic relations between the Fourier transforms of the kernels.

Thus suppose:

$$\begin{aligned} &f(t, t_1) = f_1(t - t_1), \qquad g(t, t_1) = g_1(t - t_1), \qquad h(t, t_1) = h_1(t - t_1), \\ &f(t;\, t_1, t_2) = f_1(t - t_1, t - t_2), \qquad \dots, \qquad \dots, \\ &\dots \end{aligned}$$

Without loss of generality, the kernels of f and g may be assumed symmetrical so that, e.g. $f(t;\, t_1, t_2) = f(t;\, t_2, t_1)$.

Corresponding to eqn. (E) is the equation:

$$h_1(t-\alpha_1, t-\alpha_2, t-\alpha_3) = \iiint g_1(t-\beta_1, t-\beta_2, t-\beta_3) f_1(\beta_1-\alpha_1) f_1(\beta_2-\alpha_2) \times f_1(\beta_3-\alpha_3)\, d\beta_1\, d\beta_2\, d\beta_3$$
$$+ \iint g_1(t-\beta_1, t-\beta_2) f_1(\beta_1-\alpha_1) f_1(\beta_2-\alpha_2, \beta_2-\alpha_3)\, d\beta_1\, d\beta_2$$
$$+ \iint g_1(t-\beta_1, t-\beta_2) f_1(\beta_1-\alpha_1, \beta_1-\alpha_2) f_1(\beta_2-\alpha_3)\, d\beta_1\, d\beta_2$$
$$+ \int g_1(t-\beta) f_1(\beta-\alpha_1, \beta-\alpha_2, \beta-\alpha_3)\, d\beta. \quad \text{(E')}$$

Taking the Fourier transform of both sides, we have:

$$H(\omega_1, \omega_2, \omega_3) = G(\omega_1, \omega_2, \omega_3) F(\omega_1) F(\omega_2) F(\omega_3) + G(\omega_1, \omega_2+\omega_3) F(\omega_1) F(\omega_2, \omega_3) + G(\omega_1+\omega_2, \omega_3) F(\omega_1, \omega_2) F(\omega_3) + G(\omega_1+\omega_2+\omega_3) F(\omega_1, \omega_2, \omega_3), \quad \text{(F)}$$

where $F(\omega) = \int f_1(\tau) \exp(-i\omega\tau)\, d\tau,$

$$F(\omega_1, \omega_2) = \iint f_1(\tau_1, \tau_2) \exp[-i(\omega_1\tau_1 + \omega_2\tau_2)]\, d\tau_1\, d\tau_2,$$

etc. (with the same notation for g's and h's) are the Fourier transforms of the various kernels.

It is clear that, since the kernels corresponding to f and g are symmetrical in their variables, so are the corresponding Fourier transforms F and G. Since the kernels of h are not, in general, symmetrical, the Fourier transforms may not be expected to be symmetrical.

Thus the right-hand side of (F) appears unsymmetrical. By linearity, it is seen that the Fourier transform of the symmetrical kernel corresponding to $h_1(t-t_1, t-t_2, t-t_3)$ is:

$$H(\omega_1, \omega_2, \omega_3) = G(\omega_1, \omega_2, \omega_3) F(\omega_1) F(\omega_2) F(\omega_3) + \tfrac{2}{3} \begin{bmatrix} G(\omega_1, \omega_2+\omega_3) F(\omega_1) F(\omega_2, \omega_3) \\ + G(\omega_2, \omega_3+\omega_1) F(\omega_2) F(\omega_3, \omega_1) \\ + G(\omega_3, \omega_1+\omega_2) F(\omega_3) F(\omega_1, \omega_2) \end{bmatrix} + G(\omega_1+\omega_2+\omega_3) F(\omega_1, \omega_2, \omega_3).$$

§ 4.

4.1. *Inverse Functional*

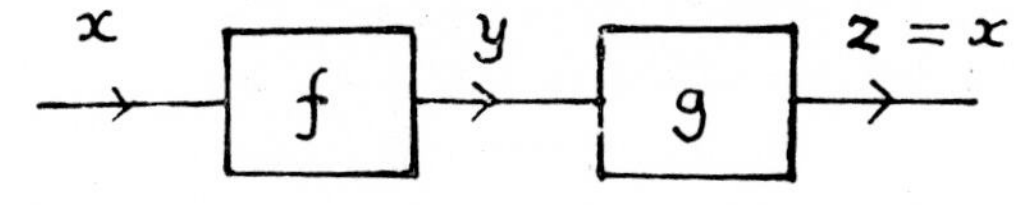

Two functionals, f and g, will be said to be inverse to one another if, $f \circ g$ and $g \circ f$ are both the identity. In this case, if a device representing

f is put in series with one represented by g, it will restore the input of this element from its output—and vice versa.

The functional relational between z and x is thus:

$$z(t) = x(t)$$
$$= \int \delta(t, \alpha) x(\alpha)\, d\alpha, \quad \text{(A)}$$

where $\delta(t, \alpha) = \delta(t-\alpha)$ is the δ-function kernel.

4.1.1. *Linear case*

It is easily seen that if a linear functional has an inverse, this must be linear too. Suppose that the functionals f and g have kernels $f(t, t_1)$ and $g(t, t_1)$. Then if they are inverse to one another, it follows from § 3.2.1 that

$$\left.\begin{aligned} \int f(t, t_1) g(t_1, t_2)\, dt_2 = \delta(t, t_2), \\ \int g(t, t_1) f(t_1, t_2)\, dt_1 = \delta(t, t_2). \end{aligned}\right\} \quad \text{(B)}$$

From this it is seen that f and g cannot both be proper functions since the integral of their product is a δ-function.

A case of practical importance is when

$$f(t, t_1) = \delta(t, t_1) + k(t, t_1), \quad \text{(C)}$$

where $k(t, t_1)$ is a proper function. In this case the equation

$$y(t) = \int f(t, t_1) x(t_1)\, dt_1 \quad \text{(D)}$$

is

$$y(t) = x(t) + \int k(t, t_1) x(t_1)\, dt_1; \quad \text{(E)}$$

i.e. a linear integral equation which will be of the Volterra type when f acts only on the past. It is known (Lovitt 1950) that an equation of this type has a unique solution of the form:

$$x(t) = y(t) - \int k^{\dagger}(t, t_1) y(t_1)\, dt_1, \quad \text{(F)}$$

where $k^{\dagger}(t, t_1)$ is related to $k(t, t_1)$ by an equation:

$$k^{\dagger}(t, t_1) = k(t, t_1) - \int k(t, t_2) k(t_2, t_1)\, dt_2$$
$$+ \int\int k(t, t_2) k(t_2, t_3) k(t_3, t_1)\, dt_2\, dt_3 - \dots$$

(Neumann–Liouville Series). (G)

Moreover the inversion of eqn. (F) gives eqn. (E).

Thus the inverse kernel corresponding to (C) is:

$$g(t, t_1) = \delta(t, t_1) - k^{\dagger}(t, t_1). \quad \text{(H)}$$

Time-invariant case: In this case $f(t, t_1)$ and $g(t, t_1)$ are functions only of the time difference $t - t_1$. Suppose $f(t, t_1) = f_1(t-t_1)$, $g(t, t_1) = g_1(t-t_1)$, then eqns. (B) both give, on a Fourier transformation of both sides:

$$F(\omega)G(\omega)=1, \qquad \text{(B}'\text{)}$$

where $$F(\omega)=\int f_1(\tau)\exp(-i\omega\tau)\,d\tau,\quad G(\omega)=\int g_1(\tau)\exp(-i\omega\tau)\,d\tau$$

are the Fourier transforms of $f_1(t)$ and $g_1(t)$.

Thus

$$G(\omega)=\frac{1}{F(\omega)} \qquad \text{(I)}$$

and so $g_1(t)$ may be found by an inverse Fourier transformation. It has yet to satisfy certain conditions for stability: $G(\omega)$ must have no poles in the region $\mathscr{I}(\omega)<0$ of the complex ω-plane. This means that $F(\omega)$ must have no zeros in this region. Also if $F(\omega)$ corresponds to a stable system, it too must have no poles in the region $\mathscr{I}(\omega)<0$. Therefore it must have no poles or zeros in this region. This condition is satisfied by minimum-phase networks (Bode 1945).

4.1.2. *Analytic case*

If f is an analytic functional with an invertible linear part then f has an analytic inverse. This follows from eqns. (D) of § 3.2.2 and relation (A) above. Thus, considering $g\circ f$ it follows that, in contracted notation:

$$\left.\begin{aligned}
\delta_{t,\alpha} &= g_{t\beta}f_{\beta\alpha},\\
0 &= g_{t\beta_1\beta_2}f_{\beta_1\alpha_1}f_{\beta_2\alpha_2}+g_{t\beta}f_{\beta\alpha_1\alpha_2},\\
0 &= g_{t\beta_1\beta_2\beta_3}f_{\beta_1\alpha_1}f_{\beta_2\alpha_2}f_{\beta_3\alpha_3}\\
&\quad+g_{t\beta_1\beta_2}f_{\beta_1\alpha_1}f_{\beta_2\alpha_2\alpha_3}\\
&\quad+g_{t\beta_1\beta_2}f_{\beta_1\alpha_1\alpha_2}f_{\beta_2\alpha_3}\\
&\quad+g_{t\beta}f_{\beta\alpha_1\alpha_2\alpha_3}.
\end{aligned}\right\} \qquad \text{(J)}$$

Suppose that the linear part of f, viz. the linear functional determined by the kernel $f(\beta,\alpha)$, is invertible. Then it is possible to find a kernel $g(t,\beta)$ satisfying the first of these equations and also

$$\delta_{t,\alpha}=f_{t\beta}g_{\beta\alpha} \qquad \text{(K)}$$

and it is then possible to solve for all the higher kernels.

Thus from the second of the set of eqns. (J):

$$\begin{aligned}
0 &= (g_{t\beta_1\beta_2}f_{\beta_1\alpha_1}f_{\beta_2\alpha_2}+g_{t\beta}f_{\beta\alpha_1\alpha_2})g_{\alpha_1r_1}g_{\alpha_2r_2}.\\
&= g_{t\beta_1\beta_2}f_{\beta_1\alpha_1}g_{\alpha_1r_1}f_{\beta_2\alpha_2}g_{\alpha_2r_2}+g_{t\beta}f_{\beta\alpha_1\alpha_2}g_{\alpha_1r_1}g_{\alpha_2r_2}\\
&= g_{t\beta_1\beta_2}\delta_{\beta_1r_1}\delta_{\beta_2r_2}+g_{t\beta}f_{\beta\alpha_1\alpha_2}g_{\alpha_1r_1}g_{\alpha_2r_2} \text{ from eqn. (K)}\\
&= g_{tr_1r_2}+g_{t\beta}f_{\beta\alpha_1\alpha_2}g_{\alpha_1r_1}g_{\alpha_2r_2};\\
\therefore\ g_{tr_1r_2} &= -g_{t\beta}f_{\beta\alpha_1\alpha_2}g_{\alpha_1r_1}g_{\alpha_2r_2}.
\end{aligned}$$

This solves for $g_{tr_1r_2}$ in terms of kernels which are assumed known. The same procedure of using eqn. (J) to isolate the required kernel will also solve the equations for the higher kernels which may be found by this method one by one. The next equation gives:

$$g_{tr_1r_2r_3} = +g_{t\delta}f_{\delta\epsilon_1\epsilon_2}g_{\epsilon_1r_1}g_{\epsilon_2\beta_2}f_{\beta_2\alpha_2\alpha_3}g_{\alpha_2r_2}g_{\alpha_3r_3}$$
$$+g_{t\delta}f_{\delta\epsilon_1\epsilon_2}g_{\epsilon_1\beta_1}g_{\epsilon_2r_3}f_{\beta_1\alpha_2\alpha_3}g_{\alpha_2r_1}g_{\alpha_3r_2}$$
$$-g_{t\beta}f_{\beta\alpha_1\alpha_2\alpha_3}g_{\alpha_1r_1}g_{\alpha_2r_2}g_{\alpha_3r_3}.$$

Time-invariant case: Here it is possible, by taking Fourier transforms of the appropriate dimension, to convert the set of eqns. (J) into a set of algebraic equations between the Fourier transforms of kernels. To avoid expressions of great complexity this method will be illustrated in the case when the functional f is odd. The power series expansion of f has then only odd powers. Let it be:

$$f_{t\alpha}x_\alpha + f_{t\alpha_1\alpha_2\alpha_3}x_{\alpha_1}x_{\alpha_2}x_{\alpha_3} + f_{t\alpha_1\alpha_2\alpha_3\alpha_4\alpha_5}x_{\alpha_1}x_{\alpha_2}\ldots x_{\alpha_5} + \ldots.$$

Equations (J) for this case are:

$$\delta_{t\alpha} = g_{t\beta}f_{\beta\alpha},$$
$$0 = g_{t\beta_1\beta_2\beta_3}f_{\beta_1\alpha_1}f_{\beta_2\alpha_2}f_{\beta_3\alpha_3} + g_{t\beta}f_{\beta\alpha_1\alpha_2\alpha_3}$$
$$0 = g_{t\beta_1\beta_2\beta_3\beta_4\beta_5}f_{\beta_1\alpha_1}f_{\beta_2\alpha_2}f_{\beta_3\alpha_3}f_{\beta_4\alpha_4}f_{\beta_5\alpha_5}$$
$$+g_{t\beta_1\beta_2\beta_3}f_{\beta_1\alpha_1\alpha_2\alpha_3}f_{\beta_2\alpha_4}f_{\beta_3\alpha_5}$$
$$+g_{t\beta_1\beta_2\beta_3}f_{\beta_1\alpha_1}f_{\beta_2\alpha_2\alpha_3\alpha_4}f_{\beta_3\alpha_5}$$
$$+g_{t\beta_1\beta_2\beta_3}f_{\beta_1\alpha_1}f_{\beta_2\alpha_2}f_{\beta_3\alpha_3\alpha_4\alpha_5}$$
$$+g_{t\beta}f_{\beta\alpha_1\alpha_2\alpha_3\alpha_4\alpha_5}, \quad \ldots\ldots \quad (J')$$

the corresponding Fourier relations being:

$$1 = G(\omega)F(\omega),$$
$$0 = G(\omega_1, \omega_2, \omega_3)F(\omega_1)F(\omega_2)F(\omega_3) + G(\omega_1+\omega_2+\omega_3)F(\omega_1, \omega_2, \omega_3),$$
$$0 = G(\omega_1, \omega_2, \omega_3, \omega_4, \omega_5)F(\omega_1)F(\omega_2)\ldots F(\omega_5)$$
$$+G(\omega_1+\omega_2+\omega_3, \omega_4, \omega_5)F(\omega_1,\omega_2,\omega_3)F(\omega_4)F(\omega_5)$$
$$+G(\omega_1, \omega_2+\omega_3+\omega_4, \omega_5)F(\omega_1)F(\omega_2,\omega_3,\omega_4)F(\omega_5)$$
$$+G(\omega_1, \omega_2, \omega_3+\omega_4+\omega_5)F(\omega_1)F(\omega_2)F(\omega_3,\omega_4,\omega_5)$$
$$+G(\omega_1+\omega_2+\omega_3+\omega_4+\omega_5)F(\omega_1, \omega_2, \omega_3, \omega_4, \omega_5),$$

which may be solved to give:

$$G(\omega) = 1/F(w),$$

$$G(\omega_1, \omega_2, \omega_3) = \frac{-F(\omega_1, \omega_2, \omega_3)}{F(\omega_1+\omega_2+\omega_3)F(\omega_1)F(\omega_2)F(\omega_3)},$$

$$G(\omega_1, \omega_2, \ldots, \omega_5) = \frac{1}{F(\omega_1+\omega_2+\ldots+\omega_5)\prod_{i=1}^{5}F(\omega_i)}$$
$$\times\left\{\frac{F(\omega_1+\omega_2+\omega_3,\omega_4,\omega_5)F(\omega_1,\omega_2,\omega_3)}{F(\omega_1+\omega_2+\omega_3)}\right.$$
$$+\frac{F(\omega_1,\omega_2+\omega_3+\omega_4,\omega_5)F(\omega_2,\omega_3,\omega_4)}{F(\omega_2+\omega_3+\omega_4)}$$
$$\left.+\frac{F(\omega_1,\omega_2,\omega_3+\omega_4+\omega_5)F(\omega_3,\omega_4,\omega_5)}{F(\omega_3+\omega^4+\omega_5)} - F(\omega_1,\omega_2,\omega_3,\omega_4,\omega_5)\right\}.$$

4.2. *A Servomechanism Problem*

To illustrate the results of the preceding section on the inversion of functionals it will be shown how expressions may be obtained for error and output of a simple servomechanism (James *et al.* 1947) (see diagram). The input, error and output signals are denoted by the letters x, e and y, the functional giving the relation between error and output being denoted by h. h is assumed to be an odd functional of error, as is usually the case in systems of this nature.

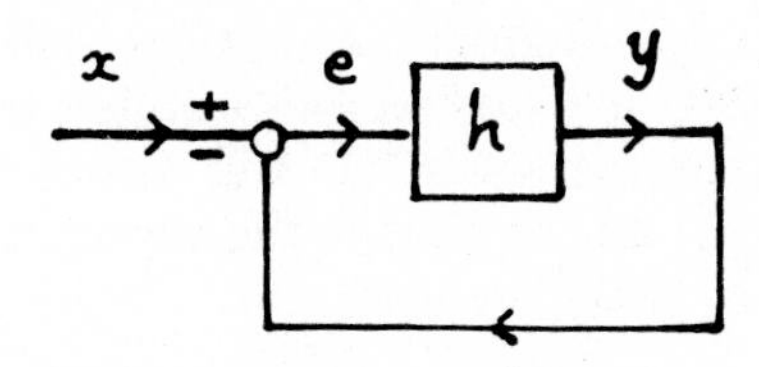

The equations describing the system are then:

$$e(t) = x(t) - y(t), \qquad \text{(A)}$$

$$y(t) = h(t\,;\, e(t'), t' \leqslant t). \qquad \text{(B)}$$

To study the behaviour of the system it is necessary to consider the relation of error and output to input. From the above equations it follows by elimination of $y(t)$ that:

$$x(t) = e(t) + h(t\,;\, e(t)', t' \leqslant t). \qquad \text{(C)}$$

It is therefore necessary to invert this functional relation to find $e(t)$ as a functional of input. Once this has been done, it is easy to find $y(t)$ as a functional of input from eqn. (A).

4.2.1. *The linear case*

Here, eqn. (B) may be written:

$$y(t) = \int h(t, t_1) e(t_1)\, dt_1 \qquad \text{(B')}$$

and eqn. (C) becomes:

$$x(t) = e(t) + \int h(t, t_1) e(t_1)\, dt_1. \qquad \text{(C')}$$

The inversion problem for this equation has already been considered (see § 4.1.1), the solution being:

$$e(t) = x(t) - \int h^{\dagger}(t, t_1) x(t_1)\, dt_1.$$

The output is given by eqn. (A) as:

$$y(t) = \int h^{\dagger}(t, t_1) x(t_1)\, dt_1.$$

Consequently $h^{\dagger}(t, t_1)$ is seen to have a simple interpretation. It is the weighting function (impulse response function) for the whole servo-mechanism.

In the time-invariant case the method of Fourier transformation may be used. Put $h(t, t_1) = h_1(t - t_1)$, $h^\dagger(t, t_1) = h_1^\dagger(t - t_1)$ and

$$H(\omega) = \int h_1(t) \exp(-i\omega t)\, dt, \quad H^\dagger(\omega) = \int h_1^\dagger(t) \exp(-i\omega t)\, dt.$$

Then

$$1 - H^\dagger(\omega) = \frac{1}{1 + H(\omega)} \qquad \text{cf. § 4.1.1, eqn. (I),}$$

$$H^\dagger(\omega) = \frac{H(\omega)}{1 + H(\omega)},$$

a well-known formula. Stability conditions still have to be fulfilled. These are very familiar in linear servomechanism theory; assuming $H(\omega)$ corresponds to a stable system, $H^\dagger(\omega)$ will correspond to a stable system if and only if $1 + H(\omega)$ has no zeros in the region $\mathscr{I}(\omega) \leqslant 0$ of the complex ω-plane (James *et al.* 1947).

4.2.2. *Case where h is analytic*

Since h has been assumed odd it will have an expansion:

$$h(t; e(t'), t' \leqslant t) = \int h(t; t_1) e(t_1)\, dt_1 + \iiint h(t; t_1, t_2, t_3) e(t_1) e(t_2) e(t_3)\, dt_1\, dt_2\, dt_3 + \dots$$

involving only terms of odd order.

Equation (C) is consequently, in abbreviated notation:

$$\begin{aligned} x_t &= e_t + h_{t\alpha_1} e_{\alpha_1} + h_{t\alpha_1\alpha_2\alpha_3} e_{\alpha_1} e_{\alpha_2} e_{\alpha_3} + \dots \\ &= \delta_{t\alpha_1} e_{\alpha_1} + h_{t\alpha_1} e_{\alpha_1} + h_{t\alpha_1\alpha_2\alpha_3} e_{\alpha_1} e_{\alpha_2} e_{\alpha_3} + \dots \\ &= (\delta_{t\alpha_1} + h_{t\alpha_1}) e_{\alpha_1} + h_{t\alpha_1\alpha_2\alpha_3} e_{\alpha_1} e_{\alpha_2} e_{\alpha_3} + \dots . \end{aligned}$$

It has been seen that the invertibility of this equation is dependent on the invertibility of the linear part, viz.

$$x_t = (\delta_{t\alpha_1} + h_{t\alpha_1}) e_{\alpha_1},$$

$$x(t) = e(t) + \int h(t, t_1) e(t_1)\, dt_1.$$

This is the equation of the linear case already considered, the solution being:

$$e(t) = x(t) - \int h^\dagger(t, t_1) x(t_1)\, dt_1;$$

i.e.

$$\begin{aligned} e_t &= x_t - h_{t\alpha}^\dagger x_\alpha \\ &= (\delta_{t\alpha} - h_{t\alpha}^\dagger) x_\alpha \\ &= g_{t\alpha} x_\alpha . \end{aligned}$$

e may now be calculated as an odd power series in x:

$$e_t = g_{t\alpha} x_\alpha + g_{t\alpha_1\alpha_2\alpha_3} x_{\alpha_1} x_{\alpha_2} x_{\alpha_3} + \dots,$$

where the higher g-kernels may be worked out as in § 4.1.2;

e.g.

$$g_{t\alpha_1\alpha_2\alpha_3} = -g_{tr} h_{r\beta_1\beta_2\beta_3} g_{\beta_1\alpha_1} g_{\beta_2\alpha_2} g_{\beta_3\alpha_3}.$$

If the functional h is time-invariant, Fourier transforms may be used to give the Fourier transforms of the g-kernels in terms of those of the h-kernels.

For example, consider the case where

$$y(t) = \int k_1(t-t_1) f(e(t_1))\, dt_1, \quad f(e) = a_1 e + a_3 e^3 + a_5 e^5 + \dots .$$

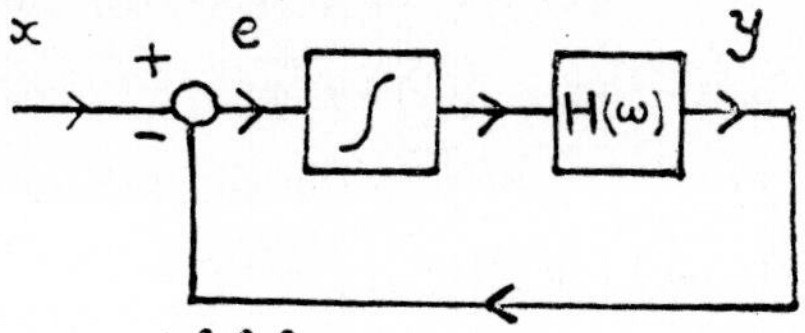

Then

$$y(t) = \int h_1(t-t_1) e(t_1)\, dt_1 + \iiint h_1(t-t_1, t-t_2, t-t_3) e(t_1) e(t_2) e(t_3)\, dt_1\, dt_2\, dt_3 + \dots$$

on putting

$$\begin{aligned} h_1(\tau) &= a_1 k_1(\tau), \\ h_1(\tau_1, \tau_2, \tau_3) &= a_3 k_1(\tau_1) \delta(\tau_2 - \tau_1) \delta(\tau_3 - \tau_1), \\ h_1(\tau_1, \tau_2, \dots, \tau_5) &= a_5 k_1(\tau_1) \delta(\tau_2 - \tau_1) \dots \delta(\tau_5 - \tau_1) \end{aligned}$$

... etc.

the corresponding Fourier transforms are

$$\begin{aligned} H_1(\omega) &= a_1 K(\omega), \\ H_1(\omega_1, \omega_2, \omega_3) &= a_3 K(\omega_1 + \omega_2 + \omega_3), \\ H_1(\omega_1, \omega_2, \dots, \omega_5) &= a_5 K(\omega_1 + \omega_2 + \dots + \omega_5), \end{aligned}$$

... etc.

this gives

$$G(\omega) = \frac{1}{1 + a_1 K(\omega)}$$

$$G(\omega_1, \omega_2, \omega_3) = -a_3 \frac{1}{(1 + a_1 K(\omega_1 + \omega_2 + \omega_3)) \prod_{i=1}^{3} (1 + a_1 K(w_i))}$$

for the Fourier transforms of the time-invariant kernels relating e to input x.

§ 5. The Class of Functionals with Finite Mean Square

So far, mainly functionals of analytic type have been considered. The class of analytic functionals includes the linear and polynomial functionals as special cases. A class of functionals will now be introduced which arises naturally in theory based on mean square values of stochastic processes such as occurs in Wiener's theory of least square smoothing and prediction. Analytic functionals are included as special cases as also are certain discontinuous functionals, in particular the simple type of functional (function) occurring in relay systems ($f(x) = +1$, $x > 0$, $f(x) = -1$, $x < 0$).

5.1. *Functionals and Functions with Finite Mean Square*

Consider first the analogous class of functions. Let a bar denote any linear averaging operation of functions, satisfying the axioms:

$$\text{(i) } \overline{f_1(x)+f_2(x)} = \overline{f_1(x)} + \overline{f_2(x)},$$

$$\text{(ii) } \overline{kf(x)} = k\overline{f(x)} \text{ for any constant } k,$$

$$\text{(iii) } f(x) \geqslant 0 \text{ for all } x \text{ implies } \overline{f(x)} \geqslant 0.$$

Then, relative to this operation, the class of functions of finite mean square will be defined as the class of functions for which $\overline{f(x)^2}$ is finite. A common case is where the averaging operation is given by the formula:

$$\overline{f(x)} = \int_a^b f(x)w(x)\,dx,$$

where $w(x)$ is a non-negative weighting function. If $w(x)=1$, the class of function with finite mean square becomes the class $L^2(a,b)$ of functions of Lebesgue integrable square in (a,b). The case of special interest at present is that in which the bar denotes statistical averaging: this may be written in integral form as:

$$\overline{f(x)} = \int f(x)p(x)\,dx,$$

where $p(x)$ is the probability density function of x. In this case the practical significance of functions of finite mean square is as follows: suppose that the output of a simple instantaneous device is related to the input by

$$y(t) = f(x(t)),$$

$x(t)$ and $y(t)$ representing input and output quantities at time t. Suppose x has a constant one-dimensional distribution with probability density function $p(x)$ and the function f is of finite mean square relative to $p(x)$ as weighting function. Then

$$\int f^2(x)p(x)\,dx,$$

i.e. the mean square output (variance of output) is finite. Hence it may be argued that only functions of this class have physical significance†.

Functionals of finite mean square may be defined similarly, relative to any linear averaging operation.

If $x(t)$ is a stationary ergodic stochastic process, for any functional $f(t; x(t'), t' \leqslant t)$ the mean square value

$$\overline{f(t; x(t'), t' \leqslant t)^2}$$

may be considered. If this is finite, f will be said to be of finite mean square relative to this process. If $x(t)$ is not stationary, the mean square value may be taken at a particular time. Certain difficulties arise in attempting

† Strictly speaking, the process should be ergodic for this argument to apply.

to define an integral form for this averaging operation. In the case where $x(t)$ is a purely random process, the integral has been defined by Wiener (1930).

5.2. *Mean Square Approximation*

The method of mean square approximation may sometimes be used to get approximate solutions of engineering problems where exact methods fail. Thus Booton *et al.* (1953) were able to obtain results about the behaviour of a non-linear servomechanism by using a mean square approximation method which linearized the system. Applications to filtering and prediction theory are given below.

Consider the problem of finding the best mean square approximation to a function $f(x)$ of finite mean square by a polynomial of degree n. This is the problem of minimizing

$$\overline{(f(x)-a_0-a_1x-\ldots-a_nx^n)^2} \qquad \text{(A)}$$

by variation of the coefficients $a_0, a_1, \ldots, a_n$.

This is a special case of the more general problem of finding the best mean square approximation to a function $f(x)$ of finite mean square by linear combinations of given functions $f_0(x), \ldots f_1(x), \ldots, f_n(x)$, i.e. of minimizing

$$\overline{(f(x)-a_0f_0(x)-a_1f_1(x)-\ldots-a_nf_n(x))^2} \qquad \text{(B)}$$

by variation of the parameters $a_0, a_1, \ldots, a_n$. For the expression to be finite, it is necessary that all functions appearing should be of finite mean square. It may be assumed without any loss of generality that the approximating functions are linearly independent, i.e. no one of them is expressible as a linear sum of the others†. This condition may be shown equivalent to the condition that

$$\begin{vmatrix} \overline{f_0^2} & \overline{f_0f_1} & & \overline{f_0f_n} \\ \overline{f_1f_0} & \overline{f_1^2} & & \overline{f_1f_n} \\ & & & \\ \overline{f_nf_0} & \overline{f_nf_1} & & \overline{f_n^2} \end{vmatrix} \neq 0. \qquad \text{(C)}$$

Here $\overline{f_if_j}$ is a contracted notation for $\overline{f_i(x)f_j(x)}$. This notation will be retained.

The calculation of the minimizing values of the parameters $a_0, a_1, \ldots, a_n$ is straightforward.

Denote the expression (B) by $F(a_0, a_1, \ldots, a_n)$:

$$F(a_0, a_1, \ldots, a_n) = \overline{f^2} - 2\sum_{i=0}^{n} a_i\overline{ff_i} + \sum_{i,j=0}^{n} a_ia_j\overline{f_if_j}. \qquad \text{(D)}$$

A necessary condition that $F(a_0, a_1, \ldots, a_n)$ should be a minimum is that

$$\frac{\partial F}{\partial a_0} = \frac{\partial F}{\partial a_1} = \ldots = \frac{\partial F}{\partial a_n} = 0. \qquad \text{(E)}$$

† More precisely, it is required that the set of values of x for which any one of the functions is equal to a linear combination of the others should have zero probability measure.

Now

$$\frac{\partial F}{\partial a_k} = -2\overline{f_k f} + 2\sum_{i=0}^{n} a_i \overline{f_i f_k}, \quad k=0,1,\ldots,n$$

$$= 2\left\{ \sum_{i=0}^{n} a_i \overline{f_i f_k} - \overline{f_k f} \right\}. \qquad \text{(F)}$$

∴ eqns. (E) are:

$$\left.\begin{array}{l} a_0\overline{f_0^2} + a_1\overline{f_0 f_1} + \ldots + a_n\overline{f_0 f_n} = \overline{f_0 f} \\ a_0\overline{f_1 f_0} + a_1\overline{f_1^2} + \ldots + a_n\overline{f_1 f_n} = \overline{f_1 f} \\ \ldots \quad \ldots \quad \ldots \quad = \ldots \\ a_0\overline{f_n f_0} + a_1\overline{f_n f_1} + \ldots + a_n\overline{f_n^2} = \overline{f_n f} \end{array}\right\} \qquad \text{(G)}$$

The determinant of the a's is non-zero by condition (C) and so these equations have a unique solution for the a's. It is easily seen that these values must correspond to a minimum value of $F(a_0, a_1, \ldots, a_n)$. The approximating function, $a_0 f_0(x) + a_1 f_1(x) + \ldots + a_n f(x)$

$$= -\begin{vmatrix} 0 & f_0(x) & f_1(x) & \ldots & f_n(x) \\ \overline{f_0 f} & \overline{f_0^2} & \overline{f_0 f_1} & \ldots & \overline{f_0 f_n} \\ \overline{f_1 f} & \overline{f_1 f_0} & \overline{f_1^2} & \ldots & \overline{f_1 f_n} \\ \ldots & \ldots & \ldots & \ldots & \ldots \\ \overline{f_n f} & \overline{f_n f_0} & \overline{f_n f_1} & \ldots & \overline{f_n^2} \end{vmatrix} \div D, \qquad \text{(H)}$$

where D is the determinant on the left-hand side of condition (C).

The value of the expression (A), i.e. the mean square error of the approximation is found as:

$$\begin{vmatrix} \overline{f^2} & \overline{f f_0} & & \overline{f f_n} \\ \overline{f_0 f} & \overline{f_0^2} & & \overline{f_0 f_n} \\ & & & \\ \overline{f_n f} & \overline{f_n f_0} & & \overline{f_n^2} \end{vmatrix} \div D. \qquad \text{(I)}$$

This calculation is the same if the f's represent functionals instead of functions. The expression $f(x)$ may be considered an abbreviation for $f(t; x(t'), t' \leqslant t)$. With functionals the result may be interpreted in engineering terms as follows.

It is required to simulate, or construct a system with functional relation f between output and input when the input has a known distribution. Using a finite number of elements with relations between output and input given by functionals $f_1, f_2, \ldots, f_n$, the arrangement of these elements in parallel which will approximate with least mean square output discrepancy is that shown in the diagram. The multiplication factors have the values found above.

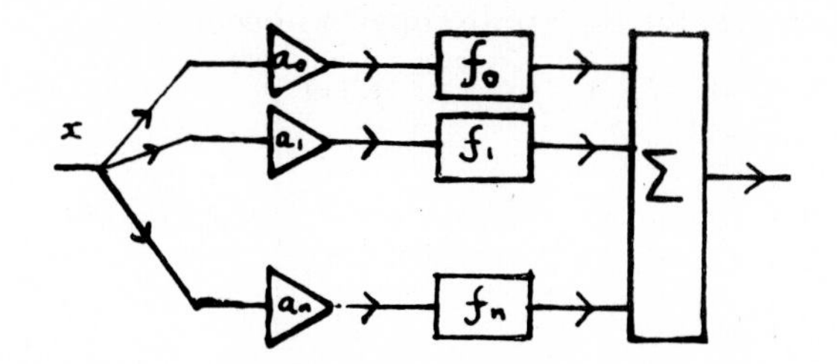

By suitable choice of what might be called coordinate functionals $f_1, f_2, \ldots, f_n$ it should by this means be possible to reproduce quickly to within permissible error a wide range of functional relationships†.

5.3. *The Expansion of a Functional of Finite Mean Square by a Complete Orthonormal System*

Two functionals $f_1(t; x(t'), t' \leqslant t)$ and $f_2(t; x(t''), t'' \leqslant t)$ will be said to be *orthogonal* relative to a stationary ergodic stochastic process if

$$\overline{f_1(t; x(t'), t' \leqslant t) f_2(t; x(t''), t'' \leqslant t)} = 0.$$

If $f_1, f_2, \ldots$ is a finite or infinite set of functionals every two of which are orthogonal, the set will be called an orthogonal set. The norm of a functional relative to a stationary ergodic stochastic process may be defined as

$$\|f\|_x = \sqrt{[\overline{f^2(t; x(t'), t' \leqslant t)}]}.$$

It is thus the root mean square value of the output of any system with functional f when the input is the stochastic process mentioned. If $f_1, f_2, \ldots$ is an orthogonal set of functionals each with norm one relative to some particular stochastic process, the set will be called orthonormal relative to this process.

In this case

$$\overline{f_i(t; x(t'), t' \leqslant t) f_j(t; x(t''), t'' \leqslant t)} = \begin{cases} 1 & i = j, \\ 0 & i \neq j. \end{cases}$$

For finite orthonormal set $f_0, f_1, \ldots, f_n$, the above equations dealing with least mean square approximation take on a very much simpler form.

The determinant

$$D = \begin{vmatrix} 1 & 0 & \ldots & 0 \\ 0 & 1 & \ldots & 0 \\ \ldots & & \ldots & . \\ 0 & 0 & \ldots & 1 \end{vmatrix} = 1.$$

Equations (G) become:

$$\begin{aligned} a_0 \qquad\qquad\qquad &= \overline{f_0 f} \\ a_1 \qquad\quad &= \overline{f_1 f} \\ \ldots \qquad & \\ a_n &= \overline{f_n f}, \end{aligned}$$

giving immediately the values of the a's.

† This arrangement has been used by Lampard (1955) to synthesis linear time-invariant filters. See also Lee (1932).

The expression (H) for the approximation becomes:

$$-\begin{vmatrix} 0 & f_0(x) & f_1(x)\ldots & f_n(x) \\ \overline{f_0 f} & 1 & \ldots & \ldots \\ \overline{f_1 f} & & 1\ \ldots & \ldots \\ \ldots & \ldots & \ldots & \ldots \\ \overline{f_n f} & \ldots & \ldots\ \ldots & 1 \end{vmatrix} = \sum_{i=0}^{n} \overline{ff_i}\, f_i(x)$$

and the minimum mean square value is:

$$\begin{vmatrix} \overline{f^2} & \overline{ff_0} & \ldots & \overline{ff_n} \\ \overline{f_0 f} & 1 & \ldots & 0 \\ \ldots & & \ldots & \ldots \\ \overline{f_n f} & 0 & \ldots & 1 \end{vmatrix} = \overline{f^2} - \sum_{i=0}^{n} (\overline{f_i f})^2.$$

It may happen with an infinite orthonormal set that every functional with finite mean square can be approximated with any prescribed accuracy by finite linear combinations of functionals of this set. Such a set will be said to be *complete*. A complete set of orthonormal functionals for the case where the input stochastic process is completely random has been considered by Cameron and Martin (1947).

In this case it is possible to write:

$$f(x) = \sum \overline{ff_i}\, f_i(x), \qquad \text{(H'')}$$

$$\overline{f^2} = \sum (\overline{ff_i})^2. \qquad \text{(I'')}$$

(H″) thus gives an expansion for any functional of finite mean square in terms of the basic functionals f_i.

5.4. *Formation of Orthogonal and Orthonormal Sets*

Given any linearly independent set of functions or functionals, it is possible to form from these an orthogonal set by first arranging these function(al)s into a sequence:

$$f_0(x), f_1(x), \ldots, f_n(x), \ldots$$

and then forming the sequence (Szegö 1939)†:

$$D_0(x) = f_0(x),$$

$$D_1(x) = \begin{vmatrix} f_1(x) & f_0(x) \\ \overline{f_0 f_1} & \overline{f_0^2} \end{vmatrix},$$

$$D_2(x) = \begin{vmatrix} f_2(x) & f_1(x) & f_0(x) \\ \overline{f_1 f_2} & \overline{f_1^2} & \overline{f_1 f_0} \\ \overline{f_0 f_2} & \overline{f_0 f_1} & \overline{f_0^2} \end{vmatrix}$$

$$\ldots \qquad \ldots \qquad \ldots$$

† The process given here is equivalent to the Gram–Schmidt process.

$$D_n(x) = \begin{vmatrix} f_n(x) & f_{n-1}(x) & \cdots & f_0(x) \\ \overline{f_{n-1}f_n} & \overline{f_{n-1}^2} & \cdots & \overline{f_{n-1}f_0} \\ \cdots & \cdots & \cdots & \cdots \\ \overline{f_0 f_n} & \overline{f_0 f_{n-1}} & \cdots & \overline{f_0^2} \end{vmatrix}$$

$$\cdots = \quad \cdots \quad \text{etc.}$$

If any one of these determinants vanishes identically then there will be a linear relation between some of the function(al)s f. Suppose $D_r(x)$ is the first determinant to vanish identically. Then $f_r(x)$ is linearly dependent on its predecessors and may be excluded from the sequence, f_{r+1} being taken in its place.

The determinants $D_n(x)$ are mutually orthogonal: $D_n(x)$ is by its construction, orthogonal to $f_{n-1}(x), \ldots, f_0(x)$. Hence it is orthogonal to any linear combination of these, in particular, to any preceding determinant. Since, of any two determinants, one is of lower order than the other, the result follows; i.e.

$$\overline{D_m(x)D_n(x)} = 0, \quad m \neq n. \qquad \text{(A)}$$

If $m = n$,

$$\overline{D_m(x)D_n(x)} = \overline{D_n^2(x)}$$

$$= \overline{\begin{vmatrix} f_n(x) & f_{n-1}(x) & \cdots & f_0(x) \\ \overline{f_{n-1}f_n} & \overline{f_{n-1}^2} & \cdots & \overline{f_{n-1}f_0} \\ \cdots & \cdots & \cdots & \cdots \\ \overline{f_0 f_n} & \overline{f_0 f_{n-1}} & & \overline{f_0^2} \end{vmatrix} \times (f_n(x)\Delta_{n-1} + R(x))},$$

where

$$\Delta_{n-1} = \begin{vmatrix} \overline{f_{n-1}^2} & \cdots & \overline{f_{n-1}f_0} \\ \cdots & \cdots & \cdots \\ \overline{f_0 f_{n-1}} & \cdots & \overline{f_0^2} \end{vmatrix}$$

$$\Delta_{-1} = 1$$

and $R(x)$ is a linear combination of $f_{n-1}(x), \ldots, f_0(x)$.

On taking mean values, the term involving $R(x)$ vanishes by the orthogonality properties of $D_n(x)$.

$$\therefore \overline{D_n^2(x)} = \overline{\begin{vmatrix} f_n(x) & f_{n-1}(x) & \cdots & f_0(x) \\ \overline{f_{n-1}f_n} & \overline{f_{n-1}^2} & \cdots & \overline{f_{n-1}f_0} \\ \cdots & \cdots & \cdots & \cdots \\ \overline{f_0 f_n} & \overline{f_0 f_{n-1}} & \cdots & \overline{f_0^2} \end{vmatrix} \times f_n(x)\Delta_{n-1}}$$

$$= \Delta_n \Delta_{n-1};$$

i.e.

$$\overline{D_n^2(x)} = \Delta_n \Delta_{n-1}, \; n = 0, 1, \ldots. \qquad \text{(B)}$$

It follows that the quantities

$$\chi_n(x) = \frac{D_n(x)}{\sqrt{(\Delta_n \Delta_{n-1})}}$$

form an orthonormal set. (It may be shown that $\Delta_n > 0$ and so the square root is always real and positive. This is obvious from (B) since the left-hand side is positive.)

Orthogonal polynomials

Important sets of orthogonal functions (polynomials) are obtained by taking the sequence $f_0(x), f_1(x), \ldots$ to be the sequence of powers of x, viz. $1, x, x^2, \ldots$.

In this case the sequence of D's is:

$$1, \quad \begin{vmatrix} x & 1 \\ \overline{x} & 1 \end{vmatrix}, \quad \begin{vmatrix} x^2 & x & 1 \\ \overline{x^3} & \overline{x^2} & \overline{x} \\ \overline{x^2} & \overline{x} & 1 \end{vmatrix}, \quad \begin{vmatrix} x^3 & x^2 & x & 1 \\ \overline{x^5} & \overline{x^4} & \overline{x^3} & \overline{x^2} \\ \overline{x^4} & \overline{x^3} & \overline{x^2} & \overline{x} \\ \overline{x^3} & \overline{x^2} & \overline{x} & 1 \end{vmatrix}, \ldots .$$

In particular, if the x-distribution is symmetrical with zero mean, all odd moments vanish and this sequence becomes:

$$1, \quad x, \quad \begin{vmatrix} x^2 & 1 \\ \overline{x^2} & 1 \end{vmatrix}, \quad \begin{vmatrix} x^3 & x \\ \overline{x^4} & \overline{x^2} \end{vmatrix}, \ldots,$$

a sequence of alternately even and odd polynomials.

It is convenient to introduce the polynomials:

$$P_n(x) = \frac{D_n(x)}{\Delta_{n-1}}, \qquad n = 0, 1, \ldots$$

which have the orthogonality properties of the D's and, in addition, unit coefficient for the highest power of x.

From eqns. (A) and (B), it follows that:

$$\left.\begin{aligned} \overline{P_m(x)P_n(x)} &= 0, \quad m \neq n, \\ \overline{P_n{}^2(x)} &= \frac{\Delta_n}{\Delta_{n-1}} \end{aligned}\right\} \quad \ldots \ldots \quad \text{(C)}$$

In the case when the x-distribution is normal with unit variance and zero mean, the polynomials $P_n(x)$ become the Hermite polynomials (Cramér 1946) $H_n(x)$ defined by the equation:

$$\frac{d^n}{dx^n} \exp(-x^2/2) = (-1)^n H_n(x) \exp(-x^2/2).$$

5.5. *Multi-dimensional Orthogonal Polynomials: Symmetric Partially Orthogonal Systems*

Multi-dimensional orthogonal polynomials will be mentioned here for they lead towards the definition of the partially orthogonal systems of functionals, which will be used in the applications of the present theory to least square filtering and prediction.

Consider functions in n variables $x_1, x_2, \ldots, x_n$. It is possible, by starting with the sequence:

$$1;\ x_1, x_2, \ldots, x_n;\ x_1^2, x_1, x_2, \ldots, x_1 x_n;\ x_2 x_1, x_2^2, \ldots, x_2 x_n; \ldots$$
$$\ldots x_n x_1 \ldots x_n^2;\ x_1^3, \ldots,$$

and using the preceding method to construct a set of orthogonal or orthonormal polynomials in the n variables $x_1, x_2, \ldots, x_n$. However, this procedure is unsymmetrical in the variables and unsuitable for certain purposes. It is possible to construct a sequence of *symmetrical* polynomials with the property that any two polynomials of different degrees are orthogonal.

The polynomials are, denoting $(x_1, x_2, \ldots, x_n)$ by $\mathbf{x}$:

$$\left.\begin{aligned}
&Q^{(0)}(\mathbf{x}) = 1,\\
&Q_i^{(1)}(\mathbf{x}) = \begin{vmatrix} x_i & 1 \\ \overline{x_i} & 1 \end{vmatrix}, \quad i = 1, 2, \ldots, n,\\
&Q_{ij}^{(2)}(\mathbf{x}) = \begin{vmatrix} x_i x_j & x_1 & x_2 & \ldots & x_n & 1 \\ \overline{x_1 x_i x_j} & \overline{x_1^2} & \overline{x_1 x_2} & \ldots & \overline{x_1 x_n} & \overline{x_1} \\ \ldots & \ldots & \ldots & \ldots & \ldots & \ldots \\ \overline{x_i x_j} & \overline{x_1} & \overline{x_2} & \ldots & \overline{x_n} & 1 \end{vmatrix} i, j = 1, 2, \ldots\\
&\ldots \quad \ldots \quad \ldots \quad \text{etc.}
\end{aligned}\right\} \quad \text{(D)}$$

The orthogonality property of polynomials of different degrees is seen as above.

The values of the averaged products of polynomials of the same degree can be evaluated simply by following the previous method. Thus

$$\overline{Q_{ij}^{(2)}(\mathbf{x}) Q_{kl}^{(2)}(\mathbf{x})} = \begin{vmatrix} \overline{x_k x_l x_i x_j} & \overline{x_k x_l x_1} & \ldots & \overline{x_k x_l} \\ \overline{x_1 x_i x_j} & \overline{x_1^2} & \ldots & \overline{x_1} \\ \ldots & \ldots & \ldots & \ldots \\ \overline{x_i x_j} & \ldots \ \overline{x_1} & \ldots & 1 \end{vmatrix} \times C_2,$$

where (E)

$$C_2 = \begin{vmatrix} \overline{x_1^2} & \overline{x_1 x_2} & \ldots & \overline{x_1} \\ \overline{x_2 x_1} & \overline{x_2^2} & \ldots & \overline{x_2} \\ \ldots & \ldots & \ldots & \ldots \\ \overline{x_1} & \overline{x_2} & \ldots & 1 \end{vmatrix}.$$

As in the one-dimensional case, it may be convenient to work with polynomials having the coefficient of the term of highest degree unity. These polynomials are:

$$\left.\begin{aligned}
&P^{(0)}(\mathbf{x}) = Q^{(0)}(\mathbf{x}),\\
&P_i^{(1)}(\mathbf{x}) = Q_i^{(1)}(\mathbf{x}), \quad i = 0, 1 \ldots\\
&P_{ij}^{(2)}(\mathbf{x}) = \frac{Q_{ij}^{(2)}(\mathbf{x})}{C_2}, \quad i, j = 0, 1, \ldots\\
&\ldots \quad \ldots \quad \ldots \quad \text{etc.}
\end{aligned}\right\} \quad . \quad . \quad . \quad . \quad \text{(F)}$$

The polynomials $P^{(n)}(\mathbf{x})$ in the case where the variables $x_1, x_2, \ldots, x_n$ are independently and normally distributed with zero mean and unit variance, have been proposed by Grad (1949) as giving a symmetrical set of multi-dimensional Hermite polynomials. In this case, the polynomials of the same degree are orthogonal in the sense that

$$\overline{P_{i_1 i_2 \ldots i_n}^{(n)}(\mathbf{x})\, P^{(n)}_{j_1 j_2 \ldots j_n}(\mathbf{x})} = \begin{cases} n! \text{ if } j_1, j_2, \ldots, j_n \text{ is a permutation} \\ \text{of } i_1, i_2, \ldots, i_n \\ 0 \text{ otherwise.} \end{cases} \qquad \text{(G)}$$

In this case, the first few polynomials have the values (Grad 1949) (changing to Grad's notation):

$$\begin{aligned} \mathscr{H}^{(0)}(\mathbf{x}) &= 1, \\ \mathscr{H}_i^{(1)}(\mathbf{x}) &= x_i, \\ \mathscr{H}_{ij}^{(2)}(\mathbf{x}) &= x_i x_j - \delta_{ij}, \\ \mathscr{H}_{ijk}^{(3)}(\mathbf{x}) &= x_i x_j x_k - (x_i \delta_{jk} + x_j \delta_{ik} + x_k \delta_{ij}), \end{aligned}$$

where δ_{rs} is the Kronecker δ-function: $\delta_{rs} = 0$, $r \neq s$, $\delta_{rs} = 1$, $r = s$; $r, s = 1, 2, \ldots, n$.

In the more general case where the x's may be correlated, the Kronecker δ-function is replaced by the covariance $\overline{x_r x_s}$ between x_r and x_s. Thus

$$\left.\begin{aligned} \mathscr{H}^{(0)}(\mathbf{x}) &= 1, \\ \mathscr{H}_i^{(1)}(\mathbf{x}) &= x_i, \\ \mathscr{H}_{ij}^{(2)}(\mathbf{x}) &= x_i x_j - \overline{x_i x_j}, \\ \mathscr{H}_{ijk}^{(3)}(\mathbf{x}) &= x_i x_j x_k - (x_i \overline{x_j x_k} + x_j \overline{x_k x_i} + x_k \overline{x_i x_j}). \end{aligned}\right\} \qquad \text{(H)}$$

5.6. *Symmetric Partially Orthogonal Systems of Functionals*

The multi-dimensional polynomials just considered are now applied to find a partially orthogonal system of functionals.

As in § 3 (p. 65), a functional may be regarded as a function of a continuous infinity of variables, $x(t)$ where t runs over the appropriate range of time. Thus, to take a definite example, the Hermite polynomials (H) of § 5.5 will become (Friedrichs 1953):

$$\begin{aligned} \mathscr{H}^{(0)}(x) &= 1, \\ \mathscr{H}^{(1)}(t_1; x) &= x(t_1), \\ \mathscr{H}^{(2)}(t_1, t_2; x) &= x(t_1)x(t_2) - \overline{x(t_1)x(t_2)}, \\ \mathscr{H}^{(3)}(t_1, t_2, t_3; x) &= x(t_1)x(t_2)x(t_3) - x(t_1)\overline{x(t_2)x(t_3)} - x(t_2)\overline{x(t_3)x(t_1)} - x(t_3)\overline{x(t_1)x(t_2)}. \end{aligned}$$

It may be verified that for any values of $s_1, s_2, \ldots, s_m$ and $t_1, t_2, \ldots, t_n$, $\mathscr{H}^{(m)}(s_1, s_2, \ldots, s_m; x)$ and $\mathscr{H}^{(n)}(t_1, t_2, \ldots, t_n; x)$ are orthogonal if $m \neq n$. From this it follows that the functionals

$$\iint \cdots \int h(s_1, s_2, \ldots, s_m) \mathscr{H}^{(m)}(s_1, s_2, \ldots, s_m; x)\, ds_1 ds_2 \ldots ds_m$$

$$\iint \cdots \int k(t_1, t_2, \ldots, t_n) \mathscr{H}^{(n)}(t_1, t_2, \ldots, t_n; x)\, dt_1\, dt_2 \ldots dt_n$$

are orthogonal for any weighting functions (kernels) h and k. These form the required partially orthonormal system in this case. The general case, for any distribution (stochastic process) will correspond to a system of functionals:

$$\iint \cdots \int k(t_1, t_2, \ldots, t_n) P^{(n)}(t_1, t_2, \ldots, t_n\,;\, x)\, dt_1\, dt_2 \ldots dt_n,$$

where the kernel $k(t_1, t_2, \ldots, t_n)$ is arbitrary and $P^{(n)}(t_1, t_2, \ldots, t_n)$ is the partially orthogonal system of the stochastic process.

The values of the partially orthogonal system of P's does not follow from a simple limiting procedure applied to the polynomials $P_{i_1 i_2 \ldots i_n}{}^{(n)}(\mathbf{x})$ of the multi-nomial case because it is not clear how the determinants will behave, but, taking note of the form of the polynomial P's and the Hermite case just considered, it is possible to write down the expressions $P^{(n)}(t_1, t_2, \ldots, t_n; x)$ as:

$$P^{(0)}(x) = 1,$$

$$P^{(1)}(t_1\,;\, x) = x(t_1) - g_{10} P^{(0)}(x),$$

$$P^{(2)}(t_1, t_2\,;\, x) = x(t_1)x(t_2) - g_{20}(t_1, t_2) P^{(0)}(x) - \int g_{21}(t_1, t_2\,;\, s) P^{(1)}(s\,;\, x)\, ds,$$

$$\begin{aligned} P^{(3)}(t_1, t_2, t_3\,;\, x) = {} & x(t_1)x(t_2)x(t_3) - g_{30}(t_1, t_2, t_3) P^{(0)}(x) \\ & - \int g_{31}(t_1, t_2, t_3\,;\, s) P^{(1)}(s\,;\, x)\, ds \\ & - \iint g_{32}(t_1, t_2, t_3\,;\, s_1, s_2) P^{(2)}(s_1, s_2\,;\, x)\, ds_1\, ds_2 \end{aligned}$$

... ... etc.,

and determine the g's by the orthogonality conditions. This leads to a set of simultaneous integral equations for the g's. In the case where the stochastic process is stationary, the kernels g will involve only time differences (see the remark in the next section).

When the x-process is stationary with zero mean and symmetrical distribution the p's become:

$$P^{(0)}(x) = 1,$$

$$P^{(1)}(t_1\,;\, x) = x(t_1),$$

$$P^{(2)}(t_1, t_2\,;\, x) = x(t_1)x(t_2) - \overline{x(t_1)x(t_2)},$$

$$P^{(3)}(t_1, t_2, t_3\,;\, x) = x(t_1)x(t_2)x(t_3) - \int G(s - t_1, s - t_2, s - t_3) x(s)\, ds,$$

... ... etc.,

where G satisfies

$$\overline{x(t_1)x(t_2)x(t_3)x(t_4)} = \int G(s - t_1, s - t_2, s - t_3)\overline{x(t_4)x(s)}\, ds \text{ (all } t_1, t_2, t_3 \text{ and } t_4).$$

This only involves time differences and may be solved by taking a 4-dimensional Fourier transform of both sides.

§ 6. Generalized Wiener–Hopf Equations

In this section the orthogonal expansions just discussed will be used to derive a set of equations of the Wiener–Hopf type applicable to the problem of least squares filtering and prediction. In the linear case when the associated stochastic processes are stationary ergodic, the set of equations reduces to one, namely, the Wiener Hopf equation.

The following general problem will be considered ('regression problem').

Given two dependent stochastic processes with values $x(t)$, $\xi(t)$ at time t, what is the best mean square estimate of $\xi(t)$, based on observation of past values of x, i.e. the values $x(t'), t' \leqslant t$? This includes the filtering problem (ξ = signal, x = signal + noise), and the prediction problem ($\xi(t) = x(t+\alpha)$, $\alpha > 0$) and various others, e.g. combined filtering and prediction, approximate differentiation in the presence of noise (Wiener 1949), etc.

This regression problem may be stated in the following way: What Volterra-type functional of finite mean square relative to the x-process minimizes

$$\delta = \overline{(\xi(t) - f(t;\, x(t'), t' \leqslant t))^2},$$

where the bar denotes ensemble average?

It is not difficult to show (Doob) that there always exists a unique minimizing functional f of the class considered. From the uniqueness property it follows that if the x- and ξ-processes are stationary then the minimizing functional f is time-invariant. For if f minimizes the expression, so do all its time-translates and so these must coincide. In this stationary case, it follows that if the given expression δ is minimized at one time, it is minimized for all time.

If the x- and ξ-processes are stationary ergodic, the ensemble mean may be interpreted as a time-mean

$$\lim_{T\to\infty} \frac{1}{2T} \int_{-T}^{T} (\xi(t) - f(x(t'), t' \leqslant t))^2\, dt$$

(omitting t from f because f is time-invariant).

6.1. *Use of Orthogonal Expansions*

To derive the above-mentioned equations, f will be developed in a partially orthogonal set of functionals relative to the x-process†. To illustrate the method, the analogous problem will be considered between two real variables x and ξ.

Assuming that x and ξ are real dependent random variables, consider the problem of minimizing $\overline{(\xi - f(x))^2}$ for functions f of finite mean square relative to the x-distribution.

† x is the input to the filter to be optimized.

Expand f by the set of orthogonal polynomials $P_n(x)$ of the x-distribution. Assume this set complete:

$$f(x) = a_0 P_0(x) + a_1 P_1(x) + \ldots \quad \ldots + a_n P_n(x) + \ldots \qquad \text{(A)}$$

$$\begin{aligned}\delta &= \overline{(\xi - f(x))^2}\\ &= \overline{\xi^2} - 2\sum_{n=0}^{\infty} a_n \overline{\xi P_n(x)} + \sum_{m,n=0}^{\infty} a_m a_n \overline{P_m(x) P_n(x)}\\ &= \overline{\xi^2} - 2\sum_{n=0}^{\infty} a_n \overline{\xi P_n(x)} + \sum_{n=0}^{\infty} a_n^2 \overline{P_n^2(x)} \qquad \text{(B)}\end{aligned}$$

since $\overline{P_m(x)P_n(x)} = 0$ if $m \neq n$. If δ is a minimum, $\partial\delta/\partial a_r = 0$, $r = 0, 1, \ldots$:

$$\begin{aligned}\frac{\partial \delta}{\partial a_r} &= -2\overline{\xi P_r(x)} + 2a_r \overline{P_r^2(x)}\\ &= 0 \quad \text{if}\end{aligned}$$

$$\overline{P_r^2(x)}\, a_r = \overline{\xi P_r(x)}; \qquad \text{(C)}$$

i.e. $$\text{if } a_r = \frac{\overline{\xi P_r(x)}}{\overline{P_r^2(x)}} \quad \text{(the denominator does not vanish)} \qquad \text{(C}')$$

or $$f(x) = \sum_{n=0}^{\infty} \frac{\overline{\xi P_n(x)} P_n(x)}{\overline{P_n^2(x)}}. \qquad \text{(D)}$$

This will correspond to the minimum value.

The partial sums of $f(x)$ may be expressed by determinants of the form

$$f(x) = -\begin{vmatrix} 0 & 1 & x & \ldots & x^m \\ \overline{\xi x} & \overline{x} & \overline{x^2} & \ldots & \overline{x^{m+1}} \\ \ldots & \ldots & \ldots & \ldots & \ldots \\ \overline{\xi x^m} & \overline{x^n} & \overline{x^{n+1}} & \ldots & \overline{x^{2n}} \end{vmatrix} \div D \qquad \text{(E)}$$

where D is the leading minor. This follows as in § 5.2.

Now return to the main problem. *It is assumed that there exists a complete set of partially orthogonal functional polynomials* $P^{(0)}(x)$, $P^{(1)}(t; x)$, $P^{(2)}(t_1, t_2; x), \ldots$ *relative to the stochastic process* x. Expand f in terms of these:

$$\begin{aligned}f(t; x) &= f(t; x(t'), t' \leqslant t)\\ &= a_0(t) P^{(0)}(x) + \int a_1(t; t_1) P^{(1)}(t_1; x)\, dt_1\\ &\quad + \iint a_2(t; t_1, t_2) P^{(2)}(t_1, t_2; x)\, dt_1\, dt_2 \ldots \qquad \text{(A}')\end{aligned}$$

The problem is to find the optimal values of the kernels a.

$$\begin{aligned}\delta &= \overline{(\xi(t) - f(t; x))^2}\\ &= \overline{\xi^2(t)} - 2\sum_{n=0}^{\infty} \iint \ldots \int a_n(t; t_1, t_2, \ldots, t_n)\\ &\qquad \overline{\xi(t) P^{(n)}(t_1, t_2, \ldots, t_n; x)}\, dt_1 \ldots dt_n\\ &\quad + \sum_{n=0}^{\infty} \iint \ldots \int a_n(t; s_1, s_2, \ldots, s_n) a_n(t; t_1, t_2, \ldots, t_n)\\ &\qquad \overline{P^{(n)}(s_1, s_2, \ldots, s_n) P^{(n)}(t_1, t_2, \ldots, t_n)}\, ds_1 \ldots ds_n\, dt_1 \ldots dt_n\end{aligned}$$

the cross-terms vanishing as in the dimensional case.

If this expression is minimized by a particular set of kernels a, then the first variation when one of these kernels, say $a_r(t; s_1, s_2, \ldots, s_r)$ is varied by $\delta a_r(t; s_1, s_2, \ldots, s_r)$ will vanish. This gives, as in the usual Wiener–Hopf derivation, the equation:

$$-2\int\int\cdots\int \delta a_r(t; t_1, t_2, \ldots, t_n)\Big[\overline{\xi(t)P^{(r)}(t_1, t_2, \ldots, t_r; x)}$$

$$-\int\int\cdots\int a_r(t; s_1, s_2, \ldots, s_r)\overline{P^{(r)}(s_1, s_2, \ldots, s_r; x)P^{(r)}(t_1, t_2, \ldots, t_r; x)}$$

$$\times\, ds_1\, ds_2 \ldots ds_r\Big]\, dt_1\, dt_2 \ldots dt_r = 0.$$

$\therefore$ since $\delta a_r(t; t_1, t_2, \ldots, t_r)$ is an arbitrary function in the region $t_1, t_2, \ldots, t_r \leqslant t$ it follows that:

$$\int\int\cdots\int a_r(t; s_1, s_2, \ldots, s_r)\overline{P^{(r)}(s_1, s_2, \ldots, s_r; x)P^{(r)})(t_1, t_2, \ldots, t_r; x)}$$
$$ds_1\, ds_2 \ldots ds_r = \overline{\xi(t)P^{(r)}(t_1, t_2, \ldots, t_r; x)}$$
$$\text{for } t_1, t_2, \ldots, t_r \leqslant t \quad r = 0, 1, \ldots \qquad \text{(C}'\text{)}$$

This is the required set of equations to be solved for the kernel a_r:

$P^{(0)}(x) = 1$ Therefore the first equation is

$a_0(t) = \overline{\xi(t)}$ (Compare (C′) with $r = 0$).

$P^{(1)}(x) = x(t)$ Thus the second equation is

$$\int a_1(t; s_1)\overline{x(s_1)x(t_1)}\, ds_1 = \overline{\xi(t)x(t_1)};$$

i.e. the Wiener–Hopf equation in the non-stationary case. This has been considered by Dolph and Woodbury (1952).

In most applications to filtering and prediction problems, the x- and ξ-processes have a symmetrical distribution about $x = 0$, $\xi = 0$. In this case, only the equations corresponding to r odd will remain. For $P^{(r)}(t_1, t_2, \ldots, t_r; x)$ is an even polynomial functional of x in this case if r is even.

Therefore $$\overline{\xi(t)P^{(r)}(t_1, t_2, \ldots, t_r; x)} = 0$$

by the symmetry of the distribution.

When the x- and ξ-processes are stationary, all the functions occurring in eqns. (C′) depend only on time differences. It would seem likely that these equations could be solved by an extension of the methods used for the Wiener–Hopf equation.

References

Bellman, R., 1954, Rand Corporation Report.

Bochner, S., and Chandrasekharan, K., 1949, *Fourier Transforms*. Annals of Mathematics Studies (Princeton).

Bode, H. W., 1945, *Network Analysis and Feedback Amplifier Design* (D. van Nostrand Co. Inc.).

Bode, H. W., and Shannon, C. E., 1950, *Proc. Inst. Radio Engrs, N.Y.*, **38**, 417.

Booton, R. C., Mathews, M. V., and Seifert, W., 1953, *Report No.* 70, D.A.C.L., M.I.T.

Bourbaki, N., 1952, *Éléments de Mathématique. Actualitiés Scientifiques et Industrielles*, No. 1175 (Paris : Hermann & Cie).

Cameron, R. H., and Martin, W. T., 1947, *Ann. Math.*, **48**, 385.

Cherry, C., 1951, *Phil. Mag.*, **42**, 1161.

Courant, R., and Hilbert, D., 1953, *Methods of Mathematical Physics* (Interscience Publishers).

Cramér, H., 1946, *Mathematical Methods of Statistics* (Princeton).

Dolph, C. L., and Woodbury, M. A., 1952, *Trans. Amer. Math. Soc.*, **72**, 519.

Doob, J. L., *Stochastic Processes* (Chapman & Hall and John Wiley & Sons Inc.).

Friedrichs, K. O., 1953, *Mathematical Aspects of the Quantum Theory of Fields* (Interscience Publishers), p. 52.

Gabor, D., 1946, *J. Instn. elect. Engrs*, **93**, 429.

Goursat, E., 1904, *A Course in Mathematical Analysis* (Hedrick Ginn & Co.).

Grad, H., 1949, *Commun. Pure appl. Math.*, **2**, 325.

Hille, E., 1948, *Functional Analysis and Semi-Groups*. American Mathematical Society Colloquium Publication. Vol. XXXI, Chap. IV.

Ikehara, S., 1951, Report No. 217. Research Laboratory of Electronics, M.I.T.

James, H. M., Nichols, N. B., and Phillips, R. S., 1947, *Theory of Servomechanisms* (McGraw-Hill).

Lafleur, Ch., and Namias, V., 1954, *Proc. U.R.S.I.* (The Hague).

Lampard, D. G., 1955, *Proc. Instn. elect. Engrs*, C, **102**, 34.

Lee, Y. W., 1932, *J. Math. Phys.*, M.I.T., 83.

Lévy, P., 1951, *Problèmes concret's d'Analyse Fonctionelle* (Paris : Gauthier-Villars).

Lovitt, W. V., 1950, *Linear Integral Equations* (Dover).

Millar, W., 1951, *Phil. Mag.*, **42**, 1150.

Schwartz, L., (1950, 1951), *Theorie des Distributions*. 2 volumes. Actualitiés Scientifiques et Industrielles, 1091, 1122. Publications de l'institute de mathematique de l'Université de Strasbourg.

Shannon, C. E., and Weaver, W., 1949, *The Mathematical Theory of Communication* (University of Illinois Press).

Szegö, G., 1939, *Orthogonal Polynomials*. American Mathematical Society Colloquium Publication, Vol. XXIII.

Trimmer, J. D., 1950, *The Response of Physical Systems* (Chapman & Hall and John Wiley & Sons Inc.).

Turing, A. M., 1936, *Proc. Lond. Math. Soc.*, **42**, 230.

Volterra, V., 1930, *Theory of Functionals* (Blackie).

Wiener, N., 1930, *Acta Math., Stockh.*, **55**, 117 ; 1942, Report No. 129, Radiation Laboratory, M.I.T. Available as U.S. Department of Commerce report PB-58087 ;
1949, *Extrapolation, Interpolation and Smoothing of Stationary Time Series* John Wiley & Sons Inc. and Technology Press, M.I.T.).

Zadeh, L. S., 1952, *J. Franklin Inst.*, **253**, 310.

D1

Hermite Functional Expansions and the Calculation of Output Autocorrelation and Spectrum for any Time-invariant Non-linear System with Noise Input†

By J. F. Barrett
Department of Mechanical Engineering, Birmingham University

[Received March 29, 1963]

Abstract

By expansion of the input–output relation of an arbitrary non-linear system into a series of Hermite functionals, it is shown that if the input is Gaussian noise, it is possible to derive convenient theoretical expressions for the output autocorrelation function and spectrum.

§ 1. Introduction

The value of functional expansions as a method of analysis and synthesis of non-linear systems in communication engineering was first shown by Wiener and his colleagues at M.I.T. (Wiener 1942). His more recent work is summarized in his *Nonlinear Problems in Random Theory* (Wiener 1958). The present paper is closely related with this work.

In connection with noise problems, expansions using Hermite polynomials are of special importance because of the orthogonality of these polynomials with respect to Gaussian distributions. A practical method of realizing such an expansion is given by Bose (1956). Wiener, in his book *Nonlinear Problems in Random Theory*, gives another procedure for expanding a given functional into terms orthogonal with respect to a white noise process (see Wiener 1958, lecture 3). This expansion can be considered as a Hermite expansion of a different type. The appropriate Hermite polynomials for this case were briefly described in the writer's report (Barrett 1955), (see also Zadeh 1955, 1957). They were suggested by Grad's (1949) treatment of multi-dimensional Hermite polynomials. This type of Hermite expansion was used also by Friedrichs (1953) in a different application (to the quantum theory of fields).

The present paper gives a description of this type of Hermite expansion and then uses it to calculate the output autocorrelation function and spectrum of an arbitrary time-invariant non-linear system with white

† Communicated by Dr. A. T. Fuller.

noise input. The calculations generalize the known method for using Hermite polynomials to analyse the passage of Gaussian noise through an instantaneous (zero-memory) non-linear element (see e.g. Thompson 1954, Barrett and Lampard 1955).

The results of this paper were obtained during the author's research for Ph.D. (1953–8) at Cambridge University Engineering Laboratory under the direction of J. F. Coales and at Columbia University Electrical Engineering Laboratory under the direction of Professor L. A. Zadeh. The author wishes to thank Mr. Coales and Professor Zadeh for their help and encouragement.

§ 2. Hermite Functionals

Let $x(t)$, $-\infty<t<\infty$ be a stationary white Gaussian noise process having zero mean and autocorrelation function:

$$\overline{x(t_1)x(t_2)}=N\delta(t_2-t_1) \quad \quad (1)$$

($\delta(t_2-t_1)=$ Dirac δ-function).

Associated with this process, the following Hermite functionals are defined:

$$\left.\begin{aligned}\mathscr{H}^{(0)}(x)&=1,\\ \mathscr{H}^{(1)}(x;t)&=x(t),\\ \mathscr{H}^{(2)}(x;t_1,t_2)&=x(t_1)x(t_2)-N\delta(t_2-t_1),\\ \mathscr{H}^{(3)}(x;t_1,t_2,t_3)&=x(t_1)x(t_2)x(t_3)-Nx(t_1)\delta(t_3-t_2)-Nx(t_2)\delta(t_1-t_3)\\ &\quad -Nx(t_3)\delta(t_2-t_1),\end{aligned}\right\} \quad (2)$$

etc. They are defined generally by the formula:

$$\mathscr{H}^{(n)}(x;t_1,t_2\ldots,t_n)w(x)=(-1)^nN^n\frac{\partial^n w(x)}{\partial x(t_1)\partial x(t_2)\ldots\partial x(t_n)}, \quad (3)$$

where

$$w(x)=\exp\left[-\frac{1}{2N}\int_{-\infty}^{\infty}x^2(t)\,dt\right], \quad \quad (4)$$

or equivalently by the expansion:

$$\exp\left[-\frac{1}{2N}\int_{-\infty}^{\infty}u^2(t)\,dt+\frac{1}{N}\int_{-\infty}^{\infty}u(t)x(t)\,dt\right]$$

$$=\sum_{n=0}^{\infty}\frac{1}{N^n n!}\iint_{-\infty}^{\infty}\cdots\int\mathscr{H}^{(n)}(x;t_1,t_2,\ldots,t_n)u(t_1)u(t_2)\ldots u(t_n)\,dt_1\,dt_2\ldots dt_n. \quad (5)$$

These polynomials are generalizations of the Hermite polynomials defined by the formula:

$$H_n(x)w(x)=(-1)^n\frac{d^n}{dx^n}w(x), \quad \quad (3')$$

where

$$w(x)=\exp\left(-\tfrac{1}{2}x^2\right), \quad \quad (4')$$

or equivalently by:

$$\exp\left(-\tfrac{1}{2}u^2+ux\right)=\sum_{n=0}^{\infty}\frac{1}{n!}H_n(x)u^n \quad . \quad . \quad . \quad . \quad . \quad . \qquad (5')$$

(see e.g. Cramèr 1946, p. 133). The generalization was suggested in the following way. Grad (1949) defined Hermite polynomials of a vector variable $x=(x_1,x_2,\ldots,x_M)$ in which formulae (3′), (4′), (5′) would read (making minor changes in Grad's notation for convenience):

$$\mathscr{H}^{(n)}{}_{i_1 i_2 \ldots i_n}(x)w(x)=(-1)^n\frac{\partial^n}{\partial x_{i_1}\partial x_{i_2}\ldots\partial x_{i_n}}w(x), \quad . \quad . \qquad (3'')$$

where

$$w(x)=\exp\left(-\tfrac{1}{2}\sum_{i=1}^{M}x_i^2\right) \quad . \quad . \quad . \quad . \quad . \quad . \quad . \quad . \qquad (4'')$$

and

$$\exp\left(-\tfrac{1}{2}\sum_{i=1}^{M}u_i^2+\sum_{i=1}^{M}u_ix_i\right)=\sum_{n=0}^{\infty}\frac{1}{n!}\left(\sum_{i_1,i_2,\ldots,i_n}\mathscr{H}^{(n)}{}_{i_1,i_2,\ldots i_n}(x)u_{i_1}u_{i_2}\ldots u_{i_n}\right) \qquad (5'')$$

(indices $i_1, i_2, \ldots, i_n$, etc., vary over $1, 2, \ldots, M$.).

The definitions (3), (4), (5) then follow by treating x not as a finite dimensional vector but as a vector in a continuous infinity of dimensions, i.e. a function $x=(x(t):-\infty<t<\infty)$. The expressions (2) are also made homogeneous in the noise intensity N when this is not unity.

Grad, in his paper, developed a series of formulae for the multi-dimensional polynomials (3″) which are completely analogous to the corresponding formulae for the ordinary Hermite polynomials (3′). Such formulae are easily carried over by analogy to the polynomials (3). However, the proofs of these formulae are not so easily carried over. The main difficulty is the absence of an integration process $\int\ldots dx$ with respect to a function x. This difficulty is partially avoided by suitable interpretation of $\int\ldots w(x)\,dx$ as a mean value but this is not enough; for example, the formula for integration by parts is required. Other difficulties are caused by the occurrence of δ-functions and functional derivation.

The present paper will overlook such difficulties because there seems to be no immediate solution for them†. The aim of the paper is to present some interesting formulae and their application in communication engineering.

The following two results which will be needed, are derived by analogy with the corresponding formulae for Grad's polynomials.

2.1. *The Orthogonality Property*

Two polynomials of different degree are orthogonal:

$$\overline{\mathscr{H}^{(m)}(x;s_1,s_2,\ldots,s_m)\mathscr{H}^{(n)}(x;t_1,t_2,\ldots,t_n)}$$
$$=\begin{cases}0 \text{ if } m\neq n\\ \displaystyle\sum_{\binom{p_1\,p_2\ldots\,p}{1\;\,2\;\ldots\;n^n}}\delta(s_1-t_{p_1})\delta(s_2-t_{p_2})\ldots\delta(s_n-t_{p_n}) \text{ if } m=n,\end{cases} \quad . \quad . \qquad (6)$$

† The correct mathematical basis may come from the theory of group representations (cf. Kakutani 1950).

the bar denoting mean value and $\sum_{\binom{p_1 p_2 \dots p_n}{1\ 2\ \dots\ n}}$ meaning that summation is over all permutations $p_1, p_2, \dots, p_n$ of $1, 2, \dots, n$.

2.2. *The Expansion Property*

Any functional $\phi(x(t); -\infty < t < \infty)$ of finite mean square

$$\overline{\phi^2(x(t); -\infty < t < \infty)} < \infty \qquad (7)$$

has a mean-square expansion:

$$\phi(x(t); -\infty < t < \infty)$$

$$= \sum_{n=0}^{\infty} \frac{1}{n!} \int\int \dots \int_{-\infty}^{\infty} k^{(n)}(t_1, t_2, \dots, t_n) \mathscr{H}^{(n)}(x; t_1, t_2, \dots, t_n)\, dt_1\, dt_2 \dots dt_n, \qquad (8)$$

where

$$k^{(n)}(t_1, t_2, \dots, t_n) = \overline{\mathscr{H}^{(n)}(x; t_1, t_2, \dots, t_n) \phi(x(t); -\infty < t < \infty)}$$

$$= \overline{\frac{\partial^n}{\partial x(t_1) \partial x(t_2) \dots \partial x(t_n)} \phi(x(t); -\infty < t < \infty)}. \qquad (9)$$

Since the individual terms of this expansion have a mean square value proportional to $\int\int \dots \int_{-\infty}^{\infty} [k^{(n)}(t_1, t_2, \dots, t_n)]^2\, dt_1\, dt_2 \dots, dt_n$, $k^{(n)}$ is assumed square integrable.

The expansion (8) is a slightly different form from that given by Wiener (1958, lecture 3). The connection is as follows: e.g. for $n = 2$:

$$\int\int_{-\infty}^{\infty} k^{(2)}(t_1, t_2) \mathscr{H}^{(2)}(x;\ t_1, t_2)\, dt_1\, dt_2$$

$$= \int\int_{-\infty}^{\infty} k^{(2)}(t_1, t_2) [x(t_1) x(t_2) - N\delta(t_2 - t_1)]\, dt_1\, dt_2$$

$$= \int\int_{-\infty}^{\infty} k^{(2)}(t_1, t_2) x(t_1) x(t_2)\, dt_1\, dt_2 - N \int_{-\infty}^{\infty} k^{(2)}(t, t)\, dt. \qquad (10)$$

Next, write $x(t)$ as the (generalized) derivative of a Brownian motion $X(t)$, so that $x(t)\, dt = dX$. Then, e.g.,

$$\int\int_{-\infty}^{\infty} k^{(2)}(t_1, t_2) x(t_1) x(t_2)\, dt_1\, dt_2 = \int\int_{-\infty}^{\infty} k^{(2)}(t_1, t_2)\, dX(t_1)\, dX(t_2). \qquad (11)$$

Hence,

$$\int\int_{-\infty}^{\infty} k^{(2)}(t_1, t_2) \mathscr{H}^{(2)}(x; t_1, t_2)\, dt_1\, dt_2$$

$$= \int\int_{-\infty}^{\infty} k^{(2)}(t_1, t_2)\, dX(t_1)\, dX(t_2) - N \int_{-\infty}^{\infty} k^{(2)}(t, t)\, dt. \qquad (12)$$

The expression (12) is then that of Wiener (with $N = 1$). (Compare the right-hand side of (12) with the expression of Wiener (1958) marked (3.15), p. 31.)

2.3. *Application of Non-linear Systems*

Let us consider an arbitrary time-invariant non-linear system (diagram) with output $y(t)$ and input $x(t)$ which is Gaussian white noise having zero mean and autocorrelation function:

$$\overline{x(t_1)x(t_2)} = N\delta(t_2 - t_1). \qquad (13)$$

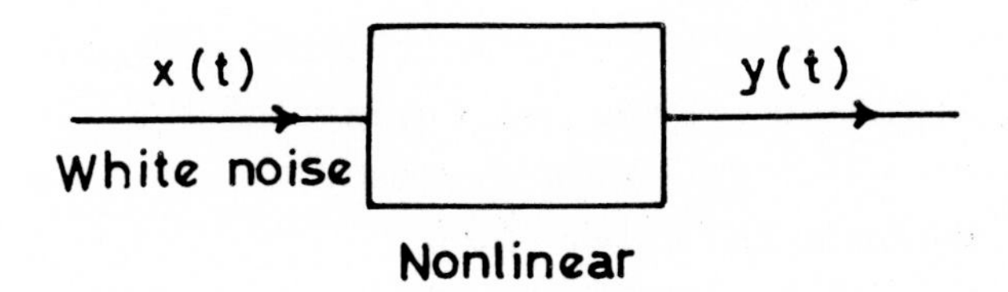

$y(t)$ is a general functional of the input values $x(t'), t' \leqslant t$ and the time t:

$$y(t) = f(t; x(t'), t' \leqslant t). \qquad (14)$$

The only restriction placed on $f(\ )$ will be that it is of finite mean square (giving finite output variance). Expand $f(\ )$ in terms of the Hermite functionals associated with the white noise input:

$$y(t) = f(t; x(t'), t' \leqslant t)$$

$$= \sum_{n=0}^{\infty} \frac{1}{n!} \int\int \overset{\infty}{\underset{-\infty}{\cdots}} \int k^{(n)}(t_1, t_2, \ldots, t_n) \mathscr{H}^{(n)}(x; t_1, t_2, \ldots, t_n)\, dt_1\, dt_2 \ldots dt_n. \qquad (15)$$

The kernels $k^{(n)}(\ )$ will be dependent on t. The assumption of time-invariance implies that the relations:

$$k^{(n)}(t+\tau; t_1+\tau, t_2+\tau, \ldots, t_n+\tau) = k^{(n)}(t; t_1, t_2, \ldots, t_n) \qquad (16)$$

must be true for any value of τ. Hence $k^{(n)}$ must be of the form:

$$k^{(n)}(t; t_1, t_2, \ldots, t_n) = g^{(n)}(t-t_1, t-t_2, \ldots, t-t_n). \qquad (17)$$

So (15) takes the form:

$$y(t) = \sum_{n=0}^{\infty} \frac{1}{n!} \int\int \overset{\infty}{\underset{-\infty}{\cdots}} \int g^{(n)}(t-t_1, t-t_2, \ldots, t-t_n) \times \mathscr{H}^{(n)}(x; t_1, t_2, \ldots, t_n)\, dt_1\, dt_2 \ldots dt_n, \qquad (18)$$

where, by (9),

$$g^{(n)}(t-t_1, t-t_2, \ldots, t-t_n) = \overline{\frac{\partial^n}{\partial x(t_1)\partial x(t_2) \ldots \partial x(t_n)} f(t; x(t'), t' < t)}, \qquad (19)$$

the bar denoting mean value. Now form the correlation of y:

$$\overline{y(t)y(t+\tau)} = \sum_{m=0}^{\infty} \sum_{n=0}^{\infty} \frac{1}{m!\,n!} \int\int \cdots \int g^{(m)}(t-s_1, t-s_2, \ldots, t-s_m)$$

$$g^{(n)}(t+\tau-t_1, t+\tau-t_2, \ldots, t+\tau-t_n) \overline{\mathscr{H}^{(m)}(x; s_1, s_2, \ldots, s_m)}$$

$$\overline{\mathscr{H}^{(n)}(x; t_1, t_2, \ldots, t_n)}\, ds_1\, ds_2 \ldots ds_m\, dt_1\, dt_2 \ldots dt_n. \qquad (20)$$

Using the orthogonality property (6) and the symmetry of the kernels $g^{(n)}(\ )$ in their arguments, this equation transforms to

$$\overline{y(t)y(t+\tau)} = \sum_{n=0}^{\infty} \frac{N^n}{n!} \iint_{-\infty}^{\infty} \cdots \int g^{(n)}(t-t_1, t-t_2, \ldots, t-t_n)$$
$$g^{(n)}(t+\tau-t_1, t+\tau-t_2, \ldots, t+\tau-t_n)\, dt_1\, dt_2 \ldots dt_n. \quad (21)$$
$$= \sum_{n=0}^{\infty} \frac{N^n}{n!} \iint_{-\infty}^{\infty} \cdots \int g^{(n)}(\tau_1, \tau_2, \ldots, \tau_n) g^{(n)}(\tau_1+\tau, \tau_2+\tau, \ldots, \tau_n+\tau)$$
$$d\tau_1\, d\tau_2 \ldots d\tau_n. \quad (22)$$

Now, taking mean values in eqn. (18), it is found that

$$\bar{y}(t) = g^{(0)}(t) \quad (=\text{const.}). \quad (23)$$

Hence the autocorrelation function of y is:

$$R_{yy}(\tau) = \overline{y(t)y(t+\tau)} - (\bar{y})^2$$
$$= \sum_{n=1}^{\infty} \frac{N^n}{n!} \iint_{-\infty}^{\infty} \cdots \int g^{(n)}(\tau_1, \tau_2, \ldots, \tau_n) g^{(n)}(\tau_1+\tau, \tau_2+\tau, \ldots, \tau_n+\tau)$$
$$d\tau_1\, d\tau_2 \ldots d\tau_n. \quad (24)$$

This is the required expression for output autocorrelation function. The kernels $g^{(n)}$ are, by (19), mean values of the functional derivatives of $f(\ )$.

Putting $\tau = 0$ gives the output variance:

$$\sigma_y{}^2 = R(0) = \sum_{n=1}^{\infty} \frac{\|g^{(n)}\|^2}{n!} N^n, \quad (25)$$

where

$$\|g^{(n)}\| = \sqrt{\iint_{-\infty}^{\infty} \cdots \int [g^{(n)}(\tau_1, \tau_2, \ldots, \tau_n)]^2\, d\tau_1\, d\tau_2 \ldots d\tau_n}$$
$$= L^2 \text{ norm of the kernel } g^{(n)}. \quad (26)$$

To find output spectrum, it is necessary to pass to Fourier transforms in eqn. (24). Apply Parseval's relation to the terms on the right-hand side of eqn. (24). Then

$$\iint_{-\infty}^{\infty} \cdots \int g^{(n)}(\tau_1, \tau_2, \ldots, \tau_n) g^{(n)}(\tau_1+\tau, \tau_2+\tau, \ldots, \tau_n+\tau)\, d\tau_1\, d\tau_2 \ldots d\tau_n$$
$$= \frac{1}{(2\pi)^n} \iint_{-\infty}^{\infty} \cdots \int |G^{(n)}(i\omega_1, i\omega_2, \ldots, i\omega_n)|^2$$
$$\exp\,[i(\omega_1+\omega_2+\ \ldots\ +\omega_n)\tau] d\omega_1 d\omega_2 \ldots d\omega_n, \quad (27)$$

where

$$G^{(n)}(p_1, p_2, \ldots, p_n)$$
$$= \iint_{-\infty}^{\infty} \cdots \int g^{(n)}(\tau_1, \tau_2, \ldots, \tau_n) \exp\,[-(p_1\tau_1 + p_2\tau_2 + \ \ldots\ + p_n\tau_n)]$$
$$d\tau_1\, d\tau_2 \ldots d\tau_n. \quad (28)$$

Further,

$$\text{right-hand side of (27)} = \frac{1}{2\pi} \int_{-\infty}^{\infty} \Gamma_n(\omega) \exp\,(i\omega\tau)\, d\omega, \quad (29)$$

where

$$\Gamma_n(\omega)=\frac{1}{(2\pi)^{n-1}}\iint\limits_{\omega_1+\omega_2+\ldots+\omega_n=\omega}\cdots\int|G^{(n)}(i\omega_1,i\omega_2,\ldots,i\omega_n)|^2\,d\omega_1\,d\omega_2\ldots,d\omega_n. \qquad (30)$$

Then (24) becomes, using (27), (30):

$$R_{yy}(\tau)=\frac{1}{2\pi}\int_{-\infty}^{\infty}\left(\sum_{n=1}^{\infty}\frac{N^n}{n!}\Gamma_n(\omega)\right)\exp(i\omega\tau)\,d\omega. \qquad (31)$$

Comparing this with the Wiener–Khinchin formula:

$$R_{yy}(\tau)=\frac{1}{2\pi}\int_{-\infty}^{\infty}\Phi_{yy}(\omega)\exp(i\omega\tau)\,d\omega, \qquad (32)$$

we see that (since Fourier transform is one-to-one):

$$\Phi_{yy}(\omega)=\sum_{n=1}^{\infty}\frac{N^n}{n!}\Gamma_n(\omega). \qquad (33)$$

This is the expression for the output spectrum. The successive terms of this expansion give a generalization of the intensity of harmonics for the case of sinusoidal forcing. The expansion (18) into Hermite functionals may be thought of as expansion into ' harmonics of the white noise '.

§ 3. Generalizations

The calculations in this paper could be generalized to where some or all of the following conditions apply.

(i) input or output are multi-dimensional,
(ii) input is non-stationary,
(iii) the system is not time-invariant,
(iv) input is Gaussian but not white.

Of course, autocorrelation function must be replaced by autocovariance function for non-stationary processes. Also, in this case, the formula for spectrum will not apply unless there is a spectrum in a generalized sense; e.g. there occurs a process with a stationary nth derivative.

References

Barrett, J. F., 1955, unpublished report. At present under publication in *J. Electron. Contr.*; 1958, Ph.D. Thesis, Cambridge University.

Barrett, J. F., and Lampard, D. G., 1955, *Proc. Inst. Radio Engrs, N.Y.*, IT–1, 10.

Bose, Amar G., 1956, *Tech. Rep. Electron. Mass. Inst. Tech.*, 309, May 15.

Cramèr, H., 1946, *Mathematical Methods of Statistics* (Princeton).

Friedrichs, K. O., 1953, *Mathematical Aspects of the Quantum Theory* (Interscience).

Grad, H., 1949, *Commun. pure appl. Math.*, **2,** 325.

Kakutani, S., 1950, *Proc. nat. Acad. Sci., Wash.*, **36,** 319.

Thompson, W. E., 1954, *Proc. Inst. elect. Engrs*, C.

Wiener, N., 1942, *Rep. U.S. Dep. Commerce*, PB–1–58087; 1958, *Nonlinear Problems in Random Theory* (Technology Press and John Wiley).

Zadeh, L. A., 1955, *I.R.E. Trans.*, IT–1 ; 1957, *I.R.E. WESCON Convention*, Part 2, August, 105.

Part II.

Optimal nonlinear deterministic control

INTRODUCTION TO PART II

Part II is concerned with the optimization of nonlinear (saturating) deterministic control systems. As was discussed in the Preface, this is a preliminary to the more difficult study of nonlinear stochastic optimization problems.

Paper 4 gives an introduction to the theory of phase space, or state space as it is normally called now. State space theory has undergone elaborations in recent years; however, Paper 4 retains its utility as an elementary account and as a guide to the historical background.

Paper 5 shows that the optimal controller generates a control signal which is a certain function of the current values of the state coordinates. Papers 6–9 discuss methods of calculating this nonlinear control function when the performance index is the integral of a quadratic form, e.g. integral-square-error. Incidentally Paper 8 gives a self-contained account of and rigorization of the theory of dimensions, which is useful in simplifying optimization problems in special cases.

Paper 10 treats the nonlinear optimization problem when the plant contains a pure delay. Sub-optimal techniques are also discussed, and these give a method for extending optimization techniques for low-order plants to high-order plants.

Paper 4

Reprinted from JOURNAL OF ELECTRONICS AND CONTROL, Vol. 8, No. 5, pp. 381–400, May, 1960.

Paper 5

Reprinted from JOURNAL OF ELECTRONICS AND CONTROL, Vol. 8, No. 6, pp. 465–479, June, 1960.

Paper 6

Reprinted from JOURNAL OF ELECTRONICS AND CONTROL, Vol. 15, No. 1, pp. 63–71, July, 1963.

Paper 7

Reprinted from INTERNATIONAL JOURNAL OF CONTROL, Vol. 2, No. 1, pp. 33–73, July, 1965.

Paper 8

Reprinted from INTERNATIONAL JOURNAL OF CONTROL, Vol. 3, No. 4, pp. 359–394, April, 1966.

Paper 9

Reprinted from INTERNATIONAL JOURNAL OF CONTROL, Vol. 8, No. 1, pp. 65–87, July, 1968.

Paper 10

Reprinted from INTERNATIONAL JOURNAL OF CONTROL, Vol. 8, No. 2, pp. 145–168, August, 1968.

Phase Space in the Theory of Optimum Control†

By A. T. Fuller
Department of Engineering, University of Cambridge

[Received April 26, 1960]

Abstract

Phase space is an essential concept in the theory of optimum non-linear control, but has been given only cursory exposition hitherto. This paper attempts to bridge the gap between the control engineer who uses phase space heuristically and the mathematician who adopts phase space as an arbitrary starting point. A history of the phase-space concept is given. It is shown how to define and measure the phase coordinates of linear and non-linear systems. A special treatment of electrical networks is given.

§ 1. Introduction

In the theory of optimum relay control systems and optimum saturating control systems, which was begun by Ivachnenko (1948) and Feldbaum (1949), and is at present being actively developed, the concept of phase space plays an important part. However, this concept has as yet received no adequate exposition in the context of control theory. On the one hand control engineers have adapted phase-space ideas piece-meal from older sciences, and used them tentatively and heuristically. On the other hand, control mathematicians have tended to adopt the phase-space equations as an arbitrary and unexplained starting point. As a result, many papers on optimum systems have seemed obscure. The present paper attempts to bridge the gap by giving the historical origins of the phase-space concept, by laying down precise definitions, and by showing how they may be applied to typical physical systems.

The present paper is introductory in character. The ideas in it are applied to control optimization problems in the writer's thesis (Fuller 1959 a). It is hoped to publish the results in the latter in subsequent papers.

§ 2. A Control Problem

In this section we give a simple formulation of a control optimization problem when saturation is present. This formulation will motivate and lend some concreteness to our subsequent discussion of phase space.

In its simplest form a control system consists of three basic parts: a *main process* which has the ‘controlled variable’ as output, an *amplifier* which drives the main process, and a *controller* which drives the amplifier. (The term ‘main process’ as here used, embraces the terms ‘plant’,

† Communicated by J. F. Coales.

'load', 'motor-and-load', 'dynamics', 'fixed elements', which are sometimes used in the literature.) The main process and the amplifier usually operate at high-power level and their design is largely fixed by such considerations as efficiency, economics, safety, tradition, etc. The output of the amplifier is at high-power level and usually tends to saturate when the amplifier input is large, as shown in fig. 1. Obviously, this saturation

Fig. 1

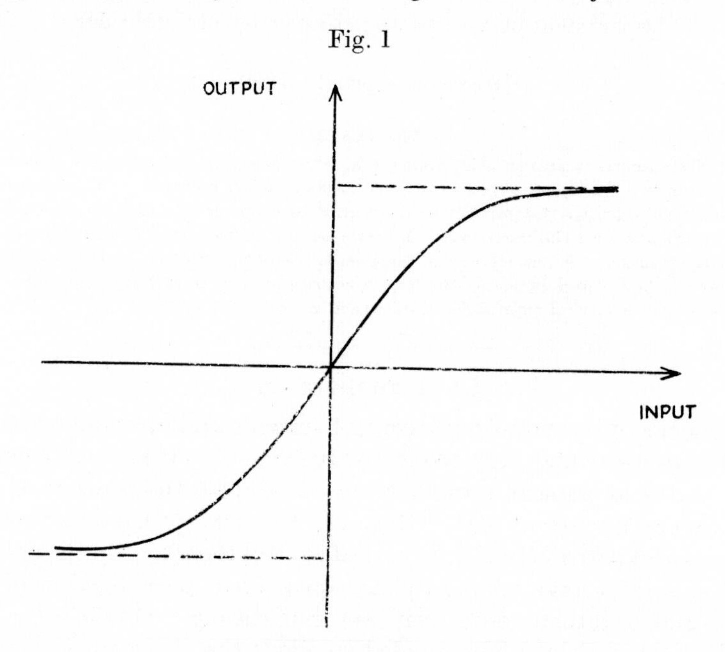

Saturation characteristic.

Fig. 2

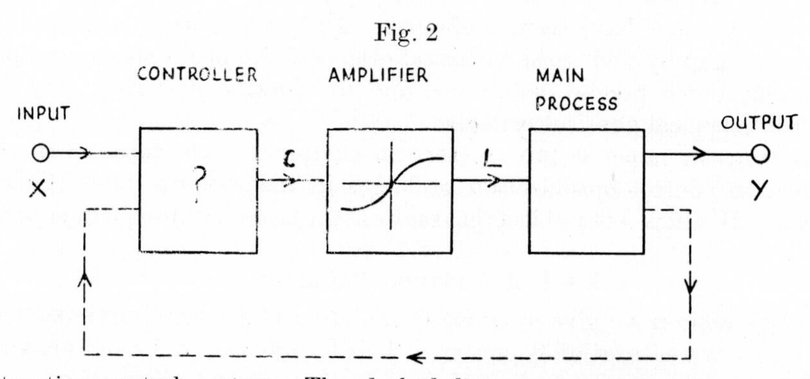

Saturating control system. The dashed line represents possibly multiple connections from the main process to the controller.

drastically limits the ability of the system to respond quickly or fully to a given input signal. The controller operates at low power level, and its design is relatively unrestricted. The block diagram of the system is shown in fig. 2.

We now make the following assumptions concerning the system of fig. 2. The main process is assumed to be fixed, i.e. specified by some given set of linear or non-linear differential equations. The nature of these differential equations will be discussed in later sections. The amplifier is assumed to be free from lag and completely specified by its given input–output characteristic. This characteristic is assumed for definiteness to be single-valued, monotonic, and bounded by positive and negative saturation levels of equal amplitude, as shown in fig. 1. The system input is assumed to be a single-valued function (deterministic or random) of time t.

The problem is to design the controller so as to minimize some given performance criterion which is a time-integral or a time-average of some given function of the error $E(t)$ where

$$E \equiv X - Y, \qquad (1)$$

Y being the system output. In order to do this, the controller may make any physically realizable measurements on the main process and on the system input (as indicated in fig. 2) and may perform any physically realizable operations (linear or non-linear) on these measurements.

For example, if the system input X is a simple transient, such as a step-change applied at $t=0$, it is often appropriate to adopt as a performance criterion M_ν where

$$M_\nu \equiv \int_0^\infty |E|^\nu \, dt, \quad \nu \geqslant 0, \qquad (2)$$

where ν is some chosen constant. Thus M_2 is the familiar criterion integral-square-error, and M_0 can be shown (Fuller 1959 b) to be often equivalent to the well-known settling time criterion. Alternatively, if X is a random input, it is often appropriate to adopt as a performance criterion M_ν' where

$$M_\nu' \equiv \lim_{T \to \infty} \frac{1}{2T} \int_{-T}^{T} |E|^\nu \, dt, \quad \nu \geqslant 0. \qquad (3)$$

These various criteria have been discussed in a previous paper (Fuller 1959 b).

A special case of the saturating amplifier described above is the *ideal relay*. Its input–output characteristic is shown in fig. 3, and is given by the relation

$$L = \operatorname{sgn} C \qquad (4)$$

where C is the relay input and L is the relay output. Thus the ideal relay is assumed free from hysteresis, dead-zone, dead-time, etc. We term a control system with this characteristic and subject to the above assumptions an *ideal relay control system*.

The key to the solution of the control optimization problems discussed above is a thorough understanding of phase space. The present paper is mainly an exposition of phase-space principles; the application of these principles will be dealt with in subsequent papers.

§ 3. General History of Phase-space Concepts

The concepts of phase space evolved mainly from the classical theory of the dynamics of particles and rigid bodies. An excellent account of this theory has been given by Whittaker (1937). According to Lanczos (1949), an adequate history of the subject has not yet been written; and completeness cannot be claimed for the following survey.

Fig. 3

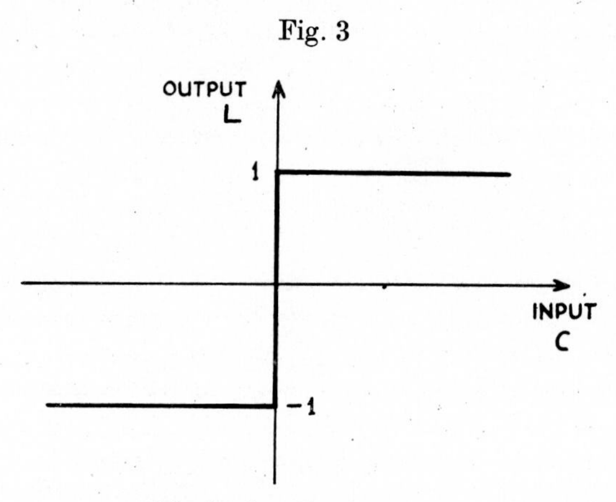

Ideal relay characteristic.

The theory of the dynamics of particles and rigid bodies is concerned basically with determining the motion of a system of particles when these are subject to certain constraints, such as rigid inter-connections, sliding on perfectly smooth surfaces and rolling on perfectly rough surfaces. The theory is based on Newton's (1687) laws of motion. The instantaneous position of the system is expressed by r independent *generalized coordinates*, $q_1, q_2, \ldots q_r$, which are determined (in terms of the Cartesian coordinates of the particles) by the geometry of the constraints. If the system is *holonomic*, i.e. if there are no additional constraint relations between the time-derivatives of the coordinates, the generalized coordinates are termed degrees of freedom. Lagrange (1778) showed that the equations of motion of a conservative holonomic system could be simplified to:

$$\frac{d}{dt}\left(\frac{\partial \mathscr{L}}{\partial q_i'}\right) - \frac{\partial \mathscr{L}}{\partial q_i} = 0 \quad (i=1, 2, \ldots r) \qquad (5)$$

where

$$\mathscr{L} \equiv T - V, \qquad (6)$$

T being the kinetic energy and V the potential energy of the system, and where

$$q_i' \equiv \frac{dq_i}{dt}. \qquad (7)$$

At first sight eqns. (5) appear to be a rather formidable set of simultaneous partial differential equations; however, it is important to realize that both $\partial\mathscr{L}/\partial q_i'$ and $\partial\mathscr{L}/\partial q_i$ are known functions of the variables

$$q_1, q_2, \ldots q_r, q_1', q_2', \ldots q_r'.$$

Thus eqns. (5) can be written

$$\frac{d}{dt}A_i(q_1, q_2, \ldots q_r, q_1', q_2', \ldots q_r') - B_i(q_1, q_2, \ldots q_r, q_1', q_2', \ldots q_r') = 0 \quad (i = 1, 2, \ldots r). \qquad (8)$$

In eqns. (8) the differentiation of A_i with respect to time yields terms involving $q_1'', q_2'', \ldots q_r''$. Hence eqns. (8), and therefore Lagrange's eqns. (5), reduce to simply a set of simultaneous second-order ordinary differential equations. The advantage of eqns. (5) over those obtained by a straightforward application of Newton's laws to each component of the system is that eqns. (5) do not involve the reaction forces which maintain the given constraints.

Originally attention was concentrated on finding the position $(q_1, q_2, \ldots q_r)$ of the system at a given instant (especially for problems in celestial mechanics). Somewhat later the more general problem of finding the 'complete state' of the system at a given instant must have been recognized. The 'complete state' is characterized by the instantaneous values not only of $q_1, q_2, \ldots q_r$ but also of their first derivatives with respect to time, namely $q_1', q_2', \ldots q_r'$. This follows heuristically from the fact that, as we have seen, Lagrange's r equations are each of second order, so that the future motion depends on $2r$ 'initial conditions'. Hamilton (1835) showed that the equations of motion can be further simplified by taking as the extra coordinates needed to specify the 'complete state', not $q_1', q_2', \ldots q_r'$ but $p_1, p_2, \ldots p_r$ where

$$p_i \equiv \frac{\partial\mathscr{L}}{\partial q_i'} \quad (i = 1, 2, \ldots r). \qquad (9)$$

Since $\mathscr{L}$ is a known function of $q_1, q_2, \ldots q_r, q_1', q_2', \ldots q_r'$ eqn. (9) means merely that p_i is a known function of the same arguments:

$$p_i = A_i(q_1, q_2, \ldots q_r, q_1', q_2', \ldots q_r'). \qquad (10)$$

On substituting (9) in (5), Hamilton found that the r second-order equations of Lagrange could be replaced by the following $2r$ equations, each of first order:

$$\left.\begin{aligned} q_i' &= \frac{\partial H}{\partial p_i} \\ p_i' &= -\frac{\partial H}{\partial q_i} \end{aligned}\right\} \quad (i = 1, 2, \ldots r) \qquad (11)$$

where H is (for a conservative holonomic system) the total energy:

$$H = T + V. \qquad (12)$$

H is thus a known function of $q_1, q_2, \ldots q_r, q_1', q_2', \ldots q_r'$, and hence, in view of relations (10), H is a known function of $q_1, q_2, \ldots q_r, p_1, p_2, \ldots p_r$. Therefore eqns. (11) reduce to the following simple set of first-order ordinary differential equations:

$$\left.\begin{aligned} q_i' &= C_i(q_1, q_2, \ldots q_r, p_1, p_2, \ldots p_r), \\ p_i' &= D_i(q_1, q_2, \ldots q_r, p_1, p_2, \ldots p_r), \end{aligned}\right\} (i = 1, 2, \ldots r). \quad . \quad . \quad (13)$$

These $2r$ equations emphasize that $2r$ variables are needed to specify the instantaneous state of the system†.

A simple generalization of eqns. (13) is as follows:

$$\frac{dy_i}{dt} = f_i(y_1, y_2, \ldots y_n, t) \quad (i = 1, 2, \ldots n). \quad . \quad . \quad . \quad (14)$$

(These equations are more general than eqns. (13) because the functions f_i are not assumed to be $\pm$ the partial derivatives of some function H and because t appears explicitly on the right-hand sides.) Cauchy (1819) had already encountered equations similar to eqns. (14), in dealing with the general theory of partial differential equations‡. Thus, after Hamilton's work, it became clear that eqns. (14) constitute a class of systems of differential equations of importance in both pure and applied mathematics. Equations of the general form of eqns. (14) are sometimes called *canonical equations*.

Cauchy (1842) proved that, subject to certain assumptions, eqns. (14) have a unique solution in terms of the 'initial conditions', i.e. in terms of the given values of $y_1, y_2, \ldots y_n$ at any given time t_0. Although this result has the appearance of being merely an abstract uniqueness theorem, it is of great practical importance because it establishes that

$$y_1(t), y_2(t), \ldots y_n(t)$$

characterize completely the 'state' at time t of the system represented by eqns. (14). Another uniqueness proof also due to Cauchy, was published by Moigno (1844).

Moigno also pointed out that the nth order differential equation in two variables

$$\frac{d^n y}{dt^n} = F\left(y, \frac{dy}{dt}, \ldots \frac{d^{n-1}y}{dt^{n-1}}, t\right) \quad . \quad . \quad . \quad . \quad . \quad . \quad (15)$$

is reduced to a set of first order differential equations of the same canonical form as (14), by making the substitutions:

† Thus the term 'degrees of freedom', which refers to the coordinates $q_1, q_2, \ldots q_r$ and ignores the coordinates $p_1, p_2, \ldots p_r$ is almost a misnomer, and is best avoided in control theory (compare MacLellan 1957). It gives rise to confusion especially when electro-mechanical analogies are considered.

‡ According to Cayley (1890), in 1831 Cauchy was already familiar with the Hamiltonian eqns. (11) though Cauchy did not publish their derivation.

$$y_1 = y, \quad y_2 = \frac{dy}{dt}, \ldots y_n = \frac{d^{n-1}y}{dt^{n-1}}. \quad . \quad . \quad . \quad . \quad . \quad . \quad (16)$$

Cauchy's theorem applied to the resulting set of equations showed that the solution of eqn. (15) is uniquely determined by the initial values

$$(y)_0, \left(\frac{dy}{dt}\right)_0, \ldots \left(\frac{d^{n-1}y}{dt^{n-1}}\right)_0$$

of the derivatives at $t = t_0$. Hitherto this result had been assumed without proof on the basis of the known special cases in which eqn. (15) could be solved explicitly.

Since the work of Cauchy, his proofs have been perfected and extended to more general cases by Briot and Bouquet, Weierstrass, Darboux, Méray, Riquier, Kovalevski, Lipschitz, and many others. Goursat (1917) gives a general account of the theory, for both ordinary and partial differential equations. We shall state one of the relevant theorems in § 7.

Poincaré (1881) pointed out that it is sometimes helpful to consider the variables $y_1, y_2, \ldots y_n$ as the coordinates of a point in n-dimensional space, and applied his own topological methods to the study of the trajectory which the point describes as time increases. Around this time Lie discussed some abstract group properties of eqns. (14) in a great number of papers and in his treatise Lie (1888) (see Goursat (1917) for an introduction to this theory). Liapunov (1892), and Bendixson (1901) contributed to the stability theory of eqns. (14). By this time mathematicians felt that eqns. (14) were an appropriate standard form of ordinary differential equations, and adopted them as the starting point of their theories without preamble.

The physicist Gibbs (1902) considered the application of the Hamiltonian eqns. (11) to gas molecules, in the science of 'statistical mechanics'. He thought of the instantaneous positions and velocities of the K molecules as representing the coordinates of a point in space. He called the position of the point the *phase* of the system, and the $6K$-dimensional space the *phase space*. This terminology has remained, except that it is often nowadays applied more generally, e.g. to the $(y_1, y_2, \ldots y_n)$ space of Poincaré†.

Around 1930 Van der Pol and a number of Russians‡ began adapting the methods of celestial mechanics to the study of non-linear phenomena in applied science, i.e. to what is now called 'non-linear mechanics'. The work of these Russians (notably Mandelstam, Papalexi, Andronov, Chaikin, Lochakov, Krylov, Bogoliubov) was made more accessible by Minorsky's (1947) (English language) reports which were published in book form. This work stimulated the English language translation of a

---

† We shall give a precise general definition of phase space in § 6.

‡ An interesting history of Russian contributions to differential equations is given by Lefschetz (1953).

similar book by Andronov and Chaikin (1949), originally published in Russian in 1937. Phase space concepts were used for discussing the stability of simple electrical and mechanical systems whose differential equations could be put in form (14) with $n = 2$ (in this case the phase space is simply a phase plane). The emphasis at this stage was on using the phase plane as a convenient medium for depicting the totality of all possible motions of a given system.

Various other papers applying the phase plane to engineering problems are scattered throughout the engineering literature, but we shall make no attempt to assess them here. Many references are given in the textbook of Ku (1958). Papers dealing with control problems will be discussed in the next section.

§ 4. History of Phase Space in Control Theory

We shall not give here a history of phase space in general control theory, as we wish to concentrate attention on *optimum* control theory. However, it is interesting to note that, according to Tsypkin (1955), Léauté (1885, 1891) was the first to use what was effectively the phase plane in control system studies. The recent revival of interest in phase space seems to be due, at least in part, to Andronov (1956) and his collaborators around 1944. We should also mention Ashby's *Design for a brain* which appeared in 1952. This book provides a purely mechanistic model of biological adaptation and control and relies heavily on phase space concepts. Some attempt is made to explain and justify these concepts, however Ashby's argument seems incomplete.

Let us now turn to the history of phase space in optimum control theory, i.e. in the solution of problems like those defined in § 2. Some of the relevant literature is very heuristic, and some is very abstract; for this reason it would not be profitable at this stage to attempt a detailed assessment. Instead we give little more than a list of published works in the present section; a few general comments will also be given in the next section.

Serious work† on optimum relay control systems was begun in Russia by Ivachnenko (1948) and Feldbaum (1949), in America by McDonald (1950), and in England by Uttley and Hammond (1952). These early papers were exploratory, dealing only with simple linear main processes of second order, and with the problem of minimizing the settling time after application of a step-input. The arguments used were mainly heuristic. Hopkin (1951) and West (1952) made further contributions along these lines. In 1953 three American reports appeared which rigorized and generalized the theory for linear main processes. These reports, by Bushaw (1953), Lasalle (1953) and Rose (1953) were unfortunately not published, and much subsequent work seems to have been done in ignorance

† A. J. Williams (1938) had already touched on the subject.

of them†. Bogner and Kazda (1954) again treated higher-order systems, but only tentatively and heuristically, and this approach was further developed for third-order systems by Hopkin and Iwama (1956), Passera and Willoh (1956), Doll and Stout (1957), and Hagin and Fett (1957). Meanwhile somewhat similar work was being done in Russia by Ivachnenko (1951), Lerner (1952), and Feldbaum (1953, 1955). Bellman *et al.* (1956) gave a generalization of some of Lasalle's and Rose's results, using an abstract topological argument. Pontryagin (1956) gave a very general method for minimizing the settling time of both linear and non-linear high-order main processes. This method has subsequently been developed in several papers by Boltyanskii, Gamkrelidze and Pontryagin (1956), Boltyanskii (1956), Gamkrelidze (1958 a, b, 1959), and Pontryagin (1959). Kalman (1957), Kalman and Bertram (1958), and Kalman and Koepcke (1958) applied phase-space ideas to the optimization of sampled-data control systems. Their methods are related to those of 'dynamic programming', a technique due to Bellman (1957). Dynamic programming uses implicitly phase space, except that the system is assumed to be described by difference equations rather than differential equations. Recently Desoer (1959) has shown how classical techniques from the calculus of variations may be used in minimizing settling time. The above list of references is not exhaustive‡, but will serve to illustrate the remarks in the next section.

§ 5. Commentary on the History of Phase Space

Sections 3 and 4 show that the concepts of phase space have evolved slowly in a long and tortuous history. This history is not given in text-books, which tend rather to adopt eqns. (14) as an arbitrary and unexplained starting point, or at most to show that a few simple examples of physical systems are described by such equations. Moreover many textbooks put most emphasis on (i) the use of phase space as a convenient medium for depicting the totality of possible solutions of differential equations (as in Poincaré's topology), (ii) the use of special graphical techniques for computing the solution of differential equations (method of isoclines, etc.). The impression is given that phase-space techniques constitute a rather miscellaneous set of mathematical tools (of mysterious origin) which happen to yield solutions to certain complicated problems of dynamics.

In control engineering, however, it is becoming recognized that phase space is not merely an occasionally convenient tool of analysis, but is a fundamental concept in the synthesis of optimum systems. In fact it can

---

† Some of Bushaw's work was reported in the book by Tsien (1954), and Bushaw (1958) published a condensed version of his report. The present writer has seen only an abstract of Lasalle's (1954) work. Rose (1956) has published a summary of his results.

‡ Kalman (1957) stated that the literature then consisted of well over 50 titles.

be shown that *an optimum controller simply measures a certain phase state, and forms an instantaneous function of the resulting measurements*. This remark will be made precise and substantiated in a later paper (Fuller 1960). The proof depends on the property that *the entire future of a system is (in a sense) determined by its present phase state*. This property, which, as we have seen, is implied for eqns. (14) by Cauchy's results, is of great importance, and in the next section we shall make it the basis of a precise generalized definition of phase state. The property seems to be under-emphasized in present textbooks, which treat it as an abstract theorem on the existence and uniqueness of trajectories in phase space.

The papers referred to in § 4 roughly fall into two classes. On the one hand there are the heuristic papers written by engineers who had previously obtained experience of phase space by applying the above methods (i) and (ii) to the analysis of non-optimum control systems. On the other hand there are the abstract papers written by mathematicians who take phase-space equations as an unexplained starting point. Both classes of paper are much easier to understand when the importance of the property underlined above is appreciated. It is a purpose of the present paper and a companion paper (Fuller 1960) to emphasize the importance of this property†.

§ 6. Phase-space Definitions

Associated with any dynamic system there are infinitely many time-variables. For example, in an electrical network there are the voltages at the nodes and their first, second, etc., integrals with respect to time. In practice, however, only a few variables are of direct interest in any given problem. For example, in a control system often the error $E(t)$ is the only variable that needs to be known if the problem is to assess the system performance.

Let us suppose that in a general system there is a particular set of variables $y_1, y_2, \ldots y_m$ which are required to be evaluated as functions of time. We define the system as *determinate* if the futures of $y_1, y_2, \ldots y_m$ are completely determined by the present values of a set of time-variables $z_1, z_2, \ldots z_n$, by certain specified constants and by the specified futures of a set of time-variables $x_1, x_2, \ldots x_q$. (Thus if the system is represented by a set of ordinary differential equations, $z_1, z_2, \ldots z_n$ represent 'initial conditions' and $x_1, x_2, \ldots x_q$ represent 'forcing functions'.) We regard the n quantities $z_1, z_2, \ldots z_n$ as coordinates of a point in an n-dimensional space, termed the *phase space*. The point is termed the *representative point*, and its position the *phase state*.

Thus associated with any one determinate system there are many different phase spaces. Which of these are relevant depends on the problem in hand and on the presence or absence of forcing functions. Moreover,

† Kalman's papers show that he recognized the importance of this property; however, he did not explain it, as he was mainly concerned with the development of special sampled-data techniques.

even when the latter considerations are allowed for, the relevant phase space is not unique, for we may take as phase-space coordinates any set of functions

$$\bar{z}_1(z_1, z_2, \ldots z_n), \quad \bar{z}_2(z_1, z_2, \ldots z_n), \ldots \bar{z}_n(z_1, z_2, \ldots z_n),$$

provided that for every set of values $\bar{z}_1, \bar{z}_2, \ldots \bar{z}_n$ there is a unique set of values $z_1, z_2, \ldots z_n$ and vice versa. However, once a phase space has been selected in any given problem it is customary to speak loosely of it as 'the phase space of the system'.

Note that according to our definition, the set of phase coordinates $z_1, z_2, \ldots z_n$ need not all be independent, in other words the set may include redundant members. For reasons of practical and mathematical economy it is usual to choose a minimal set, i.e. a set without redundant members. Sometimes however it is not clear how to make this choice.

The above definition of phase space is similar to definitions by Ashby (1952) and Kalman and Koepcke (1958). Note that our definition is not restricted to systems described by eqns. (14). Note also that the term ' space ' in phase space is used purely for linguistic convenience; it is not necessary for our purposes to visualize the corresponding geometric space (for $n \leqslant 3$ it is sometimes helpful to do so, but for $n > 3$ an attempt at such visualization tends to be more confusing than helpful).

§ 7. Continuous Systems

It is clear from § 3 that many dynamic systems satisfy equations of the form

$$\frac{dy_i}{dt} = f_i(y_1, y_2, \ldots y_n, t) \quad (i = 1, 2, \ldots n) \quad . \quad . \quad . \quad . \quad (18)$$

and that when the f's are such that Cauchy's theorems apply, the coordinates $y_1, y_2, \ldots y_n$ constitute a phase space in the sense defined in § 6. This is the case when the f_i are continuous and possess continuous first partial derivatives with respect to $y_1, y_2, \ldots y_n, t$; or more generally when the f_i satisfy the Lipschitz condition of the following theorem (see e.g. Birkhoff (1927)):

Theorem I (Cauchy-Lipschitz): *If for every i and for every pair of points*† $(y_1^1, y_2^1, \ldots y_n^1, t_1)$ *and* $(y_1^2, y_2^2, \ldots y_n^2, t_2)$ *the* f_i *satisfy the following Lipschitz condition*

$$|f_i(y_1^1, y_2^1, \ldots y_n^1, t_1) - f_i(y_1^2, y_2^2, \ldots y_n^2, t_2)| \leqslant \sum_{j=1}^{n} K_j |y_j^1 - y_j^2| + K_{n+1}|t_1 - t_2|, \quad . \quad . \quad . \quad . \quad . \quad (19)$$

the quantities $K_1, K_2, \ldots K_{n+1}$ *being fixed positive quantities, then there is only one solution* $y(t) \equiv [y_1(t), y_2(t), \ldots y_n(t)]$ *such that* $y(t_0) = y^0$.

This theorem cannot be applied immediately to relay control systems, because for such systems the f_i of eqns. (18) are discontinuous. Moreover, it often turns out that control systems which are optimum in the sense of § 2 behave like relay control systems. Such discontinuous systems require the discussion given in the next section.

† Here superscripts are used for labelling purposes only.

§ 8. Discontinuous Systems

Suppose the f_i in eqns. (18) are discontinuous functions of their arguments. As long as the system remains in a finite region where the f_i are continuous we may apply Theorem I and say that the motion is uniquely determined. What happens when the system reaches a point at which an f_i is discontinuous? It is important to realize that the answer to this question cannot be deduced from eqns. (18). In fact eqns. (18) are, strictly speaking, meaningless at this point. Equations (18) are differential equations and are only meaningful within each continuous region of the f_i's. The discontinuities of the f_i occur at the boundaries of these regions, and the motion of the system cannot be determined until the boundary conditions have been assigned.

The boundary conditions must be chosen to make the resulting mathematical model appropriate to the physical system being studied. In many cases of control systems it is realistic to assume that when the system reaches a discontinuity it will cross the boundary into the neighbouring continuous region, and will start motion there with initial conditions which are the same as the end conditions in the previous region. Systems which have this behaviour will be called systems with *identified internal boundary conditions.* We thus have the result that for systems described by eqns. (18), in which the f_i are piece-wise continuous and satisfy a Lipshitz condition in each region of continuity, and which have identified internal boundary conditions, the coordinates $y_1, y_2, \ldots y_n$ constitute a phase space.

Bushaw (1953, 1958) pointed out that the solution at a discontinuity is a matter of definition. Other writers have tended to ignore this point and to assume implicitly identified internal boundary conditions.

If in a relay system, the system is given impulses whenever the relay switches over, the system may still satisfy eqns. (18) (with discontinuous f_i), but will not necessarily have identified internal boundary conditions. Nevertheless $y_1, y_2, \ldots y_n$ will still constitute a phase space if a rule is given relating the initial conditions of the trajectory after a point of discontinuity to the end conditions before this point. In such a phase space the trajectories are discontinuous, and this turns out to complicate appreciably the discussion of optimum systems (compare, e.g. Flügge-Lotz 1960). For this reason it is better, when possible (as is sometimes the case), to select an alternative phase space with continuous trajectories.

§ 9. Linear Systems

The main process of a control system can often be represented, with reasonable approximation, by a set of linear differential equations with constant coefficients. Consequently the main process can often be characterized by a transfer function between its input and its output. Control engineers are very familiar with this characterization, and will wish to know how the phase coordinates of the main process are related to its transfer function.

9.1. *Transfer Function without Zeros*

Let us consider first the simple case when the transfer function is

$$\frac{1}{a_0+a_1p+\ldots+a_np^n}. \qquad (20)$$

This corresponds to a system satisfying the differential equation

$$(a_0+a_1D+\ldots+a_nD^n)y=x, \qquad (21)$$

where $x(t)$ is a known forcing function (finite but possibly discontinuous), y is the output, and D represents d/dt. Equation (21) may be replaced by the set of canonical equations

$$\left.\begin{aligned} Dy_1&=y_2,\\ Dy_2&=y_3,\\ &\cdots\\ Dy_{n-1}&=y_n,\\ Dy_n&=-a_n^{-1}(a_0y_1+a_1y_2+\ldots+a_{n-1}y_n-x), \end{aligned}\right\} \qquad (22)$$

by making the substitutions

$$y=y_1, \quad Dy=y_2, \ldots, \quad D^{n-1}y=y_n. \qquad (23)$$

Thus the phase coordinates corresponding to a system with transfer function (20) are the output and its first, second... $(n-1)$th time-derivatives. Notice also that according to eqns. (22), these coordinates are continuous functions of time when $x(t)$ is continuous, and are still continuous when $x(t)$ is discontinuous if we postulate identified internal boundary conditions. (This postulate is implicit in the use of the transfer function, in any case.)

9.2. *Transfer Function with Zeros*

Next let us consider the more general transfer function

$$\frac{b_0+b_1p+\ldots+b_mp^m}{a_0+a_1p+\ldots+a_np^n} \quad (n>m\geqslant 0). \qquad (24)$$

This corresponds to a system satisfying the differential equation

$$\left.\begin{aligned}(a_0+a_1D+\ldots+a_nD_n)y=(b_0+b_1D+\ldots+b_mD^m)x,\\ (n>m\geqslant 0).\end{aligned}\right\} \qquad (25)$$

Strictly speaking, eqn. (25) with $m>0$ is meaningless when $x(t)$ is discontinuous, since then the right side does not exist. However, whenever $x^{(m)}(t)$ is continuous it follows as before that $y, y', \ldots y^{(n-1)}$ are phase coordinates. When $x^{(m)}(t)$ is discontinuous, the resulting jumps in this phase space can be calculated by assuming that the right side of eqn. (25) contains delta-functions. But the use of $(y, y', \ldots y^{(n-1)})$ phase space in this case raises two difficulties: (i) the instruments measuring

$$y, y', \dots y^{(n-1)}$$

may tend to overload when $x(t)$ changes discontinuously, (ii) the jumps in the phase space complicate considerably the design of optimum controllers (see § 8). These difficulties can be partially alleviated by attempting to cancel the impulse functions arriving at the controller with other impulse functions obtained by differentiating $x(t)$. The method of Hung and Chang (1957) amounts to this, and is unsatisfactory inasmuch as it requires the mutual cancellation of infinite quantities.

Thus we still require a method for choosing continuous phase coordinates for a system with transfer function (24). One method is as follows. Let us recall that the transfer function is obtained by performing operations on the differential equations describing the component parts of the main process. These differential equations, at least for components such as inertias, inductors, capacitors, are simple equations of the same form as eqn. (21), for which we know the phase coordinates. Thus we can take as phase coordinates of the complete main process the phase coordinates of the component parts. Loosely speaking, the phase coordinates are signals stored in the energy-storage elements, and are readily available when the internal structure of the main process is accessible. This point will be illustrated in § 11, dealing with electrical networks. As a matter of fact theoretical papers on linear systems often take what are in effect the component differential equations, written in canonical form, as the starting point: i.e. they assume that the main process is described by

$$\frac{dy_i}{dt} = a_{i1}y_1 + a_{i2}y_2 + \dots + a_{in}y_n + x_n \quad (i = 1, 2, \dots n). \quad . \quad . \quad (26)$$

Here $y_1, y_2, \dots y_n$ are clearly the phase coordinates.

We saw that the continuous phase coordinates of system (20) are related to the output by linear operations, namely differentiations. We shall now obtain a somewhat similar result for system (24). Equation (25) may be written

$$(a_0 + a_1 D + \dots + a_n D^n)z = x \quad . \quad . \quad . \quad . \quad . \quad . \quad (27)$$

where

$$z = \frac{y}{b_0 + b_1 D + \dots + b_m D^m}; \quad . \quad . \quad . \quad . \quad . \quad . \quad (28)$$

i.e. where

$$y = (b_0 + b_1 D + \dots + b_m D^m)z. \quad . \quad . \quad . \quad . \quad . \quad (29)$$

Equation (27) is of the same form as eqn. (21), and consequently $z, Dz, \dots D^{n-1}z$ constitute a set of continuous phase coordinates. (When these are known, y may be found immediately from eqn. (29).) Using eqn. (28), these continuous phase coordinates for system (24) may be written symbolically as

$$\left.\begin{array}{c}\dfrac{y}{b_0+b_1D+\ldots+b_mD^m},\quad \dfrac{Dy}{b_0+b_1D+\ldots+b_mD^m},\ldots \\ \dfrac{D^{n-1}y}{b_0+b_1D+\ldots+b_mD^m}\end{array}\right\} . \qquad (30)$$

Expressions (30) indicate conveniently the relations between the output and the phase coordinates. However, it is important to realize that expressions (30) (with $m>0$) cannot be used to calculate the phase coordinates explicitly when only $y(t)$ is known. (This calculation would require the solution of eqn. (29) for $z(t)$, which in turn would require knowledge of the 'initial values' of z, $Dz, \ldots D^{m-1}z$, which would be some of the phase coordinates being sought.) We shall use expressions (30) in the next section. They are also useful in the theory of control systems with random inputs (Fuller 1959 a).

§ 10. Measurement of Phase State of a Linear System

As mentioned above, the phase coordinates are usually directly measurable quantities appearing within the system. Sometimes however the internal structure of the system is inaccessible, and the problem becomes that of determining the phase state from measurements on the output, y. Expressions (30) show that if y is connected to the inputs of n filters having transfer functions

$$\left.\begin{array}{c}\dfrac{1}{b_0+b_1p+\ldots+b_mp^m},\quad \dfrac{p}{b_0+b_1p+\ldots+b_mp^m},\ldots \\ \dfrac{p^{n-1}}{b_0+b_1p+\ldots+b_mp^m}\end{array}\right\} . \qquad (31)$$

then the phase state of the system is represented by the n filter outputs, provided

(i) at some time $t=t_0$ the main system and each filter all start with zero 'initial conditions',

(ii) each filter is stable.

Proviso (i) can often be waived if the system is in operation long enough for the initial stored energies in the filters and the system to decay to insignificance. Proviso (ii) implies that all the zeros of

$$(b_0+b_1p+\ldots+b_mp^m)$$

must be in the left half plane, i.e. that the main system must be a 'minimum-phase' structure in the well-known sense of Bode.

We shall call the $(z, z', \ldots z^{(n-1)})$ space of § 9.2 the *master* phase space, and that specified by the outputs $z_c, z_c', \ldots z_c^{(n-1)}$ of filters (31) the *copy* phase space.

Example. Consider the electrical network shown in fig. 4. The differential equation relating the output voltage y to the input voltage x is of the form

$$(1+a_1D+a_2D^2+a_3D^3)y=(1+b_1D)x \quad . \quad . \quad . \quad . \quad (32)$$

where

$$b_1=R_3C_3. \quad . \quad . \quad . \quad . \quad . \quad (33)$$

Fig. 4

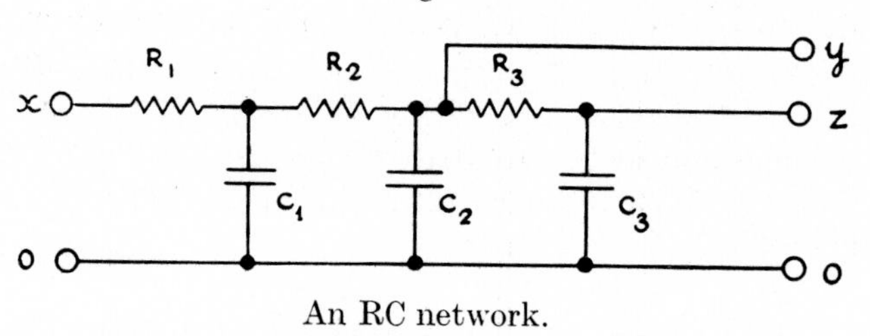

An RC network.

Also the relation between the voltages y and z is

$$(1+b_1D)z=y. \quad . \quad . \quad . \quad . \quad . \quad . \quad . \quad . \quad (34)$$

Equations (32) and (34) correspond to eqns. (25) and (29) respectively. Hence the master phase coordinates are z, z', z''. If $z(t)$ is not accessible for measurement, we measure the copy coordinates. Figure 5 shows a filter suitable for this purpose, the copy coordinates being z_c, z_c', z_c''. It is obvious that if $R_4C_4=R_3C_3$, and if any initial discrepancy between the charges on C_3 and C_4 has had time to disappear, then the coordinates z_c, z_c' and z_c'' will be the same as z, z' and z''.

Fig. 5

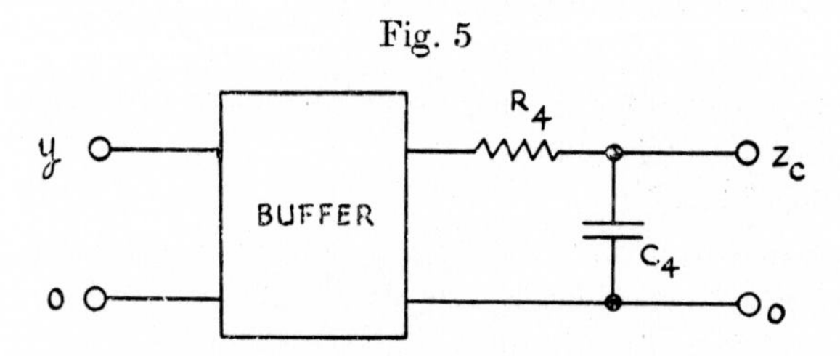

Filter for measuring the phase state of the network of fig. 4.

The above example shows that measurement of copy coordinates virtually involves construction of an analogue of part of the main process. This idea may be extended, in fact the whole main process can be simulated and copy coordinates measured within the simulator. Kalman and Bertram (1958) discuss this and related methods.

§ 11. Phase State of Electrical Networks

Many main processes are analogues of electrical networks (see e.g. Olson (1958), 2nd ed.). In this section we shall show how the phase coordinates can be determined directly as signals appearing within the network, without recourse to linear operations such as differentiation. We consider a linear RLC network, i.e. a network containing only time-invariant resistors, self-inductors and capacitors, together with voltage generators and current generators. The variables which are to be determined are taken to be the voltages (V) across and the currents (I) through the network components.

For the capacitors, the following equations hold (generalized Ohm's law):

$$\frac{dV_{cj}}{dt} = \frac{1}{C_j} I_{cj} \quad (j = 1, 2, \ldots n_c) \quad \ldots \ldots \quad (35)$$

in which the meaning of the symbols is self-evident. Similarly for the inductors:

$$\frac{dI_{Lj}}{dt} = \frac{1}{L_j} V_{Lj} \quad (j = 1, 2, \ldots n_L) \quad \ldots \ldots \quad (36)$$

and for the resistors:

$$I_{Rj} = \frac{1}{R_j} V_{Rj} \quad (j = 1, 2, \ldots n_R). \quad \ldots \ldots \quad (37)$$

Furthermore the currents and voltages must satisfy the constraints imposed by Kirchhoff's laws. For each independent loop there is an algebraic relation of the form

$$\pm V_\alpha \pm V_\beta \pm \ldots = V_A \quad \ldots \ldots \quad (38)$$

where V_α, V_β, etc. are those of V_{cj}, V_{Lj} and V_{Rj} which occur in the loop, and V_A represents the sum of the applied e.m.f's (forcing functions) round the loop. Similarly for each independent node there is an algebraic relation of the form

$$\pm I_\alpha \pm I_\beta \pm \ldots = I_A. \quad \ldots \ldots \quad (39)$$

It follows from relations (37), (38) and (39) that not necessarily all the V's and I's appearing in eqns. (36) and (37) are independent. When the dependent variables (if any) are eliminated, eqns. (36) and (37), which are already nearly in canonical form, reduce to a smaller (or equal) set of canonical equations, having as variables some of V_{cj} and I_{Lj}. Hence we may take this sub-set of variables as the coordinates of the phase space. Thus we have the result that the phase coordinates are voltages across certain capacitors and currents in certain inductors.

In many cases we do not have to take *all* the energy-storage elements as providing phase coordinates. To find a non-redundant set of coordinates we may of course carry out the elimination technique mentioned above; but in complicated networks it is quicker to apply the following simple topological rule. This rule was first proved by Bryant (1959); a shorter but less complete proof was given in the writer's thesis (Fuller 1959). A similar but more complicated rule was given by Bashkow (1957).

Rule

(i) Replace voltage generators by short-circuits and current generators by open-circuits.

(ii) Form a capacitor-only network by open-circuiting all the resistors and inductors. The resulting network will consist of one or more sub-networks, each of which is entirely separate from (i.e. disconnected from) the others. In each such sub-network form a tree†. Take as the voltage phase coordinates the voltages across all the capacitors of all the trees.

(iii) Form an inductor-only network in the network of step (i) by short-circuiting all the resistors and capacitors. Form a tree in each separate resulting sub-network. Take as the current phase coordinates the currents in all the inductors which do not belong to the trees.

Example 1. In the network of fig. 6, application of the rule shows that the five phase coordinates are: the voltage across C_4; any two of the voltages across C_1, C_2 and C_3; the current through L_1; either of the currents through L_2 and L_3.

Fig. 6

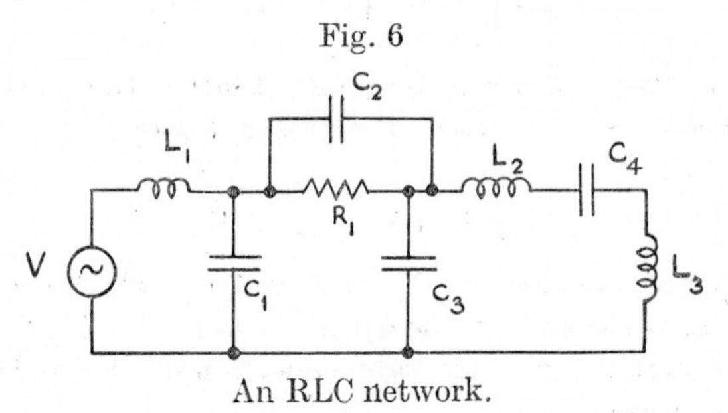

An RLC network.

Example 2. In the network of fig. 4, the three phase coordinates are simply the voltages across C_1, C_2 and C_3. (As a check one may show that z, z' and z'' are linear combinations of the voltages across C_1, C_2 and C_3.)

§ 12. Conclusions

This paper has given an exposition of phase space with emphasis on fundamentals. It is necessary to know these fundamentals in order fully to understand and justify some of the heuristic papers and abstract papers that have been written on optimum non-linear control theory. The principles outlined here will be applied, in a companion paper (Fuller 1960), to a specific theory of optimum control.

† A tree T in a network N is a sub-network of N such that (*a*) T contains all the nodes of N, (*b*) T contains no closed loops, (*c*) T is fully connected (i.e. T does not consist of two or more parts which are entirely separate from one another).

References

Andronov, A. A., 1956, *Collected works* (Moscow: A.N.U.S.S.R.).

Andronov, A. A., and Chaikin, C. E., 1949, *Theory of Oscillations*, English language version, ed. by S. Lefschetz, of Russian edition, Moscow, 1937 (Princeton : University Press).

Ashby, W. R., 1952, *Design for a Brain* (London : Chapman and Hall).

Bashkow, T. R., 1957, *Inst. Radio Engrs, Trans. Circuit Theory, CT*-4, 117.

Bellman, R., 1957, *Dynamic programming* (Princeton : University Press).

Bellman, R., Glicksberg, I., and Gross, O., 1956, *Quart. appl. Math.*, **14,** 11.

Bendixson, I., 1901, *Acta Math.*, **24,** 1.

Birkhoff, G. D., 1927, *Dynamical systems* (New York : Amer. Math. Soc.).

Bogner, I., and Kazda, L. F., 1954, *Trans. Amer. Inst. elect. Engrs*, **73,** 118.

Boltyanskii, V. G., 1958, *Dokl. Akad. Nauk. U.S.S.R.*, **119,** 1070.

Boltyanskii, V. G., Gamkrelidze, R. V., and Pontryagin, L. S., 1956, *Dokl. Akad. Nauk U.S.S.R.*, **110,** 7.

Bryant, P. R., 1959, *Proc. Inst. elect. Engrs*, C, **106,** 174.

Bushaw, D. W., 1953, *E.T.T. Report* 469 (Hoboken, New Jersey : Stevens Inst. of Technology) ; 1958, *Contributions to the Theory of Non-linear Oscillations*, **4,** 29, ed. by S. Lefschetz (Princeton : University Press).

Cauchy, A., 1819, *Bull. Soc. Philo., Paris*, p. 10; 1842, *C.R. Acad. Sci., Paris*, **14,** 1020; **15,** 14.

Cayley, A., 1890, *Collected mathematical papers*, **3,** 156 ; **4,** 513 (Cambridge : University Press).

Desoer, C. A., 1959, *Information and Control*, **2,** 333.

Doll, H. G., and Stout, T. M., 1957, *Trans. Amer. Soc. mech. Engrs*, **79,** 513.

Feldbaum, A. A., 1949, *Avt. i Tel.* (U.S.S.R.), **10,** 249 ; 1953, *Ibid.*, (U.S.S.R.), **14,** 712 ; 1955, *Ibid.*, (U.S.S.R.), **16,** 129.

Flügge-Lotz, I., 1960, *Proc. I.F.A.C. Conf. Moscow*.

Fuller, A. T., 1959 a, Ph.D. Thesis, Cambridge ; 1959 b, *J. Electron. Contr.*, **7,** 456; 1960, *Ibid* (to be published).

Gamkrelidze, R. V., 1958 a, *Dokl. Akad. Nauk. U.S.S.R.*, **123,** 223 ; 1958 b, *Izv. Akad. Nauk. U.S.S.R.*, **22,** 449 ; 1959, *Dokl. Akad. Nauk. U.S.S.R.*, **125,** 475.

Gibbs, J. W., 1902, *Elementary Principles in Statistical Mechanics* (New York : Scribner).

Goursat, E., 1917, *A course in mathematical analysis*, Part II of Vol. II : *Differential equations*. Trans. by E. R. Hedrick and O. Dunkel (Boston : Ginn).

Hagin, E. J., and Fett, G. H., 1957, *Proc. nat. Electron. Conf.* (Chicago), **13,** 537.

Hamilton, W. R., 1835, *Phil. Trans. roy. Soc.*, p. 95.

Hopkin, A. M., 1951, *Trans. Amer. Inst. elect. Engrs*, **70,** 631.

Hopkin, A. M., and Iwama, M., 1956, *Trans. Amer. Inst. elect. Engrs*, **75,** 1.

Hung, J. C., and Chang, S. S. L., 1957, *Inst. Radio Engrs, Nat. Conv. Rec.*, Part 4, 22.

Ivachnenko, A. G., 1948, *Sbornik. Inst. Elektrotechn. A.N.U.S.S.R.*, **4,** 47; 1951, *Ibid.*, **7,** 53.

Kalman, R. E., 1957, *Inst. Radio Eng. WESCON Conv. Rec.*, Part 4, 130.

Kalman, R. E., and Bertram, J. E., 1958, *Trans. Amer. Inst. elect. Engrs*, **77,** 602.

Kalman, R. E., and Koepcke, R. W., 1958, *Trans. Amer. Soc. mech. Engrs*, **80,** 1820.

Ku, Y. H., 1958, *Analysis and Control of Non-linear Systems* (New York : Ronald).

Lagrange, J. L., 1778, *Mécanique analytique*, Part 2.

LANCZOS, C., 1949, *The Variational Principles of Mechanics* (Toronto: University Press).
LASALLE, J. P., 1953, *Goodyear Aircraft Corp. Report* (Akron, Ohio); 1954, *Bull. Amer. math. Soc.*, **60,** 154.
LÉAUTÉ, H., 1885, *J. Éc. polyt., Paris*, **55,** 1; 1891, *Ibid.*, **61,** 1.
LEFSCHETZ, S., 1953, *Proc. Symp. Polytechn. Inst. Brooklyn*, p. 68.
LERNER, A. YA., 1952, *Avt. i Tel.* (U.S.S.R.), **13,** 134, 429.
LIAPUNOV, A. M., 1907, *Ann. Fac. Sci. Toulouse*, **9** (reprinted by Princeton University Press, 1946). French trans. of the Russian version of 1892.
LIE, S., 1888, *Theorie der Transformationsgruppen* (Leipzig).
MACLELLAN, G. D. S., 1957, *Trans. Soc. Instrum. Tech.*, **9,** 62.
McDONALD, D. C., 1950, *Proc. nat. Electron. Conf.* (Chicago), **6,** 400.
MINORSKY, N., 1947, *Introduction to Non-linear Mechanics* (Ann Arbor, Michigan: Edwards).
MOIGNO, F. N. M., 1844, *Leçons de calcul d'après les méthodes de Cauchy*, **2,** 513, 547, (Paris).
NEWTON, I., 1687, *Principia mathematica.*
OLSON, H. F., 1958, *Dynamical Analogies*, 2nd ed. (New York: Van Nostrand).
PASSERA, A. L., and WILLOH, R. G., 1956, *N.A.C.A. Report* TN 3743.
POINCARÉ, H., 1881, *J. Math. pures appl.*, **7,** 375; 1882, *Ibid.*, **8,** 251. (Also in his *Oeuvres*, **1**, 3–222.)
PONTRYAGIN, L. S., 1956, *Session of A.N.U.S.S.R. on fundamental problems of automatic controls*, **2,** 107 (Moscow: A.N.U.S.S.R., 1957), 1959, *Uspeki Mat. Nauk.*, **14,** 3. (Also in *Proc. Internat. Congr. Math.*, Edinburgh, 1958, published by Cambridge University Press, 1960).
ROSE, N. J., 1953, *E.T.T. Report* 459 (Hoboken, New Jersey: Stevens Inst. of Technology); 1956, *Inst. Radio Engrs, Conv. Rec.*, p. 61.
TSIEN, H. S., 1954, *Engineering Cybernetics* (New York: McGraw-Hill).
TSYPKIN, YA. Z., 1955, *Theory of Relay Systems of Automatic Control* (Moscow). German trans. by W. Hahn and R. Herschel, 1958 (Munich: Oldenbourg).
UTTLEY, A. M., and HAMMOND, P. H., 1952, *Automatic and Manual Control*, ed. by A. Tustin, p. 285 (London: Butterworths).
WEST, J. C., 1952, *Automatic and Manual Control*, ed. by A. Tustin, p. 300 (London: Butterworths).
WILLIAMS, A. J., 1938, *Trans. Amer. Inst. elect. Engrs*, **57,** 565.
WHITTAKER, E. T., 1937, *Analytical Dynamics*, 4th ed. (Cambridge: University Press).

Optimization of Non-linear Control Systems with Transient Inputs†

By A. T. Fuller
Department of Engineering, University of Cambridge

[Received April 26, 1960]

ABSTRACT

The nature of the optimum controller is investigated for control systems which are subject to saturation. It is shown that for a wide class of performance criteria the optimum controller simply generates an instantaneous non-linear function of the input phase coordinates and of the output phase coordinates. For relay control systems, the optimum controller is characterized by a switching surface in the phase space. For special cases, the dimensions of the phase space can be reduced in number by using error coordinates. These results systematize and generalize several known results, and explain the starting points of some recent abstract papers.

§ 1. Introduction

Many papers have been written using phase space in the optimization of non-linear control systems (see Fuller 1960 b for a bibliography). These papers fall roughly into two classes. On the one hand there are papers written by engineers, in which phase space is used heuristically and tentatively; on the other hand there are papers written by mathematicians, which take phase space as an arbitrary and abstract starting point. Consequently the essential property of phase space, the property which makes phase space relevant to the optimization problem, has been obscured. In a previous paper (Fuller 1960 b) the writer attempted to remove this obscurity by giving a systematic exposition of phase space. The present paper continues the argument by applying the phase space concepts of the previous paper to the problem of control optimization.

It is now well known that when saturation exists in a control system, the optimum controller is non-linear. What is the *nature* of the non-linearity, or in other words, to which class of non-linear filters does the controller belong? This is the question we shall be concerned with. It can be answered very simply, without involving a discussion of the sometimes difficult problem of the exact numerical or algebraic specification of the optimum controller.

---

† Communicated by J. F. Coales.

§ 2. Formulation of the Problem

The block diagram of the system to be considered is shown in the figure. The justification of this block diagram has been given in the previous paper (Fuller 1960 b).

2.1. *Main Process*

The main process (i.e. the plant) is assumed to be a specified *determinate* process with phase coordinates $Y_1, Y_2, \ldots Y_n$. As defined by Fuller (1960 b), this means that if the present values of $Y_1, Y_2, \ldots Y_n$ (which vary with time t) are known, and if the future of the input $L(t)$ to the main process is known, then the future of the output $Y(t)$ of the main process is completely determined. For example if the main process is represented by the following *canonical* set of differential equations

$$\frac{dY_i}{dt} = f_i(Y_1, Y_2, \ldots Y_n, L) \quad (i=1, 2, \ldots n) \quad . \quad . \quad . \quad (1)$$

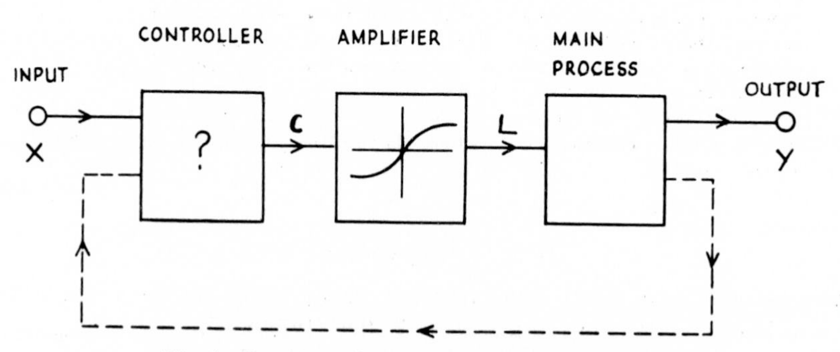

Block diagram of saturating control system.
Note : The dashed line represents possibly multiple connections from the main process to the controller.

where the f_i are piece-wise continuous functions with piece-wise continuous first partial derivatives with respect to all the arguments, and if

$$Y_1 \equiv Y, \quad . \quad . \quad . \quad . \quad . \quad . \quad . \quad . \quad . \quad (2)$$

then $Y_1, Y_2, \ldots Y_n$ are indeed a relevant set of phase coordinates (Fuller 1960 b). Again, the main process may have more complicated non-linearity than that allowed by eqns. (1), e.g. it may have backlash, or hysteresis (in such cases some ingenuity may be called for in deciding what are the relevant phase coordinates $Y_1, Y_2, \ldots Y_n$). Questions of the physical meaning and measurement of the phase coordinates have already been discussed by Fuller (1960 b).

2.2. *Amplifier*

The amplifier, which drives the main process, is assumed to be completely specified by its given input–output characteristic

$$L = g(C), \quad . \quad . \quad . \quad . \quad . \quad . \quad . \quad . \quad . \quad (3)$$

i.e. the amplifier is free from time-lags, hysteresis, etc. $g(C)$ is assumed to be a single-valued monotonic function with a finite positive maximum and a finite negative minimum. This saturation characteristic obviously restricts the control of the main process and implies that the optimum systems we shall discuss are not trivial in the sense of being perfect. If, in a practical case, the amplifier contains lags, it is often possible to regard these lags as belonging to the main process, and then our ensuing argument still applies.

2.3. *Controller*

The controller, which drives the amplifier, is assumed to be physically realizable (i.e. operates only on the present and past of available signals), and is otherwise unrestricted. It is to be designed so as to minimize one of the performance criteria discussed later.

2.4. *Input*

The control system input, $X(t)$, is assumed to be a generalized transient input applied at $t=0$. More precisely, $X(t)$, $t \geqslant 0$, is assumed to be a determinate input with phase coordinates, $X_1, X_2, \ldots X_m$, which coordinates are directly available for measurement. For example, if $X(t)$ is a step plus a ramp, it satisfies

$$\frac{d^2X}{dt^2}=0, \qquad (4)$$

which corresponds to the canonical equations

$$\left.\begin{aligned} \frac{dX}{dt} &= X', \\ \frac{dX'}{dt} &= 0. \end{aligned}\right\} \qquad (5)$$

Equations (5) show that $X(t)$ and $X'(t)$ are the phase coordinates. Kalman and Koepcke (1958) suggested that a general class of transient inputs can be defined, in a similar way, by a general linear differential equation with constant coefficients. This idea can be taken a little further by assuming that X satifies the following general canonical set of differential equations

$$\frac{dX_i}{dt}=h_i(X_1, X_2, \ldots X_m) \quad (i=1, 2, \ldots m) \qquad (6)$$

where

$$X_1 \equiv X, \qquad (7)$$

and where the h_i satisfy continuity restrictions of the type mentioned above for eqns. (1). The phase coordinates $X_1, X_2, \ldots X_m$ of eqns. (6) thus define a further example of the generalized transient input defined above.

The input is assumed to be undefined for $t<0$. This implies that $X(t)$, $t<0$, has no known relation with $X(t)$, $t \geqslant 0$, and thus that the controller, being physically realizable, has no advance knowledge of the impending transient.

2.5. *Performance Criterion*

The error $E(t)$ of the control system is defined by

$$E \equiv X - Y. \qquad (8)$$

A typical and well-known performance criterion is

$$\int_0^\infty E^2 \, dt.$$

A more general criterion is M_ν where

$$M_\nu \equiv \int_0^\infty |E|^\nu \, dt \quad (\nu \geqslant 0). \qquad (9)$$

It has been shown (Fuller 1959 b), that the criterion M_0 (suitably defined) is usually equivalent to the well-known settling time criterion. Criteria (9) can be further generalized to

$$\int_0^\infty q(E) \, dt, \qquad (10)$$

where $q(E)$ is any single-valued function for which the integral exists. Finally, great generality can be obtained by assuming (compare Pontryagin 1959) a performance criterion Q where

$$Q \equiv \int_0^\infty q(X_1, X_2, \dots X_m, Y_1, Y_2, \dots Y_n) \, dt. \qquad (11)$$

The criterion Q allows us, if we wish, to penalize excess values of the output coordinates (e.g. output speed), and also to attach more weight to certain inputs. Moreover criterion Q includes criteria (9) and (10) as special cases.

2.6. *The Problem*

The control optimization problem we shall consider is to find the controller which minimizes some given performance criterion Q as defined by eqn. (11). (It is assumed that the criterion chosen can exist when a suitable controller is incorporated in the system.) This formulation of the problem could still be made much more general, e.g. by considering main processes with multiple inputs ; however such generality might make the ensuing argument lose clarity.

§ 3. Nature of the Optimum Controller

Since the controller is physically realizable, its output $C(t)$ can in general be any non-linear or linear operation, however complicated, on the pasts of all the available signals ; i.e. $C(t)$ is a general *functional* of these signals.

We shall now show that when the controller is optimum in the sense of § 2, $C(t)$ simplifies to a *function* of the present values of the input phase coordinates and the output phase coordinates. (We use the term 'function' here in its legitimate mathematical sense; in engineering the same meaning is sometimes reserved for the term 'instantaneous function'.) This result is of great practical and theoretical importance, for the following reasons. The practical synthesis of instantaneous functions, though difficult, is much simpler than the synthesis of functionals. Moreover, the empirical synthesis of the optimum controller on an analogue computer becomes feasible in simple cases; and in more complicated cases it is simple to envisage self-optimizing processes for finding the optimum function (see Fuller 1959 a). The theoretical importance of the result is that it explains and justifies the starting point of many abstract papers which derive expressions for optimum controllers. This starting point is, in fact, the *assumption* that the optimum controller must generate simply a function of the phase coordinates (see e.g. Bushaw 1953, Pontryagin 1956).

The proof of the result is basically as follows (see the Appendix for a precise proof). From the definition of phase coordinates (Fuller 1960 b) it follows that $q(t)$ for $t \geqslant 0$ is completely determined by (i) the initial values $X_1(0), X_2(0), \ldots X_m(0)$ of the input phase coordinates; (ii) the initial values $Y_1(0), Y_2(0), \ldots Y_n(0)$ of the output phase coordinates; (iii) the known constants (i.e. non-time-varying parameters) of the input, of the main process, and of the performance criterion; (iv) the input to the main process, $L(t)$, for $t \geqslant 0$. We assume that there exists† an optimum $L(t)$, i.e. an $L(t)$ for which Q is not greater than the Q obtained with any other $L(t)$. Although this optimum $L(t)$ is not necessarily unique, for linguistic convenience we shall refer to it as *the* optimum $L(t)$, and designate it $L_{\text{opt}}(t)$. Items (i), (ii) and (iii) constitute all the relevant given data of the problem, and item (iv) is to be chosen on the basis of this data, i.e.

$$\left.\begin{aligned} L_{\text{opt}}(t) = F[X_1(0), X_2(0), \ldots X_m(0), Y_1(0), Y_2(0), \ldots Y_n(0), t] \\ (t \geqslant 0). \end{aligned}\right\} \qquad (12)$$

Here F is some unknown function of arguments (i), (ii) (iii) and time. We have not written arguments (iii) explicitly in (12) as they are not needed for present purposes.

Consider the system at time $t = t_1 > 0$, assuming that L has been optimum for $0 \leqslant t < t_1$. The performance criterion may be written

$$Q = \int_0^{t_1} q\,dt + \int_{t_1}^{\infty} q\,dt. \qquad \ldots\ldots\ldots \qquad (13)$$

† We assume that the optimum $L(t)$ exists in some well-defined general mathematical sense, e.g. $L(t)$ may be restricted to piece-wise continuous functions, or to measurable functions. However, for our purposes it is not necessary to specify any particular class of functions to which $L(t)$ must belong.

At time t_1 the first integral is completely determined and will be unchanged whatever the subsequent choice of L. Therefore at time t_1 the problem is to choose $L(t)$ for $t \geqslant t_1$ so that

$$Q_1 \equiv \int_{t_1}^{\infty} q(t)\,dt$$

is minimized. The data from which this choice is to be made can be taken as the new values of the phase coordinates, namely

$$X_1(t_1), X_2(t_1), \ldots X_m(t_1), Y_1(t_1), Y_2(t_1), \ldots Y_n(t_1).$$

Thus the problem is now the same as that with which we started† at $t=0$, except that we are now starting at a new time origin t_1. Hence the solution is given by the formula of (12) if in the right side we replace t by $t-t_1$ and $X_1(0), \ldots X_m(0), Y_1(0), \ldots Y_n(0)$ by

$$X_1(t_1), \ldots X_m(t_1), Y_1(t_1), \ldots Y_n(t_1)$$

respectively, i.e.

$$\left.\begin{aligned} L_{\mathrm{opt}}(t) = F[X_1(t_1), X_2(t_1), \ldots X_m(t_1), Y_1(t_1), Y_2(t_1), \ldots Y_n(t_1), (t-t_1)] \\ (t \geqslant t_1). \end{aligned}\right\} \quad (14)$$

In eqn. (14) put $t=t_1$:

$$L_{\mathrm{opt}}(t_1) = F[X_1(t_1), X_2(t_1), \ldots X_m(t_1), Y_1(t_1), Y_2(t_1), \ldots Y_n(t_1), \quad 0] \quad (15)$$

$$= G[X_1(t_1), X_2(t_1), \ldots X_m(t_1), Y_1(t_1), Y_2(t_1), \ldots Y_n(t_1)], \text{ say.} \quad (16)$$

Since this equation holds for all $t_1 \geqslant 0$, we can drop the suffix and write

$$L_{\mathrm{opt}}(t) = G[X_1(t), X_2(t), \ldots X_m(t), Y_1(t), Y_2(t), \ldots Y_n(t)], \quad . \quad . \quad (17)$$

or, more briefly,

$$L_{\mathrm{opt}} = G[X_1, X_2, \ldots X_m, Y_1, Y_2, \ldots Y_n]. \quad . \quad . \quad . \quad (18)$$

This is the required result, stating that the optimum control input to the main process is an instantaneous non-linear function of the input phase coordinates and the output phase coordinates.

§ 4. Relay Control Systems

It has been shown that in many circumstances $L_{\mathrm{opt}}(t)$ takes only the maximum and minimum values allowed by the saturation‡ (see Feldbaum 1953, 1955, Bellman *et al.* 1956, Pontryagin 1956, 1959, Desoer 1959, Lasalle 1960). In other words $L_{\mathrm{opt}}(t)$ is a discontinuous function of time, jumping alternately from one extreme of L to the other. This type of operation can be achieved by inserting a high-gain amplifier in front of the saturating amplifier, or by using a relay as the saturating amplifier. In this section for definiteness, we shall confine attention to the *ideal relay control system* in which

$$L = \operatorname{sgn} C. \quad . \quad . \quad . \quad . \quad . \quad . \quad . \quad . \quad . \quad (19)$$

---

† This is an application of Bellman's (1957) 'principle of optimality' which states that the continuation of an optimal policy is optimal.

‡ This holds except possibly during a certain end-regime which we shall discuss below in § 6.

It follows that, for such a system, L_{opt} as given by eqn. (18) is a discontinuous function of the phase coordinates $X_1, X_2, \dots X_m, Y_1, Y_2, \dots Y_n$. For a given value of each of the latter coordinates, L_{opt} is either $+1$ or -1. If L_{opt} is $+1$, the corresponding point in the phase space is termed a *P-point*. Similarly if L_{opt} is -1, the point is termed an *N-point*. The whole phase space† consists of regions of P-points and regions of N-points. The boundary between these regions constitutes a hypersurface‡. When, the representative point in the phase space travels through the hypersurface, the relay switches over, and consequently the hypersurface is called a *switching surface*. We thus have the result that *for an ideal relay control system with transient input X and performance criterion Q, the optimum controller is characterized by a switching surface in* $(X_1, X_2, \dots X_m, Y_1, Y_2, \dots Y_n)$ *phase space.* This result is well known in various special cases (notably when Q is the settling time criterion, see e.g. Rose 1956), but in many papers it tends to be obscured by the technical detail of calculating formulae for the switching surface.

If the hypersurface is represented by the equation

$$C(X_1, X_2, \dots X_m, Y_1, Y_2, \dots Y_n) = 0 \quad . \quad . \quad . \quad . \quad (20)$$

where C is some known function, then C is positive on one side of the hypersurface and negative on the other. The side on which C is positive can be made to correspond to the region of P-points (if necessary by multiplying throughout eqn. (20) by (-1) and taking the resulting left side of the equation as a new C). Therefore a controller which generates the instantaneous non-linear function C of the phase space coordinates will give the correct output for optimum relay switching. C is termed the *switching function*§.

The above argument indicates clearly the nature of the optimum controller. One or two practical matters however require further comment.

(*a*) If the representative point is precisely on the switching surface, C as given by eqn. (20) is exactly zero. The relay output is then indeterminate, or equals zero if we interpret sgn (0) as zero, and is thus non-optimum. In practice this difficulty does not matter, because owing to instrumental inaccuracies, small parasitic delays, etc., the representative point cannot remain in the switching surface for a finite time (see Feldbaum 1955).

† Strictly, the whole of the region in which L_{opt} can exist. (In some control systems with unstable main processes, there are regions of phase space in which the error $E(t)$ always diverges whatever $L(t)$ is chosen, so that in these regions L_{opt} does not exist.)

‡ A hypersurface is the analogue in general-dimensional space of an ordinary surface in three-dimensional space.

§ The switching function C representing any one switching surface is not unique. In fact there are infinitely many such switching functions. These functions have the same sign at any given point of the phase space, but that is all they have in common. For example, there is a well-known switching function in (E, E') phase space, namely $E + \frac{1}{2}a_2 E'|E'|$ (compare eqn. (48)); and the same switching surface is obtained with the well-known SERME switching function of West, namely $\text{sgn}(E)\sqrt{(|E|)} + \sqrt{(\frac{1}{2}a_2)}E'$.

(*b*) If the settling time performance criterion is used, the error $E(t)$ eventually reaches a régime in which it is always precisely zero. This régime often corresponds to a state of equilibrium of the main process, which is preserved if we make $L=0$ (using a relay with an infinitesimal dead-zone). Thus in this régime we need not make $L=\pm 1$. Even with other performance criteria there is usually an end-régime in which $E=0$ and in which L_{opt} can be theoretically other than ± 1 (for some examples see Fuller 1960 a). However, we can still make $L=\pm 1$ during the end-régime by postulating a small (infinitesimal) delay in the system which will make L switch at infinite frequency, causing E to oscillate infinitesimally about zero (compare Feldbaum 1955).

§ 5. Error Coordinates

In view of the relation

$$E=X_1-Y_1 \qquad (21)$$

we may use E as a phase coordinate instead of X_1. More generally if we write

$$E_i \equiv X_i-Y_i \qquad (22)$$

we can adopt E_i as a phase coordinate in place of X_i. (In general any one–one transformation of a complete set of phase coordinates results in a new complete set of phase coordinates (see Fuller 1960 b)). It is theoretically advantageous to make this transformation of coordinates when both the main process and the 'input generator' include pure integrators, as we shall now show. We term the E_i *error coordinates.*

We assume that the input X is generated by integrating r times a signal X_{r+1} which is itself generated by a determinate system with phase coordinates $X_{r+1}, X_{r+2}, \ldots X_m$. Thus

$$\frac{d^rX}{dt^r}=X_{r+1}. \qquad (23)$$

Equation (23) may be written as the following set of first-order equations:

$$\left.\begin{aligned} \frac{dX_1}{dt} &= X_2, \\ \frac{dX_2}{dt} &= X_3, \\ &\cdots\cdot \\ \frac{dX_r}{dt} &= X_{r+1}, \end{aligned}\right\} \qquad (24)$$

where

$$X_1 \equiv X,\ X_2 \equiv X', \ldots X_r \equiv X^{(r-1)}. \qquad (25)$$

Equations (24) are in canonical form, and so represent a system with phase coordinates $X_1, X_2, \ldots X_r$ and forcing function $X_{r+1}(t)$. The latter signal has phase coordinates $X_{r+1}, X_{r+2}, \ldots X_m$ by definition, hence the complete set of phase coordinates of the signal X is $X_1, X_2, \ldots X_m$ (agreeing with our previous notation). Note that if $X_{r+1}=0$, it follows from eqn. (23) that X is of the polynomial form

$$X = k_0 + k_1 t + \ldots + k_{r-1} t^{r-1} \quad (t \geqslant 0), \qquad (26)$$

so that X is then an ordinary transient input of the type often discussed in control theory.

Similarly we assume that the main process consists of a determinate system followed by r integrators, i.e.

$$\frac{d^r Y}{dt^r} = Y_{r+1} \qquad (27)$$

where Y_{r+1} is the output of a system with phase coordinates $Y_{r+1}, Y_{r+2}, \ldots Y_n$. Equation (27) may be written in canonical form as

$$\left.\begin{aligned} \frac{dY_1}{dt} &= Y_2, \\ \frac{dY_2}{dt} &= Y_3, \\ &\ldots \\ \frac{dY_r}{dt} &= Y_{r+1}, \end{aligned}\right\} \qquad (28)$$

where

$$Y_1 \equiv Y, \quad Y_2 \equiv Y', \quad \ldots \quad Y_r = Y^{(r-1)}. \qquad (29)$$

It follows as above that the complete set of phase coordinates of the signal Y are $Y_1, Y_2, \ldots Y_n$ (again agreeing with our previous notation).

We also assume throughout the present section that the performance criterion is of the special type (10), i.e. depends only on E.

We shall show that $E, E', \ldots E^{(r-1)}, X_{r+1}, X_{r+2}, \ldots X_m, Y_{r+1}, Y_{r+2}, \ldots Y_n$ constitute a complete set of phase coordinates for E. Subtract eqns. (28) from eqns. (24) and substitute from eqns. (22), for $i = 1, 2, \ldots r$. The result is

$$\left.\begin{aligned} \frac{dE_1}{dt} &= E_2, \\ \frac{dE_2}{dt} &= E_3, \\ &\ldots \\ \frac{dE_r}{dt} &= X_{r+1} - Y_{r+1}, \end{aligned}\right\} \qquad (30)$$

where

$$E_1 \equiv E, \quad E_2 \equiv E', \ldots, \quad E_r \equiv E^{(r-1)}. \qquad (31)$$

Equations (30) are in canonical form, with phase coordinates $E_1, E_2, \ldots E$ and forcing functions $X_{r+1}(t)$ and $Y_{r+1}(t)$. The phase coordinates of the latter signals are by definition $X_{r+1}, X_{r+2}, \dots X_m$ and $Y_{r+1}, Y_{r+2}, \dots Y_n$. Hence the complete set of phase coordinates for the determination of E_1 is

$$E_1, E_2, \dots E_r, X_{r+1}, X_{r+2}, \dots X_m, Y_{r+1}, Y_{r+2}, \dots Y_n. \quad . \quad (32)$$

Substitution from eqns. (31) now shows that the complete set of phase coordinates for the determination of E is

$$E, E', \dots E^{(r-1)}, X_{r+1}, X_{r+2}, \dots X_m, Y_{r+1}, Y_{r+2}, \dots Y_n. \quad . \quad (33)$$

On applying the argument of § 3 we see that the optimum controller must be an instantaneous function of coordinates (33). These coordinates are $(m+n-r)$ in number, whereas the combined input and output coordinates are $(m+n)$ in number. Hence we have the result:
When the input generating system and the main process each include r pure integrators, by the use of error coordinates the optimum controller can be expressed as an instantaneous function of only $(m+n-r)$ arguments; moreover, for relay control systems the dimensions of the phase space in which the optimum switching surface lies are reduced in number from $(m+n)$ to $(m+n-r)$.

This result can easily be generalized as follows. Let the r integrators be replaced by a general filter described by an rth order canonical set of linear differential equations with constant coefficients (the filter in the input generating system must be the same as that in the main process). Then coordinates (32) are all that are necessary for specification of the optimum controller. However this generalization is of little practical interest as it describes a rather coincidental situation.

In the particular case when the input X is a simple transient of the form (26) the following stronger statement can be made. *When the main process includes r pure integrators, let the optimum controller for zero system input ($X=0$) be represented by*

$$C = R(Y, Y', \dots Y^{(r-1)}, Y_{r+1}, Y_{r+2}, \dots Y_n). \quad . \quad . \quad . \quad (34)$$

Then for an input

$$X = k_0 + k_1 t + \dots + k_{r-1} t^{r-1} \quad (k\text{'s constant}, t \geqslant 0) \quad . \quad . \quad . \quad (35)$$

the optimum controller is represented by

$$C = R(-E, -E', \dots -E^{(r-1)}, Y_{r+1}, Y_{r+2}, \dots Y_n). \quad . \quad . \quad (36)$$

Proof: when $X=0$ the system is described by eqns. (28) and the performance criterion is

$$\int_0^\infty q(-Y_1)\, dt.$$

When X is given by eqn. (35), the system is described by eqns. (30) with $X_{r+1}=0$, and the performance criterion is

$$\int_0^\infty q(E_1)\, dt.$$

Comparison of these two cases thus shows that the second is the same as the first with $-E_i$ written for Y_i $(i=1,2,\ldots r)$. Hence all mathematical properties of the second case are the same as those of the first case, with $-E_i$ written for Y_i, that is with $-E^{(i-1)}$ written for $Y^{(i-1)}$ $(i=1,2,\ldots r)$. This proves result (36).

Various special cases of the above results are well known; see for example Feldbaum (1955), Kalman and Bertram (1958).

§ 6. Examples

In this section we shall give various examples of optimum relay control systems. We wish to bring out the qualitative nature of the optimum controller; however for interest we shall also occasionally quote the quantitative expression for the switching function.

Example 1. Let the input X be a step input and let the main process satisfy the simple linear differential equation

$$(a_0+a_1D+a_2D^2+a_3D^3)Y=L \quad \left(D\equiv\frac{d}{dt}\right) \quad . \quad . \quad . \quad . \quad (37)$$

where the a's are constants. Then the only phase coordinate of the input is $X_1=X$, and the phase coordinates of the output are Y, Y', Y'' (see Fuller 1960 b). Hence the optimum switching function is of the form $C(X, Y, Y', Y'')$.

Example 2. Let the input X satisfy

$$X=k_0+k_1t+\ldots+k_{m-1}t^{m-1} \quad . \quad . \quad . \quad . \quad . \quad . \quad (38)$$

where the k's are constants, and let the main process have a transfer function

$$\frac{b_0+b_1p+\ldots+b_lp^l}{a_0+a_1p+\ldots+a_np^n}. \quad . \quad . \quad . \quad . \quad . \quad . \quad (39)$$

Then the input phase coordinates are $X, X', \ldots X^{(m-1)}$ and the output phase coordinates are represented by (Fuller 1960 c):

$$\frac{Y}{b_0+b_1D+\ldots+b_lD^l}, \frac{DY}{b_0+b_1D+\ldots+b_lD^l}, \ldots \frac{D^{n-1}Y}{b_0+b_1D+\ldots+b_lD^l}. \quad (40)$$

Hence the optimum switching function is

$$C\left(X, \ldots D^{m-1}X, \frac{Y}{b_0+\ldots+b_lD^l}, \ldots \frac{D^{n-1}Y}{b_0+\ldots+b_lD^l}\right). \quad . \quad (41)$$

In principle the coordinates (40) can be measured directly within the main process (Fuller 1960 b). Alternatively it is sometimes feasible to measure instead the *copy coordinates* (Fuller 1960 b) obtained by feeding Y into filters having transfer functions corresponding to the operators in (40). In such cases the controller output $C(t)$ is a *functional of* $X(t)$ *and* $Y(t)$, though it is still of course an *instantaneous function of the phase coordinates.*

Example 3. Let the input X be

$$X=A\exp(-\alpha t) \quad (t\geqslant 0), \quad . \quad . \quad . \quad . \quad . \quad . \quad (42)$$

where α is a known constant and A is an unknown constant. Let the main process satisfy

$$Y^{(n)}=f(Y, Y', \dots Y^{(n-1)}, L) \quad \dots \dots \quad (43)$$

where f is continuous and with continuous first partial derivatives. Then X satisfies the differential equation

$$(\alpha+D)X=0, \quad \dots \dots \quad (44)$$

hence the input phase coordinate is simply X. Also the output phase coordinates are (Fuller 1960 b) $Y, Y', \dots Y^{(n-1)}$. Therefore the optimum switching function is of the form $C(X, Y, Y', \dots Y^{(n-1)})$.

Example 4. Let the input be zero, and let the main process have a transfer function $1/a_2p^2$. This is a special case of Examples 1 and 2. The optimum switching function is of the form $C(Y, Y')$.

If the performance criterion is settling time, it is well known that C is given by

$$C(Y, Y')=-(Y+\tfrac{1}{2}a_2Y'|Y'|). \quad \dots \quad (45)$$

If, on the other hand, the performance criterion is

$$M_2\equiv\int_0^\infty E^2\,dt \quad \dots \dots \quad (46)$$

the optimum C is given by (Fuller 1960 a):

$$C(Y, Y')=-\left[Y+\tfrac{1}{2}\left(\frac{(33)^{1/2}-1}{6}\right)^{1/2} a_2Y'|Y'|\right]. \quad \dots \quad (47)$$

Example 5. Let the problem be the same as in Example 4 except that the input is now a step plus a ramp. Using error coordinates we can say immediately that the optimum switching function is of the form $C(E, E')$. Moreover it follows from the results of Example 4 that for minimum settling time

$$C(E, E')=E+\tfrac{1}{2}a_2E'|E'|, \quad \dots \dots \quad (48)$$

and for minimum M_2

$$C(E, E')=E+\tfrac{1}{2}\left(\frac{(33)^{1/2}-1}{6}\right)^{1/2} a_2E'|E'|. \quad \dots \quad (49)$$

Example 6. Let the input be a step plus a ramp plus a parabola. Let the main process have a transfer function $1/a_3p^3$. Then the optimum switching function is of the form $C(E, E', E'')$.

Feldbaum (1955) has shown that if the performance criterion is settling time, C is given by

$$C=E+\tfrac{1}{3}a_3{}^2(E'')^3+[\operatorname{sgn} B][a_3E'E''+a_3{}^{1/2}\{\tfrac{1}{2}a_3(E'')^2+E'\operatorname{sgn} B\}^{3/2}] \quad (50)$$

where

$$B=E'+\tfrac{1}{2}a_3E''|E''|. \quad \dots \dots \quad (51)$$

§ 7. Generalizations

The results in §§ 3 and 4 can be readily generalized in various directions.

If there are several controllable inputs L_i to the main process, each will require a separate controller, and the principles of § 3 will apply to each controller output C_i. For such a *multi-loop* system, when optimum, each C_i will be an instantaneous function of the phase coordinates of X and the phase coordinates of the main process. Bellman *et al* (1956), and Pontryagin (1956, 1959), among others, have investigated optimum multi-loop systems.

If there are several unwanted transient disturbances, i.e. *load variations*, acting on the main process, each C_i will be an instantaneous function of the phase coordinates of X, the phase coordinates of the main process, and the phase coordinates of each disturbance.

If the main process is time-variant, time will be one of the phase coordinates of the main process, and the above results will then be immediately applicable.

If the performance criterion is time weighted, time must be taken as an additional coordinate of the phase space (if time does not already appear as such a coordinate).

In the writer's thesis (Fuller 1959 a), generalizations to cases of random inputs have been treated. These will be the subject of a subsequent paper (Fuller 1960 c).

§ 8. Conclusions

The central result of the present paper is that the optimum controller is an instantaneous function of the phase coordinates. Hitherto this result has been assumed explicitly or implicitly, conjectured, asserted, or proved, in special cases; but has apparently not been presented with the clarity and generality it deserves.

It can be argued that this result provides us with 'design in the large', and that the details of calculation of the optimum C_i can be left to analogue or digital computers, using, say, the techniques of dynamic programming (Bellman 1957, Kalman and Koepcke 1958), or perhaps self-optimizing techniques.

APPENDIX

Nature of the Optimum Controller

In this appendix a precise proof is given of the result that the optimum controller generates simply an instantaneous function of the phase coordinates.

The definition of the phase coordinates implies that

$$q(t) = U\left[X_1(0), \ldots X_m(0), Y_1(0), \ldots Y_n(0), t, \begin{pmatrix} L(u) \text{ for all} \\ 0 \leqslant u < t \end{pmatrix}\right], \tag{52}$$

i.e. $q(t)$ is some unique function of the past of $L(t)$, which functional depends only on the arguments written explicitly in (52) (and on known constants). Thus the problem is to choose $L(t)$ for all $t \geqslant 0$ so that

$$Q = \int_0^\infty U\left[X_1(0), \ldots X_m(0), Y_1(0), \ldots Y_n(0), t. \begin{pmatrix} L(u) \text{ for all} \\ 0 \leqslant u < t \end{pmatrix}\right] dt \quad (53)$$

is minimized. It follows that the optimum $L(t)$ depends only on $X_1(0), \ldots X_m(0), Y_1(0), \ldots Y_n(0)$ and t:

$$L_{\text{opt}}(t) = F[X_1(0), \ldots X_m(0), Y_1(0), \ldots Y_n(0), t] \quad (t \geqslant 0). \quad . \quad (54)$$

Suppose

$$L(t) = L_{\text{opt}}(t) \text{ for all } 0 \leqslant t < t_1. \quad . \quad . \quad . \quad . \quad . \quad (55)$$

Then the problem reduces to minimizing

$$Q_1 \equiv \int_{t_1}^\infty q(t)\, dt. \quad . \quad . \quad . \quad . \quad . \quad . \quad . \quad . \quad (56)$$

Put

$$\tau = t - t_1. \quad . \quad . \quad . \quad . \quad . \quad . \quad . \quad . \quad . \quad (57)$$

Then

$$Q_1 = \int_0^\infty q(t_1 + \tau)\, d\tau. \quad . \quad . \quad . \quad . \quad . \quad . \quad . \quad . \quad (58)$$

The definition of the phase coordinates implies that

$$q(t_1 + \tau) = U\left[X_1(t_1), \ldots X_m(t_1), Y_1(t_1), \ldots Y_n(t_1), \tau, \begin{pmatrix} L\{u\} \text{ for all} \\ 0 \leqslant u < \tau \end{pmatrix}\right] \quad (59)$$

where

$$L\{u\} \equiv L(t_1 + u). \quad . \quad . \quad . \quad . \quad . \quad . \quad . \quad . \quad (60)$$

From (58) and (59) the problem is now to choose $L\{\tau\}$ for all $\tau \geqslant 0$ so that

$$Q_1 = \int_0^\infty U\left[X_1(t_1), \ldots X_m(t_1), Y_1(t_1), \ldots Y_n(t_1), \tau, \begin{pmatrix} L\{u\} \text{ for all} \\ 0 \leqslant u < \tau \end{pmatrix}\right] d\tau \quad (61)$$

is minimized. Comparing (61) and (53) we see that the problem is the same as before except that $X_i(t_1)$ is written for $X_i(0)$, $Y_i(t_1)$ for $Y_i(0)$, τ for t, and $L\{u\}$ for $L(u)$. Hence the solution is the same as in eqn. (54), except for the latter modifications, i.e.

$$L_{\text{opt}}\{\tau\} = F[X_1(t_1), \ldots X_m(t_1), Y_1(t_1), \ldots Y_n(t_1), \tau], (\tau \geqslant 0). \quad (62)$$

From eqns. (57), (60) and (62)

$$L_{\text{opt}}(t) = F[X_1(t_1), \ldots X_m(t_1), Y_1(t_1), \ldots Y_n(t_1), (t - t_1)], (t \geqslant t_1). \quad (63)$$

Equation (63) is the same as eqn. (14). The proof now follows through eqns. (15) to (18).

REFERENCES

BELLMAN, R., GLICKSBERG, I., and GROSS, O., 1956, *Quart. appl. Math.*, **14,** 11.

BELLMAN, R., 1957, *Dynamic programming* (Princeton : University Press).

BUSHAW, D. W., 1953, *E.T.T. Report* 469 (Hoboken, New Jersey : Stevens Institute of Technology).

DESOER, C. A., 1959, *Information and Control*, **2,** 333.

FELDBAUM, A. A., 1953, *Avt. i Tel.*, **14,** 712 ; 1955, *Ibid.*, **16,** 129.

FULLER, A. T., 1959 a, Ph.D. Thesis, Cambridge ; 1959 b, *J. Electron. Contr.*, **7,** 456 ; 1960 a, *Proc. I.F.A.C. Conf.*, Moscow ; 1960 b, *J. Electron. Contr.*, **8,** 381; 1960 c, *Ibid.* (in the press).

KALMAN, R. E., and BERTRAM, J. E., 1958, *Trans. Amer. Inst. elect. Engrs*, **77,** 602.

KALMAN, R. E., and KOEPCKE, R. W., 1958, *Trans. Amer. Soc. mech. Engrs*, **80,** 1820.

LASALLE, J. P., 1960, *Proc. I.F.A.C. Conf.*, Moscow.

PONTRYAGIN, L. S., 1956, *Session of A.N.U.S.S.R. on fundamental problems of automatic controls*, **2,** 107 (Moscow, 1957) ; 1959, *Usp. Nat. Nauk*, **14,** 3.

ROSE, N. J., 1956, *Inst. Radio Engrs. Convention Record*, p. 61.

Study of an Optimum Non-linear Control System†

By A. T. Fuller
Department of Engineering, University of Cambridge

[Received December 10, 1962]

Abstract

The system studied has a plant which consists of two integrators with a saturable control input. The command signal input is a step plus a ramp. The performance criterion which is to be minimized is integral-error-squared. The optimum controller for this problem was found previously by a very specialized method. In the present paper this optimum controller is verified by means of (i) Pontryagin's maximum principle, (ii) a special method involving variation of the switching curve.

---

§ 1. Introduction

The optimum switching curves have been previously calculated for a double-integrator plant with saturable input, with various performance criteria (Fuller 1959, 1960 a). The methods used were very specialized but yielded the solutions in closed form. Fitsner (1960) considered effectively the same problem and gave an approximate solution. Subsequently Pearson (1961) and Brennan and Roberts (1962) showed that one of the exact solutions of Fuller (1960a) was in close agreement with digital and analogue computational solutions obtained from the equations resulting from Pontryagin's maximum principle. In the present paper it will be proved algebraically that the known optimum system satisfies Pontryagin's maximum principle. The example thus turns out to be one of the few cases of Pontryagin's method which have a closed-form solution for the optimum controller. We shall also verify the result by a method involving variation of the switching curve.

A further verification of the optimum system, obtained from the Bellman–Hamilton–Jacobi partial differential equation, is given in a recent paper by Wonham (1963).

§ 2. The Problem

We shall consider a system in which the main process, or plant, has a transfer function $1/ap^2$ $(a>0)$ between input u and output x. The desired output, or command signal, is zero. The input to the main process is subject to the saturation constraint :

$$|u|\leqslant 1. \qquad (1)$$

---

† Communicated by Mr. J. F. Coales.

The problem is to design the controller so that the integral-error-squared, namely

$$\int_0^\infty x^2\,dt,$$

is minimized, for any initial phase state of the main process at $t=0$, where t is time.

Note that in the present example it is a trivial matter to generalize the results for the case of zero command signal to the case of a command signal consisting of a step plus a ramp (Fuller 1960 a, b). For brevity however we shall confine our working to the zero command signal case.

It turns out that the optimum controller is characterized by a *switching curve* in the $(x, dx/dt)$ phase plane (Fuller 1960 a). See e.g. Fuller (1960 b) for a discussion of switching curves. The problem thus reduces to calculating the optimum switching curve. In fact the optimum switching curve was shown in Fuller (1960 a) by special methods to be given by the equation :

$$x+hax'|x'|=0, \qquad (2)$$

where

$$x'\equiv\frac{dx}{dt}, \qquad (3)$$

and where

$$h=\sqrt{\left(\frac{\sqrt{(33)}-1}{24}\right)}\simeq 0{\cdot}4446. \qquad (4)$$

Our first object is to show that the system with controller (2) satisfies Pontryagin's maximum principle.

§ 3. Formulation of the Hamiltonian Equations

Various expositions of Pontryagin's method have been given, e.g., Rozonoer (1959), Boltyanski *et al.* (1960). The application of the method to the example in hand involves the following steps (see Pearson 1961 and Brennan and Roberts 1962). The differential equations of the main process are written as a set of first-order equations as follows :

$$\begin{cases} x_1'=x_2, & (5) \\ x_2'=\dfrac{u}{a}, & (6) \end{cases}$$

where

$$x_1=x \text{ and } x_2=x'=\frac{dx}{dt}. \qquad (7)$$

A new variable x_0 is defined as having a derivative equal to the integrand in the performance criterion

$$\int_0^\infty x^2\,dt.$$

Thus

$$x_0' = x_1^2. \qquad (8)$$

Equations (5), (6) and (8) can be written as :

$$x_i' = f_i, \quad i = 0,1,2. \qquad (9)$$

The Hamiltonian H is defined as :

$$H = \psi_0 f_0 + \psi_1 f_1 + \psi_2 f_2, \qquad (10)$$

where the ψ_i are new variables, which satisfy the following equations :

$$\psi_i' = -\frac{\partial H}{\partial x_i}, \quad i = 0,1,2. \qquad (11)$$

From eqns. (5)–(10), the equation for H can be written more explicitly as :

$$H = \psi_0 x_1^2 + \psi_1 x_2 + \psi_2 \frac{u}{a}. \qquad (12)$$

From eqns. (11) and (12) the equations for the ψ_i can be written more explicitly as :

$$\begin{cases} \psi_0' = 0, & (13) \\ \psi_1' = -2\psi_0 x_1, & (14) \\ \psi_2' = -\psi_1. & (15) \end{cases}$$

Rozonoer (1959) and Katz (1960) have shown that when the performance criterion is of the form

$$\int_0^\tau f_0\,dt$$

with τ fixed, certain boundary conditions on the $\psi_i(\tau)$ may be assigned. With ∞ substituted for τ, these conditions are (we assume that Rozonoer's argument remains valid for the case $\tau = \infty$) :

$$\begin{cases} \psi_0(\infty) = -1, & (16) \\ \psi_1(\infty) = 0, & (17) \\ \psi_2(\infty) = 0. & (18) \end{cases}$$

From eqns. (13) and (16) :

$$\psi_0(t) = -1, \quad 0 \leqslant t < \infty. \qquad (19)$$

The Pontryagin maximum principle states that, subject to constraint (1), u must be chosen as a function of the x_i and ψ_i, so that H is maximized. Therefore, from (12) :

$$u = \operatorname{sgn} \psi_2. \qquad (20)$$

With substitution for ψ_0 from (19) and for u from (20), the eqns. (5), (6), (14), (15) for the x_i and ψ_i reduce to :

$$\begin{cases} x_1' = x_2, & (21) \\ x_2' = \dfrac{1}{a} \operatorname{sgn} \psi_2, & (22) \\ \psi_1' = 2x_1, & (23) \\ \psi_2' = -\psi_1, & (24) \end{cases}$$

with boundary conditions

$$\begin{cases} x_1(0) = x_1^0, & (25) \\ x_2(0) = x_2^0, & (26) \\ \psi_1(\infty) = 0, & (27) \\ \psi_2(\infty) = 0. & (28) \end{cases}$$

The solution of differential eqns. (21)–(24), with boundary conditions (25)–(28), traces out the optimum trajectory in (x_1, x_2) phase space, starting from arbitrary initial conditions (x_1^0, x_2^0). The switch points (where u changes sign) on this trajectory can be recognized as 'corners' and the locus of switch points when (x_1^0, x_2^0) is varied represents the required switching curve.

§ 4. The Switch Points of the Hamiltonian Equations

Our aim is now to show that there is a solution of the Pontryagin eqns. (21)–(28), which gives switching whenever the trajectory in the (x_1, x_2) phase space crosses curve (2). This will confirm the validity of the optimum switching curve (2), and at the same time demonstrate a closed-form solution for the locus of switching points of Pontryagin's equations.

As a preliminary let us show that it is possible to choose initial values ψ_1^0 and ψ_2^0 so that the *first two* switch points occur on curve (2), i.e. on the curve :

$$x_1 + hax_2|x_2| = 0. \qquad (29)$$

Without real loss of generality† we choose $x_2^0 > 0$ and x_1^0 to be such that the starting point is on curve (29) :

$$x_1^0 = -ha(x_2^0)^2. \qquad (30)$$

The initial value of ψ_2 must then be

$$\psi_2^0 = 0 \qquad (31)$$

for switching to occur at $t = 0$ when the representative point is on curve (29) (eqn. (20) shows that switching occurs whenever ψ_2 changes sign).

Let $\psi_2(t)$ next reach zero at $t = t_1$. For $0 \leqslant t \leqslant t_1$ it is straightforward to integrate eqns. (22), (21), (23), (24) (in that order), with initial conditions satisfying $x_2^0 > 0$, $\psi_1^0 > 0$, and eqns. (30) and (31). We obtain, for $0 \leqslant t \leqslant t_1$:

$$\begin{cases} x_2 = -\dfrac{t}{a} + x_2^0, & (32) \\[2ex] x_1 = -\dfrac{t^2}{2a} + x_2^0 t - ha(x_2^0)^2, & (33) \\[2ex] \psi_1 = -\dfrac{t^3}{3a} + x_2^0 t^2 - 2ha(x_2^0)^2 t + \psi_1^0, & (34) \\[2ex] \psi_2 = \dfrac{t^4}{12a} - \dfrac{x_2^0 t^3}{3} + ha(x_2^0)^2 t^2 - \psi_1^0 t. & (35) \end{cases}$$

† A trajectory with a starting point not on curve (29) can be regarded as a continuation of a trajectory which had started earlier on curve (29).

It is straightforward algebra (see Appendix 1) to show from eqns. (32), (33), (35) and the equation $\psi_2(t_1)=0$, that the condition for $x_1(t_1)$ and $x_2(t_1)$ to be on curve (29) is :

$$\psi_1{}^0 = h^2a^2(x_2{}^0)^3. \quad . \quad . \quad . \quad . \quad . \quad . \quad . \quad . \quad (36)$$

Thus when $\psi_1{}^0$ and $\psi_2{}^0$ are chosen to satisfy (31) and (36), the first two switch points are indeed on curve (29).

Let us next consider $t=t_1$ as a new starting time, with initial conditions $x_2(t_1)=x_2{}^1$ and $\psi_1(t_1)=\psi_1{}^1$. If these conditions obey the same relation as that obeyed by $x_2(0)$ and $\psi_1(0)$, namely relation (36), then it will follow that the third switch point will also be on curve (29). Thus we have to show that:

$$\psi_1{}^1 = h^2a^2(x_2{}^1)^3. \quad . \quad . \quad . \quad . \quad . \quad . \quad . \quad . \quad (37)$$

But it is elementary to calculate $\psi_1{}^1$ and $x_2{}^1$. This is done in Appendix 2 and it turns out that relation (37) is in fact satisfied. We conclude that the third switch point is indeed on curve (29).

The above proves that if any two successive switch points are on curve (29), then so is the next switch point. It follows by induction that all the switch points of the trajectory satisfying initial conditions (30), (31) and (36) are on curve (29).

It only remains to be shown that the above solution of the Pontryagin equations satisfies boundary conditions (27) and (28). From eqn. (32) and the expression for t_1 (eqn. (52) in Appendix 1) :

$$x_2{}^1 = -x_2{}^0\sqrt{k}. \quad . \quad . \quad . \quad . \quad . \quad . \quad . \quad . \quad . \quad (38)$$

Since $k<1$, eqn. (38) implies that successive switch points approach closer and closer to the origin of (x_1, x_2) space. From (36) the value of ψ_1 at a switch point at the origin of (x_1, x_2) space is zero ; and moreover the value of ψ_2 at any switch point is zero. Hence the trajectory approaches the origin not merely of (x_1, x_2) space but also of $(x_1, x_2, \psi_1, \psi_2)$ space. Actually, as shown in Fuller (1960 a), the origin is reached in a finite time T, although the number of switches is infinite. For $t>T$, the representative point stays at the origin, because eqns. (21)–(24) imply that the origin is a point of equilibrium†. Hence $\psi_1(\infty)=\psi_2(\infty)=0$, i.e. boundary conditions (27) and (28) are indeed satisfied.

§ 5. The Vanishing of the Hamiltonian

A corollary of the Pontryagin theory in some cases when the performance criterion is an improper integral is that along an optimum trajectory H is identically zero (Boltyanski *et al.* 1960). Let us confirm this result for the example in hand. Substituting eqns. (19), (32), (33), (34), (35) and (36) in eqn. (12) we find :

$$H(t)=0, \quad 0 \leqslant t \leqslant t_1. \quad . \quad . \quad . \quad . \quad . \quad . \quad . \quad . \quad (39)$$

Result (39) holds for $0 \leqslant t \leqslant t_1$, i.e. for the initial segment of the trajectory between the first two switch points. But since any similar segment of the

† In eqn. (22) we interpret sgn ψ_2 when $\psi_2=0$ as zero. That this is the correct interpretation follows from the fact that with $u\ (t)=0$ for $t>T$ there is no further contribution to the performance criterion after $t=T$.

trajectory can be regarded as an initial segment, $H(t)=0$ holds for all t until the origin is reached at $t=T$. For $t>T$ the result $H(t)=0$ is a trivial consequence of eqn. (12). Hence the Hamiltonian H vanishes† for all $t \geqslant 0$.

§ 6. Method of Variation of the Switching Curve

As further verification of the validity of switching curve (2), let us investigate the effect of varying this curve slightly.

For simplicity consider a trajectory starting on the x_1 axis at $(S, 0)$ in the (x_1, x_2) phase plane for the system with switching curve (2). As shown in eqn. (32) of Fuller (1960 a), the performance criterion

$$M=\int_0^\infty x_1{}^2\,dt$$

for this trajectory is given by

$$M=\frac{(23-14k+23k^2)(1+k)^{1/2}a^{1/2}|S|^{5/2}}{30(1-k^{5/2})}, \quad . \quad . \quad . \quad (40)$$

where $(-k)$ is the (constant) ratio of successive overshoot peaks. A further result from Fuller (1960 a) that we shall need is that the contribution of the initial part of the trajectory, up to the first overshoot peak, to the performance criterion, is‡:

$$M^*=\frac{(23-14k+23k^2)(1+k)^{1/2}a^{1/2}|S|^{5/2}}{30}. \quad . \quad . \quad . \quad (41)$$

Suppose, to begin with, that the switching curve is varied slightly only in the neighbourhood of the first switch point. Then the first overshoot peak (point at which the x_1–axis is first crossed) will be no longer at $x_1=-kS$ but at $x_1=-(k+\Delta)S$, where Δ is some small quantity.

Thus the first overshoot ratio will be $-(k+\Delta)$ and subsequent overshoot ratios will remain at $-k$. We shall show that this special variation results in a worsened performance, if $|\Delta|$ is sufficiently small.

The contribution of the initial part of the varied trajectory, up to the first overshoot peak, to the performance criterion is obtained by substituting $(k+\Delta)$ for k in the right side of (41). The remaining contribution to the performance criterion is obtained by substituting $-(k+\Delta)S$ for S in the right side of (40). Adding these two contributions we obtain the following expression for the performance criterion $M(\Delta)$ of the varied system :

$$M(\Delta)=\left[\{23-14(k+\Delta)+23(k+\Delta)^2\}(1+k+\Delta)^{1/2} + \frac{(23-14k+23k^2)(1+k)^{1/2}(k+\Delta)^{5/2}}{1-k^{5/2}}\right]\frac{a^{1/2}|S|^{5/2}}{30}. \quad . \quad . \quad (42)$$

---

† We could have shortened the working in § 4 by using the fact that $H=0$ at the outset, but this might have entailed a sacrifice of clarity.

‡ Equation (41) above follows immediately from eqns. (27) and (31) of Fuller (1960 a).

We have to show that $M'(0)=0$ and $M''(0)>0$. Straightforward evaluation from (42) shows in fact that

$$M'(0) = -\frac{(k^{7/2}-10k^{5/2}-23k^2-23k^{3/2}-10k+1)a^{1/2}|S|^{5/2}}{12(1-k^{5/2})(1+k)^{1/2}}. \quad . \quad (43)$$

But according to eqn. (33) of Fuller (1960 a), k satisfies the following equation:

$$k^{7/2}-10k^{5/2}-23k^2-23^{3/2}-10k+1=0. \quad . \quad . \quad . \quad . \quad (44)$$

From eqns. (43) and (44) :

$$M'(0)=0. \quad . \quad . \quad . \quad . \quad . \quad . \quad . \quad . \quad . \quad (45)$$

Further straightforward evaluation from (42) and use of (44) yields :

$$M''(0) \doteq \frac{(1+k^{1/2})^2(9+14k^{1/2}+9k)}{8(1+k)^{3/2}(1+k^{1/2}+k+k^{3/2}+k^2)} a^{1/2}|S|^{5/2}. \quad . \quad . \quad (46)$$

Since $k>0$, eqn. (46) implies :

$$M''(0)>0. \quad . \quad . \quad . \quad . \quad . \quad . \quad . \quad . \quad . \quad (47)$$

Results (45) and (47) show that a sufficiently small change in the first switch point increases

$$\int_0^\infty x_1^2\,dt.$$

Now suppose a sufficiently small change in the switching curve is additionally made in the neighbourhood of the second switch point. This will, by the same argument, increase the contribution of the trajectory beyond the first overshoot peak to the performance criterion (without affecting the contribution of the remaining initial part of the trajectory†). Similarly, sufficiently small additional changes in the neighbourhoods of the 3rd, 4th, . . . etc. switch points will result in an additionally increased performance criterion. Therefore curve (2) is at least locally optimum.

The same method may be used to check the validity of the optimum switching curve given in Fuller (1960 a) for the performance criterion integral-modulus-error.

§ 7. Conclusions

In this paper the optimum switching curve for a double-integrator plant with integral-error-squared performance criterion has been verified by Pontryagin's method and by a method involving variation of the switching curve.

The example treated provides a direct illustration (without the need for computing) of Pontryagin's method and so lends concreteness to the otherwise rather abstract theory of the maximum principle.

The variation method uses only elementary algebraic steps, but is specialized.

Note that we have not actually derived the optimum switching curve by the maximum principle and the variation method ; we have only

† The second switch point occurs after the first overshoot peak (Fuller 1960 a).

verified it, taking its known form as a starting point. Thus the derivation in Fuller (1960 a) is not rendered redundant by the present paper. Note also that the present paper shows only that the switching curve is locally optimum.

APPENDIX 1

Calculation of ψ_1^0

In this Appendix it is shown that the condition for the second switch point of eqns. (21)–(24) to lie on curve (29) is that the initial value of ψ_1 should be given by eqn. (36).

The condition for the trajectory to meet curve (29) at $t=t_1$ is :

$$x_1(t_1)-hax_2^2(t_1)=0, \qquad (48)$$

i.e. with substitution for x_1 and x_2 from eqns. (32) and (33),

$$t_1^2-2ax_2^0t_1+\frac{4h(ax_2^0)^2}{1+2h}=0 \qquad (49)$$

or

$$t_1=\left[1+\sqrt{\left(\frac{1-2h}{1+2h}\right)}\right]ax_2^0. \qquad (50)$$

Using the fact that

$$h=\tfrac{1}{2}\left(\frac{1-k}{1+k}\right), \qquad (51)$$

which is a consequence of eqn. (52) of Fuller (1960 a), we can write condition (50) as :

$$t_1=(1+\sqrt{k})ax_2^0. \qquad (52)$$

t_1 is also determined by the switching condition (see eqn. (20)) :

$$\psi_2(t_1)=0, \qquad (53)$$

i.e. from eqn. (35),

$$\psi_1^0=\frac{t_1^3}{12a}-\frac{x_2^0t_1^2}{3}+ha(x_2^0)^2t_1. \qquad (54)$$

From eqns. (51), (52) and (54), ψ_1^0 must satisfy :

$$\frac{\psi_1^0}{a^2(x_2^0)^3}=\frac{(1+\sqrt{k})^3}{12}-\frac{(1+\sqrt{k})^2}{3}+\frac{(1-k)(1+\sqrt{k})}{2(1+k)}. \qquad (55)$$

The right side of (55) simplifies to h^2. One way of showing this is to subtract the square of eqn. (51) from (55), then multiply throughout by $12(1+k)^2(1+\sqrt{k})$.

This yields :

$$12(1+k)^2(1+\sqrt{k})\left(\frac{\psi_1^0}{a^2(x_2^0)^3}-h^2\right)$$
$$=k^{1/2}(k^{7/2}-10k^{5/2}-23k^2-23k^{3/2}-10k+1). \qquad (56)$$

From eqn. (44) the right side of (56) is zero. Hence (56) yields :

$$\psi_1^0 = h^2a^2(x_2^0)^3, \qquad (57)$$

which is the required condition.

APPENDIX 2

Calculation of ψ_1^1

In this Appendix it is shown that when ψ_1^0 is chosen to satisfy eqn. (36), $\psi_1(t_1)$ satisfies eqn. (37).

From eqns. (34) and (36) :

$$\psi_1^1 = -\frac{t_1^3}{3a} + x_2^0t_1^2 - 2ha(x_2^0)^2t_1 + h^2a^2(x_2^0)^3. \qquad (58)$$

Substitution for t_1 and h from (52) and (51) yields :

$$\frac{\psi_1^1}{a^2(x_2^0)^3} = -\frac{(1+\sqrt{k})^3}{3} + (1+\sqrt{k})^2 - \frac{(1-k)(1+\sqrt{k})}{1+k} + \frac{(1-k)^2}{4(1+k)^2}. \qquad (59)$$

The right side of (59) simplifies to $-h^2k^{3/2}$. One way of showing this is to add $k^{3/2}$ times the square of eqn. (51) to (59), then multiply throughout by $12(1+k)^2$. This yields :

$$12(1+k)^2\left(\frac{\psi_1^1}{a^2(x_2^0)^3} + h^2k^{3/2}\right)$$
$$= -k^{7/2} + 10k^{5/2} + 23k^2 + 23k^{3/2} + 10k - 1. \qquad (60)$$

From eqn. (44) the right side of (60) is zero. Hence (60) yields :

$$\frac{\psi_1^1}{a^2(x_2^0)^3} = -h^2k^{3/2}. \qquad (61)$$

From eqns. (38) and (61) :

$$\psi_1^1 = h^2a^2(x_2^1)^3 \qquad (62)$$

which is the required result.

REFERENCES

BOLTYANSKI, V. G., GAMKRELIDZE, R. V., and PONTRYAGIN, L. S., 1960, *Bull. Acad. Sci. U.R.S.S.*, **24,** 3.

BRENNAN, P. J., and ROBERTS, A. P., 1962, *J. Electron. Contr.*, **12,** 345.

FITSNER, L. N., 1960, *Avtomat. Telemech., Moscow*, **21,** 1115.

FULLER, A. T., 1959, Ph.D. Thesis, Cambridge ; 1960 a, *Proceedings of the International Federation of Automatic Control Congress, Moscow* (London : Butterworths), **1,** 510; 1960 b, *J. Electron. Contr.*, **8,** 465.

KATZ, S., 1960, *Ann. N.Y. Acad. Sci.*, **84,** 441.

PEARSON, J. D., 1961, *J. Electron. Contr.*, **10,** 323.

ROZONOER, L. I., 1959, *Avtomat. Telemech., Moscow*, **20,** 1320, 1441, 1561.

WONHAM, W. M., 1963, *J. Electron. Contr.*, **15,** 59.

Minimization of Integral-square-error for Non-linear Control Systems of Third and Higher Order

By P. E. W. Grensted and A. T. Fuller
Engineering Department, University of Cambridge

[Received November 13, 1964]

Abstract

The paper deals principally with a control system which has a plant consisting of three integrators with a saturable control input. The command signal input is a step plus a ramp plus a parabola. The switching surface which minimizes integral-square-error is found, partly algebraically and partly numerically, by methods which start from Pontryagin's maximum principle. All optimum trajectories have an infinite number of switches before the origin is reached, except for two trajectories which have no switches. Some optimum trajectories have the property that the ratio of any two successive switching intervals is constant. All other optimum trajectories (apart from the two exceptional cases) converge rapidly towards these constant ratio trajectories. Thus, when finding optimum trajectories by backwards numerical computation from near the origin of the Hamiltonian system of equations, it is necessary to adjust the 'initial' values to be, not only small, but also close to a constant ratio trajectory.

The above behaviour is similar to that for an analogous linear system. Approximate analysis of the non-linear Hamiltonian system by means of describing function techniques also indicates this behaviour; in fact, the constant ratio trajectories were only discovered as a result of preliminary linear and quasi-linear studies.

It is conjectured that similar behaviour occurs for other plants and other performance criteria.

§ 1. Introduction

The optimization of saturating control systems for integral-square-error performance criterion and for similar criteria, has been treated by Fuller (1960 a, 1963, 1964 a, b), Pearson (1961) Brennan and Roberts (1962), Eggleston (1963) and Wonham (1963). The systems considered were all of first and second orders. In the present paper a system of third order is optimized with respect to integral-square-error; and systems of higher order are discussed.

Pearson (1961) and Brennan and Roberts (1962) found that computational optimization of second-order systems could be obtained by tracing backwards solutions of the differential equations resulting from Pontryagin's maximum principle. The extension of this technique to systems of higher order is not straightforward. For higher-order systems the choice of the boundary values associated with the Pontryagin equations becomes critical. Much of the present paper is concerned with explanation of and circumvention of this difficulty.

† Communicated by the Authors.

§ 2. The Problem

The system to be considered has a plant with a transfer function $1/(ap^3)$, between control input u and system output x. The differential equation of this plant is:

$$ax''' = u, \tag{1}$$

where primes denote differentiation with respect to time t, and a is a positive constant. The control input is subject to the saturation constraint:

$$|u| \leqslant 1. \tag{2}$$

The desired output, or command signal, is considered to be zero, but it is a trivial matter to generalize the results for this case to the case of a command signal consisting of a step plus a ramp plus a parabola (see, e.g., Fuller 1960 b for the technique).

The problem is to design the controller so that the integral-square-error, namely

$$\int_0^\infty x^2\,dt, \tag{3}$$

is minimized for any initial state of the plant at $t = 0$. It will be shown that the optimum controller is characterized by a *switching surface* in (x, x', x'') phase space. A discussion of switching surfaces has been given by Fuller (1960 b).

§ 3. The Hamiltonian System

3.1. *Statement of the Maximum Principle*

For expositions of Pontryagin's maximum principle see, e.g., Rozonoer (1959) and Pontryagin *et al.* (1961). Here we state formally Pontryagin's maximum principle for a moderately general problem, and then specialize it to the problem in hand.

Given an nth-order plant described by

$$x_i' = f_i(x_1, x_2, \ldots x_n, u) \quad (i = 1, 2, \ldots n) \tag{4}$$

with

$$|u| \leqslant 1, \tag{5}$$

suppose it is required to minimize

$$\int_0^\infty f_0(x_1, x_2, \ldots x_n, u)\,dt. \tag{6}$$

The Hamiltonian H is defined as:

$$H(\psi_0, \ldots \psi_n, x_0, \ldots x_n, u) = \psi_0 f_0 + \psi_1 f_1 + \ldots + \psi_n f_n, \tag{7}$$

where the ψ_i are new variables defined as satisfying:

$$\psi_i' = -\frac{\partial H}{\partial x_i} \quad (i = 0, 1, \ldots n). \tag{8}$$

The maximum principle states† that, if $(x_1(t)\ldots x_n(t))$ is an optimum trajectory, then there exist $\psi_0(t),\ldots\psi_n(t)$ satisfying (8) such that H is maximized with respect to $u(t)$, for all $t\geqslant 0$.

3.2. *Application of the Maximum Principle*

To apply Pontryagin's maximum principle to the problem in hand, we write the plant equations as:

$$\left.\begin{aligned} x_1' &= x_2, \\ x_2' &= x_3, \\ x_3' &= u/a, \end{aligned}\right\} \tag{9}$$

where

$$x_1=x, \quad x_2=x', \quad x_3=x'' \tag{10}$$

are taken to be the state variables of the plant. From (3) and (6):

$$f_0=x_1^2. \tag{11}$$

Corresponding to (7), we have, therefore;

$$H=\psi_0 x_1^2+\psi_1 x_2+\psi_2 x_3+\psi_3 u/a. \tag{12}$$

Equations (8) now become:

$$\left.\begin{aligned} \psi_0' &= 0, \\ \psi_1' &= -2\psi_0 x_1, \\ \psi_2' &= -\psi_1, \\ \psi_3' &= -\psi_2. \end{aligned}\right\} \tag{13}$$

Now Rozonoer (1959) has shown that when the performance criterion is of the form $\int_0^\tau f_0\,dt$ with τ fixed and finite, the following boundary conditions on the ψ's may be assigned:

$$\left\{\begin{aligned} &\psi_0(\tau)=-1, \\ &\psi_i(\tau)=0 \quad (i=1,2,\ldots n). \end{aligned}\right.$$

We shall assume Rozonoer's argument remains valid‡ for the case $\tau=\infty$,

† For a precise statement of the maximum principle, including the restrictions on the nature of the f's and u under which it is valid, see Pontryagin *et al.* (1961).

‡ Rozonoer's boundary conditions with $\tau=\infty$ are certainly valid for the simpler case where the plant has a transfer function $1/(ap^2)$. In this case the use of Rozonoer's boundary conditions has been shown to lead to unique and absolutely optimum solutions (Fuller 1964 b).

and assign boundary conditions in our example as follows:

$$\left.\begin{aligned} \psi_0(\infty) &= -1, \\ \psi_1(\infty) = \psi_2(\infty) &= \psi_3(\infty) = 0. \end{aligned}\right\} \quad (14)$$

From (13) and (14):

$$\psi_0(t) = -1 \quad (0 \leqslant t < \infty). \quad (15)$$

Also, since H is a maximum with respect to u, it follows from (2) and (12) that†

$$u = \operatorname{sgn} \psi_3. \quad (16)$$

Thus the optimum control is bang-bang. With substitution of (15) and (16) in (9) and (13), the differential equations for the x's and ψ's reduce to:

$$\left.\begin{aligned} x_1' &= x_2, \\ x_2' &= x_3, \\ x_3' &= (\operatorname{sgn} \psi_3)/a, \\ \psi_1' &= 2x_1, \\ \psi_2' &= -\psi_1, \\ \psi_3' &= -\psi_2. \end{aligned}\right\} \quad (17)$$

We shall refer to the system of eqns. (17) as the *Hamiltonian system*‡. It is equivalent to a system containing six integrators and a relay in a closed loop as shown in fig. 1.

Fig. 1

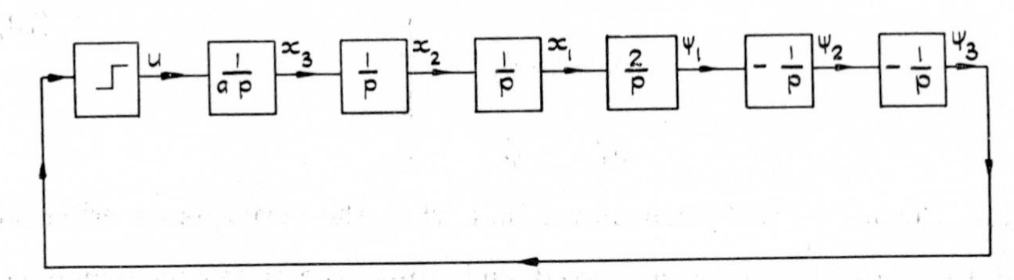

Block diagram of the Hamiltonian system.

At first sight there is a superficial similarity to a conventional control loop, but it is very unstable as it contains 540 degrees of phase lag, and, moreover, the feedback is positive, not negative.

† It can be proved that $\psi_3(t)$ does not remain zero for a positive interval of time, at least until $x_1, x_2, x_3, \psi_1, \psi_2, \psi_3$ are all simultaneously zero. For the method see Wonham (1963) or Fuller (1964 b). It follows that in (16) we can treat sgn ψ_3 as zero when ψ_3 is zero.

‡ From (12), (15) and (16), $H = -x_1^2 + \psi_1 x_2 + \psi_2 x_3 + |\psi_3|/a$. With this expression for H, eqns. (17) follow directly from

$$x_i' = \frac{\partial H}{\partial \psi_i} \text{ and } \psi_i' = -\frac{\partial H}{\partial x_i} \; (i = 1, 2, 3),$$

which is a set of equations in the standard Hamiltonian form.

3.3. *Boundary Conditions*

The control signal $u = \operatorname{sgn} \psi_3$ for an optimum trajectory is determined by the solutions of eqns. (17), together with the boundary conditions of (14):

$$\psi_1(\infty) = \psi_2(\infty) = \psi_3(\infty) = 0, \tag{18}$$

and those given by the initial state of the plant, $x_1(0)$, $x_2(0)$ and $x_3(0)$. It can be proved rigorously (see Fuller 1964b for the method) that all optimum trajectories also satisfy:

$$x_1(\infty) = x_2(\infty) = x_3(\infty) = 0. \tag{19}$$

The origin is thus a singular point† in $(x_1, x_2, x_3, \psi_1, \psi_2, \psi_3)$ space of the system of eqns. (17) through which all optimum trajectories pass. It corresponds to a point of unstable equilibrium.

3.4. *Change of Variables*

It is convenient to take as new variables certain multiples of x_1, x_2, x_3, ψ_1, ψ_2, ψ_3 in order to avoid the variety of coefficients which appear in eqns. (17). Let

$$\left.\begin{aligned} y_1 &= \tfrac{1}{2}a\psi_3, \\ y_2 &= -\tfrac{1}{2}a\psi_2, \\ y_3 &= \tfrac{1}{2}a\psi_1, \\ y_4 &= ax_1, \\ y_5 &= ax_2, \\ y_6 &= ax_3. \end{aligned}\right\} \tag{20}$$

Then in place of eqns. (17) we obtain the tidier set:

$$\left.\begin{aligned} y_1' &= y_2, \\ y_2' &= y_3, \\ y_3' &= y_4, \\ y_4' &= y_5, \\ y_5' &= y_6, \\ y_6' &= \operatorname{sgn} y_1, \end{aligned}\right\} \tag{21}$$

with boundary conditions for all optimum trajectories, corresponding to (18) and (19):

$$y_i(\infty) = 0 \quad (i = 1, 2, \ldots 6). \tag{22}$$

† It is the only singular point, as inspection of (17) shows. (We take $\operatorname{sgn} 0 = 0$, see footnote to eqn. (16).)

We shall refer to the normalized system (as well as the non-normalized system) as the Hamiltonian system†.

Equations (21) can be expressed alternatively as the single non-linear differential equation:

$$y^{(6)} = \operatorname{sgn} y, \tag{23}$$

where $y = y_1$ and $y^{(r)} = y_{r+1}$ $(r = 1, 2, \dots 5)$.

3.5. *Behaviour of the Hamiltonian System*

The values of x_1, x_2, and x_3 when ψ_3 is zero (or y_4, y_5 and y_6 when y_1 is zero) for optimum trajectories define the switching surface of the original problem. The surface is completely determined if all the appropriate solutions of the Hamiltonian system (21) have been found. In principle one can compute approximately such solutions by starting near the unstable singular point at the origin, and tracing y-space trajectories backwards in time. Pearson (1961) and Brennan and Roberts (1962) have shown that this procedure works for some second-order plants, using small but otherwise arbitrary starting points for the computation. But it has been found that for the third-order plant of the present example one obtains only trajectories which are devoid of switches or, occasionally, with one or two switches so close to the arbitrary starting point that they are of doubtful validity for determining the optimum switching surface‡.

To explain this phenomenon by an exact analysis of the Hamiltonian system would seem to involve difficulties owing to the non-linearity of eqns. (21). Instead we first investigate, in § 4, an analogous linear system, which turns out to yield a similar phenomenon. Then, in § 5, we analyse the non-linear system approximately by a linearization technique. These preliminary studies provide a heuristic understanding of the phenomenon, and suggest how to find trajectories which do have switches.

§ 4. An Analogous Linear Problem

The system studied in this section is the same as that described in § 2, except that the saturation constraint (2) is removed, and instead an 'energy' constraint on the control signal is imposed by changing the performance criterion (3) to

$$\int_0^\infty (x^2 + \lambda u^2)\, dt, \tag{24}$$

---

† If K is defined as $\frac{1}{2}a^2 H$,

$$K = -\tfrac{1}{2}y_4^2 + y_3 y_5 - y_2 y_6 + |y_1|.$$

Hence (21) are equivalent to the Hamiltonian equations:

$$y'_{4-2i} = \frac{\partial K}{\partial y_{3+2i}} \quad \text{and} \quad y'_{3+2i} = -\frac{\partial K}{\partial y_{4-2i}} \quad (i = -1, 0, 1).$$

‡ We are indebted to J. Lipscombe and D. Stainton who discovered this phenomenon on attempting to solve the problem on an analogue computer.

where λ is fixed. It is well known that for this criterion the optimal controller is linear if the plant is linear. Of the several available methods of analysis we shall use Pontryagin's maximum principle.

4.1. *The Linear Hamiltonian System*

The general statement of Pontryagin's maximum principle in §3.1 covers the present problem. The Hamiltonian defined in (7) becomes:

$$H = \psi_0(x_1^2 + \lambda u^2) + \psi_1 x_2 + \psi_2 x_3 + \psi_3 u/a. \tag{25}$$

The boundary conditions (14) remain valid, and so does the expression (15), $\psi_0 = -1$. Maximizing H with respect to u now gives:

$$u = \psi_3/(2\lambda a), \tag{26}$$

so eqns. (8) applied to (25) yield the linear system:

$$\left\{\begin{aligned} x_1' &= x_2, \\ x_2' &= x_3, \\ x_3' &= \psi_3/(2\lambda a^2), \\ \psi_1' &= 2x_1, \\ \psi_2' &= -\psi_1, \\ \psi_3' &= -\psi_2. \end{aligned}\right.$$

On introducing normalized variables defined by (20) we obtain:

$$\left.\begin{aligned} y_1' &= y_2, \\ y_2' &= y_3, \\ y_3' &= y_4, \\ y_4' &= y_5, \\ y_5' &= y_6, \\ y_6' &= y_1/(\lambda a^2). \end{aligned}\right\} \tag{27}$$

This Hamiltonian system for the linear problem is the same as the Hamiltonian system (21) except that the relay is now replaced by an amplifier of gain $1/(\lambda a^2)$.

4.2. *General Solution of the Linear Hamiltonian System*

The characteristic equation of system (27) is:

$$p^6 = 1/(\lambda a^2). \tag{28}$$

Thus the characteristic roots are distributed uniformly round a circle centred on the origin in the complex plane, as shown in fig. 2, and with radius R, where

$$R = (\lambda a^2)^{-1/6}. \tag{29}$$

Fig. 2

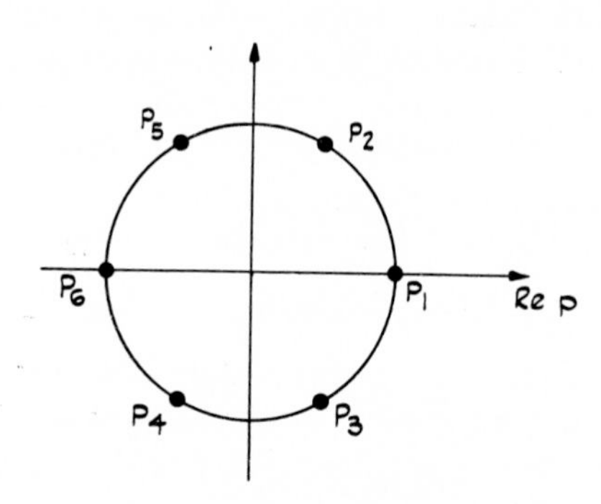

Characteristic roots of the linear Hamiltonian system.

Specifically the roots are

$$\left.\begin{aligned} p_1 &= R, \\ p_2, p_3 &= \tfrac{1}{2}(1 \pm j\sqrt{3})R, \\ p_4, p_5 &= -\tfrac{1}{2}(1 \pm j\sqrt{3})R, \\ p_6 &= -R. \end{aligned}\right\} \qquad (30)$$

The general solution for y_1 can be written:

$$y_1 = \sum_{k=1}^{6} A_k \exp(p_k t), \qquad (31)$$

where the A_k are six arbitrary constants. Using (31) and eqns. (27) in succession, the general vector solution of (27) is:

$$y_i = \sum_{k=1}^{6} p_k^{i-1} A_k \exp(p_k t) \quad (i = 1, 2, \ldots 6). \qquad (32)$$

We will call the component of the general solution associated with a real characteristic root a *monotonic mode*; and we will call the sum of the pair of components associated with a complex conjugate pair of characteristic roots an *oscillatory mode*. Then inspection of fig. 2, or of (30), shows that the linear Hamiltonian system has one convergent monotonic mode, one convergent oscillatory mode, one divergent monotonic mode and one divergent oscillatory mode.

4.3. *Optimum Trajectories*

The boundary conditions (22) still apply for optimum trajectories in (y_4, y_5, y_6) space. Hence the divergent modes of the linear Hamiltonian system must not be excited, i.e.

$$A_1 = A_2 = A_3 = 0. \qquad (33)$$

Hence (32) becomes:

$$y_i = p_4^{i-1} A_4 \exp(p_4 t) + p_5^{i-1} A_5 \exp(p_5 t) + p_6^{i-1} A_6 \exp(p_6 t) \qquad (34)$$
$$(i = 1, 2, \ldots 6).$$

If the arbitrary constants A_4, A_5 and A_6 are eliminated between the six equations (34), y_1, y_2 and y_3 may be expressed in terms of the plant coordinates y_4, y_5 and y_6. This can be done by solving the last three equations for $A_4 \exp(p_4 t)$, $A_5 \exp(p_5 t)$, $A_6 \exp(p_6 t)$, and then substituting into the first three equations. Using the values of p_4, p_5, p_6 the result is:

$$\left.\begin{aligned} y_1 &= -R^{-3}y_4 - 2R^{-4}y_5 - 2R^{-5}y_6, \\ y_2 &= 2R^{-2}y_4 + 3R^{-3}y_5 + 2R^{-4}y_6, \\ y_3 &= -2R^{-1}y_4 - 2R^{-2}y_5 - R^{-3}y_6. \end{aligned}\right\} \tag{35}$$

On reverting to the non-normalized coordinates, using (20) and (29), eqns. (35) become:

$$\left.\begin{aligned} \psi_1 &= -4(\lambda a^2)^{1/6}x_1 - 4(\lambda a^2)^{1/3}x_2 - 2(\lambda a^2)^{1/2}x_3, \\ \psi_2 &= -4(\lambda a^2)^{1/3}x_1 - 6(\lambda a^2)^{1/2}x_2 - 4(\lambda a^2)^{2/3}x_3, \\ \psi_3 &= -2(\lambda a^2)^{1/2}x_1 - 4(\lambda a^2)^{2/3}x_2 - 4(\lambda a^2)^{5/6}x_3. \end{aligned}\right\} \tag{36}$$

Thus along an optimum trajectory the ψ's are linear functions of the x's.

Finally, using (26), the last of eqns. (36) gives the optimum control input to be a linear function of the plant phase coordinates:

$$u/a = -(\lambda a^2)^{-1/2}x - 2(\lambda a^2)^{-1/3}x' - 2(\lambda a^2)^{-1/6}x''. \tag{37}$$

Result (37) is the analytic solution of the problem of finding the optimum controller for the linear system defined at the beginning of § 4.

4.4. *Solution by Computation*

Suppose, instead of analysing the linear Hamiltonian system explicitly as above, we attempt to compute the relevant solutions by the procedure suggested in § 3.5. By starting the computation of the solution of eqns. (27) near a point at the origin and computing backwards in time, the divergent modes in forward time appear as convergent modes in backward time, and so die out as the computation proceeds. The computation will find only those modes which are convergent in forward time, i.e. optimum trajectories. But the magnitude of the characteristic root for the monotonic convergent mode is twice the magnitude of the real parts of the characteristic roots for the oscillatory convergent mode. This means that the monotonic mode will grow much faster during the computation than the oscillatory mode, and so that by the time the unwanted modes have become insignificant it is likely that only the monotonic convergent mode will be significant in the computed solution. Thus in order not to lose the valid oscillatory mode it is necessary to start the computation at a point very close to this mode. We conclude that if all optimum trajectories are to be computed for the linear third-order system then the starting point for the computation cannot be chosen completely arbitrarily.

4.5. *Comparison of Linear and Non-linear Systems*

The above argument shows that a phenomenon will occur in the computation of optimum trajectories for the linear Hamiltonian system which is similar to the phenomenon described in § 3.5 in the computation of the non-linear Hamiltonian system. In the latter case oscillatory solutions are required so that zeros of the control variable, which define the switching surface, are obtained. We are led to the conjecture that the non-linear system has types of motion which are analogous to the modes of the linear system, and that the computational difficulty in the non-linear system is due to a monotonic motion, divergent in backward time, becoming rapidly dominant over an oscillatory motion, divergent in backward time.

In support of this conjecture, we note from Appendices 8 and 9 that in simpler cases, where the plant transfer function is $1/(ap)$ or $1/(ap^2)$, the behaviour of the non-linear Hamiltonian systems is very similar to the behaviour of the analogous linear Hamiltonian systems.

§ 5. Describing Function Analysis of the Non-linear Hamiltonian System

We can regard the linear Hamiltonian system as a crude first approximation to the non-linear Hamiltonian system, obtained by replacing the relay characteristic by a linear characteristic. This procedure is somewhat similar to the use of the describing function approximation in calculating the frequency response of non-linear systems. In the present case, however, we are concerned with transient response, so it is necessary to refine the describing function method by replacing the relay by a linear characteristic with *time varying* slope. A general technique for extending the describing function technique in this way has been given by Grensted (1955, 1962).

5.1. *Derivation of the Approximate Equation*

We now apply the transient describing function technique to the non-linear system (23):

$$\frac{d^6y}{dt^6} = \operatorname{sgn} y. \qquad (38)$$

We assume that the input to the relay can be written:

$$y(t) = a(t) \sin \phi(t), \qquad (39)$$

where the *phase* $\phi(t)$ is a monotonic function increasing with time, and the *amplitude* $a(t)$ is a positive function of time. Then $\operatorname{sgn} y$ is a square waveform of unit amplitude, periodic in ϕ. The output of the relay is therefore:

$$\operatorname{sgn} y = \frac{4}{\pi}(\sin\phi + \tfrac{1}{3}\sin 3\phi + \tfrac{1}{5}\sin 5\phi + \dots). \qquad (40)$$

The input to the relay is the six-fold integration with respect to time of (40). Now because all terms after the first on the right-hand side of (40) are not only of smaller amplitude, but are also varying more rapidly, than the first term, we suppose that only the first term becomes significant at the input to the relay†. With this approximation, eqn. (38), with (39), becomes:

$$\frac{d^6}{dt^6} a(t) \sin \phi(t) = \frac{4}{\pi} \sin \phi(t). \tag{41}$$

5.2. *Introduction of Complex Variables*

It is convenient to treat (41) as the imaginary part of

$$\frac{d^6}{dt^6} [a \exp(j\phi)] = \frac{4}{\pi} \exp(j\phi). \tag{42}$$

It is also convenient to introduce the *frequency* $\omega(t)$ and *damping* $\mu(t)$ of the transient solution, defined by

$$\omega = \frac{d\phi}{dt}, \quad \text{whence} \quad \phi = \int \omega \, dt \tag{43}$$

and

$$\mu = -\frac{1}{a}\frac{da}{dt}, \quad \text{whence} \quad a = \exp\left(-\int \mu \, dt\right). \tag{44}$$

Then (42) becomes:

$$\frac{d^6}{dt^6} \exp\left[\int (-\mu + j\omega) \, dt\right] = \frac{\pi}{4} \exp\left(j \int \omega \, dt\right), \tag{45}$$

which can be simplified, by using the identity (Grensted 1955):

$$P\left(\frac{d}{dt}\right)\left[\exp\left(\int z(t) \, dt\right) f(t)\right] = \exp\left[\int z(t) \, dt\right] P\left(\frac{d}{dt} + z(t)\right)[f(t)],$$

where $P(d/dt)$ is a polynomial in the operator d/dt, to yield:

$$\left(\frac{d}{dt} - \mu + j\omega\right)^6 = \frac{4}{\pi} \exp\left(\int \mu \, dt\right). \tag{46}$$

5.3. *Solution of the Approximate Equation*

If we try to satisfy (46) by

$$-\mu = \frac{\alpha}{t} \quad \text{and} \quad \omega = \frac{\beta}{t}, \tag{47}$$

† This corresponds to the argument justifying the assumption of the conventional describing function method used to analyse oscillations of constant frequency and amplitude. In that case a numerical assessment of the relevant magnitude of the neglected terms at the input to the non-linear element can be given. In the transient case, $\phi(t)$ being a fairly general function, a reliable guide as to the validity of the approximation is not directly available. In the present example it turns out that the approximation yields answers of remarkable accuracy. See also Appendix 1 and the footnote to eqn. (53).

where α and β are real constants to be determined, we obtain directly† from (46):

$$\frac{(\alpha+j\beta)(\alpha-1+j\beta)\dots(\alpha-5+j\beta)}{t^6}=\frac{4}{\pi\xi t^\alpha}, \tag{48}$$

where ξ is a constant arising from the integration on the right-hand side of (46).

From (48) we must have:

$$\alpha=6. \tag{49}$$

With this value of α the imaginary and real parts of (48) yield, respectively,

$$\beta(\beta^4-35\beta^2+84)=0 \tag{50}$$

and

$$-\beta^6+175\beta^4-1624\beta^2+720=\frac{4}{\pi\xi}. \tag{51}$$

The roots of (50) are

$$\left.\begin{aligned}\beta_1&=0,\\ \beta_2,\beta_3&=\pm\sqrt{\left(\frac{35+\sqrt{889}}{2}\right)},\\ \beta_4,\beta_5&=\pm\sqrt{\left(\frac{35-\sqrt{889}}{2}\right)}.\end{aligned}\right\} \tag{52}$$

We shall deal first with the root $\beta_1=0$. This corresponds to a non-oscillatory solution. This case is not strictly covered by the initial hypothesis of the present section, but appropriate modifications of the technique would show it to be a valid solution. It is analogous to a monotonic mode of the linear Hamiltonian system. We do not pursue if further here, as it is easier to obtain the corresponding exact solution of the non-linear Hamiltonian system (see § 6).

For the remaining roots, substitution of (47) into (43) and (44) yields the approximate solution‡:

$$y=\xi t^6\sin\left(\beta\log\frac{t}{t_0}\right), \tag{53}$$

† The identity $\left(\frac{d}{dt}+\frac{\gamma}{t}\right)^r=\gamma(\gamma-1)\dots(\gamma-r+1)/t^r$ is easily proved by induction.

‡ Justification of the correctness of this approximate solution can be reduced to the justification of the conventional describing function method of analysis. The substitution of $\tau=\log(t/t_0)$, t_0 arbitrary, and $\eta(\tau)=t^{-6}y(t)$ into the exact equation (38) yields

$$\left(\frac{d}{d\tau}+1\right)\left(\frac{d}{d\tau}+2\right)\dots\left(\frac{d}{d\tau}+6\right)\eta(\tau)=\operatorname{sgn}\eta(\tau).$$

The conventional describing function technique applied to this equation yields the following as the only *periodic* solution, and it is unstable:

$$\eta(\tau)=\xi\sin\beta\tau,$$

where β and ξ are given by (54) and (55). Substitution back to the original variables, y and t, yields eqn. (53).

where t_0 is an arbitrary constant, provided it has the same sign as t. But we find that

$$\beta = \beta_2 = \sqrt{\left(\frac{35 + \sqrt{889}}{2}\right)} \tag{54}$$

with

$$\xi = \xi_2 = \frac{1}{60(199 + 7\sqrt{889})\pi} \tag{55}$$

provides the only significant case. The case $\beta = \beta_3$ turns out to be the same as that above, while β_4 and β_5 yield, from (51), a negative value of ξ, so that the amplitude is negative, contrary to hypothesis.

5.4. *Discussion of the Approximate Solution*

If t is positive, result (53) represents an oscillatory solution diverging from the starting point $t=0$, $y=0$, with amplitude increasing as the sixth power of time, and frequency decreasing as time increases. If t is negative, result (53) represents an oscillatory solution converging from negative time to the point $t=0$, $y=0$, with amplitude decreasing as the sixth power of the time remaining before $t=0$ is reached, and frequency increasing as this time reduces. These two solutions are respectively analogous to the divergent and convergent oscillatory modes of the linear Hamiltonian system.

There are certain further properties of these solutions which provide useful clues as to the nature of the corresponding exact solutions of the non-linear Hamiltonian system. From (53) the zeros of y occur at $t = t_r$, where

$$t_r = t_0 \exp(-r\pi/\beta_2) \quad (r \text{ integral}). \tag{57}$$

Hence the ratio of a pair of consecutive intervals between zeros is the constant

$$\exp(-\pi/\beta_2). \tag{58}$$

Furthermore, it is shown in Appendix 1 that for any solution (53)

$$(-1)^r y^{(i)}(t_r)(t_{r+1} - t_r)^{-i} \quad (i = 1, 2, \ldots 5)$$

are constants independent of r.

§ 6. Trajectories without Switches

It was pointed out in § 3.5 that backwards computation from near the origin of the non-linear Hamiltonian system yields trajectories without switches. In this section we obtain exact expressions for these switchless trajectories. A proof that these trajectories are absolutely optimum, not merely locally optimum, is given in Appendix 2.

6.1. *Convergent Switchless Trajectories*

We consider first the case when the relay output is positive during the forward motion until the origin is reached†. For convenience we

† After the origin is reached the relay output becomes zero. For linguistic convenience we do not count this change to zero as a switch.

take the time origin as the time at which the y-space origin is reached. Then eqns. (21) become:

$$\left.\begin{aligned} y_i' &= y_{i+1} \quad (i=1,2,\ldots 5), \\ y_6' &= \begin{cases} 1 & (t<0), \\ 0 & (t\geqslant 0), \end{cases} \end{aligned}\right\} \tag{59}$$

with boundary conditions:

$$y_i(0)=0 \quad (i=1,2,\ldots 6). \tag{60}$$

The solution of (59) with (60) is:

$$y_i = \left\{\begin{array}{ll} \dfrac{t^{7-i}}{(7-i)!} & (t\leqslant 0) \\ 0 & (t\geqslant 0) \end{array}\right\} \quad (i=1,2,\ldots 6). \tag{61}$$

Similarly, taking the case when the relay output is negative during the forward motion until the origin is reached, we find:

$$y_i = \left\{\begin{array}{ll} -\dfrac{t^{7-i}}{(7-i)!} & (t\leqslant 0) \\ 0 & (t\geqslant 0) \end{array}\right\} \quad (i=1,2,\ldots 6). \tag{62}$$

The solutions (61) and (62) of the non-linear Hamiltonian system are both† analogous to the convergent monotonic mode of the linear Hamiltonian system.

6.2. *Divergent Switchless Trajectories*

To find the analogue of the divergent monotonic mode, we consider that the non-linear Hamiltonian system is at the origin, with zero relay output, until $t=0$; then, due to some slight disturbance, the relay output is made positive. The eqns. (21) are now:

$$\left.\begin{aligned} y_i' &= y_{i+1} \quad (i=1,2,\ldots 5), \\ y_6' &= \begin{cases} 0 & (t\leqslant 0), \\ 1 & (t>0), \end{cases} \end{aligned}\right\} \tag{63}$$

with boundary conditions:

$$y_i(0)=0 \quad (i=1,2,\ldots 6). \tag{64}$$

† Solution (61) is analogous to the linear solution with $A_6>0$ and solution (62) is analogous to the linear solution with $A_6<0$.

The solution of (63) with (64) is:

$$y_i = \begin{cases} 0 & (t \leqslant 0) \\ \dfrac{t^{7-i}}{(7-i)!} & (t \geqslant 0) \end{cases} \quad (i=1,2,\dots 6). \tag{65}$$

Similarly, taking the case when the relay output is made negative after $t=0$, we find:

$$y_i = \begin{cases} 0 & (t \leqslant 0) \\ -\dfrac{t^{7-i}}{(7-i)!} & (t \geqslant 0) \end{cases} \quad (i=1,2,\dots 6). \tag{66}$$

The solutions (65) and (66) of the non-linear Hamiltonian system are both analogous to the divergent monotonic mode of the linear Hamiltonian system.

§ 7. Trajectories with Switches

In this section and the following we investigate solutions of the non-linear Hamiltonian system which involve switches. It is shown in Appendix 3 that any switch on an optimum trajectory must be followed by an infinite number of switches.

7.1. *Switch Points of the Hamiltonian System*

Consider a solution of (21) which starts at time t_0, and has switches (zeros of y_1) at times $t_1, t_2, \dots t_r, \dots$. During the interval between t_r and t_{r+1}, eqns. (21) are:

$$\left.\begin{aligned} y_i' &= y_{i+1} \quad (i=1,2,\dots 5), \\ y_6' &= \operatorname{sgn} y_1(t_r+). \end{aligned}\right\} \tag{67}$$

The solution of (67), in terms of the values $y_j(t_r)$ at the beginning of the interval, is:

$$y_i(t) = \sum_{j=i}^{6} y_j(t_r)\frac{(t-t_r)^{j-i}}{(j-i)!} + \frac{(t-t_r)^{7-i}}{(7-i)!}\operatorname{sgn} y_1(t_r+) \quad (i=1,2,\dots 6). \tag{68}$$

In fact, (68) is the Taylor series for $y_i(t)$, $t_r \leqslant t \leqslant t_{r+1}$.

Let the interval between t_r and t_{r+1} be denoted by T_r:

$$T_r = t_{r+1} - t_r. \tag{69}$$

From (68) and (69), the y_i at the end of the interval are:

$$y_i(t_{r+1}) = \sum_{j=i}^{6} y_j(t_r)\frac{T_r^{\,j-i}}{(j-i)!} + \frac{T_r^{\,7-i}}{(7-i)!}\operatorname{sgn} y_1(t_r+) \quad (i=1,2,\dots 6). \tag{70}$$

Since, from the definition of the t_r, $y_1(t_r)=y_1(t_{r+1})=0$, the first of eqns. (70) reduces to:

$$0=y_2(t_r)+\frac{1}{2!}y_3(t_r)T_r+\frac{1}{3!}y_4(t_r)T_r^2+\frac{1}{4!}y_5(t_r)T_r^3+\frac{1}{5!}y_6(t_r)T_r^4$$

$$+\frac{1}{6!}\operatorname{sgn} y_1(t_r+)T_r^5. \quad (71)$$

By definition, T_r is the smallest positive root of eqn. (71). When this value of T_r is substituted into the remaining equations of (70), the latter give the values of the y_i at the beginning of a switching interval in terms of the values of the y_i at the beginning of the previous interval†.

7.2. *A Property of the Switch Points*

The switch points of the Hamiltonian system have the following property, which we shall make use of later. Suppose we have found a particular solution of (70) and (71), say

$$\left.\begin{aligned} T_r&=\tau_r \quad (r=1,2,\ldots),\\ y_i(t_r)&=\theta_{ir} \quad (r=1,2,\ldots;\ i=1,2,\ldots 6), \end{aligned}\right\} \quad (72)$$

where τ_r and θ_{ir} are certain numbers. Then another solution is:

$$\left.\begin{aligned} T_r&=k\tau_r \quad (r=1,2,\ldots),\\ y_i(t_r)&=k^{7-i}\theta_{ir} \quad (r=1,2,\ldots;\ i=1,2,\ldots 6), \end{aligned}\right\} \quad (73)$$

where k is an arbitrary positive constant.

The result can be verified by noting that if (73) is substituted into (70) and (71) then k cancels out, leaving (70) and (71) with (72) substituted.

7.3. *Change of Variables*

To simplify eqns. (70) and (71), we define new variables as follows:

$$\left.\begin{aligned} b_r&=y_2(t_r)T_r^{-5}\operatorname{sgn} y_1(t_r+),\\ c_r&=y_3(t_r)T_r^{-4}\operatorname{sgn} y_1(t_r+),\\ d_r&=y_4(t_r)T_r^{-3}\operatorname{sgn} y_1(t_r+),\\ e_r&=y_5(t_r)T_r^{-2}\operatorname{sgn} y_1(t_r+),\\ f_r&=y_6(t_r)T_r^{-1}\operatorname{sgn} y_1(t_r+), \end{aligned}\right\} \quad (74)$$

and also

$$\rho_r=T_{r+1}/T_r. \quad (75)$$

† It may be noted, from (21), that the y_i are continuous functions of time, even at the switching instants.

Then eqns. (70) and (71) become:

$$\left.\begin{aligned}
0 &= \frac{1}{1!}b_r + \frac{1}{2!}c_r + \frac{1}{3!}d_r + \frac{1}{4!}e_r + \frac{1}{5!}f_r + \frac{1}{6!}, \\
-\rho_r^{5}b_{r+1} &= b_r + \frac{1}{1!}c_r + \frac{1}{2!}d_r + \frac{1}{3!}e_r + \frac{1}{4!}f_r + \frac{1}{5!}, \\
-\rho_r^{4}c_{r+1} &= c_r + \frac{1}{1!}d_r + \frac{1}{2!}e_r + \frac{1}{3!}f_r + \frac{1}{4!}, \\
-\rho_r^{3}d_{r+1} &= d_r + \frac{1}{1!}e_r + \frac{1}{2!}f_r + \frac{1}{3!}, \\
-\rho_r^{2}e_{r+1} &= e_r + \frac{1}{1!}f_r + \frac{1}{2!}, \\
-\rho_r^{1}f_{r+1} &= f_r + \frac{1}{1!}.
\end{aligned}\right\} \tag{76}$$

These are six simultaneous non-linear difference equations in the six unknowns $\rho_r, b_r, c_r, d_r, e_r, f_r$.

Given the general solution of (76), the general solution for $y_i(t_r)$ could be obtained using (74) and (75). Since ρ_r defines only the ratio T_{r+1}/T_r this general solution would contain an additional arbitrary constant, say T_1, equivalent to the arbitrary constant k of § 7.2.

§ 8. Constant Ratio Trajectories

It follows from the definition of ρ_r, b_r, c_r, d_r, e_r, and f_r that, if all these quantities are constant, independent of r, then the corresponding solution of the non-linear Hamiltonian system will have the properties, described in § 5.4, of the oscillatory solution obtained by the describing function method. For such trajectories, successive time intervals between switches decrease in geometric progression, and, accordingly, we describe them as *constant ratio* trajectories. Exact expressions for all constant ratio trajectories are derived in this section.

8.1. *Equations of Constant Ratio Trajectories*

Denoting ρ_r, b_r, c_r, d_r, e_r and f_r by ρ, b, c, d, e and f, eqns. (76) become:

$$\left.\begin{aligned}
\frac{1}{1!}b + \frac{1}{2!}c + \frac{1}{3!}d + \frac{1}{4!}e + \frac{1}{5!}f + \frac{1}{6!} &= 0, \\
(1+\rho^5)b + \frac{1}{1!}c + \frac{1}{2!}d + \frac{1}{3!}e + \frac{1}{4!}f + \frac{1}{5!} &= 0, \\
(1+\rho^4)c + \frac{1}{1!}d + \frac{1}{2!}e + \frac{1}{3!}f + \frac{1}{4!} &= 0, \\
(1+\rho^3)d + \frac{1}{1!}e + \frac{1}{2!}f + \frac{1}{3!} &= 0, \\
(1+\rho^2)e + \frac{1}{1!}f + \frac{1}{2!} &= 0, \\
(1+\rho)f + \frac{1}{1!} &= 0.
\end{aligned}\right\} \tag{77}$$

If we regard ρ as a parameter, (77) is a system of six linear equations in the five unknowns b, c, d, e, f. It will be soluble only if the determinant of the coefficients is zero, i.e. ρ must satisfy:

$$\begin{vmatrix} \frac{1}{1!} & \frac{1}{2!} & \frac{1}{3!} & \frac{1}{4!} & \frac{1}{5!} & \frac{1}{6!} \\ 1+\rho^5 & \frac{1}{1!} & \frac{1}{2!} & \frac{1}{3!} & \frac{1}{4!} & \frac{1}{5!} \\ \cdot & 1+\rho^4 & \frac{1}{1!} & \frac{1}{2!} & \frac{1}{3!} & \frac{1}{4!} \\ \cdot & \cdot & 1+\rho^3 & \frac{1}{1!} & \frac{1}{2!} & \frac{1}{3!} \\ \cdot & \cdot & \cdot & 1+\rho^2 & \frac{1}{1!} & \frac{1}{2!} \\ \cdot & \cdot & \cdot & \cdot & 1+\rho & \frac{1}{1!} \end{vmatrix} = 0. \tag{78}$$

We will denote the determinant in (78) as D_6, and the minor of ith order formed from a square in the bottom right-hand corner as D_i. If we solve eqns. (77) by means of Cramer's rule, we find, after manipulation of the determinants involved,

$$\left.\begin{aligned} b &= \frac{-D_5}{(1+\rho)(1+\rho^2)(1+\rho^3)(1+\rho^4)(1+\rho^5)}, \\ c &= \frac{D_4}{(1+\rho)(1+\rho^2)(1+\rho^3)(1+\rho^4)}, \\ d &= \frac{-D_3}{(1+\rho)(1+\rho^2)(1+\rho^3)}, \\ e &= \frac{D_2}{(1+\rho)(1+\rho^2)}, \\ f &= \frac{-D_1}{(1+\rho)}. \end{aligned}\right\} \tag{79}$$

The D_i are polynomials of $\frac{1}{2}i(i-1)$th degree in ρ. By expanding D_i in terms of the elements of its first row, it is seen to satisfy the recurrence relation:

$$D_i = D_{i-1} - \frac{1}{2!}(1+\rho^{i-1})D_{i-2} + \frac{1}{3!}(1+\rho^{i-1})(1+\rho^{i-2})D_{i-3} - \dots + (-1)^{i-1}\frac{1}{i!}(1+\rho^{i-1})(1+\rho^{i-2})\dots(1+\rho)D_0, \qquad (80)$$

where $D_0 = 1$. Successive applications of (80) yield:

$$\left.\begin{aligned}
D_1 &= \frac{1}{1!}\\
D_2 &= \frac{1}{2!}(1-\rho)\\
D_3 &= \frac{1}{3!}(1-2\rho-2\rho^2+\rho^3)\\
D_4 &= \frac{1}{4!}(1-3\rho-5\rho^2+5\rho^4+3\rho^5-\rho^6)\\
D_5 &= \frac{1}{5!}(1-4\rho-9\rho^2-3\rho^3+12\rho^4+22\rho^5+12\rho^6-3\rho^7-9\rho^8-4\rho^9+\rho^{10})\\
D_6 &= \frac{1}{6!}(1-5\rho-14\rho^2-9\rho^3+21\rho^4+61\rho^5+70\rho^6+31\rho^7-31\rho^8-70\rho^9\\
&\qquad -61\rho^{10}-21\rho^{11}+9\rho^{12}+14\rho^{13}+5\rho^{14}-\rho^{15}).
\end{aligned}\right\} \qquad (81)$$

8.2. *Zeros of a Polynomial*

The polynomials on the right-hand sides of eqns. (81) are *reciprocal* polynomials, and may be factorized by standard techniques (see, e.g. Burnside and Panton (1904)). In particular, it is shown in Appendix 4 that the factors of $D_6(\rho)$ are:

$$D_6(\rho) = \frac{1}{6!}(1-\rho)(1+\rho)^2(1+\rho^2)(1+\rho+\rho^2)(1-\sigma_1\rho+\rho^2)(1-\sigma_2\rho+\rho^2)(1-\sigma_3\rho+\rho^2)(1-\sigma_4\rho+\rho^2), \qquad (82)$$

where σ_1, σ_2, σ_3, and σ_4 are the roots of the auxiliary equation:

$$\sigma^4 - 7\sigma^3 - 6\sigma^2 + 29\sigma + 23 = 0. \qquad (83)$$

By computation, the roots of (83) are:

$$\left.\begin{aligned}
\sigma_1 &= -0{\cdot}797994928,\\
\sigma_2 &= -0{\cdot}172780181,\\
\sigma_3 &= 2{\cdot}31264295,\\
\sigma_4 &= 7{\cdot}21315379.
\end{aligned}\right\} \qquad (84)$$

From (82) and (84) the zeros of $D_6(\rho)$, and thus the roots of (78), are:

$$\left.\begin{aligned} \rho_1 &= 1, \\ \rho_2 &= 1/\rho_3 = -1, \\ \rho_4 &= 1/\rho_5 = j, \\ \rho_6 &= 1/\rho_7 = -\tfrac{1}{2}+j\tfrac{1}{2}\sqrt{3}, \\ \rho_8 &= 1/\rho_9 = -0{\cdot}398997464+j0{\cdot}916952029, \\ \rho_{10} &= 1/\rho_{11} = -0{\cdot}863900907+j0{\cdot}503661814, \\ \rho_{12} &= 1/\rho_{13} = 0{\cdot}141407793, \\ \rho_{14} &= 1/\rho_{15} = 0{\cdot}575736118. \end{aligned}\right\} \quad (85)$$

8.3. *Selection of Relevant Zeros*

We will now show that the necessary and sufficient conditions for a ρ_k in (85) to be relevant solution of (77) are:

$$\rho_k \text{ is real and positive} \quad (86)$$

and

$$b(\rho_k) > 0, \quad (87)$$

where $b(\rho_k)$ is given by the first of eqns. (79).

The necessity of condition (86) follows immediately from the definition of ρ (see (75)). The necessity of condition (87) is proved by noting, from definitions (74), that

$$b = y_2(t_r)T_r^{-5} \operatorname{sgn} y_1(t_r+). \quad (88)$$

But $y_2(t_r)$ has the same sign as $y_1(t_r+)$, since, by definition, $y_2(t) = y_1'(t)$ and $y_1(t_r) = 0$. Hence b must be positive.

It remains to show that conditions (86) and (87) are sufficient. We have to show that the root ρ_k, when substituted back, yields a solution $y_1(t)$ which is of constant sign during the interval t_r to t_{r+1}, i.e. we have to verify that the corresponding value of T_r is the *smallest* positive root of (71). A proof that this is the case if conditions (86) and (87) are satisfied is given in Appendix 5.

From (85), the only roots which satisfy condition (86) are ρ_1, ρ_{12}, ρ_{13}, ρ_{14}, ρ_{15}. By substitution, it is found that the only ones of these remaining roots which satisfy condition (87) are ρ_{14} and ρ_{15}†.

8.4. *Discussion of Constant Ratio Trajectories*

The value of ρ_{14} and corresponding values of b, c, d, e and f define (to within an arbitrary constant‡) a trajectory of the Hamiltonian system

† It can be shown that ρ_1, ρ_{12} and ρ_{13} correspond to valid solutions of the equation

$$\frac{d^6y}{dt^6} = -\operatorname{sgn} y.$$

‡ See § 7.2.

which is such that the ratio of any two switching intervals is the constant ρ_{14}: moreover the ratio of $|y_i|$ at the beginning of a switching interval to the ith power of the interval is a constant for all intervals of the trajectory. Since $\rho_{14} < 1$, the trajectory approaches the origin of y-space, reaching it in finite time. The value ρ_{15} defines a similar type of trajectory, but since $\rho_{15} > 1$ this trajectory diverges from the origin.

The constant ratio trajectories with $\rho = \rho_{14}$ are analogous to the trajectories of the convergent oscillatory mode of the linear Hamiltonian system. The constant ratio trajectories with $\rho = \rho_{15}$ are analogous to the trajectories of the divergent oscillatory mode of the linear Hamiltonian system. Taking account of the switchless trajectories (§ 6), we have now found analogues of all the modes of the linear Hamiltonian system.

§ 9. Computation of the Optimum Switching Surface

By analogy with the linear case (§ 4.4) we may expect to obtain general trajectories with switches by starting close to the origin and near to a constant ratio trajectory, and computing backwards in time. This turns out to be the case.

9.1. *Nature of the Optimum Switching Surface*

In § 7.2 it was shown that if $y_i = \theta_{ir}$ $(i = 1, 2, \ldots 6;\ r = 1, 2, \ldots)$ is a sequence of switch points of the Hamiltonian system, then so is $y_i = k^{7-i}\theta_{ir}$ $(i = 1, 2, \ldots 6;\ r = 1, 2, \ldots)$. It follows that if the point (x_1, x_2, x_3) is in the optimum switching surface, so is the point (k^3x_1, k^2x_2, kx_3). Here k is an arbitrary positive constant.

If (x_1, x_2, x_3) is a fixed point, and k varies, the point (k^3x_1, k^2x_2, kx_3) describes a curve in the switching surface. This curve sweeps over the switching surface if (x_1, x_2, x_3) varies along any suitable locus in the switching surface. Thus to determine the switching surface, we need only specify it along a suitable locus. Further discussion of this result is given in Appendix 6.

Suppose we take as coordinates of the phase space, not x_1, x_2 and x_3, but z_1, z_2 and z_3, where

$$\left.\begin{aligned} z_1 &= |ax_1|^{1/3}\,\mathrm{sgn}\,x_1, \\ z_2 &= |ax_2|^{1/2}\,\mathrm{sgn}\,x_2, \\ z_3 &= ax_3. \end{aligned}\right\} \tag{89}$$

Then, if (z_1, z_2, z_3) is a point on the optimum switching surface, so is (kz_1, kz_2, kz_3), where k is a positive parameter. Thus in z-space the optimum switching surface is a cone made up of straight line generators through the origin.

9.2. *Computational Results*

Rather than compute complete trajectories, it is simpler to compute sequences of switch points. We use, in effect, the recurrence relations (76), but, since we wish to trace backwards in time, we write (76) as:

$$\left.\begin{aligned}
0 &= \frac{1}{1!}b_r\rho_{r-1}^{\,5} - \frac{1}{2!}c_r\rho_{r-1}^{\,4} + \frac{1}{3!}d_r\rho_{r-1}^{\,3} - \frac{1}{4!}e_r\rho_{r-1}^{\,2} + \frac{1}{5!}f_r\rho_{r-1} + \frac{1}{6!}\\
b_{r-1} &= -b_r\rho_{r-1}^{\,5} + \frac{1}{1!}c_r\rho_{r-1}^{\,4} - \frac{1}{2!}d_r\rho_{r-1}^{\,3} + \frac{1}{3!}e_r\rho_{r-1}^{\,2} - \frac{1}{4!}f_r\rho_{r-1} - \frac{1}{5!}\\
c_{r-1} &= -c_r\rho_{r-1}^{\,4} + \frac{1}{1!}d_r\rho_{r-1}^{\,3} - \frac{1}{2!}e_r\rho_{r-1}^{\,2} + \frac{1}{3!}f_r\rho_{r-1} + \frac{1}{4!}\\
d_{r-1} &= -d_r\rho_{r-1}^{\,3} + \frac{1}{1!}e_r\rho_{r-1}^{\,2} - \frac{1}{2!}f_r\rho_{r-1} - \frac{1}{3!}\\
e_{r-1} &= -e_r\rho_{r-1}^{\,2} + \frac{1}{1!}f_r\rho_{r-1} + \frac{1}{2!}\\
f_{r-1} &= -f_r\rho_{r-1} - \frac{1}{1!}
\end{aligned}\right\} \qquad (90)$$

(The last of eqns. (90) is obtained directly from the last of (76), then the remaining equations are obtained successively in upwards order.) If b_r, c_r, d_r, e_r and f_r are given, ρ_{r-1} can be computed from the largest root of the first of eqns. (90); then $b_{r-1}, c_{r-1}, d_{r-1}, e_{r-1}$ and f_{r-1} can be computed from the rest of eqns. (90).

For a constant ratio trajectory which converges (in forward time) to the origin, the values of b, c, d, e, f are found by substituting $\rho=\rho_{14}$ in (79), and are:

$$\left.\begin{aligned}
b &= 0{\cdot}0053895997,\\
c &= -0{\cdot}025271563,\\
d &= 0{\cdot}041597667,\\
e &= 0{\cdot}10110916,\\
f &= -0{\cdot}63462403.
\end{aligned}\right\} \qquad (91)$$

In calculating a sequence of switch points, the initial values b_0, c_0, d_0, e_0 and f_0 are chosen close to the values (91), then recurrence relations (90) are used.

This procedure was found to work well in practice, using a digital computer (EDSAC). The initial values† b_0, c_0, d_0, e_0 and f_0 were given fractional discrepancies of about 10^{-4} from the values (91). It was then

† The particular set of values used was based on the exact solution of (90) linearized about the stationary solution (91). Details are given in Appendix 7.

found that about six switch points could be obtained before the trajectory became dominated by the switchless 'mode'. The results of the computations are shown in fig. 3 and the table. These give the spherical coordinates of the switching surface whose cartesian coordinates are $|ax_1|^{1/3}\,\mathrm{sgn}\,x_1$, $|ax_2|^{1/2}\,\mathrm{sgn}\,x_2$ and ax_3 (see § 9.1).

There is one particular solution of (90) appropriate to any step change in the desired value of the plant output. The principal properties (value of the performance criterion, settling time and first overshoot) for this step response are derived in Appendix 10, and displayed in fig. 4. For comparison, the step responses for plants with transfer functions $1/(ap)$ and $1/(ap^2)$ are also shown in fig. 4, with numerical details.

Fig. 3

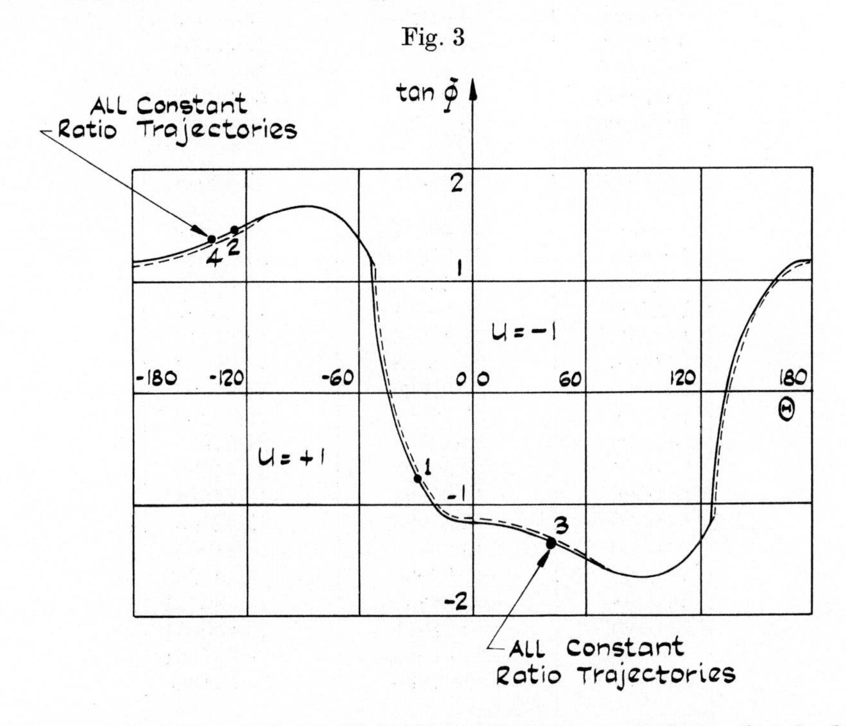

Computed switching surface. If figure is rolled into a cylinder so that $\Theta = 180$ coincides with $\Theta = -180$ for all Φ, the switching surface with

$$|ax_1|^{1/3}\,\mathrm{sgn}\,x_1,\ |ax_2|^{1/2}\,\mathrm{sgn}\,x_2 \text{ and } x_3$$

as cartesian coordinates to equal scales is generated by straight lines joining (Θ, Φ) to $(\Theta \pm 180, -\Phi)$.

——— minimum integral-square-error.

– – – minimum settling time.

$$\Phi = \tan^{-1}\frac{ax_3}{(|ax_2| + |ax_1|^{2/3})^{1/2}};$$

$$\Theta = \tan^{-1}\frac{|ax_2|^{1/2}\,\mathrm{sgn}\,x_2}{|ax_1|^{1/3}\,\mathrm{sgn}\,x_1}.$$

Numbers 1, 2, . . indicate 1st, 2nd, . . . switches of step response.

Polar coordinates of switching surface

ρ_r/ρ	Θ degrees	Φ degrees
0·610000	125·70332	−50·43700
0·700000	123·89160	−51·79533
0·760000	121·29062	−53·30374
0·800000	117·83488	−54·82640
0·840000	108·75029	−57·34493
0·850000	100·06922	−58·54433
0·860000	72·537497	−57·93608
0·870000	67·462613	−57·31711
0·880000	64·011867	−56·83821
0·900000	58·923544	−56·06955
0·920000	54·939880	−55·43505
0·940000	51·499661	−54·87785
0·960000	48·365219	−54·37178
0·980000	45·407378	−53·90226
1·000000	42·542922	−53·46014
1·05000	35·406496	−52·43879
1·10000	27·612300	−51·49682
1·15000	17·566120	−50·60448
1·20000	−10·85511	−48·70189
1·25000	−21·11676	−44·61216
1·30000	−26·28483	−40·96624
1·40000	−32·21626	−34·64293
1·50000	−35·72759	−29·26286
1·60000	−38·12402	−24·57918
1·70000	−39·89401	−20·44085
1·80000	−41·26994	−16·74614
1·90000	−42·37898	−13·42105
2·00000	−43·29712	−10·41009
3·00000	−47·97900	9·130263
3·99998	−49·88259	19·186621
4·99999	−50·94011	25·295571
5·99996	−51·61384	29·397161
6·99997	−52·07860	32·341860

$\rho=0{\cdot}575736118$.
ρ_r = ratio of next but one switching interval to next interval.

9.3. *Accuracy*

There is evidence to suggest that the computed switching surface of § 9.2 is highly accurate.

In the first place one can calculate the accuracy of the computational solution of the analogous linear Hamiltonian system (§ 4.4). It is found that if, in tracing backwards from an initial point near the divergent oscillatory mode, one obtains as many as six zeros of y, the fractional discrepancy† in the coefficients of the controller is within about 10^{-15}.

† This assumes that the data on which the controller coefficients are computed is based on the sixth zero of y.

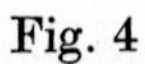

Fig. 4

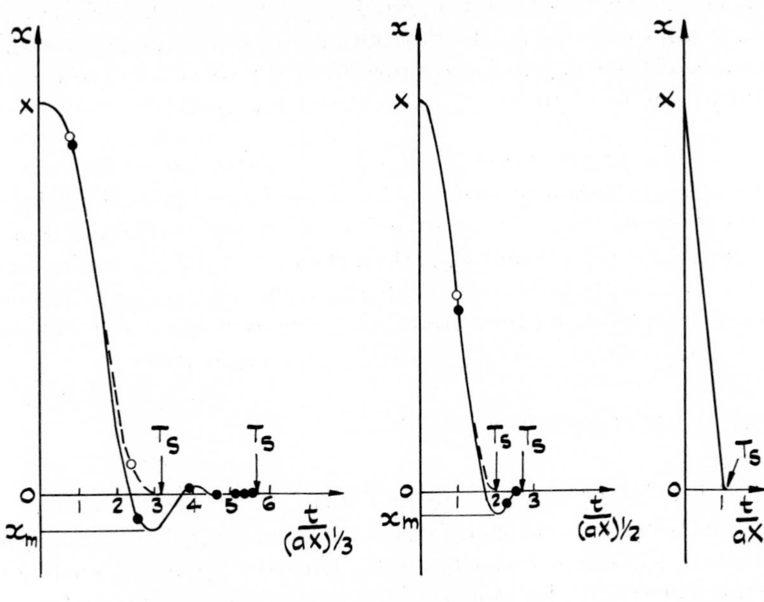

	$1/ap^3$	$1/ap^2$	$1/ap$
i.s.e.	$1{\cdot}2521(aX)^{1/3}X^2$ $(1{\cdot}2665(aX)^{1/3}X^2)$	$0{\cdot}76402(aX)^{1/2}X^2$ $(0{\cdot}76667(aX)^{1/2}X^2)$	$\frac{1}{3}aX^3$ $(\frac{1}{3}aX^3)$
T_s	$5{\cdot}6873(aX)^{1/3}$ $(3{\cdot}1748(aX)^{1/3})$	$2{\cdot}7152(aX)^{1/2}$ $(2(aX)^{1/2})$	aX (aX)
$-x_m$	$0{\cdot}0965X$ (0)	$0{\cdot}0586X$ (0)	0 (0)
n	∞ (2)	∞ (1)	0 (0)

Step response of systems under optimal control.
——— minimum integral-square-error (i.s.e.); —●— switch point.
– – – minimum settling time (T_s); – –○– – switch point.
n = number of switches except start or finish. Numerical data are for minimum i.s.e. trajectories, with data for minimum T_s trajectories in brackets. Saturation constraint on control input u is $|u| \leqslant 1$.

Secondly, Eggleston (1963) has, in effect, traced backward solutions of the Hamiltonian system for the case when the plant consists of two integrators, and found a remarkably accurate computed switching curve. He found the accuracy increased with the number of switches obtained along the backwards trajectory. After only four switches, the switching curve was in error by about one part in 10^{10}. Eggleston was dealing with a second-order plant, and, moreover, his performance criterion was integral-modulus-error, not integral-square-error. Nevertheless his results suggest that out results are also highly accurate.

An analysis of optimum trajectories which lie close to constant ratio trajectories is given in Appendix 7. From this analysis, estimates can be given of the rapidity with which errors (introduced by the arbitrary starting point of the computation) decay during the first few computed switch points, i.e. while the computed trajectory is still fairly close to a constant ratio trajectory. In this initial region, fractional errors in the differences of the computed trajectory from the constant ratio trajectory decay by a factor of about 0·008 at each switch point.

9.4. *Comparison with Minimum Settling Time Case*

When the performance criterion to be minimized is settling time, i.e. time to reach the origin of x-space (the plant and constraint being the same as before, see (1) and (2)), the optimum switching surface is well known. It consists of the set of single switch trajectories to the origin. The explicit equation for this surface was given by Feldbaum (1955), and is:

$$0 = x_1 + \tfrac{1}{3}a^2x_3^3 + [ax_2x_3 + a^{1/2}\{\tfrac{1}{2}ax_3^2 + x_2 \operatorname{sgn}(x_2 + \tfrac{1}{2}ax_3|x_3|)\}^{3/2}] \operatorname{sgn}(x_2 + \tfrac{1}{2}ax_3|x_3|). \quad (92)$$

In z-space (see (89)) this surface is:

$$0 = z_1^3 + \tfrac{1}{3}z_3^3 + [z_2|z_2|z_3 + \{\tfrac{1}{2}z_3^2 + z_2|z_2| \operatorname{sgn}(z_2|z_2| + \tfrac{1}{2}z_3|z_3|)\}^{3/2}] \operatorname{sgn}(z_2|z_2| + \tfrac{1}{2}z_3|z_3|). \quad (93)$$

Direct substitution shows that if values z_1, z_2 and z_3 satisfy (93) so do values kz_1, kz_2 and kz_3, where k is arbitrary. Thus the minimum settling time switching surface is also composed of straight lines through the origin of z-space. It is represented by the broken line in fig. 3. Figure 3 shows that the switching surfaces for minimum integral-square-error and minimum settling time are close to one another.

The step responses for third, second and first-order plants under optimum control for minimum settling time are shown in fig. 4. On comparing these responses with those based on the integral-square-error criterion, it appears that there is only a small increase in the numerical value of the integral-square-error when the minimum settling time criterion is used.

§ 10. Extension to the General Problem

It is conjectured that the techniques of the previous sections extend to more general problems as follows.

For a plant with transfer function $1/(ap^n)$, and for a performance criterion integral-square-error, there will be $(n-1)/2$ distinct types of constant ratio trajectory and one type of switchless trajectory if n odd, and $n/2$ distinct types of constant ratio trajectory if n even†. These

† The authors have confirmed this conjecture for $n=4$. Using the obvious extension of the notation of earlier sections, the two valid and distinct types of constant ratio trajectory are defined by:

$$\begin{aligned}
\rho &= +0{\cdot}10018783, & &+0{\cdot}74026837\\
b &= +0{\cdot}000032739, & &+0{\cdot}00053248\\
c &= -0{\cdot}00043982, & &-0{\cdot}00207564\\
d &= +0{\cdot}00380036, & &+0{\cdot}00133518\\
e &= -0{\cdot}02433724, & &+0{\cdot}01655976\\
f &= +0{\cdot}11695296, & &-0{\cdot}05153327\\
g &= -0{\cdot}40487175, & &-0{\cdot}04820681\\
h &= +0{\cdot}90893570, & &+0{\cdot}57462401
\end{aligned}$$

will correspond to the trajectories of the modes of the analogous linear system. The calculation of the constant ratio trajectories will proceed in the same way as in § 8. When tracing trajectories backwards, in order to find the optimum switching surface, it will be necessary to start near a constant ratio trajectory.

A plant with transfer function $1/(a_np^n+a_{n-1}p^{n-1}+\ldots+a_0)$ behaves like a pure integrator plant $1/(a_np^n)$ when the phase point is near the origin. To see this, note that the differential equation of the plant is:

$$a_n\frac{d^nx}{dt^n}+a_{n-1}\frac{d^{n-1}x}{dt^{n-1}}+\ldots+a_0x=u. \tag{94}$$

On the right-hand side, $u=\pm 1$. Thus if

$$\frac{d^{n-1}x}{dt^{n-1}},\quad \frac{d^{n-2}x}{dt^{n-2}},\ldots,x$$

are small compared with $a_{n-2}^{-1}, a_{n-1}^{-1},\ldots,a_0^{-1}$, respectively, eqn. (94) becomes approximately:

$$a_n\frac{d^nx}{dt^n}=u, \tag{95}$$

i.e. the equation of a pure integrator plant. It follows that we can compute the optimum switching surface for a plant with a transfer function without zeros by tracing backwards the solution of the Hamiltonian system, starting from points near a constant ratio trajectory of a pure integrator plant.

Many of the techniques of the present paper will apply for the more general performance criterion $\int_0^\infty |x|^\nu dt$, $\nu>0$. There will be constant ratio trajectories, and if ν is an integer the corresponding values of ρ will be given by a high order polynomial equation. For still more general performance criteria $\int_0^\infty g(x)\,dt$ provided $g(x)$ behaves like $|x|^\nu$ when $|x|$ is small, our methods will still indicate the correct starting procedure in backward tracing.

§ 11. Conclusions

In this paper the optimum switching surface has been found for a plant with transfer function $1/(ap^3)$ between saturable input $u(|u|\leqslant 1)$ and output x, with a performance criterion integral-square-error. The behaviour of the optimum non-linear system is strongly analogous to that of a corresponding linear system. The analogue of the trajectories of the oscillatory mode in the latter is the set of constant ratio trajectories in the former. An essential preliminary in the computation of the switching surface is the calculation of the parameters of the constant ratio trajectories. It is conjectured that similar techniques will yield the optimum switching surface with plants and performance criteria of greater generality.

APPENDIX 1

Properties of the Describing Function Solution

The non-linear Hamiltonian system

$$y_1^{(6)} = \operatorname{sgn} y_1, \tag{96}$$

where

$$y_i = y_1^{(i-1)} \quad (i = 2, 3, \ldots 6), \tag{97}$$

is shown, in § 5, to have an approximate solution (53) for a converging oscillatory trajectory:

$$y_1(t) = \xi_2 t^6 \sin(\beta_2 \log t/t_0), \quad t < 0, \tag{98}$$

where the numerical values of β_2 and ξ_2 are given by (54) and (55). Here, we derive some properties of this solution and, also, compare its agreement with the corresponding exact solution derived in § 8.

At successive 'switch points' (i.e. where $y_1 = 0$) let $t = t_r$, $r = 0, 1, 2, \ldots$, and let the duration of successive switching intervals be $T_r = t_{r+1} - t_r$. We note, first, that

$$t_r = t_0 \exp(-r\pi/\beta_2), \tag{99}$$

whence

$$T_r = (1 - \rho)(-t_r), \tag{100}$$

where

$$\rho = \frac{T_{r+1}}{T_r} = \frac{t_{r+1}}{t_r} = \exp(-\pi/\beta_2) = 0{\cdot}57588171. \tag{101}$$

In § 8, an exact analysis of 'constant ratio' trajectories shows the true numerical value of the ratio $\rho = T_{r+1}/T_r$ to be $0{\cdot}575736118$. Hence the describing function method obtains this constant with a fractional error of less than 3×10^{-4}.

Approximations for $y_i(t)$ $(i = 2, 3, \ldots 6)$ are obtained by differentiating (98) $(i-1)$ times w.r.t. t. This may be done by writing (98) in the form:

$$y_1(t) = \xi_2 t^6 \operatorname{Im} \left(\frac{t}{t_0}\right)^{j\beta_2} \tag{102}$$

to obtain, using (48):

$$y_i(t) = \frac{4}{\pi} \operatorname{Im} \frac{t^{7-i}(t/t_0)^{j\beta_2}}{(1 + j\beta_2)(2 + j\beta_2) \ldots (7 - i + j\beta_2)}. \tag{103}$$

We now derive expressions for y_i at the switch points. From (99)

$$(t_r/t_0)^{j\beta_2} = (-1)^r. \tag{104}$$

Also, since $t_{r+1}/t_0 < t_r/t_0$ and ξ_2 and β_2 are positive, inspection of (98) shows that

$$\operatorname{sgn} y_1(t_r +) = -(-1)^r. \tag{105}$$

Using (105), (104) and (100) in (103), we obtain:

$$y_i(t_r) = \left\{\frac{4}{\pi} \operatorname{Im} \frac{(-1)^i}{(1-\rho)^{7-i}(1 + j\beta_2)(2 + j\beta_2) \ldots (7 - i + j\beta_2)}\right\} T_r^{7-i} \operatorname{sgn} y_1(t_r +)$$

$$(i = 1, 2, \ldots 6). \tag{106}$$

Thus

$$y_i(t_r) = K_i T_r^{7-i} \operatorname{sgn} y_1(t_r+) \quad (i=1,2,\ldots 6), \tag{107}$$

where the K_i are constants equal to the expression in the curly brackets of (106). This establishes that an optimum trajectory derived by the describing function method has the property that $|y_1(t_r)|/T_r^{7-i}$ is a constant at all switch points.

The exact solution for a constant ratio trajectory is defined at the switch points by eqns. (74), which are identical with (107), but with K_2, K_3, K_4, K_5 and K_6 replaced by the constants b, c, d, e and f (see § 8.1). By using a partial fraction expansion of the products in (106), we obtain:

$$K_i = \frac{4}{\pi} \frac{(-1)^i}{(1-\rho)^{7-i}} \sum_{q=1}^{7-i} \frac{(-1)^q}{(q-1)!(7-i-q)!} \frac{\beta_2}{(q^2+\beta_2{}^2)}. \tag{108}$$

On inserting numerical values (from (52) and (101)) for β_2 and ρ into (108) it is found that K_2, K_3, K_4, K_5, K_6 are greater than the values (91) of b, c, d, e, f by fractional errors of about 8×10^{-4}, $3{\cdot}5 \times 10^{-3}$, -3×10^{-2}, $-0{\cdot}017$ and $0{\cdot}24$ respectively. The magnitude of these errors increases with i, because there is a corresponding increase in the number of differentiations of the approximate solution (98) involved.

These errors can be greatly reduced by avoiding differentiation of the approximate solution. If the approximate solution (102) is substituted into the right-hand side of (96), which is then expanded as a Fourier series, we obtain:

$$y_1^{(6)}(t) = \frac{4}{\pi} \sum_{s=1,3,5,\ldots} \frac{1}{s} \operatorname{Im} \left(\frac{t}{t_0}\right)^{j\beta_2 s}. \tag{109}$$

Integrating (109) $(7-i)$ times w.r.t. t yields:

$$y_i(t) = \frac{4}{\pi} \sum_{s=1,3,5,\ldots} \operatorname{Im} \left[\frac{t^{7-i}(t/t_0)^{js\beta_2}}{s(1+js\beta_2)(2+js\beta_2)\ldots(7-i+js\beta_2)}\right] \quad (i=1,2,\ldots 6). \tag{110}$$

Following the same procedure as before to obtain $y_i(t_r)$ (eqns. (102) to (108)), we find:

$$y_i(t_r) = L_i T_r^{7-i} \operatorname{sgn} y_1(t+) \quad (i=1,2,\ldots 6), \tag{111}$$

where the constants L_i are:

$$L_i = \frac{4}{\pi} \frac{(-1)^i}{(1-\rho)^{7-i}} \sum_{s=1,3,5,\ldots} \sum_{q=1}^{7-i} \frac{(-1)^q}{(q-1)!(7-i-q)!} \frac{\beta_2}{q^2+s^2\beta_2} \quad (i=1,2,\ldots 6). \tag{112}$$

After reversing the order of summation, use of the expansion†

$$\tanh\left(\frac{\pi x}{2}\right) = \frac{4x}{\pi}\left(\frac{1}{1+x^2} + \frac{1}{3^2+x^2} + \frac{1}{5^2+x^2} + \ldots\right) \tag{113}$$

† (113) can be derived by replacing x by jx in the well-known expansion:

$$\tan\left[\frac{\pi x}{2}\right] = \frac{4x}{\pi}\left[\frac{1}{1-x^2} + \frac{1}{3^2-x^2} + \frac{1}{5^2-x^2} + \ldots\right].$$

yields

$$L_i = \frac{(-1)^i}{(1-\rho)^{7-i}} \sum_{q=1}^{7-i} \frac{(-1)^q}{q!(7-i-q)!} \frac{1-\rho^q}{1+\rho^q} \quad (i=1,2,\ldots 6), \tag{114}$$

where ρ is given by (101). On inserting numerical values, it is found that L_2, L_3, L_4, L_5, L_6 are greater than the values (91) of the exact solution b, c, d, e, f by fractional errors of $-9{\cdot}6\times10^{-4}$, $1{\cdot}2\times10^{-4}$, $4{\cdot}3\times10^{-4}$, $-4{\cdot}7\times10^{-4}$ and $1{\cdot}1\times10^{-4}$ respectively. We conclude that the approximate solution for constant ratio trajectories, derived by the describing function method, is not only qualitatively correct, but is also remarkably accurate.

It may be verified that the functions of ρ on the right-hand side of (114) are the same as the functions of ρ which define b, c, d, e, and f, (79) with (81). If, as a further refinement to the approximation, ρ is chosen in (114) such that $L_1=0$, then (111) becomes in complete agreement with the exact solution.

APPENDIX 2

Absolute Optimality of the Convergent Switchless Trajectories

The convergent switchless trajectories (61) and (62) satisfy boundary conditions (22) and so correspond to optimum trajectories in (y_4, y_5, y_6) space, and thus in x-space. However, our use of Pontryagin's maximum principle proves only that these trajectories have a kind of local optimality. Let us prove that they are in fact absolutely optimum.

The trajectory in x-space corresponding to (61) is:

$$x_i = \frac{1}{a} \frac{q^{4-i}}{(4-i)!} \quad (q \leqslant 0;\ i=1,2,3), \tag{115}$$

where q is a non-positive parameter. We consider the phase point as starting on this trajectory at the point with parameter value

$$q = q_0 < 0, \tag{116}$$

at time $t=0$ (thus we are using a different time origin from that used in § 6). If the control $u(t)$ is

$$u^*(t) = \begin{cases} 1 & (0 \leqslant t < -q_0), \\ 0 & (-q_0 \leqslant t < \infty), \end{cases} \tag{117}$$

the phase point will follow trajectory (115) to the origin and then stay there. We shall show that any control different† from (117) yields a greater value of the performance criterion $\int_0^\infty x_1^2(t)\,dt$.

From (115) and (116), x_1 begins negative. From (1) and (117), the rate of change of acceleration of x_1 during $0 \leqslant t < -q_0$ is the maximum possible only if $u = u^*$ then; moreover x_1 remains negative then. Therefore

† i.e. a control producing a measurable difference in $x_1(t)$.

during $0 \leqslant t < -q_0$, $x_1(t)$ has the minimum possible magnitude only if $u = u^*$ then. Also, during $-q_0 \leqslant t < \infty$, x_1 is zero, i.e. has the minimum possible magnitude, only if $u = u^*$. Therefore, for all $t \leqslant 0$, $x_1(t)$ has the minimum possible magnitude only if $u = u^*$. Therefore $u^*(t)$ absolutely minimizes

$$\int_0^\infty x_1^2(t)\, dt.$$

Similarly the trajectory in x-space corresponding to (62), namely

$$x_i = -\frac{1}{a}\frac{q^{4-i}}{(4-i)!} \quad (q \leqslant 0; \quad i = 1, 2, 3), \tag{118}$$

is an absolutely optimum trajectory.

APPENDIX 3

Infiniteness of the Number of Switches of an Optimum Trajectory with Switches

We show that an optimum trajectory, i.e. a solution of (21) which leads to the origin of y-space, must have an infinite number of switches, unless it is one of the two switchless trajectories.

Suppose that, if possible, an optimum trajectory has only a finite number $N > 0$ of switches, the last switch being at t_N. Let the interval between the last switch and arrival of the phase point at the origin be T_N. Since a switch is not counted as such if it is at the origin, we have:

$$T_N > 0. \tag{119}$$

y_1 at t_N can be expressed in terms of the y's at $t_N + T_N$ by means of Taylor's series:

$$y_1(t_N) = \sum_{i=0}^{5} y_{i+1}(t_N + T_N)\frac{(-T_N)^i}{i\,!} + \operatorname{sgn}\,[y_1(t_N+)]\frac{T_N{}^6}{6\,!} \tag{120}$$

By definition $y_1(t_N)$ and $y_{i+1}(t_N + T_N)$ $(i = 0, 1, \ldots 5)$ are all zero, and $y_1(t_N+)$ is non-zero. Hence (120) yields:

$$T_N = 0. \tag{121}$$

(121) contradicts (119). Therefore our initial hypothesis that the optimun trajectory has only a finite number of switches is invalid. It follows that if an optimum trajectory has even one switch, then in the neighbourhood of the origin it has an infinite number of switches.

APPENDIX 4

Factors of $D_6(\rho)$

In this Appendix it is shown that the factors of

$$\begin{aligned} 6\,!\,D_6(\rho) = 1 - 5\rho - 14\rho^2 - 9\rho^3 + 21\rho^4 + 61\rho^5 + 70\rho^6 + 31\rho^7 - 31\rho^8 \\ - 70\rho^9 - 61\rho^{10} - 21\rho^{11} + 9\rho^{12} + 14\rho^{13} + 5\rho^{14} - \rho^{15} \end{aligned} \tag{122}$$

are as given by (82) and (83).

G

The factor $(1-\rho)$ is readily found by inspection and divided out. Then the factor $(1+\rho)$ is found by inspection and divided out. Similarly another factor $(1+\rho)$ is found by inspection and divided out. The polynomial remaining is:

$$1-6\rho-7\rho^2-7\rho^3+15\rho^4+32\rho^5+46\rho^6+32\rho^7+15\rho^8-7\rho^9 -7\rho^{10}-6\rho^{11}+\rho^{12}. \tag{123}$$

This is a reciprocal polynomial. Following standard procedure (Burnside and Panton, 1904, p. 90) we substitute:

$$\sigma=\rho+\frac{1}{\rho}. \tag{124}$$

Then (123) becomes:

$$\rho^6(\sigma^6-6\sigma^5-13\sigma^4+23\sigma^3+52\sigma^2+23\sigma). \tag{125}$$

In (125) the factors σ and $(\sigma+1)$ are readily found by inspection and divided out; this yields (125) as:

$$\rho^6\sigma(\sigma+1)(\sigma^4-7\sigma^3-6\sigma^2+29\sigma+23), \tag{126}$$

i.e.

$$\rho^6\sigma(\sigma+1)(\sigma-\sigma_1)(\sigma-\sigma_2)(\sigma-\sigma_3)(\sigma-\sigma_4), \tag{127}$$

where σ_1, σ_2, σ_3, σ_4 are the roots of (83). Finally substitution of (124) in (127) shows that the latter is:

$$(1+\rho^2)(1+\rho+\rho^2)(1-\sigma_1\rho+\rho^2)(1-\sigma_2\rho+\rho^2)(1-\sigma_3\rho+\rho^2)(1-\sigma_4\rho+\rho^2) \tag{128}$$

Thus all the factors given in (82) are accounted for.

APPENDIX 5

Validity of roots ρ_{14} and ρ_{15}

It is shown in § 8.3 that (86) and (87) are necessary conditions for a solution of eqns. (77) to be a valid solution of the Hamiltonian system. These conditions arise from the nature of the definitions (74) and (75), and the requirement that T_r be positive. The conditions (86) and (87) would be sufficient if they also implied compliance with the only remaining restriction on T_r, i.e. that T_r is the *smallest* positive root of eqn. (71). We now show that this is the case, and it then follows that (86) and (87) are sufficient as well as necessary conditions for a valid solution of (77).

Let $P(v)$ be the polynomial of sixth degree defined by

$$P(v)=\frac{1}{1!}bv+\frac{1}{2!}cv^2+\frac{1}{3!}dv^3+\frac{1}{4!}ev^4+\frac{1}{5!}fv^5+\frac{1}{6!}v^6, \tag{129}$$

where b, c, d, e, f are solutions of (77). Let

$$v=(t-t_r)/T_r, \tag{130}$$

where $T_r(T_r>0)$ is derived (to within an arbitrary factor) from the solution of (77), using (75). Then, using definitions (74), we obtain from (130) and (129):

$$\bar{y}_1(t) = P(v) T_r^{\,6} \operatorname{sgn} y_1(t_r+), \tag{131}$$

where

$$\bar{y}_1(t) = \sum_{j=1}^{6} y_j(t_r) \frac{(t-t_r)^{j-1}}{(j-1)!} + \frac{(t-t_r)^6}{6!} \operatorname{sgn} y_1(t_r+). \tag{132}$$

From (129) it is seen that $P(v)$ has a zero at $v=0$. Also, using the first of eqns. (77), $P(v)$ has a zero at $v=1$. Hence, from (130) and (131), $\bar{y}_1(t)$ has zeros at $t=t_r$ and $t=t_r+T_r$. We now show that $\bar{y}_1(t)$ has no zero for $t_r<t<t_r+T_r$; this is equivalent to showing that $P(v)$ has no zero for $0<v<1$.

The theorem of Fourier and Budan (see, e.g. Burnside and Panton, 1904, p. 189) states that if $P(v)$ is any polynomial of degree n, and if $S(v)$ is the number of changes of sign in the sequence

$$P(v), P'(v), \ldots P^{(n)}(v), \tag{133}$$

and Z is the number of zeros of $P(v)$ between $v=\alpha$ and $v=\beta$ $(\alpha<\beta)$, then

$$Z \leqslant S(\alpha) - S(\beta). \tag{134}$$

Now, at $v=0+$, and for $P(v)$ defined by (129), the sequence (133) is:

$$(0+)b, b, c, d, e, f, 1. \tag{135}$$

At $v=1-$, the sequence (133) is:

$$(0+)\rho^5 b, -\rho^5 b, -\rho^4 c, -\rho^3 d, -\rho^2 e, -\rho f, 1, \tag{136}$$

using the first two of eqns. (77) to obtain the first term of the sequence, and using the ith of eqns. (77) to obtain the ith term of the sequence, $i=2,3,\ldots 6$. Now ρ is positive by condition (86). Also f is negative by condition (86) and the last of eqns. (77). Thus the end terms contribute just one change of sign out of a possible two in each of the sequences (135) and (136). Also the number of changes of sign between the remaining terms is the same for each sequence, irrespective of the signs of b, c, d, e, f. Hence $S(0+) - S(1-) = 0$; and so from, (134), $P(v)$ defined by (129) has no zeros for $0<v<1$.

We have now shown that the right-hand side of (132) has no zeros for $t_r<t<t_r+T_r$ and that $t=t_r$ and $t=t_r+T_r$ are zeros. Since the right-hand side of (132) is the same as the right-hand side of the first of eqns. (68) and also, at $t=t_r+T_r$ is the same as the right-hand side of (71), we have proved that T_r, derived from a solution of (77), is the smallest positive root of (71) provided conditions (86) and (87) are satisfied. Hence these conditions are sufficient for a solution of (77) to be valid.

APPENDIX 6

Nature of the Optimum Switching Surface

In this Appendix it is shown that the optimum switching surface can be expressed by a relation between only two variables.

Suppose that the equation of the optimum switching surface is:

$$F(z_1, z_2, z_3) = 0, \tag{137}$$

where the z's are defined by (89). Then, in view of the remark following eqn. (89)

$$F(kz_1, kz_2, kz_3) = 0, \tag{138}$$

where k is an arbitrary positive number. Result (138) can be generalized to

$$F(lz_1, lz_2, lz_3) = 0, \tag{139}$$

where l is an arbitrary positive or negative number. To prove (139), let us note that, from symmetry†

$$F(-z_1, -z_2, -z_3) = 0. \tag{140}$$

Hence also

$$F(-kz_1, -kz_2, -kz_3) = 0, \tag{141}$$

where k is an arbitrary positive number. Result (139) now follows from (138) and (141).

In (139) put

$$l = \frac{1}{z_3}. \tag{142}$$

Then

$$F\left(\frac{z_1}{z_3}, \frac{z_2}{z_3}, 1\right) = 0, \tag{143}$$

i.e.

$$G\left(\frac{z_1}{z_3}, \frac{z_2}{z_3}\right) = 0. \tag{144}$$

(144) is valid for all points (z_1, z_2, z_3) on the switching surface, and is thus the equation of the switching surface. The latter is thus specified by a relation between only two (dimensionless) variables‡,

$$\frac{z_1}{z_3} \text{ and } \frac{z_2}{z_3}.$$

APPENDIX 7

Trajectories near Constant Ratio Trajectories

All optimum trajectories with switches of the non-linear Hamiltonian system are defined by solutions of the non-linear difference equations (76) for $\rho_r, b_r, c_r, d_r, e_r$ and f_r. In this Appendix we obtain solutions of (76) linearized about the stationary solution, ρ, b, c, d, e, and f, derived in § 8. Let

$$\rho_r = \rho + \rho_r', \quad b_r = b + b_r', \quad c_r = c + c_r', \ldots f_r = f + f_r', \tag{145}$$

where ρ, b, c, d, e and f are the constants appropriate to a constant ratio

† The equations specifying the problem, namely (2), (3) and (9) are unchanged if we write $-x_i$ for x_i $(i = 1, 2, 3)$ and $-u$ for u. Thus the solution (137) is unchanged if we write $-z_i$ for z_i $(i = 1, 2, 3)$.

‡ This result has also been obtained by M. M. Newmann (private note), on the basis of dimensional considerations.

trajectory, and the primed symbols are small. Ignoring second-order terms, eqns. (76) become:

$$\left.\begin{aligned}
0 &= \frac{1}{1!}b_r' + \frac{1}{2!}c_r' + \frac{1}{3!}d_r' + \frac{1}{4!}e_r' + \frac{1}{5!}f_r' \\
-\rho^5 b_{r+1}' &= b_r' + \frac{1}{1!}c_r' + \frac{1}{2!}d_r' + \frac{1}{3!}e_r' + \frac{1}{4!}f_r' + 5b\rho^4\rho_r' \\
-\rho^4 c_{r+1}' &= c_r' + \frac{1}{1!}d_r' + \frac{1}{2!}e_r' + \frac{1}{3!}f_r' + 4c\rho^3\rho_r' \\
-\rho^3 d_{r+1}' &= d_r' + \frac{1}{1!}e_r' + \frac{1}{2!}f_r' + 3d\rho^2\rho_r' \\
-\rho^2 e_{r+1}' &= e_r' + \frac{1}{1!}f_r' + 2e\rho\rho_r' \\
-\rho f_{r+1}' &= f_r' + f\rho_r'
\end{aligned}\right\} \tag{146}$$

These are six linear simultaneous equations in the six unknowns ρ_r', b_r', c_r', d_r', e_r' and f_r'. Their solution will be the sum of four modes of the form:

$$\left.\begin{aligned}
\rho_r' &= \epsilon_k q_k^r \\
b_r' &= B_k \epsilon_k q_k^r \\
c_r' &= C_k \epsilon_k q_k^r \\
d_r' &= D_k \epsilon_k q_k^r \\
e_r' &= E_k \epsilon_k q_k^r \\
f_r' &= F_k \epsilon_k q_k^r,
\end{aligned}\right\} \quad (k=1,2,3,4), \tag{147}$$

where B_k, C_k, D_k, E_k, F_k are constants which define the kth mode shape, ϵ_k is arbitrary, and q_k is a root of the quartic equation:

$$\begin{vmatrix}
\frac{1}{1!} & \frac{1}{2!} & \frac{1}{3!} & \frac{1}{4!} & \frac{1}{5!} & 0 \\
1+q\rho^5 & \frac{1}{1!} & \frac{1}{2!} & \frac{1}{3!} & \frac{1}{4!} & 5b\rho^4 \\
0 & 1+q\rho^4 & \frac{1}{1!} & \frac{1}{2!} & \frac{1}{3!} & 4c\rho^3 \\
0 & 0 & 1+q\rho^3 & \frac{1}{1!} & \frac{1}{2!} & 3d\rho^2 \\
0 & 0 & 0 & 1+q\rho^2 & \frac{1}{1!} & 2e\rho \\
0 & 0 & 0 & 0 & 1+q\rho & f
\end{vmatrix} = 0. \tag{148}$$

On inserting the numerical values of ρ, b, c, d, e and f from (85) and (91) into (148), the four roots q_k are found to be:

$$q_1 = -0{\cdot}2120139\ldots,\quad q_2 = 27{\cdot}5\ldots,\quad q_3 = 47{\cdot}7\ldots,\quad q_4 = -225\ldots. \tag{149}$$

Only q_1 yields a converging mode. When computing solutions of (76) by *backwards* computation, an arbitrary starting point will contain components of all four modes, but only the mode derived from q_1 will diverge. This mode will grow by a factor of about 5 at each switch point. Of the remaining three modes, that corresponding to q_2 will converge to zero slowest, by a factor of about 0·04 at each switch point. Hence the ratio of the desired component to the unwanted components increases by a factor of about $5/0{\cdot}04 = 125$ at each switch point.

Initial errors due to the arbitrary starting point will be reduced if the initial deviations from the constant ratio solution correspond to the first mode, $k=1$, of (147). Then, the starting point for the solution (76) becomes:

$$\left.\begin{aligned}
b_0 &= b + B_1\epsilon_1 = 0{\cdot}0053895997 + 0{\cdot}0322083075\epsilon_1\\
c_0 &= c + C_1\epsilon_1 = -0{\cdot}025271563 - 0{\cdot}174907101\epsilon_1\\
d_0 &= d + D_1\epsilon_1 = 0{\cdot}041597667 + 0{\cdot}521009501\epsilon_1\\
e_0 &= e + E_1\epsilon_1 = 0{\cdot}10110916 - 0{\cdot}902723998\epsilon_1\\
f_0 &= f + F_1\epsilon_1 = -0{\cdot}63462403 + 0{\cdot}722859154\epsilon_1
\end{aligned}\right\} \tag{150}$$

where ϵ_1 is small, but otherwise arbitrary. The numerical values of B_1, C_1, D_1, E_1 and F_1 were obtained by substituting (147) into (146) and using the numerical value of q_1 from (149). The starting points (150) were used for all computations of optimum trajectories described in § 9.2.

APPENDIX 8

Results for a First-order Plant

When the plant has a transfer function $1/(ap)$ the following results are readily obtained.

The linear Hamiltonian system is:

$$\frac{d^2y}{dt^2} = \frac{1}{\lambda a^2}y \tag{151}$$

and has one convergent monotonic mode

$$\pm\exp\left(-\frac{t-\alpha_1}{a\sqrt{\lambda}}\right)$$

and one divergent monotonic mode

$$\pm\exp\left(\frac{t-\alpha_2}{a\sqrt{\lambda}}\right)$$

where α_1 and α_2 are arbitrary.

The non-linear Hamiltonian system is:

$$\frac{d^2y}{dt^2} = \operatorname{sgn} y. \tag{152}$$

This has two convergent switchless trajectories:

$$y = \begin{cases} \pm \frac{1}{2}(t-\alpha_1)^2 & (t \leqslant \alpha_1), \\ 0 & (t \geqslant \alpha_1), \end{cases} \tag{153}$$

which are analogous to the convergent monotonic mode of (151), and two divergent switchless trajectories:

$$y = \begin{cases} 0 & (t \leqslant \alpha_2), \\ \pm \frac{1}{2}(t-\alpha_2)^2 & (t \geqslant \alpha_2), \end{cases} \tag{154}$$

which are analogous to the divergent monotonic mode of (151).

The optimum non-linear controller is (Fullér 1960 a):

$$u = -\operatorname{sgn} x. \tag{155}$$

The integral-square-error for the optimum step response of the non-linear system is (Fuller 1960 a):

$$\tfrac{1}{3}a|X|^3, \tag{156}$$

where X is the value of the step command signal.

APPENDIX 9

Results for a Second-order Plant

When the plant has a transfer function $1/(ap^2)$, the following results may be obtained.

The linear Hamiltonian system is:

$$\frac{d^4y}{dt^4} = -\frac{1}{\lambda a^2}y. \tag{157}$$

The characteristic roots are:

$$(2\lambda^{1/2}a)^{-1/2}(1+j, 1-j, -1+j, -1-j). \tag{158}$$

Thus the linear Hamiltonian system has one convergent oscillatory mode and one divergent oscillatory mode. Along an optimum trajectory, only the convergent oscillatory mode is excited, and the magnitude of the ratio of successive overshoots is constant and equal to:

$$\exp(-\pi) \simeq 0{\cdot}043213918. \tag{159}$$

The non-linear Hamiltonian system is:

$$\frac{d^4y}{dt^4} = -\operatorname{sgn} y. \tag{160}$$

This has one type of convergent constant ratio trajectory, for which the ratio of successive switching intervals is constant and given by

$$\rho = \tfrac{1}{4}(3+\sqrt{33}-\sqrt{(26+6\sqrt{33})}) \simeq 0{\cdot}24212, \tag{161}$$

and one type of divergent constant ratio trajectory, for which the ratio of successive switching intervals is the reciprocal of (161). All optimum trajectories are convergent constant ratio trajectories, and the magnitude of the ratio of successive overshoots is constant and equal to:

$$\rho^2 \simeq 0{\cdot}058623. \tag{162}$$

Thus the behaviour of the non-linear Hamiltonian system is very analogous to that of the linear Hamiltonian system.

The optimum non-linear controller is (Fuller 1960 a):

$$u = -\operatorname{sgn}(x + hax'|x'|), \tag{163}$$

where

$$h = \sqrt{\left(\frac{\sqrt{33}-1}{24}\right)} \simeq 0{\cdot}4446236. \tag{164}$$

The integral-square-error for the optimum step response of the non-linear system is (Fuller 1964 a):

$$\tfrac{1}{20}\sqrt{(222+2\sqrt{33})}a^{1/2}|X|^{5/2} \simeq 0{\cdot}764017548a^{1/2}|X|^{5/2}, \tag{165}$$

where X is the value of the step command signal.

APPENDIX 10

Step Response for a Third-order Plant

Properties of the step response of the system under optimum control as defined in § 2 are derived here. If the amplitude of the step is $X (X > 0)$, the calculations are the same as if the initial state of the plant at $t = 0$ is:

$$(x_1, x_2, x_3) = (X, 0, 0),$$

and the command is zero.

Initially the control input will be $u = -1$, so, from (9), the trajectory up to the first switch at $t = t_1$ is:

$$\left.\begin{aligned} x_1 &= X - t^3/(6a) \\ x_2 &= -t^2/(2a) \\ x_3 &= -t/a \end{aligned}\right\} \quad (0 \leqslant t \leqslant t_1). \tag{166}$$

From definitions (20) and (74), (166) at $t = t_1$ yields:

$$\left.\begin{aligned} T_1{}^3 d_1 &= aX - t_1{}^3/6 \\ T_1{}^2 e_1 &= -t_1{}^2/2 \\ T_1 f_1 &= -t_1 \end{aligned}\right\} \tag{167}$$

Elimination amongst eqns. (167) yields:

$$e_1 = -\tfrac{1}{2}f_1{}^2, \quad f_1 < 0. \tag{168}$$

By computing trajectories backwards in the manner described in § 9, and systematically varying the starting point, the solution of (90) which satisfies (168) to five decimal places is found to yield:

$$\left.\begin{aligned} \rho_1 &= 0{\cdot}79172, \\ b_1 &= 0{\cdot}01436, \\ c_1 &= -0{\cdot}07172, \\ d_1 &= 0{\cdot}17242, \\ e_1 &= -0{\cdot}11234, \\ f_1 &= -0{\cdot}47402. \end{aligned}\right\} \tag{169}$$

Appropriate elimination amongst eqns. (167) yields:

$$T_1 = \left(\frac{6aX}{6d_1 - f_1^3}\right)^{1/3} \simeq 1{\cdot}73893(aX)^{1/3}. \tag{170}$$

Hence

$$t_1 = -f_1 T_1 \simeq 0{\cdot}82429(aX)^{1/3}. \tag{171}$$

The *settling time* T_s for a step response is thus†:

$$T_s = t_1 + T_1(1 + \rho_1 + \rho_1\rho_2 + \rho_1\rho_2\rho_3 + \ldots), \tag{172}$$

i.e.

$$T_s \simeq 5{\cdot}68733(aX)^{1/3}. \tag{173}$$

The *first overshoot* x_m is found to occur after the second switch and before the third switch, i.e. for $t_2 < t < t_3$. During this interval

$$x_1(t) = X - \frac{1}{6a}[t^3 - 2(t - t_1)^3 + 2(t - t_1 - T_1)^3],$$

which has a minimum x_m at $t = t_m$, where

$$t_m = 2T_1\left[1 - \sqrt{\left(\tfrac{1}{2} - \frac{t_1}{T_1}\right)}\right] \simeq 2{\cdot}91727(aX)^{1/3}$$

and

$$x_m = -0{\cdot}0965\,X. \tag{174}$$

To calculate the value of the criterion $\int_0^\infty x_1^2\,dt$, we use the first of eqns. (166) with the numerical value of t_1 to obtain the contribution:

$$I_1 \equiv \int_0^{t_1} x_1^2\,dt \simeq 0{\cdot}78684(aX)^{1/3}X^2. \tag{175}$$

The remaining contribution I_2 is derived by the following argument.

Let the minimum value of the performance criterion for an arbitrary starting point (x_1, x_2, x_3) be designated $V(x_1, x_2, x_3)$. It is known that, along an optimum trajectory Pontryagin's adjoint variables can often be identified with minus the partial derivatives of V, i.e. often

$$\psi_i = -\frac{\partial V}{\partial x_i} \tag{176}$$

† ρ_r rapidly tends to $+0{\cdot}575736$, the value for constant ratio trajectories. In fact, $\rho_2 = 0{\cdot}53428$, $\rho_3 = 0{\cdot}58457$, $\rho_4 = 0{\cdot}57387$, $\rho_5 = 0{\cdot}57613$, $\rho_6 = 0{\cdot}57565$. This enables (172) to be accurately summed.

This result is discussed by Rozonoer (1959) and Pontryagin *et al.* (1961). For it to be valid it is necessary for V to possess certain smoothness properties. We shall assume that V is sufficiently smooth in the present example for (176) to be valid. In support of this assumption, let us note that, in the simpler case when the plant is $1/(ap^2)$, result (176) (with $i=1,2$) has been shown to be valid at all points of the phase space† (Fuller 1964a).

From (176):

$$V(x_1^0, x_2^0, x_3^0) - V(0,0,0) = -\int_0^{x_1^0} \psi_1\,dx_1 - \int_0^{x_2^0} \psi_2\,dx_2 - \int_0^{x_3^0} \psi_3\,dx_3, \quad (177)$$

where the integrations are taken along any path joining (x_1^0, x_2^0, x_3^0) to the origin.

Let us take (x_1^0, x_2^0, x_3^0) to be the first switch point of the optimum step response, and the path of integration to be the curve in the switching surface corresponding to a straight line generator in z-space (see § 9.1). For points on this path, from (20), (73) and (74),

$$\left.\begin{aligned} x_3 &= kT_1 f_1/a, \\ x_2 &= k^2 T_1^2 e_1/a, \\ x_1 &= k^3 T_1^3 d_1/a, \\ \psi_1 &= 2k^4 T_1^4 c_1/a, \\ \psi_2 &= -2k^5 T_1^5 b_1/a, \\ \psi_3 &= 0, \end{aligned}\right\} \quad (178)$$

where k is a parameter which varies in the range

$$0 \leqslant k \leqslant 1. \quad (179)$$

Then (177) becomes:

$$V(x_1^0, x_2^0, x_3^0) = \frac{2T_1^7}{a^2}\int_0^1 (2b_1 e_1 - 3c_1 d_1)k^6\,dk \quad (180)$$

$$= \frac{2}{7}(2b_1 e_1 - 3c_1 d_1)T_1^7 a^{-2}. \quad (181)$$

Substituting numerical values from (169) and (170) in (181) we find

$$I_2 \equiv \int_{t_1}^{\infty} x_1^2\,dt \simeq 0{\cdot}46528(aX)^{1/3}X^2. \quad (182)$$

From (175) and (182), the minimum value of integral-square-error for a step response is:

$$I_1 + I_2 \simeq 1{\cdot}25212(aX)^{1/3}X^2. \quad (183)$$

† In this case (176) is valid when the performance criterion is integral-square-error, but not when it is settling time.

REFERENCES

BRENNAN, P. J., and ROBERTS, A. P., 1962, *J. Electron. Contr.*, **12,** 345.
BURNSIDE, W. S., and PANTON, A. W., 1904, *Theory of Equations*, 5th edition, (Dublin: University Press).
EGGLESTON, D. M., 1963, *Trans. Amer. Soc. mech. Engrs.*, **85D,** 478.
FELDBAUM, A. A., 1955, *Avtomat. Telemech., Moscow*, **16,** 129.
FULLER, A. T., 1960 a, *Proc. IFAC Congress, Moscow* (London: Butterworths 1961), **1,** 510; 1960 b, *J. Electron. Contr.*, **8,** 465; 1963, *Ibid.*, **15, 63**; 1964 a, *Ibid.*, **17,** 283; 1964 b, *Ibid.*, **17,** 301.
GRENSTED, P. E. W., 1955, *Proc. Instn. elect. Engrs*, **102C,** 244; 1962, *Progress in Control Engineering* (edited by R. H. MacMillan), **1,** 105.
PEARSON, J. D., 1961, *J. Electron. Contr.*, **10,** 323.
PONTRYAGIN, L. S., BOLTYANSKI, V. G., GAMKRELIDZE, R. V., and MISHCHENKO, E. F., 1961, *Mathematical Theory of Optimal Processes* (Moscow: Fizmatgiz). English translation (New York: Wiley, 1962).
ROZONOER, L. I., 1959, *Avtomat. Telemech., Moscow*, **20,** 1320, 1441, 1561.
WONHAM, W. M., 1963, *J. Electron. Contr.*, **15,** 59.

Optimization of some Non-linear Control Systems by means of Bellmans' Equation and Dimensional Analysis†

By A. T. Fuller
Engineering Department, Cambridge University

[Received May 18, 1966]

Abstract

For some simple control systems, controllers are found which minimize integral-square-error and similar performance criteria, subject to a constraint on control signal magnitude. The method used is that of setting up and solving Bellman's partial differential equation of continuous dynamic programming. Dimensional methods are used in the solution of Bellman's equation. The techniques are close to those of Wonham (1963), which are discussed and clarified.

It is shown incidentally that the conventional pi theorem of dimensional theory is invalid when certain variables can change sign. A modified pi theorem, applicable to variables which can change sign, is proved.

§ 1. Introduction

The problem of finding the controller which minimizes integral-square-error subject to a constraint on control signal magnitude has been treated in many papers since that of Fuller (1960 b). Wonham (1963) considered the case when the plant consists of two integrators, and derived the known optimal controller by solving Bellman's equation. This is a non-linear partial differential equation which arises in the theory of continuous dynamic programming (Bellman 1957). Wonham's technique involved reducing the number of independent variables, by what were essentially dimensional methods, so that Bellman's equation became an ordinary differential equation which could be readily solved. In the present paper a slightly different technique will be used. Bellman's equation being piece-wise linear for the problems to be considered, the general solution will be found for each linear region separately. Then dimensional methods‡ will be used to reduce the generality of these solutions.

† Communicated by the Author.

‡ Dimensional methods go back to Fourier (1822), who introduced the term 'dimension', and Maxwell (1873). During the 19th century these methods acquired a metaphysical air, which was shown to be superfluous by Bridgman (1922, 1931), See Birkhoff (1950) and Focken (1953) for references. In the context of optimal non-linear control dimensional methods have been used explicitly by Fuller (1960b), Newmann and Zachary (1965), and Woodside (1965); and implicitly by Wonham (1963) and Grensted and Fuller (1965). Bellman (1957) also used implicitly a dimensional method in his chapter 6, § 14. See also Bellman and Dreyfus (1962) chapter 7, § 7.

This technique will be applied to various problems, including a problem of time-optimality. Sun Tszyan (1960) used (in effect) Bellman's equation to treat the time-optimal system of our Example 1. However, his method was incomplete in that it required *a priori* knowledge of the solution of Bellman's equation at a certain boundary. Feldbaum (1963) also discussed our Example 1; however, he did not show how to find the solution of Bellman's equation.

Generally speaking, time-optimal problems can be treated more conveniently by Pontryagin's maximum principle (Pontryagin *et al.* 1961). However, we treat such a problem here by Bellman's equation because it provides a simple introductory example, and because the known result in this case gives a check on the validity of our methods.

In this paper various questions of rigour will be ignored. The aim is to investigate the utility of Bellman's equation as a formal device for discovering closed-form solutions to some non-linear control optimization problems. Once the solutions have been discovered, the exact sense in which they satisfy Bellman's equation can be studied (see Fuller 1964).

§ 2. Bellman's Equation

Consider a plant described by:

$$\frac{dx_i}{dt}=f_i(x_1,x_2,\ldots x_n,u) \quad (i=1,2,\ldots n), \tag{1}$$

where $x_1,x_2,\ldots,x_n$ are the phase coordinates or state coordinates, u is the control signal input to the plant and t is time. Suppose the control is subject to the saturation constraint:

$$|u(t)|\leqslant 1 \quad (0\leqslant t<\infty), \tag{2}$$

and the performance criterion to be minimized is:

$$M=\int_0^\infty f_0(x_1,x_2,\ldots x_n,u)\,dt. \tag{3}$$

The minimum value of M will be a function of the initial values

$$x_i(0)\equiv x_i^0 \quad (i=1,2,\ldots n) \tag{4}$$

of the phase coordinates. This function has been termed the best-performance function (Fuller 1964). Let us designate it as $V(x_1^0,x_2^0,\ldots x_n^0)$. Thus:

$$V(x_1^0,x_2^0,\ldots x_n^0)=\min_{u(t),0\leqslant t<\infty} M. \tag{5}$$

For notational convenience let us agree to drop the superscript in x_i^0, and write the best-performance function simply as $V(x_1,x_2,\ldots x_n)$. Then Bellman's partial differential equation for V is (Bellman 1957, 1961):

$$\min_u\left[f_0+f_1\frac{\partial V}{\partial x_1}+f_2\frac{\partial V}{\partial x_2}+\ldots+f_n\frac{\partial V}{\partial x_n}\right]=0. \tag{6}$$

For the problems treated in the present paper, the minimization with respect to u in (6) is subject to constraint (2). The meaning of eqn. (6) is as follows. Consider an optimal trajectory which starts at the point $(x_1 = X_1, x_2 = X_2, \ldots, x_n = X_n)$. Its *total* cost is $V(X_1, X_2, \ldots, X_n)$ and is thus a constant for all points of the trajectory. At an intermediate point $\{x_1(t), x_2(t), \ldots, x_n(t)\}$ of the trajectory, the rate of change of the total cost is the quantity in the square brackets in (6), being:

$$\frac{d}{dt}\left[\int_0^t f_0\{x_1(w), \ldots x_n(w), u(w)\}\,dw + V\{x_1(t), \ldots x_n(t)\}\right]. \tag{7}$$

Thus this rate of change must be zero, as (6) states. Moreover, if at time t the control u is changed to some non-optimal control, the total cost of the trajectory must increase. The quantity in the square brackets in (6) now expresses the rate of change of total cost while the trajectory is changing. This rate must be positive, as (6) implies.

Bellman's eqn. (6) is formal in the sense that it presupposes the existence of the partial derivatives

$$\frac{\partial V}{\partial x_1}, \frac{\partial V}{\partial x_2}, \ldots \frac{\partial V}{\partial x_n}.$$

However, in typical problems it turns out that these derivatives exist everywhere except possibly at the switching surface in the phase space (for examples see Fuller 1964). This switching surface is the $(n-1)$-dimensional hypersurface which is the boundary between points at which the optimal control is $u = +1$ and the points at which the optimal control is $u = -1$ (see, e.g., Fuller 1960 a for a discussion of switching surfaces).

In typical problems the f_i are all zero at the origin of the phase space and with $u = 0$.

$$f_i(0, 0, \ldots 0, 0) = 0 \quad (i = 0, 1, \ldots n). \tag{8}$$

An optimal trajectory approaches the origin of phase space, often reaching it in some finite time. Thereafter the optimal control is $u = 0$, the phase point stays at the origin , and there is no further contribution to the performance criterion. In such cases we can formulate the problem in the following alternative way. We define the performance criterion as:

$$M = \int_0^\tau f_0(x_1, x_2, \ldots x_n, u)\,dt, \tag{9}$$

and we seek the control $u(t)$ which transfers the phase point to the origin at time τ and minimizes M. Here τ is a free end time. Bellman's eqn. (6) applies also for this formulation of the problem.

§ 3. Settling Time

If in (9) we take

$$f_0 \equiv 1, \tag{10}$$

we find

$$M = \tau; \tag{11}$$

i.e. the performance criterion to be minimized is the time of transfer of the phase point to the origin of the phase space. This time is sometimes called the *settling time*, and the optimal system is sometimes called time-optimal. Thus Bellman's eqn. (6) applies also when the performance criterion is settling time (Bellman and Dreyfus, 1962, chapter 8, § 9). Another way of showing this is to substitute

$$f_0(x_1, \ldots x_n, u) = \begin{cases} 0 & \text{if } x_1 = x_2 = \ldots = x_n = 0 \\ 1 & \text{elsewhere} \end{cases} \tag{12}$$

in the first formulation of the performance criterion, namely (3). The relation between settling time and performance criteria such as integral-square-error has been discussed by Fuller (1959).

§ 4. Related Optimization Techniques

Bellman's eqn. (6) is a non-linear partial differential equation of first order. One way of treating such an equation is to attempt to solve it by Cauchy's method of characteristic strips (see, e.g., Goursat 1917 or Duff 1956). However, if we do this we arrive at equations which are the same as the equations resulting from Pontryagin's maximum principle† (see Rozonoer 1959). It is quicker, more convenient and more rigorous to start from Pontryagin's maximum principle. Since we are seeking techniques which are for the most part distinct from those of Pontryagin's maximum principle, we shall not adopt the method of characteristic strips in this paper.

After the minimization operation in Bellman's eqn. (6) is carried out, the equation becomes formally the same as the Hamilton–Jacobi equation of the classical calculus of variations (see e.g., Bliss 1946 and Berkovitz 1961). For this reason some writers treat Bellman's equation and the Hamilton–Jacobi equation as synonymous. The present writer's opinion is that Bellman's formulation of the equation gives additional insight (see remarks following (6)) and thus the term 'Bellman's equation' is retained in this paper.

§ 5. Supplementary Conditions

In all the subsequent examples it can be shown that the optimal control is bang-bang, i.e.

$$|u(t)| = 1 \quad (0 \leqslant t < \tau). \tag{13}$$

The method of proof, which involves Pontryagin's maximum principle, is indicated by Wonham (1963). Thus the optimal controller is indeed represented by a switching surface.

We assume that V is a continuous function of $x_1, x_2, \ldots, x_n$ at all points including the switching surface. A rough justification of this assumption is as follows. The theorem of continuous dependence of trajectories on initial conditions (see, e.g., Ince 1926) can (usually) be applied in a piece-

† This explains the point of view of Pearson (1961) who treated our example 3 by solving computationally the equations of the maximum principle, but called his technique dynamic programming. Compare also Newmann and Zachary (1965).

wise sense to the optimal control system. Hence two optimal trajectories which start from neighbouring initial conditions will remain close, and will incur close values of the performance criterion.

The optimal switching surface divides the phase space into regions which we denote P and N; in the region P the control is $u = +1$ and in region N the control is $u = -1$.

We assume that in the interior of region P there is no discontinuity in the first partial derivatives of V; we assume the same for region N. A rough justification for this assumption is as follows. Pontryagin's adjoint variables ψ_i can usually† be identified with $-(\partial V/\partial x_i)$ $(i = 1, 2, \ldots, n)$, except perhaps at the switching surface. Moreover, along an optimum trajectory the $\psi_i(t)$ are continuous, since they obey ordinary differential equations. This suggests that the $(\partial V/\partial x_i)$ are continuous functions of $x_1, \ldots x_n$, at least inside P and inside N.

At the switching surface it turns out that the partial derivatives $(\partial V/\partial x_i)$ can be discontinuous (see, e.g., Fuller 1964). In the examples we shall show by means of Pontryagin's maximum principle that, at the switching surface, one of the two values of one of the $(\partial V/\partial x_i)$ is zero.

All these conditions, together with considerations of dimensions and symmetry, will turn out to imply a unique solution of Bellman's eqn. (6), in our examples. Thus we bypass the difficulty of the absence of boundary conditions for Bellman's equation‡.

§ 6. Example 1. Double Integrator Plant and Settling Time Criterion

We begin with the well studied problem in which the plant has a transfer function $1/(ap^2)$ and the performance criterion is settling time. The plant eqns. (1) are:

$$\frac{dx_1}{dt} = x_2 \tag{16}$$

$$\frac{dx_2}{dt} = \frac{u}{a} \quad (a > 0), \tag{17}$$

---

† This relation is discussed by Rozooner (1959), Pontryagin *et al.* (1961) and Fuller (1964).

‡ Bellman (1957, 1961) also formulated his equation for the case where τ is fixed. In this case V depends on the additional independent variable $x_{n+1} = \tau - t$, and the following boundary condition for $V(x_1, x_2, \ldots, x_{n+1})$ holds:

$$V(x_1, x_2, \ldots x_n, 0) = 0. \tag{14}$$

Feldbaum (1963) gives the following boundary condition for $V(x_1, x_2, \ldots x_n)$ when the performance criterion is settling time:

$$V(0, 0, \ldots 0) = 0. \tag{15}$$

However, as has been pointed out to the writer by I. Gumowski, (15) is not adequate to ensure a unique solution of Bellman's equation. (E.g. (15) and (25) are satisfied by the V of (86) and by $V = ax_2$.) See also Gumowski (1965).

where a is a positive constant. The control $u(t)$ satisfies constraint (2):

$$|u| \leqslant 1. \tag{18}$$

The performance criterion is:

$$M = \int_0^\tau dt = \tau, \tag{19}$$

where τ is the time at which the origin is reached:

$$\begin{cases} x_1(\tau) = 0, & (20) \\ x_2(\tau) = 0. & (21) \end{cases}$$

6.1. *Bellman's Equation*

From (9), (16), (17) and (19):

$$\begin{cases} f_0 = 1, & (22) \\ f_1 = x_2, & (23) \\ f_2 = \dfrac{u}{a}. & (24) \end{cases}$$

Hence Bellman's eqn. (6) is:

$$\min_u \left[1 + x_2 \frac{\partial V}{\partial x_1} + \frac{u}{a} \frac{\partial V}{\partial x_2} \right] = 0. \tag{25}$$

From (18) and (25), the optimal control is given by:

$$u(x_1, x_2) = -\operatorname{sgn} \frac{\partial V}{\partial x_2}. \tag{26}$$

From (25) and (26), Bellman's equation is:

$$1 + x_2 \frac{\partial V}{\partial x_1} - \frac{1}{a} \left| \frac{\partial V}{\partial x_2} \right| = 0, \tag{27}$$

and is thus non-linear.

6.2. *Piece-wise Solution*

In the region where

$$\frac{\partial V}{\partial x_2} < 0, \quad \text{i.e. where} \quad u = +1, \tag{28}$$

Bellman's eqn. (27) is:

$$1 + x_2 \frac{\partial V}{\partial x_1} + \frac{1}{a} \frac{\partial V}{\partial x_2} = 0, \tag{29}$$

and is thus linear. It may be readily solved by the method of characteristic curves† (see e.g. Goursat 1917, Duff 1956). The procedure is as follows. We assume a solution in the form:

$$J(V, x_1, x_2) = 0, \tag{30}$$

† The method of characteristic curves applies to linear and quasi-linear partial differential equations, and is not the same as the method of characteristic strips (see § 4) which applies to non-linear partial differential equations. See Goursat (1917) for the terminology.

where J is a function to be determined. Differentiation of (30) with respect to x_1 and x_2 yields:

$$\frac{\partial J}{\partial V}\frac{\partial V}{\partial x_1}+\frac{\partial J}{\partial x_1}=0 \tag{31}$$

and

$$\frac{\partial J}{\partial V}\frac{\partial V}{\partial x_2}+\frac{\partial J}{\partial x_2}=0, \tag{32}$$

respectively. From (29), (31) and (32):

$$-\frac{\partial J}{\partial V}+x_2\frac{\partial J}{\partial x_1}+\frac{1}{a}\frac{\partial J}{\partial x_2}=0. \tag{33}$$

Thus the introduction of J has transformed the non-homogeneous eqn. (29) to the homogeneous eqn. (33). To make the notation more uniform we define:

$$x_0 \equiv -V, \tag{34}$$

so that (33) becomes:

$$\frac{\partial J}{\partial x_0}+x_2\frac{\partial J}{\partial x_1}+\frac{1}{a}\frac{\partial J}{\partial x_2}=0. \tag{35}$$

The equations of the characteristic curves for this homogeneous linear partial differential equation are:

$$\frac{dx_0}{1}=\frac{dx_1}{x_2}=\frac{dx_2}{1/a}. \tag{36}$$

(Equations (36) are formed simply by taking as the coefficient of dx_i the reciprocal of the coefficient of $\partial J/\partial x_i$ in (35).) Equations (36) are a symbolic version of the following set of ordinary differential equations:

$$\begin{cases} \dfrac{dx_0}{dx_2}=a, & (37)\\[2ex] \dfrac{dx_1}{dx_2}=ax_2. & (38)\end{cases}$$

Equations (37) and (38) may be immediately integrated, to yield:

$$\begin{cases} x_0= ax_2+\alpha, & (39)\\ x_1=\tfrac{1}{2}ax_2^2+\beta, & (40)\end{cases}$$

where α and β are arbitrary constants. Solving (39) and (40) for α and β we obtain:

$$\begin{cases} \alpha=x_0-ax_2, & (41)\\ \beta=x_1-\tfrac{1}{2}ax_2^2. & (42)\end{cases}$$

According to the theory of characteristic curves, the general solution of (35) can now be written:

$$J=G[\alpha,\beta], \tag{43}$$

where G is an arbitrary differentiable function of its two arguments, and

α and β are given by (41) and (42). From (30), (34), (41), (42) and (43) the general solution of (29) is given by:

$$G[(-V-ax_2),\quad (x_1-\tfrac{1}{2}ax_2^2)]=0, \tag{44}$$

where G is arbitrary. Solving (44) for its first argument, we can write the general solution of (29) as:

$$V_P=-ax_2+F_P(x_1-\tfrac{1}{2}ax_2^2), \tag{45}$$

where F_P is an arbitrary function of its argument. The subscript P is to remind us that (45) holds for the region (28), where u is positive.

Similarly, in the region where

$$\frac{\partial V}{\partial x_2}>0,\quad \text{i.e. where}\quad u=-1, \tag{46}$$

Bellman's eqn. (27) is:

$$1+x_2\frac{\partial V}{\partial x_1}-\frac{1}{a}\frac{\partial V}{\partial x_2}=0, \tag{47}$$

and its general solution is:

$$V_N=ax_2+F_N(x_1+\tfrac{1}{2}ax_2^2), \tag{48}$$

where F_N is an arbitrary function of its argument.

6.3. *Symmetry*

Let us use symmetry to obtain a relation between F_P and F_N. The equations of the problem, namely (16)–(21), are unchanged if we write $-x_1$, $-x_2$, $-u$ for x_1, x_2, u. It follows that if a point $(x_1=X_1, x_2=X_2)$ is in the region where $u=+1$, then the point $(x_1=-X_1, x_2=-X_2)$ is in the region where $u=-1$, and has the same value of V. Thus equating (45) and (48) for these two points, we obtain:

$$-aX_2+F_P(X_1-\tfrac{1}{2}aX_2^2)=-aX_2+F_N(-X_1+\tfrac{1}{2}aX_2^2), \tag{49}$$

i.e.

$$F_P(\xi)=F_N(-\xi) \tag{50}$$

for all ξ, where ξ is a general argument. Hence (45) and (48) can be written:

$$V=\begin{cases}-ax_2+F(-x_1+\tfrac{1}{2}ax_2^2) & (u=+1),\\ a_2x+F(x_1+\tfrac{1}{2}ax_2^2) & (u=-1).\end{cases} \tag{51}$$

where F is an arbitrary function.

6.4. *Dimensions*

Let us use dimensional methods to reduce the generality of F in (51). These methods are as expounded, for example, in Chapter 1 of Bridgman (1931). As shown by Bridgman, their justification follows from the pi theorem of the theory of dimensions. Actually the pi theorem as usually stated is incorrect if certain variables can change sign (see Appendix). However, a modified pi theorem, dealing with variables which change sign, is proved in the Appendix §§ 1.2 and 1.3. It follows that the methods of the present section are valid.

The output x_1 of the plant can have any of the conventional physical dimensions, such as length, time, mass; but these are immaterial to the present argument. We shall say x_1 has the dimensions of 'signal', S, and write this statement symbolically as:

$$D(x_1) = S. \tag{52}$$

From (16) x_2 has the dimensions of x_1 divided by time, T. Thus:

$$D(x_2) = ST^{-1}. \tag{53}$$

Constraint (18) implies that we are taking u as dimensionless, hence (17) and (53) imply that the dimensions of a are:

$$D(a) = S^{-1}T^2. \tag{54}$$

Since V is a minimum time, its dimensions are:

$$D(V) = T. \tag{55}$$

From (51) and (55) the dimensions of F are:

$$D(F) = T. \tag{56}$$

F, which has dimensions T, depends on its explicit argument, which has dimensions S (see (51) and (52)), and also on the only system parameter a, which has dimensions $S^{-1}T^2$. To produce a magnitude of dimension T from magnitudes of dimensions S and $S^{-1}T^2$ the latter magnitudes must be multiplied together to eliminate the dimension S, then the square root must be taken. Thus†:

$$F = \begin{cases} \pm\,[ka(-x_1 + \tfrac{1}{2}ax_2^2)]^{1/2} & (u = +1), \\ \pm\,[ka(\ \ x_1 + \tfrac{1}{2}ax_2^2)]^{1/2} & (u = -1), \end{cases} \tag{57}$$

where k is a dimensionless number.

In (57) and throughout the paper, the symbol $(A)^{1/2}$, where A is any positive number, means the *positive* square root of A. (51) can now be written:

$$V = \begin{cases} -ax_2 \pm k^{1/2}(-ax_2 + \tfrac{1}{2}a^2x_2^2)^{1/2} & (u = +1), \\ \ \ ax_2 \pm k^{1/2}(ax_1 + \tfrac{1}{2}a^2x_2^2)^{1/2} & (u = -1). \end{cases} \tag{59}$$

---

† The two square roots in (57) are imaginary for $k(-x_1 + \frac{1}{2}ax_2^2) < 0$ and $k(x_1 + \frac{1}{2}ax_2^2) < 0$ respectively. This means that points for which $k(-x_1 + \frac{1}{2}ax_2^2) < 0$ cannot belong to region P, and points for which $k(x_1 + \frac{1}{2}ax_2^2) < 0$ cannot belong to region N. Note that imaginary terms need not occur if we choose the following alternative to (57) and which is equally valid, as far as dimensional requirements go (see Appendix, § 1.4):

$$F = \begin{cases} \pm\,[k_1a|-x_1 + \tfrac{1}{2}ax_2^2|]^{1/2} & (u = +1,\ -x_1 + \tfrac{1}{2}ax_2^2 > 0), \\ \pm\,[k_2a|-x_1 + \tfrac{1}{2}ax_2^2|]^{1/2} & (u = +1,\ -x_1 + \tfrac{1}{2}ax_2^2 < 0), \\ \pm\,[k_1a|x_1 + \tfrac{1}{2}ax_2^2|]^{1/2} & (u = -1,\ x_1 + \tfrac{1}{2}ax_2^2 > 0), \\ \pm\,[k_2a|x_1 + \tfrac{1}{2}ax_2^2|]^{1/2} & (u = -1,\ x_1 + \tfrac{1}{2}ax_2^2 < 0), \end{cases} \tag{58}$$

where $k_1 > 0$, $k_2 > 0$. However, (58) contradicts our assumption (§ 5) that the partial derivatives of V are continuous in the interiors of regions P and N. Hence we adopt (57) rather than (58).

6.5. *Realness and Positiveness*

Substituting $x_1=0, x_2 \neq 0$ in (59) we find that since the left-hand side is real, k is real and non-negative:

$$k \geqslant 0. \tag{60}$$

Substituting $x_1 \neq 0, x_2=0$ in (59) we find that since the left-hand side is positive, the ambiguous sign in (59) is positive and k is non-zero. Hence (59) becomes:

$$V=\begin{cases} V_P=-ax_2+h(-ax_1+\frac{1}{2}a^2x_2^2)^{1/2} & (u=+1), \quad (61) \\ V_N=ax_2+h(ax_1+\frac{1}{2}a^2x_2^2)^{1/2} & (u=-1), \quad (62) \end{cases}$$

where h is some positive constant:

$$h>0. \tag{63}$$

6.6. *The Switching Curve*

The switching curve in the (x_1, x_2) phase plane consists of points for which (61) and (62) are equal (from the continuity of V, see § 5), and thus satisfies:

$$-ax_2+h(-ax_1+\tfrac{1}{2}a^2x_2^2)^{1/2}=ax_2+h(ax_1+\tfrac{1}{2}a^2x_2^2)^{1/2}. \tag{64}$$

Dividing by $a|x_2|$ and substituting

$$y=\frac{x_1}{ax_2^2}, \tag{65}$$

we write (64) as:

$$(y+\tfrac{1}{2})^{1/2}-(-y+\tfrac{1}{2})^{1/2}=-\frac{2}{h}\operatorname{sgn} x_2. \tag{66}$$

Squaring (66), rearranging, and squaring again, we obtain:

$$y^2=\frac{2}{h^2}-\frac{4}{h^4}, \tag{67}$$

the roots of which are:

$$y=\pm\left(\frac{2}{h^2}-\frac{4}{h^4}\right)^{1/2}. \tag{68}$$

To determine the ambiguous sign in (68) we argue as follows. When $y>0$, the left side of (66) is positive. Thus (66) implies that the positive sign in (68) is valid only when $x_2 \leqslant 0$. Similarly the negative sign in (68) is valid only when $x_2 \geqslant 0$. Thus (68) can be written:

$$y=-\left(\frac{2}{h^2}-\frac{4}{h^4}\right)^{1/2}\operatorname{sgn} x_2. \tag{69}$$

From (65) and (69) the equation of the switching curve is:

$$x_1+\left(\frac{2}{h^2}-\frac{4}{h^4}\right)^{1/2} ax_2|x_2|=0. \tag{70}$$

It remains to determine the constant h. We note provisionally that since (66) implies that

$$|y| \leqslant \tfrac{1}{2}, \tag{71}$$

(67) and (63) imply that h lies in the range:

$$2 \leqslant h < \infty. \tag{72}$$

6.7. *Switching Conditions*

Let us show that, at a point of the switching curve, either

$$\frac{\partial V_P}{\partial x_2} = 0 \tag{73}$$

or

$$\frac{\partial V_N}{\partial x_2} = 0. \tag{74}$$

(This condition will serve to determine the constant h.) Inside region P and inside region N we assume that $-(\partial V/\partial x_1)$ and $-(\partial V/\partial x_2)$ can be identified with Pontryagin's adjoint variables ψ_1 and ψ_2 (see § 5). Then along an optimum trajectory, (26) implies that

$$u(t) = \operatorname{sgn} \psi_2(t). \tag{75}$$

It is also known that along an optimum trajectory ψ_1 and ψ_2 are continuous functions of time. Hence if a trajectory approaches the switching curve from region P, (75) shows that $(\partial V_P/\partial x_2) \to 0$. Similarly, if a trajectory approaches the switching curve from region N, we find $(\partial V_N/\partial x_2) \to 0$. Thus at a given switch point either (73) or (74) holds (possibly both hold).

For definiteness let us consider a point of the switching curve which is approached from region P. Here (61) and (73) yield:

$$-a + \tfrac{1}{2}ha^2x_2(-ax_1 + \tfrac{1}{2}a^2x_2{}^2)^{-1/2} = 0. \tag{76}$$

Using (65) we can write (76) as:

$$(-y + \tfrac{1}{2})^{1/2} = \frac{h}{2} \operatorname{sgn} x_2. \tag{77}$$

Squaring (77), we find:

$$y = \tfrac{1}{2} - \frac{h^2}{4}. \tag{78}$$

From (67) and (78):

$$\frac{(2-h^2)^2}{16} = \frac{2}{h^4}(h^2 - 2), \tag{79}$$

i.e.

$$(h^2 - 2)(h^6 - 2h^4 - 32) = 0, \tag{80}$$

i.e.

$$(h^2 - 2)(h^2 - 4)(h^4 + 2h^2 + 8) = 0. \tag{81}$$

The last factor in (81) is positive for all real values of h. Thus the only real roots of (81) are:

$$h = \pm\sqrt{2} \quad \text{and} \quad h = \pm 2. \tag{82}$$

From (72) and (82), the only admissible value of h is:

$$h = 2. \tag{83}$$

6.8. *Results*

From (70) and (83) the equation of the switching curve is:

$$x_1 + \tfrac{1}{2}ax_2|x_2| = 0, \tag{84}$$

which is a well-known result. From (61) points for which $x_2 = 0, x_1 > 0$ cannot belong to region P. Thus points to the right of curve (84) belong to region N, and similarly points to the left of curve (84) belong to region P. Hence the optimal control satisfies:

$$u = -\operatorname{sgn}(x_1 + \tfrac{1}{2}ax_2|x_2|). \tag{85}$$

From (61), (62), (83) and (85) the best-performance function is:

$$V = \begin{cases} -ax_2 + 2(-ax_1 + \tfrac{1}{2}a^2x_2^2)^{1/2} & (x_1 + \tfrac{1}{2}ax_2|x_2| \leqslant 0), \\ ax_2 + 2(ax_1 + \tfrac{1}{2}a^2x_2^2)^{1/2} & (x_1 + \tfrac{1}{2}ax_2|x_2| \geqslant 0), \end{cases} \tag{86}$$

which agrees with results previously given by Sun Tszyan (1960) and Fuller (1964).

6.9. *Discussion*

We see that the direct solution of Bellman's equation for time-optimal problems is possible in a simple case, but is lengthy compared with the direct use of Pontryagin's maximum principle (compare § 5 of Pontryagin *et al.* (1961) where this example is treated). However, for performance criteria other than settling time, the use of the maximum principle is less straightforward and Bellman's equation is of correspondingly greater interest (see subsequent examples.)

For a discussion of the differentiability of the best-performance function (86) at the switching curve, see Fuller (1964).

§ 7. Example 2. Double Integrator Plant and Integral-modulus-error Criterion

Consider the problem of minimizing integral-modulus-error for a double-integrator plant which is the same as in Example 1. The plant equations are:

$$\begin{cases} \dfrac{dx_1}{dt} = x_2, & (87) \\[2ex] \dfrac{dx_2}{dt} = \dfrac{u}{a} \quad (a > 0). & (88) \end{cases}$$

The control constraint is:

$$|u| \leqslant 1. \tag{89}$$

The performance criterion to be minimized is:

$$M = \int_0^\infty |x_1| \, dt. \tag{90}$$

7.1. *Bellman's Equation*

For the problem specified by (87)–(90), Bellman's eqn. (6) is:

$$\min_u \left[|x_1| + x_2 \frac{\partial V}{\partial x_1} + \frac{u}{a} \frac{\partial V}{\partial x_2} \right] = 0 \tag{91}$$

From (89) and (91):

$$u(x_1, x_2) = -\operatorname{sgn} \frac{\partial V}{\partial x_2}, \tag{92}$$

so that (91) is:

$$|x_1| + x_2 \frac{\partial V}{\partial x_1} - \frac{1}{a} \left| \frac{\partial V}{\partial x_2} \right| = 0. \tag{93}$$

7.2. *Piece-wise Solution*

In region P, i.e. where

$$\frac{\partial V}{\partial x_2} < 0, \quad \text{i.e. where} \quad u = +1, \tag{94}$$

Bellman's eqn. (93) is:

$$|x_1| + x_2 \frac{\partial V}{\partial x_1} + \frac{1}{a} \frac{\partial V}{\partial x_2} = 0. \tag{95}$$

With J and x_0 defined by

$$J(V, x_1, x_2) = 0 \tag{96}$$

and

$$x_0 = -V, \tag{97}$$

(95) becomes:

$$|x_1| \frac{\partial J}{\partial x_0} + x_2 \frac{\partial J}{\partial x_1} + \frac{1}{a} \frac{\partial J}{\partial x_2} = 0. \tag{98}$$

The equations of the characteristic curves of (98) are:

$$\frac{dx_0}{|x_1|} = \frac{dx_1}{x_2} = a\,dx_2. \tag{99}$$

The general solution of (99) is:

$$\begin{cases} x_1 = \frac{1}{2} a x_2^2 + \beta, & (100) \\ x_0 = (\frac{1}{6} a^2 x_2^3 + \beta a x_2) \operatorname{sgn} x_1 + \alpha, & (101) \end{cases}$$

where α and β are arbitrary constants. From (100) and (101):

$$\begin{cases} \alpha = x_0 + (-ax_1x_2 + \tfrac{1}{3}a^2x_2^3)\operatorname{sgn} x_1, & (102) \\ \beta = x_1 - \tfrac{1}{2}ax_2^2. & (103) \end{cases}$$

Caution is necessary at this point. It would be incorrect to write the general continuous solution of (98) as $G(\alpha, \beta)$, where α and β are given by (102) and (103), and G is an arbitrary differentiable function of α and β. The reason is that α as given by (102) is a discontinuous function of x_1 and x_2. Thus $G[\alpha(x_1, x_2), \beta(x_1, x_2)]$ would be a discontinuous function of x_1 and x_2 and would not qualify as a solution of (98).

To avoid this difficulty we can solve (98) first in the regions of P where $x_1 > 0$, then in the regions where $x_1 < 0$, and finally impose continuity on the line $x_1 = 0$. Equations (102) and (103) still hold, for all x_1. Thus the general solution of (98) can be written $G(\alpha, \beta, +1)$ for $x_1 \geqslant 0$ and $G(\alpha, \beta, -1)$ for $x_1 \leqslant 0$.

$$J = G[\alpha(x_1, x_2),\ \beta(x_1, x_2),\quad \operatorname{sgn} x_1], \tag{104}$$

where the arbitrary function G must be continuous at $x_1 = 0$. From (96), (97), (102), (103) and (104):

$$G[\{-V + (-ax_1x_2 + \tfrac{1}{3}a^2x_2^3)\operatorname{sgn} x_1\}, (x_1 - \tfrac{1}{2}ax_2^2),\quad \operatorname{sgn} x_1] = 0. \tag{105}$$

Solving (105) for its first argument, and writing V_P for V in view of (94), we write the general solution of (95) as:

$$V_P = (-ax_1x_2 + \tfrac{1}{3}a^2x_2^3)\operatorname{sgn} x_1 + F[-(x_1 - \tfrac{1}{2}ax_2^2),\quad -\operatorname{sgn} x_1], \tag{106}$$

where the arbitrary function F is such that V_P is continuous at $x_1 = 0$. (The arguments in F are multiplied by -1 for later convenience.)

7.3. *Symmetry*

To obtain the expression for V_N, we merely reverse the signs of x_1 and x_2 in expression (106) for V_P (compare § 6.3). Thus:

$$V = \begin{cases} V_P = (-ax_1x_2 + \tfrac{1}{3}a^2x_2^3)\operatorname{sgn} x_1 + F[(-x_1 + \tfrac{1}{2}ax_2^2),\quad -\operatorname{sgn} x_1] \\ \qquad\qquad (u = +1) \quad (107) \\ V_N = (ax_1x_2 + \tfrac{1}{3}a^2x_2^3)\operatorname{sgn} x_1 + F[(x_1 + \tfrac{1}{2}ax_2^2),\quad \operatorname{sgn} x_1] \\ \qquad\qquad (u = -1). \quad (108) \end{cases}$$

7.4. *Dimensions*

With the notation of § 6.4, the dimensions of the quantities appearing in (107) and (108) are:

$$\begin{cases} D(x_1) = S, & (109) \\ D(x_2) = ST^{-1}, & (110) \\ D(a) = S^{-1}T^2, & (111) \\ D(\operatorname{sgn} x_1) = 1, & (112) \\ D(V) = ST, & (113) \\ D(F) = ST. & (114) \end{cases}$$

Here (112) indicates that $\operatorname{sgn} x_1$ is dimensionless. Examination of (107)–(114) shows that F (which depends on an explicit argument with the dimensions of x_1, on a, and on $\operatorname{sgn} x_1$) must satisfy†:

$$F=\begin{cases} \pm\,[k(-\operatorname{sgn} x_1)a(-x_1+\tfrac{1}{2}ax_2^2)^3]^{1/2} & (u=+1), \\ \pm\,[k(\operatorname{sgn} x_1)a(x_1+\tfrac{1}{2}ax_2^2)^3]^{1/2} & (u=-1) \end{cases} \tag{115}$$

Here $k(\operatorname{sgn} x_1)$ is a (dimensionless) function of $\operatorname{sgn} x_1$ and thus has two values $k(-1)$ and $k(+1)$ which are to be determined.

7.5. *Realness and Positiveness*

Substituting $x_1=0+$ and $0-$ in (115) with $x_2\neq 0$, we see that

$$k(-1)\geqslant 0 \quad \text{and} \quad k(+1)\geqslant 0, \tag{116}$$

since F is real. Substituting $x_1\neq 0, x_2=0$ in (107) and (108) we see that F is positive, so that the ambiguous sign in (115) is $+$ and k is non-zero. Thus (107) and (108) can be written:

$$V=\begin{cases} V_P=(-ax_1x_2+\tfrac{1}{3}a^2x_2^3)\operatorname{sgn} x_1+h(-\operatorname{sgn} x_1)a^{1/2}(-x_1+\tfrac{1}{2}ax_2^2)^{3/2} \\ \qquad (u=+1), & (117) \\ V_N=(ax_1x_2+\tfrac{1}{3}a^2x_2^3)\operatorname{sgn} x_1+h(\operatorname{sgn} x_1)a^{1/2}(x_1+\tfrac{1}{2}ax_2^2)^{3/2} \\ \qquad (u=-1), & (118) \end{cases}$$

where

$$h(+1)>0 \quad \text{and} \quad h(-1)>0. \tag{119}$$

7.6. *Continuity of V on the x_2-axis*

Since V is continuous we can substitute $x_1=0+$ and $x_1=0-$ in (117), and equate the resulting expressions. We obtain a relation between $h(-1)$ and $h(+1)$:

$$\tfrac{1}{3}a^2x_2^3+h(-1)a^2 2^{-3/2}|x_2|^3=-\tfrac{1}{3}a^2x_2^3+h(+1)a^2 2^{-3/2}|x_2|^3, \tag{120}$$

i.e.

$$\frac{2\sqrt{2}}{3}\operatorname{sgn} x_2+h(-1)=-\frac{2\sqrt{2}}{3}\operatorname{sgn} x_2+h(+1)\equiv l, \quad \text{say}. \tag{121}$$

From (121):

$$\begin{cases} h(+1)=l+\dfrac{2\sqrt{2}}{3}\operatorname{sgn} x_2, & (122) \\ h(-1)=l-\dfrac{2\sqrt{2}}{3}\operatorname{sgn} x_2. & (123) \end{cases}$$

In (122) and (123) x_2 is the ordinate of a point on the x_2-axis and in region P. Since (122) and (123) are constants (and from symmetry) it follows

† Compare the footnote to eqn. (57).

that region P contains either the positive x_2-axis or the negative x_2-axis, but not parts of both. Hence (122) and (123) are:

$$\begin{cases} h(+1) = l \pm \dfrac{2\sqrt{2}}{3}, & (124) \\ h(-1) = l \mp \dfrac{2\sqrt{2}}{3}, & (125) \end{cases}$$

where the ambiguity in sign remains to be resolved. (124) and (125) imply that

$$h(\operatorname{sgn} x_1) = l \pm \frac{2\sqrt{2}}{3} \operatorname{sgn} x_1. \tag{126}$$

Defining for brevity

$$m(\operatorname{sgn} x_1) = \pm \frac{2\sqrt{2}}{3} \operatorname{sgn} x_1, \tag{127}$$

we find from (126)

$$\begin{cases} h(\operatorname{sgn} x_1) = l + m, & (128) \\ h(-\operatorname{sgn} x_1) = l - m. & (129) \end{cases}$$

(117) and (118) now become.

$$V = \begin{cases} V_P = (-ax_1x_2 + \tfrac{1}{3}a^2x_2^3) \operatorname{sgn} x_1 + (l-m)a^{1/2}(-x_1 + \tfrac{1}{2}ax_2^2)^{3/2} \\ \qquad (u = +1), \quad (130) \\ V_N = (ax_1x_2 + \tfrac{1}{3}a^2x_2^3) \operatorname{sgn} x_1 + (l+m)a^{1/2}(x_1 + \tfrac{1}{2}ax_2^2)^{3/2} \\ \qquad (u = -1), \quad (131) \end{cases}$$

where, in view of (119) and (128):

$$l > |m|. \tag{132}$$

7.7. *Conditions at the Switching Curve*

Since V is continuous, on the switching curve expressions (130) and (131) are equal. Equating (130) and (131), dividing by $a^2|x_2|^3$, and substituting

$$y = \frac{x_1}{ax_2^2} \tag{133}$$

we obtain, for points on the switching curve:

$$(l-m)(-y+\tfrac{1}{2})^{3/2} - (l+m)(y+\tfrac{1}{2})^{3/2} = 2y \operatorname{sgn} x_1 \operatorname{sgn} x_2. \tag{134}$$

Also, at points on the switching curve which are approached from region P (compare § 6.7):

$$\frac{\partial V_P}{\partial x_2} = 0. \tag{135}$$

From (130), (133) and (135):

$$\tfrac{3}{2}(l-m)(-y+\tfrac{1}{2})^{1/2} = (y-1) \operatorname{sgn} x_1 \operatorname{sgn} x_2. \tag{136}$$

The problem is now reduced to the determination of y, l and m from (127), (134) and (136).

Let us simplify (134) and (136). Since the terms in (134) are real:

$$|y| \leqslant \tfrac{1}{2}. \tag{137}$$

From (132), the left side of (136) is positive; and from (137) the right side of (136) has the sign of $-\operatorname{sgn} x_1 \operatorname{sgn} x_2$. Hence

$$\operatorname{sgn} x_1 = -\operatorname{sgn} x_2 \tag{138}$$

for points on the switching curve. Thus (134) and (136) become:

$$(l-m)(-y+\tfrac{1}{2})^{3/2} - (l+m)(y+\tfrac{1}{2})^{3/2} = -2y \tag{139}$$

and

$$(l-m)(-y+\tfrac{1}{2})^{1/2} = -\tfrac{2}{3}(y-1). \tag{140}$$

Substitution of (140) in (139) yields:

$$(l+m)(y+\tfrac{1}{2})^{3/2} - \tfrac{2}{3}(y+\tfrac{1}{2})(y+1) = 0. \tag{141}$$

Hence either

$$y = -\tfrac{1}{2} \tag{142}$$

or

$$(l+m)(y+\tfrac{1}{2})^{1/2} = \tfrac{2}{3}(y+1). \tag{143}$$

7.8. *Elimination of Spurious Root*

(133) and (142) imply that a possible solution of the switching curve (for points approached from region P) is:

$$x_1 + \tfrac{1}{2}a{x_2}^2 = 0 \tag{144}$$

Let us show that this solution is invalid. At points on curve (144) (and not at the origin) we find from (131):

$$\frac{\partial V_N}{\partial x_2} = -a|x_1|, \tag{145}$$

so that

$$\frac{\partial V_N}{\partial x_2} < 0. \tag{146}$$

However, region N is, from its definition, the set of points for which

$$\frac{\partial V_N}{\partial x_2} > 0 \tag{147}$$

(see (92)). The contradiction between (146) and (147) shows that (142) is a spurious root of (141). Hence (143) is valid.

7.9. *Evaluation of Remaining Unknowns*

The problem now reduces to the determination of y, l and m from (127), (140) and (143). Subtracting the square of (140) from the square of (143) we find:

$$lm = -y(l^2 + m^2 - \tfrac{8}{9}). \tag{148}$$

The last two terms in the brackets in (148) cancel, in view of (127), so that (148) reduces to:

$$m = -yl. \tag{149}$$

Therefore

$$\operatorname{sgn}(m) = -\operatorname{sgn}(yl). \tag{150}$$

From (132), (133) and (150):

$$\operatorname{sgn} m = -\operatorname{sgn} x_1 \tag{151}$$

(151) determines the ambiguous sign in (127), so that (127) can be written:

$$m = -\frac{2\sqrt{2}}{3}\operatorname{sgn} x_1. \tag{152}$$

Elimination of y from (149) and the square of (140) yields:

$$l^4 - (3m^2 + \tfrac{8}{9})l^2 + 2m(m^2 - \tfrac{8}{9}l) - \tfrac{8}{9}m^2 = 0. \tag{153}$$

Substitution for m from (152) reduces (153) to:

$$l^4 - \frac{32}{9}l^2 - \frac{64}{81} = 0, \tag{154}$$

the roots of which are easily evaluated as the four values $\pm(2\sqrt{2}/3)(\pm\sqrt{5}+2)^{1/2}$. The only real root satisfying (132) is:

$$l = \frac{2\sqrt{2}}{3}(\sqrt{5}+2)^{1/2}. \tag{155}$$

From (149), (152) and (155):

$$y = (\sqrt{5}-2)^{1/2}\operatorname{sgn} x_1. \tag{156}$$

From (133), (138) and (156), the switching curve satisfies:

$$x_1 + (\sqrt{5}-2)^{1/2} a x_2 |x_2| = 0. \tag{157}$$

We have proved this for points on the switching curve which are approached from region P. Applying the symmetry argument of § 6.3, we see that (157) holds also for points which are approached from region N.

7.10. *Results.*

The optimal switching curve is (157):

$$x_1 + (\sqrt{5}-2)^{1/2} a x_2 |x_2| = 0. \tag{158}$$

Since the terms of (130) and (131) are real, the negative x_1-axis belongs to region P and the positive x_1-axis belongs to region N. Thus the optimal control is:

$$u = -\operatorname{sgn}[x_1 + (\sqrt{5}-2)^{1/2} a x_2 |x_2|]. \tag{159}$$

From (130), (131), (152), (155) and (159), the best-performance function is:

$$V=\begin{cases}(-ax_1x_2+\tfrac{1}{3}a^2x_2^3)\operatorname{sgn} x_1 \\ \quad +\tfrac{1}{3}[(\surd 5+2)^{1/2}+\operatorname{sgn} x_1]a^{1/2}(-2x_1+ax_2^2)^{3/2} \\ \qquad\qquad (x_1+(\surd 5-2)^{1/2}ax_2|x_2|\leqslant 0), & (160)\\ (ax_1x_2+\tfrac{1}{3}a^2x_2^3)\operatorname{sgn} x_1 \\ \quad +\tfrac{1}{3}[(\surd 5+2)^{1/2}-\operatorname{sgn} x_1]a^{1/2}(2x_1+ax_2^2)^{3/2} \\ \qquad\qquad (x_1+(\surd 5-2)^{1/2}ax_2|x_2|\geqslant 0). & (161)\end{cases}$$

These results are in agreement with those given previously (Fuller 1960 b, 1964). It can be directly verified that V as given by (160) and (161) is not only continuous but also with continuous first partial derivatives at all points, including points on the switching curve and points on the x_2-axis (Fuller 1964).

§ 8. Example 3. Double Integrator Plant and Integral-square-error criterion

Consider the problem of minimizing integral-square-error for a double integrator plant, which is the same as in Example 1. The plant equations are:

$$\begin{cases}\dfrac{dx_1}{dt}=x_2, & (162)\\ \dfrac{dx_2}{dt}=\dfrac{u}{a} \quad (a>0) & (163)\end{cases}$$

The control constraint is:

$$|u|\leqslant 1. \tag{164}$$

The performance criterion to be minimized is:

$$M=\int_0^\infty x_1^2\,dt. \tag{165}$$

8.1. *Bellman's Equation*

For the problem specified by (162)–(165), Bellman's eqn. (6) is:

$$\min_u\left[x_1^2+x_2\frac{\partial V}{\partial x_1}+\frac{u}{a}\frac{\partial V}{\partial x_2}\right]=0. \tag{166}$$

From (164) and (166):

$$u(x_1,x_2)=-\operatorname{sgn}\frac{\partial V}{\partial x_2}, \tag{167}$$

so that (166) is:

$$x_1^2+x_2\frac{\partial V}{\partial x_1}-\frac{1}{a}\left|\frac{\partial V}{\partial x_2}\right|=0. \tag{168}$$

8.2. *Piece-wise Solution*

In region P, i.e. where

$$\frac{\partial V}{\partial x_2} < 0, \quad \text{i.e. where} \quad u = +1, \tag{169}$$

Bellman's eqn. (168) is:

$$x_1^2 + x_2 \frac{\partial V}{\partial x_1} + \frac{1}{a}\frac{\partial V}{\partial x_2} = 0. \tag{170}$$

With J and x_0 defined by

$$J(V, x_1, x_2) = 0 \tag{171}$$

and

$$x_0 = -V \tag{172}$$

(170) becomes:

$$x_1^2 \frac{\partial J}{\partial x_0} + x_2 \frac{\partial J}{\partial x_1} + \frac{1}{a}\frac{\partial J}{\partial x_2} = 0. \tag{173}$$

The equations of the characteristic curves of (173) are:

$$\frac{dx_0}{x_1^2} = \frac{dx_1}{x_2} = a\,dx_2. \tag{174}$$

The general solution of (174) is:

$$\begin{cases} x_1 = \frac{1}{2}ax_2^2 + \beta, & (175) \\ x_0 = \frac{1}{20}a^3x_2^5 + \frac{1}{3}\beta a^2 x_2^3 + \beta^2 a x_2 + \alpha, & (176) \end{cases}$$

where α and β are arbitrary constants. From (175) and (176):

$$\begin{cases} \alpha = x_0 - ax_1^2x_2 + \frac{2}{3}a^2x_1x_2^3 - \frac{2}{15}a^3x_2^5, & (177) \\ \beta = x_1 - \frac{1}{2}ax_2^2. & (178) \end{cases}$$

The general solution of (173) is:

$$J = G(\alpha, \beta). \tag{179}$$

From (171), (172), (177), (178) and (179), the general solution of (170) is:

$$G[(-V - ax_1^2x_2 + \tfrac{2}{3}a^2x_1x_2^3 - \tfrac{2}{15}a^3x_2^5), (x_1 - \tfrac{1}{2}ax_2^2)] = 0. \tag{180}$$

Solving (180) for its first argument, and writing V_P for V in view of (169) we obtain:

$$V_P = -ax_1^2x_2 + \tfrac{2}{3}a^2x_1x_2^3 - \tfrac{2}{15}a^3x_2^5 + F(-x_1 + \tfrac{1}{2}ax_2^2), \tag{181}$$

where F is arbitrary.

8.3. *Symmetry*

By symmetry (see § 6.3) the expression for V in region N is obtained by reversing the signs of x_1 and x_2 in expression (181). Thus:

$$V = \begin{cases} V_P = -ax_1^2x_2 + \frac{2}{3}a^2x_1x_2^3 - \frac{2}{15}a^3x_2^5 + F(-x_1 + \frac{1}{2}ax_2^2) \\ \qquad\qquad (u = +1), & (182) \\ V_N = ax_1^2x_2 + \frac{2}{3}a^2x_1x_2^3 + \frac{2}{15}a^3x_2^5 + F(x_1 + \frac{1}{2}ax_2^2) \\ \qquad\qquad (u = -1). & (183) \end{cases}$$

8.4. *Dimensions*

With the notation of § 6.4, the dimensions of the quantities in (182) and (183) are:

$$\begin{cases} D(x_1) = S, & (184) \\ D(x_2) = ST^{-1}, & (185) \\ D(a) = S^{-1}T^2, & (186) \\ D(V) = S^2T, & (187) \\ D(F) = S^2T. & (188) \end{cases}$$

From (182)–(188) it follows (compare § 6.4) that F (which depends on an explicit argument with dimensions (184) and on a) satisfies:

$$F = \begin{cases} \pm\,[ka(-x_1 + \tfrac{1}{2}ax_2^2)^5]^{1/2} & (u = +1) \\ \pm\,[ka(x_1 + \tfrac{1}{2}ax_2^2)^5]^{1/2}, & (u = -1) \end{cases} \tag{189}$$

where k is a dimensionless number which is to be determined.

8.5. *Realness and Positiveness*

Substituting $x_1 = 0, x_2 \neq 0$ in (189) we find:

$$k \geqslant 0, \tag{190}$$

since F is real. Substituting $x_1 \neq 0, x_2 = 0$ in (182) and (183) we find that F is positive. Hence the ambiguous sign in (189) is $+$ and k is non-zero. Thus (182) and (183) can be written:

$$V = \begin{cases} V_P = -ax_1^2x_2 + \tfrac{2}{3}a^2x_1x_2^3 - \tfrac{2}{15}a^3x_2^5 + ha^{1/2}(-x_1 + \tfrac{1}{2}ax_2^2)^{5/2} \\ \qquad\qquad (u = +1), \quad (191) \\ V_N = ax_1^2x_2 + \tfrac{2}{3}a^2x_1x_2^3 + \tfrac{2}{15}a^3x_2^5 + ha^{1/2}(x_1 + \tfrac{1}{2}ax_2^2)^{5/2} \\ \qquad\qquad (u = -1), \quad (192) \end{cases}$$

where h is a positive constant:

$$h > 0. \tag{193}$$

8.6. *Conditions at the Switching Curve*

Since V is continuous, on the switching curve (191) and (192) are equal. Equating (191) and (192), dividing by $a^3|x_2|^5$, and substituting

$$y = \frac{x_1}{ax_2^2}, \tag{194}$$

we obtain, for points on the switching curve:

$$h(-y + \tfrac{1}{2})^{5/2} - h(y + \tfrac{1}{2})^{5/2} = 2(y^2 + \tfrac{2}{15})\,\mathrm{sgn}\,x_2. \tag{195}$$

Also, at points on the switching curve which are approached from region P (compare § 6.7):

$$\frac{\partial V_P}{\partial x_2} = 0. \tag{196}$$

H

From (191), (194) and (196):

$$h(-y+\tfrac{1}{2})^{3/2} = \tfrac{2}{5}(y^2-2y+\tfrac{2}{3})\operatorname{sgn} x_2. \tag{197}$$

The problem is now reduced to the determination of h and y from (195) and (197).

Substitution of (197) in (195) yields:

$$h(y+\tfrac{1}{2})^{5/2} + \tfrac{2}{5}(y+\tfrac{1}{2})(y^2+2y+\tfrac{2}{3})\operatorname{sgn} x_2 = 0, \tag{198}$$

Hence, either

$$y = -\tfrac{1}{2} \tag{199}$$

or

$$h(y+\tfrac{1}{2})^{3/2} = -\tfrac{2}{5}(y^2+2y+\tfrac{2}{3})\operatorname{sgn} x_2. \tag{200}$$

8.7. *Elimination of Spurious Root*

From (194) and (199), the equation

$$x_1 + \tfrac{1}{2}ax_2^{\,2} = 0 \tag{201}$$

possibly represents points on the switching curve which are approached from region P. Let us show that this possibility can be eliminated. At points on curve (201) (and not at the origin) we obtain from (192):

$$\frac{\partial V_N}{\partial x_2} = -\tfrac{1}{3}ax_1^{\,2}, \tag{202}$$

so that

$$\frac{\partial V_N}{\partial x_2} < 0. \tag{203}$$

This inequality contradicts the definition of region N, for which $(\partial V_N/\partial x_2)$ is positive (see (167)). Hence (199) is a spurious root of (198), and (200) is valid.

8.8. *Evaluation of Remaining Unknowns*

The problem now reduces to the determination of y and h from (197) and (200). Squaring these equations and eliminating h, we find:

$$y^7 + \tfrac{1}{12}y^5 - \tfrac{1}{18}y^3 = 0. \tag{204}$$

Hence either

$$y = 0 \tag{205}$$

or

$$y^4 + \tfrac{1}{12}y^2 - \tfrac{1}{18} = 0. \tag{206}$$

(205) is not valid since, on substitution in (197) and (200), it yields values of h which contradict one another. Hence (206) is valid; its roots are easily evaluated as the four values:

$$\pm\left(\frac{\pm\sqrt{(33)}-1}{24}\right)^{1/2}, \tag{207}$$

of which the only real values are:

$$y=+\left(\frac{\sqrt{(33)}-1}{24}\right)^{1/2}\equiv+g,\quad \text{say,} \tag{208}$$

and

$$y=-\left(\frac{\sqrt{(33)}-1}{24}\right)^{1/2}=-g. \tag{209}$$

If $y>0$, (200) shows that $x_2<0$. Hence, if (208) is valid, the part of the switching curve which is approached from region P is (see (194)):

$$x_1=gax_2^{\,2}\quad (x_2<0). \tag{210}$$

On the other hand, if $y<0$, (197) shows that $x_2>0$. Hence, if (209) is valid, the part of the switching curve which is approached from region P is:

$$x_1=-gax_2^{\,2}\quad (x_2>0). \tag{211}$$

(210) and (211) may both be written:

$$x_1+gax_2|x_2|=0. \tag{212}$$

Therefore, whether root (208) or root (209) is valid, the part of the switching curve approached from region P satisfies (212). By the symmetry argument (§ 6.3), the part of the switching curve approached from region N also satisfies (212).

To evaluate h we subtract the square of (197) from the square of (200) and find:

$$h^2=\frac{16}{75}\frac{12y^2+8}{4y^2+3}. \tag{213}$$

Substitution from (208) or (209) in (213) and use of (193) yields:

$$h=\tfrac{1}{20}(222+2\sqrt{(33)})^{1/2}. \tag{214}$$

8.9. *Results*

From (208) and (212) the optimal switching curve is:

$$x_1+\left(\frac{\sqrt{(33)}-1}{24}\right)^{1/2}ax_2|x_2|=0. \tag{215}$$

Since the terms of (191) and (192) are real, the negative x_1-axis belongs to region P and the positive x_1-axis belongs to region N. Thus the optimal control is:

$$u=-\operatorname{sgn}\left[x_1+\left(\frac{\sqrt{(33)}-1}{24}\right)^{1/2}ax_2|x_2|\right]. \tag{216}$$

From (191), (192), (214) and (216), the best-performance function is:

$$V = \begin{cases} -ax_1^2x_2 + \frac{2}{3}a^2x_1x_2^3 - \frac{2}{15}a^3x_2^5 \\ \quad + \frac{1}{20}(222 + 2\sqrt{(33)})^{1/2}a^{1/2}(-x_1 + \frac{1}{2}ax_2^2)^{5/2} \\ \qquad \left(x_1 + \left(\frac{\sqrt{(33)} - 1}{24}\right)^{1/2} ax_2|x_2| \leqslant 0\right) & (217) \\ ax_1^2x_2 + \frac{2}{3}a^2x_1x_2^3 + \frac{2}{15}a^3x_2^5 \\ \quad + \frac{1}{20}(222 + 2\sqrt{(33)})^{1/2}a^{1/2}(x_1 + \frac{1}{2}ax_2^2)^{5/2} \\ \qquad \left(x_1 + \left(\frac{\sqrt{(33)} - 1}{24}\right)^{1/2} ax_2|x_2| \geqslant 0\right). & (218) \end{cases}$$

These results agree with those given previously (Fuller 1960b, 1964, Wonham 1963). It can be directly verified that V as given by (217) and (218) is not only continuous but with continuous first partial derivatives at all points, including points on the switching curve (Fuller 1964).

§ 9. Another Method

Wonham (1963) treated our example 3 by techniques which are similar to those given above but differ in detail. In § 9.1 we shall reproduce his argument (but in our notation and terminology†) for solving Bellman's equation, then we shall compare techniques in § 9.2.

9.1. *Wonham's Technique*

To solve (168) first consider the transformations:

$$x_1 = \mu^2 \bar{x}_1 \quad , \quad x_2 = \mu \bar{x}_2 \quad , \quad t = \mu \bar{t}, \tag{219}$$

where μ is a parameter. It is immediately verified that under the scaling group (219) the variational problem (162)–(165) is invariant; and the new best-performance function $\bar{V}$ is related to V by the transformation:

$$\bar{V}(\bar{x}_1, \bar{x}_2) = \mu^{-5} V(x_1, x_2). \tag{220}$$

Since $\bar{V}$ and V are identical functions of their arguments, (219) and (220) imply:

$$V(x_1, x_2) = \mu^{-5} V(\mu^2 x_1, \mu x_2). \tag{221}$$

Differentiating (221) with respect to μ and setting $\mu = 1$ we obtain

$$2x_1 \frac{\partial V}{\partial x_1} + x_2 \frac{\partial V}{\partial x_2} - 5V = 0. \tag{222}$$

The general solution of (222) is (asserted to be‡):

$$V(x_1, x_2) = x_2^5 \gamma(x_1 x_2^{-2}) = a^3 x_2^5 \phi\left(\frac{x_1}{ax_2^2}\right), \tag{223}$$

† Our eqns. (219)–(227) correspond to Wonham's eqns. (14)–(22). Wonham took $a = 1$, but we leave a general.

‡ This assertion is discussed in § 9.2. A.T.F.

where the function ϕ is arbitrary. Set $a^{-1}x_1x_2^{-2}=y$ and substitute (223) in (168); the result† is:

$$\phi'(y)-|5\phi-2y\phi'(y)|+y^2=0. \tag{224}$$

Equations (167) and (223) imply:

$$u(x_1,x_2)=-\operatorname{sgn}[5\phi(y)-2y\phi'(y)]. \tag{225}$$

Denote by P and N the regions in the (x_1,x_2) plane where $u(x_1,x_2)=+1,-1$ respectively. Solving (224) and using the obvious relation $V(-x_1,-x_2)=V(x_1,x_2)$ we get:

$$\phi=\begin{cases}\phi_P(y)=-y^2+\frac{2}{3}y-\frac{2}{15}+A(\frac{1}{2}-y)^{5/2} & [(x_1,x_2)\in P],\\ \phi_N(y)=y^2+\frac{2}{3}y+\frac{2}{15}-A(\frac{1}{2}+y)^{5/2} & [(x_1,x_2)\in N],\end{cases} \tag{226}$$

where A is a constant of integration. Finally:

$$V=\begin{cases}V_P=-ax_1^2x_2+\frac{2}{3}a^2x_1x_2^3-\frac{2}{15}a^3x_2^5+ha^{1/2}(\frac{1}{2}ax_2^2-x_1)^{5/2} \\ \qquad\qquad [(x_1,x_2)\in P],\\ V_N=V_P(-x_1,-x_2) \qquad [(x_1,x_2)\in N].\end{cases} \tag{227}$$

[Wonham then goes on to determine the constant h, by methods which are a combination of those of § 8.6 above and of those used by Fuller (1960 b).]

9.2. *Comments*

Let us clarify the above technique. Examination of (184)–(187), using elementary dimensional arguments, suggests that

$$V=a^3x_2^5\phi\left(\frac{x_1}{ax_2^2}\right), \tag{228}$$

where ϕ is a dimensionless function of a dimensionless argument. In fact (228) follows from the ordinary (uncorrected) pi theorem of the elementary theory of dimensions. (228) is the same as (223). Moreover, the steps‡ used above in deriving (223) are those used in deriving the ordinary (uncorrected) pi theorem in elementary treatments of the theory of dimensions (see Appendix). We conclude that Wonham is essentially using dimensional methods. Presumably Wonham avoids specific mention of dimensional methods because of the widespread misconceptions concerning such methods.

---

† Wonham's (19) differs slightly from our (224), partly because the former has an incorrect sign, and partly because Wonham's V is half our V.

‡ (219) corresponds essentially to a simultaneous change of time scale and signal scale, such that a is unchanged. We see this if we write (219) as (with $\lambda=\mu^2$):

$$x_1=\lambda\bar{x}_1,\quad x_2=\lambda\mu^{-1}\bar{x}_2,\quad t=\mu\bar{t},\quad a=\mu^2\lambda^{-1}\bar{a}, \tag{229}$$

and compare (184)–(186). The connection between dimensional methods and group transformation methods has been shown by Birkhoff (1948, 1950) and Decius (1948). (222) is Euler's equation for a function having the homogeneity properties of (221). The above technique for deriving Euler's equation (222) from (221) is given, for example, by Courant (1936).

Wonham's technique thus involves using dimensional arguments at the outset to reduce the number of independent variables in Bellman's equation by one†. Our technique, on the other hand, involves using dimensional arguments after obtaining a general solution of Bellman's equation. There is little to choose between the two approaches; however the latter evades some minor difficulties, which are as follows.

First, in the above use of dimensional methods at the outset, it is assumed that V is differentiable everywhere (see (222)). But it is possible for V to have discontinuous derivatives at the switching curve (see our example 1). Thus (222) has to be interpreted in a piecewise sense. (However this difficulty can be avoided by using the proofs of Appendix § 1.3 rather than those of § 1.2.)

Secondly, the general solution of (222) is not actually (223), but is of the form:

$$V(x_1, x_2) = a^3 x_2{}^5 \phi \left(\frac{x_1}{a x_2{}^2}, \quad \operatorname{sgn} x_2 \right). \tag{230}$$

In fact examination of (217) and (218) confirms (230), as does the derivation of V by the use of the corrected pi theorem (see Appendix). (230) follows on solving (222) by the method of characteristics and taking account of the continuity of ϕ with respect to x_2 (compare § 7.2, where a similar difficulty arises). Thus (224) has to be solved separately for $x_2 > 0$ and $x_2 < 0$.

Thirdly, although (224) is of only first order, it is relatively complicated. Its solution is less straightforward than that of the second order system (174).

§ 10. Extension to more General Problems

In our examples there is only one plant parameter, a, and this accounts for the success of the dimensional techniques. When there is more than one plant parameter, e.g. when the plant has a transfer function $1/(ap^2 + bp)$ dimensional methods are less powerful and do not yield a direct solution of the problem. However our methods may be used to reduce the field of search for a solution.

When the plant is of higher order but has only one parameter, e.g. when the plant has a transfer function $1/(ap^3)$, again our methods do not lead directly to the solution; however they might enable the solution to be guessed.

§ 11. Conclusions

Bellman's equation can be used to solve certain simple non-linear control optimization problems. The closed form solutions obtained are

---

† The technique of using dimensional methods to reduce the number of variables in partial differential equations was introduced by Birkhoff (1948, 1950) and was followed up by Morgan (1952).

useful for checking the validity of computational methods of solving such problems, and the validity of sub-optimal control methods.

ACKNOWLEDGMENT

The writer is indebted to W. M. Wonham for several enlightening conversations and private notes.

APPENDIX

1. DIMENSIONAL METHODS

The essence of the theory of dimensions is contained in the pi theorem, which was given its first general (if vague) form by Vaschy (1892 a, b). It is often attributed to Buckingham (1914) but prior work was done by Vaschy (1890, 1892 a, b), Jeans (1905) and Riabouchinsky (1911, 1912). In § 1.1 the conventional treatment will be given. It will then be shown that the pi theorem, as usually stated, does not hold if certain variables can change sign. In § 1.2 the pi theorem and its usual proof will be modified to deal with the case when variables can change sign. In § 1.3 an alternative and more direct proof of the modified pi theorem will be given.

1.1 *Conventional Dimensional Theory*

As shown by Bridgman (1931) the dependence of a measurement y on the units of measurement can be represented by the equation:

$$y = z\mu_1^{p}\mu_2^{q}\dots\mu_m^{w} \tag{231}$$

where $\mu_1, \mu_2, \dots \mu_m$ are scale factors, by which the units of measurement are multiplied, and z is independent of the μ's. (In fact (231) must hold in order that the ratio of two measurements should be independent of the size of the units of measurement.)

Suppose then we have a set of equations (whether they represent measurements or not is actually immaterial†):

$$\begin{cases} y_1 = z_1\mu_1^{p_1}\mu_2^{q_1}\dots\mu_m^{w_1}, \\ y_2 = z_2\mu_1^{p_2}\mu_2^{q_2}\dots\mu_m^{w_2}, \\ \quad \cdot \quad \cdot \quad \cdot \quad \cdot \\ y_n = z_n\mu_1^{p_n}\mu_2^{q_n}\dots\mu_m^{w_n}, \end{cases} \tag{232}$$

where the z's, p's, q's, ... w's are independent of the μ's. Here all the symbols represent real numbers, and (since the p's, ... w's can be fractional) the μ's are therefore restricted to positive values:

$$\mu_i > 0 \quad (i = 1, 2, \dots m). \tag{233}$$

Suppose the y's satisfy:

$$\Phi(y_1, y_2, \dots y_n) = 0, \tag{234}$$

---

† Our arguments beginning with (232) are mathematical, and their validity is independent of physical interpretations.

where Φ is a fixed function, and moreover suppose (234) remains satisfied when (with fixed $z_1, z_2, \ldots z_n$) the factors $\mu_1, \mu_2, \ldots \mu_m$ are varied in the range (233). Then the pi theorem states that (234) can be expressed as:

$$\Theta(\Pi_1, \Pi_2, \ldots \Pi_{n-m}) = 0, \tag{235}$$

where

$$\left\{\begin{array}{l} \Pi_1 = y_1^{A_1} y_2^{B_1} \ldots y_n^{N_1}, \\ \Pi_2 = y_1^{A_2} y_2^{B_2} \ldots y_n^{N_2}, \\ \quad \cdot \quad \cdot \quad \cdot \quad \cdot \quad \cdot \\ \Pi_{n-m} = y_1^{A_{n-m}} y_2^{B_{n-m}} \ldots y_n^{N_{n-m}}, \end{array}\right. \tag{236}$$

and moreover the Π's are independent of the μ's. Quantities which remain constant when the μ's vary are said to be 'dimensionless'. Thus the pi theorem implies that the relation between n arguments can be simplified to a relation between only $n-m$ arguments, and moreover the latter arguments are dimensionless.

The conventional (loose) demonstration of the pi theorem runs typically as follows (see, e.g., Bridgman (1932), Duncan (1953); minor variations of the argument are introduced by these and other writers). Differentiate (234) with respect to μ_1:

$$\frac{\partial \Phi}{\partial y_1}\frac{\partial y_1}{\partial \mu_1} + \frac{\partial \Phi}{\partial y_2}\frac{\partial y_2}{\partial \mu_1} + \ldots + \frac{\partial \Phi}{\partial y_n}\frac{\partial y_n}{\partial \mu_1} = 0. \tag{237}$$

From (232):

$$\frac{\partial y_i}{\partial \mu_1} = \frac{p_i y_i}{\mu_1} \quad (i = 1, 2, \ldots n). \tag{238}$$

Substitute (238) in (237):

$$p_1 y_1 \frac{\partial \Phi}{\partial y_1} + p_2 y_2 \frac{\partial \Phi}{\partial y_2} + \ldots + p_n y_n \frac{\partial \Phi}{\partial y_n} = 0. \tag{239}$$

(Equation (239) corresponds to the Euler equation for a function Φ with the homogeneity properties implied by (232) and (234). See, e.g., Courant (1936) for a treatment of homogeneous function.)

Solve (239) by the method of characteristic curves: the equations of the latter are (compare § 6.2):

$$\frac{dy_1}{p_1 y_1} = \frac{dy_2}{p_2 y_2} = \ldots = \frac{dy_n}{p_n y_n}. \tag{240}$$

(For simplicity we assume here that none of the p's is zero.)

The integral of (240) is (allegedly†):

$$\frac{1}{p_1}\log y_1 + c_1 = \frac{1}{p_2}\log y_2 + c_2 = \ldots = \frac{1}{p_n}\log y_n + c_n, \tag{241}$$

† See discussion in Appendix § 1.2.

where the c's are arbitrary constants. From (241):

$$\begin{cases} y_1 y_n^{-p_1/p_n} = \alpha_1, \\ y_2 y_n^{-p_2/p_n} = \alpha_2, \\ \quad \cdot \quad \cdot \quad \cdot \quad \cdot \\ y_{n-1} y_n^{-p_{n-1}/p_n} = \alpha_{n-1}, \end{cases} \tag{242}$$

where $\alpha_1, \alpha_2, \ldots \alpha_{n-1}$ are arbitrary constants. The general solution of (239) is obtained by equating to zero an arbitrary function of the arbitrary constants:

$$\theta_1(\alpha_1, \alpha_2, \ldots \alpha_{n-1}) = 0, \tag{243}$$

i.e. from (242):

$$\theta_1(y_1 y_n^{-p_1/p_n}, y_2 y_n^{-p_2/p_n}, \ldots y_{n-1} y_n^{-p_{n-1}/p_n}) = 0. \tag{244}$$

In (244) the arguments $y_i y_n^{-p_i/p_n}$ are independent of μ_1, from (232). Thus we have replaced the original function Φ of n variables by a function θ, of $n-1$ variables, which variables are independent of μ_1. In the same way θ_1 can be replaced by a function θ_2 of $n-2$ variables, which are independent of μ_2. Eliminating the remaining μ's similarly, we arrive at the result given by (235) and (236).

1.2. *Modified Pi Theorem*

In § 1.1 the possibility that the y's could be negative was neglected†. Let us see how the pi theorem must be modified to take into account the signs of the y's.

First, (241) must be replaced by:

$$\frac{1}{p_1} \log |y_1| + c_1 = \ldots = \frac{1}{p_n} \log |y_n| + c_n. \tag{245}$$

(242) now becomes:

$$\begin{matrix} y_1 |y_n|^{-p_1/p_n} = \alpha_1, \\ \cdot \quad \cdot \quad \cdot \quad \cdot \\ y_{n-1} |y_n|^{-p_{n-1}/p_n} = \alpha_{n-1}, \end{matrix} \tag{246}$$

where the α's are arbitrary constants.

When the p's are positive, the α's of (246) are discontinuous functions of y_n at $y_n = 0$. Consequently the method of characteristic curves must be applied separately to the regions $y_n > 0$ and $y_n < 0$. (Compare §§ 7.2 and 9.2, where similar difficulties arise.) (244) now becomes:

$$\theta_1(y_1 |y_n|^{-p_1/p_n}, y_2 |y_n|^{-p_2/p_n}, \ldots, y_{n-1} |y_n|^{-p_{n-1}/p_n}, \operatorname{sgn} y_n) = 0. \tag{247}$$

Finally (235) and (236) become:

$$\Theta(\Pi_1, \Pi_2, \ldots \Pi_{n-m}, \operatorname{sgn} y_{n-m+1}, \operatorname{sgn} y_{n-m+2}, \ldots \operatorname{sgn} y_n) = 0, \tag{248}$$

† Riabouchinsky (1943) and Birkhoff (1950) noticed the need to take into account the signs of the variables. However they did not formulate the appropriate modification of the pi theorem.

where

$$\Pi_i = y_1^{A_i} y_2^{B_i}, \dots y_{n-m}^{H_i} |y_{n-m+1}|^{I_i} |y_{n-m+2}|^{J_i} \dots |y_n|^{N_i} \quad (i=1,2,\dots n-m). \tag{249}$$

1.3. *Precise Statement and Direct Proof of the Modified Pi Theorem*

The conventional arguments leading to the pi theorem (see §§1.1 and 1.2), via Euler's equation, are very indirect. Moreover, they implicitly assume continuity and differentiability of the functions involved. These assumptions are unnecessary, as the following direct proof shows. We also take the opportunity to formulate the modified pi theorem more precisely† than in §1.2.

Theorem (modified pi theorem)

Let $\Phi(y_1, y_2, \dots y_n)$ be a function‡ of the real variables $y_1, y_2, \dots y_n$. Whenever the equation

$$\Phi(y_1, y_2, \dots y_n) = 0 \tag{250}$$

is satisfied by a set of values:

$$y_i = z_i \quad (i=1,2,\dots n), \tag{251}$$

let it also be satisfied by all possible sets of values:

$$y_i = z_i \mu_1^{p_i} \mu_2^{q_i} \dots \mu_m^{w_i}, \quad (i=1,2,\dots n) \tag{252}$$

obtained by giving each μ_j any positive value

$$u_j > 0 \quad (j=1,2,\dots m) \tag{253}$$

with fixed real values of $p_i, q_i, \dots w_i (i=1,2,\dots n)$. Let §

$$y_i \neq 0 \quad (i=n-m+1, n-m+2, \dots n), \tag{254}$$

and let

$$\begin{vmatrix} p_{n-m+1} & q_{n-m+1} \cdots & w_{n-m+1} \\ p_{n-m+2} & q_{n-m+2} \cdots & w_{n-m+2} \\ & \cdots\cdots & \\ p_n & q_n & w_n \end{vmatrix} \neq 0. \tag{255}$$

Then the values (251) of the y_i also satisfy the equation:

$$\Phi(\Pi_1, \Pi_2, \dots \Pi_{n-m}, \operatorname{sgn} y_{n-m+1}, \operatorname{sgn} y_{n-m+2}, \dots \operatorname{sgn} y_n) = 0, \tag{256}$$

---

† Perhaps the theorem has not previously been stated and proved with this generality and precision. However, it is not far from the ideas of Vaschy, Riabouchinsky and Birkhoff.

‡ We need not restrict this function.

§ Inequalities (254) and (255) eliminate from consideration singular cases. Singular cases can usually be avoided by, if necessary, reordering the subscripts of the y_i. Otherwise singular cases are best treated individually by suitable adaptation of the proof given below. Singular cases have been discussed by Riabouchinsky (1911), Van Driest (1946), Birkhoff (1950) and Langhaar (1946, 1951) in the context of the ordinary (uncorrected) pi theorem.

where

$$\Pi_i(y_1, y_2, \dots y_n) = y_i |y_{n-m+1}|^{P_i} |y_{n-m+2}|^{Q_i} \dots |y_n|^{W_i} \quad (i = 1, 2, \dots n-m), \tag{257}$$

where the $P_i, Q_i, \dots W_i$ are constants, dependent only on the $p_j, q_j, \dots w_j$. Moreover the $\Pi_i (i = 1, 2, \dots n-m)$ are independent of the $\mu_j (j = 1, 2, \dots m)$ when (252) are substituted in (257).

Proof

Let

$$\begin{cases} y_i = z_i & (i = 1, 2, \dots n) \\ z_i \neq 0 & (i = n-m+1, \dots n) \end{cases} \tag{258, 259}$$

be a solution of (250). Then, from the hypotheses of the theorem

$$\Phi(z_1 \mu_1^{p_1} \mu_2^{p_1} \dots \mu_m^{w_1}, \quad z_2 \mu_1^{p_2} \mu_2^{p_2} \dots \mu_m^{w_2}, \\ \dots, z_n \mu_1^{p_n} \mu_2^{q_n} \dots \mu_m^{w_n}) = 0 \tag{260}$$

is valid when the μ's take any positive values, whether independently or dependently. In particular, we may assign to the μ's values which are dependent on one another and on the z's. Let us choose values of μ's which satisfy:

$$|z_i| \mu_1^{p_i} \mu_2^{q_i} \dots \mu_m^{w_i} = 1 \\ (i = n-m+1, n-m+2, \dots n). \tag{261}$$

To show that this choice is possible we take the log of (261):

$$p_i \log \mu_1 + q_i \log \mu_2 + \dots + w_i \log \mu_m = -\log |z_i| \\ (i = n-m+1, \dots n). \tag{262}$$

From (259) the right sides of (262) are finite. (262) is thus a set of m linear equations for the $\log \mu_j$ $(j = 1, 2, \dots m)$. (262) can be solved (by Cramer's rule) for the latter if the determinant of the coefficients on the left sides is non-zero, as it is by hypothesis (255). The solution of (262) gives each $\log \mu_j$ as a linear combination of the $\log |z_i|$. Thus each μ_j is a product of powers of the $|z_i|$. Substituting these μ's in the first $n-m$ arguments of (260), we find the latter are Π_i where

$$\Pi_i(z_1, z_2, \dots z_n) = z_i |z_{n-m+1}|^{P_i} |z_{n-m+2}|^{Q_i} \dots |z_n|^{W_i} \\ (i = 1, 2, \dots n-m). \tag{263}$$

Also, directly from (261), the last $n-m$ arguments of (260) are:

$$\operatorname{sgn} z_i \quad (i = n-m+1, n-m+2, \dots n). \tag{264}$$

Substituting from (251) in (263) and (264), which are the arguments of (260), we see that (260) yields (256).

It remains to be verified that the $\Pi_i(y_1, y_2, \dots y_n)$ are independent of the μ_j, when the y's of (252) are substituted in (257). Equivalently (since this amounts to a mere change of symbols) we may show that if the z's of (263) are replaced by:

$$z_i \lambda_1^{p_i} \lambda_2^{q_i} \dots \lambda_m^{w_i} \quad (i = 1, 2, \dots n), \tag{265}$$

the new Π_i resulting from (263) are independent of the λ_j. To find the new

Π_i let us make this replacement of the z_i in (260); the arguments of the latter then become:

$$z_i(\mu_1\lambda_1)^{p_i}(\mu_2\lambda_2)^{q_i}\dots(\mu_m\lambda_m)^{w_i} \quad (i=1,2,\dots n). \tag{266}$$

We apply to (266) the procedure for eliminating the μ's which was applied above to the arguments of (260). It is clear from the form of (266) that we at the same time eliminate the λ's. Hence we arrive at Π_i which are independent of the λ_j, and are in fact the same as expressions (263).

The proof of the modified pi theorem is now complete.

1.4 *Application of the Pi Theorem*

In applications the function Φ usually has the form:

$$\Phi = y_1 - \phi(y_2, y_3, \dots y_n). \tag{267}$$

In this case the pi theorem implies that

$$y_1 = |y_{n-m+1}|^{-P_1}|y_{n-m+2}|^{-Q_1}\dots|y_n|^{-W_1}\phi(\Pi_2, \Pi_3, \dots \Pi_{n-m}, \operatorname{sgn} y_{n-m+1}, \quad \operatorname{sgn} y_{n-m+2}, \dots \operatorname{sgn} y_n). \tag{268}$$

(268) is 'dimensionally homogeneous', i.e. the left and right sides have the same powers of the μ_j as factors when (252) is substituted. This fact justifies the formal method of equating dimensions which we used in our examples. Note that in using this method the presence of the sgn terms in (268) must not be overlooked (compare footnote to (57)).

In applications in physics, variables (such as viscosities or absolute temperatures) are often restricted to positive values. This explains why, although the unmodified pi theorem is not strictly correct, it has usually been applied with success.

Note that in most applications one applies, not the pi theorem itself but the technique used in proving the pi theorem. This is because, in the pi theorem itself, the constants $P_i, Q_i \dots W_i$ are not evaluated explicitly. (They could be, but would result in cumbersome formulae; it is neater to treat each application individually.)

Application 1

To emphasize the importance of the sgn terms, let us apply the technique of the pi theorem to the problem of determining the form of the optimal switching curve in Example 1. Let us represent this curve as

$$\Phi = x_1 - \phi(a, x_2) = 0. \tag{269}$$

With the signal scale multiplied by μ_1 and the time scale multiplied by μ_2, (269) becomes†:

$$\mu_1 x_1 = \phi(\mu_1^{-1}\mu_2^{2}a, \mu_1\mu_2^{-1}x_2), \tag{270}$$

† It seems intuitively obvious that since scale changes affect the problem only trivially, the ϕ of (270) is the same as the ϕ of (269). To prove this, substitute $\mu_2 t$ for t, $\mu_1 x_1$ for x_1, $\mu_1\mu_2^{-1}x_2$ for x_2, $\mu_1^{-1}\mu_2^{2}a$ for a, and $\mu_2 V$ for V, in the defining equations of the problem, and observe that the resulting equations are unchanged. (Compare Wonham's derivation of (221).)

which corresponds to (260). Corresponding to (261), we put:

$$\mu_1^{-1}\mu_2^2 a = 1 \tag{271}$$

and

$$\mu_1\mu_2^{-1}|x_2| = 1. \tag{272}$$

Solving (271) and (272) for μ_1 and μ_2, and substituting in (270) we find:

$$x_1 = ax_2^2\phi(1, \operatorname{sgn} x_2), \tag{273}$$

which corresponds to (268). (273) may be written:

$$x_1 - ax_2^2\phi_1(\operatorname{sgn} x_2) = 0. \tag{274}$$

Result (274) indicates correctly the form of the optimal switching curve (see(84)).

On the other hand, if we use the unmodified form of the pi theorem (§1.1), we arrive at, instead of (274), the incorrect result:

$$x_1 - \phi_2 ax_2^2 = 0, \tag{275}$$

where ϕ_2 is a constant.

Note also that result (274) can lead to a complete solution. Thus, applying symmetry arguments (compare § 6.3) to (274) we find:

$$\phi_1(\operatorname{sgn} x_2) = -\phi_1(-\operatorname{sgn} x_2), \tag{276}$$

so that (274) may be written:

$$x_1 - Kax_2|x_2| = 0, \tag{277}$$

where K is a constant. We have now only to optimize the system with respect to K. This approach is similar to that of Fuller (1960 b).

Application 2

Let us apply the technique of the pi theorem to prove (58). We have, for region P;

$$\Phi = F - \phi[a, (-x_1 + \tfrac{1}{2}ax_2^2)] = 0. \tag{278}$$

With the signal scale multiplied by μ_1 and the time scale multiplied by μ_2, (278) becomes:

$$\mu_2 F = \phi[\mu_1^{-1}\mu_2^2 a, \mu_1(-x_1 + \tfrac{1}{2}ax_2^2)]. \tag{279}$$

In this case eqns. (261) become:

$$\mu_1^{-1}\mu_2^2 a = 1 \tag{280}$$

and

$$\mu_1|-x_1 + \tfrac{1}{2}ax_2^2| = 1. \tag{281}$$

Solving (280) and (281) for μ_1 and μ_2, and substituting in (279) we find:

$$F = [a|-x_1 + \tfrac{1}{2}ax_2^2|]^{1/2}\,\phi[1, \operatorname{sgn}(-x_1 + \tfrac{1}{2}ax_2^2)], \tag{282}$$

which corresponds to (268). (282) verifies the first two lines of (58), and the last two lines of (58) are obtained by symmetry.

Application 3

Fuller (1960 b) treated our Example 3 by using the result that the ratio r of any two successive overshoots (intercepts of the x_1-axis in the (x_1, x_2)

phase plane) of any optimum trajectory is a constant. This result was obtained by formal dimensional reasoning; let us now confirm it by the technique of the pi theorem.

Suppose an optimum trajectory cuts the x_1-axis at successive points $(x_1, 0)$ and $(rx_1, 0)$. We have:

$$r = \phi(a, x_1), \tag{283}$$

i.e. with change of scales:

$$r = \phi(\mu_1^{-1}\mu_2^2 a, \mu_1 x_1). \tag{284}$$

Putting, as in (261),

$$\mu_1^{-1}\mu_2^2 a = 1 \tag{285}$$

and

$$\mu_1|x_1| = 1, \tag{286}$$

we find (284) becomes:

$$r = \phi(1, \quad \operatorname{sgn} x_1). \tag{287}$$

However, by symmetry (compare § 6.3), r is independent of the sign of x_1. Thus (287) implies:

$$r = \text{const.} \tag{288}$$

Application 4

Consider the problem of minimizing integral-square-error as in Example 3, but with a plant consisting of three integrators, $1/(ap^3)$. Let us apply the technique of the pi theorem to determine the form of the optimal switching surface. In the phase space with coordinates x_1 (output), $x_2[=dx_1/dt]$, and $x_3[=d^2x_1/dt^2]$, the switching surface may be represented by:

$$x_1 = \phi(x_2, x_3, a). \tag{289}$$

Thus, with change of scales:

$$\mu_1 x_1 = \phi(\mu_1\mu_2^{-1}x_2, \mu_1\mu_2^{-2}x_3, \mu_1^{-1}\mu_2^3 a). \tag{290}$$

Putting, as in (261),

$$\mu_1^{-1}\mu_2^3 a = 1 \tag{291}$$

and

$$\mu_1\mu_2^{-2}|x_3| = 1, \tag{292}$$

we find that (290) becomes:

$$\frac{x_1}{a^2|x_3|^3} = \phi\left(\frac{x_2}{ax_3^2}, \quad \operatorname{sgn} x_3, 1\right). \tag{293}$$

Therefore, in the half space $x_3 > 0$, the switching surface is represented by a relation between only two variables, $x_1/(a^2|x_3|^3)$ and $x_2/(ax_3^2)$. In the half space $x_3 < 0$, the switching surface is represented by another such relation.

By a more detailed investigation, using symmetry (compare appendix 6 of Grensted and Fuller (1965)), (293) can be simplified to;

$$\frac{x_1}{a^2x_3^3}=\phi_1\left(\frac{x_2}{ax_3|x_3|}\right), \tag{294}$$

so that the switching surface can be represented by a single relation between the two variables:

$$\frac{x_1}{a^2x_3^3} \quad \text{and} \quad \frac{x_2}{ax_3|x_3|}. \tag{295}$$

Newmann and Zachary (1965) applied the unmodified pi theorem to this problem and arrived at the incorrect result that the switching surface could be represented by a single relation between the two variables:

$$\frac{ax_3^2}{x_2} \quad \text{and} \quad \frac{x_2^2}{x_1x_3}. \tag{296}$$

To correct this result, we can represent the switching surface by two relations (one for $x_3>0$ and one for $x_3<0$) between variables (296). (Newmann and Zachary's fig. 7 must then be replaced by two separate graphs.) Alternatively we can replace variables (296) by the variables:

$$\frac{ax_3|x_3|}{x_2} \quad \text{and} \quad \frac{x_2^2}{x_1x_3} \tag{297}$$

obtained by a one-one transformation of (295). (Newmann and Zachary's fig. 7 must then be replaced by a graph with fewer branches.)

References

Bellman, R., 1957, *Dynamic Programming* (Princeton), Chaps. 6, 9; 1961, *Adaptive Control Processes: A Guided Tour* (Princeton), Chaps. 3, 4, 12.

Bellman, R., and Dreyfus, S. E., 1962, *Applied Dynamic Programming* (Princeton), Chaps. 5, 7, 8.

Berkovitz, L. D., 1961, *J. Math. Anal. Appl.*, **3,** 145.

Birkhoff, G., 1948, *Electrical Eng.*, **67,** 1185; 1950, *Hydrodynamics* (Princeton).

Bliss, G. A., 1946, *Lectures on the Calculus of Variations* (Chicago).

Bridgman, P. W., 1922, *Dimensional Analysis* (Yale); 1931, revised edition.

Buckingham, E., 1914, *Phys. Rev.*, **4,** 345.

Courant, R., 1936, *Differential and Integral Calculus* (London: Blackie), **2,** 108.

Decius, J. C., 1948, *J. Franklin Inst.*, **245,** 379.

Duff, G. F. D., 1956, *Partial Differential Equations* (Toronto).

Duncan, W. J., 1953, *Physical Similarity and Dimensional Analysis* (London: Arnold).

Feldbaum, A. A., 1963, *Principles of the Theory of Optimal Automatic Systems* (Moscow: Fizmatgiz). (In Russian.)

Focken, C. M., 1953, *Dimensional Methods and their Applications* (London: Arnold).

Fourier, J. B. J. 1822, *Théorie Analytique de la Chaleur* (Paris: Didot), p. 152; 1878, English translation (Cambridge), p. 128.

FULLER, A. T., 1959, *J. Electron. Contr.*, **7,** 456; 1960 a, *Ibid.*, **8,** 465; 1960 b, *Proc. IFAC Congress, Moscow* (London: Butterworths, 1961), **1,** 510; 1964, *J. Electron. Contr.*, **17,** 283.
GOURSAT, E., 1917, *Differential Equations*, trans. by E. R. Hedrick and O. Dunkel (Boston: Ginn).
GRENSTED, P. E. W., and FULLER, A. T., 1965, *Int. J. Contr.*, **2,** 33.
GUMOWSKI, I., 1965, *C.R. Acad. Sci., Paris*, **260,** 1096.
INCE, E. L., 1926, *Ordinary Differential Equations* (London: Longmans), Chap. 3.
JEANS, J. H., 1905, *Proc. roy. Soc.*, **76,** 545.
LANGHAAR, H. L., 1946, *J. Franklin Inst.*, **242,** 459; 1951, *Dimensional Analysis and Theory of Models* (New York: Wiley).
MAXWELL, J. C., 1873, *Treatise on Electricity and Magnetism* (Oxford), **1,** 1, 43, 90, 332; **2,** 239.
MORGAN, A. J. A., 1952, *Quart. J. Math.* **3,** 250.
NEWMANN, M. M., and ZACHARY, D. H., 1965, *Int. J. Contr.*, **2,** 149.
PEARSON, J. D., 1961, *J. Electron. Contr.*, **10,** 323.
PONTRYAGIN, L. S., BOLTYANSKI, V.G., GAMKRELIDZE, R. V., and MISHCHENKO, E. F., 1961, *Mathematical Theory of Optimal Processes* (Moscow: Fizmatgiz); 1962, English translation (New York: Wiley).
RIABOUCHINSKY, D., 1911, *L'Aérophile*, **19,** 407; reprinted in *Bull. de l'Inst. aerodynamique de Koutchino*, **4,** (1912), 50; 1943, *C.R. Acad. Sci., Paris*, **217,** 205, 220.
ROZONOER, L. I., 1959, *Avtomat. Telemech., Moscow*, **20,** 1320, 1441, 1561.
SUN TSZYAN, 1960, *Izv. A.N.S.S.S.R., Energ. i Avt.*, No. 5, 96.
VAN DRIEST, E. R., 1946, *J. appl. Mech., Trans. Amer. Soc. mech. Engrs.*, **13,** A–34.
VASCHY, A., 1890, *Traité d'électricité et de magnétisme* (Paris: Baudry), **1,** 1; 1892 a, *Annales Télégraphiques* (Paris), **19,** 25, 189; 1892 b, *C. R. Acad. Sci., Paris*, **114,** 1416.
WONHAM, W. M., 1963, *J. Electron. Contr.*, **15,** 59.
WOODSIDE, C. M., 1965, *Int. J. Contr.*, **2,** 285, 409.

Optimal control of saturating linear plants for quadratic performance indices†

By H. R. Sirisena
Engineering Department, University of Cambridge

[Received February 2, 1968]

Abstract
The problem of the optimal regulator for linear plants with bounded control variable and quadratic performance index is studied. Plants with zeros in their transfer function are included in the treatment. Some earlier results on this problem due to Wonham and Johnson (1964, 1965) and other writers are clarified and generalized.

1. Introduction

The problem of the optimal regulator for an nth order linear plant with bounded control variable and quadratic performance index‡:

$$\int_0^T (\langle x, Qx \rangle + \sigma u^2)\, dt, \tag{1}$$

posed originally by Letov (1960, 1961) has been studied more recently by Wonham and Johnson (1964, 1965). They considered plants without zeros and, for the case $\sigma = 0$, used the maximum principle to show that the optimal control is singular in a certain sub-set of state space and is normal (bang-bang) outside it. However, they limited their treatment to the class of performance indices for which this singular sub-set was of dimension $n-1$. They also required that the eigenvalues associated with the singular trajectories be distinct. The present paper aims at a unified treatment of the problem without these restrictions. A few of the results obtained are implicit in the paper by Kalman (1964) which deals with plants without saturation.

It is found that the usual requirement that the matrix Q in (1) be non-negative definite is, in general, too restrictive, e.g. when Q is also required to be diagonal. The most general non-negative definite quadratic performance index, not containing products of u and x, is shown to be

$$\int_0^T (\langle c, x \rangle^2 + \sigma u^2)\, dt. \tag{2}$$

Sufficient conditions have been given by Wonham and Johnson (1964, 1965) on the index (1) for which an nth order plant is equivalent, for optimal control purposes, to first or second-order plants. It is found that some of these conditions are not necessary. A set of necessary and sufficient conditions is derived for the more general problem of the equivalence of the nth order plant to a plant of order r ($<n$).

---

† Communicated by Dr. A. T. Fuller.
‡ $\langle x, y \rangle$ denotes the inner product of the vectors x and y.

2. The problem

Initially, only plants without zeros will be considered, the results being extended later (see § 7) to include plants with zeros as well.

The problem is as follows:

For the linear stationary plant†:

$$x' = Ax + fu, \tag{3}$$

find the control input $u(t)$ which minimizes the functional:

$$\int_0^T (\langle c, x\rangle^2 + \sigma u^2)\, dt, \tag{4}$$

subject to the terminal conditions:

$$\left.\begin{aligned} x(0) &= x_0 \quad (x_0 \text{ is unrestricted})‡, \\ x(T) &= 0 \quad (T \text{ is free}), \end{aligned}\right\} \tag{5}$$

where A is a $n \times n$ constant matrix, f, c are constant n vectors, and x is the n-dimensional state vector. The control input u is subject to the saturation constraint:

$$|u| \leqslant 1. \tag{6}$$

In general, we have, for the elements of c, the conditions:

$$\left.\begin{aligned} c_p &\neq 0, \\ c_i &= 0, \quad i = p+1, \ldots, n. \end{aligned}\right\} \tag{7}$$

Also $\sigma \geqslant 0$.

With the usual assumption that the system (3) is controllable, it has been shown by Wonham and Johnson (1964) that there is no loss of generality in taking A, f to be of the canonical forms:

$$A = \begin{bmatrix} 0 & 1 & 0 & \ldots & 0 & 0 \\ 0 & 0 & 1 & \ldots & 0 & 0 \\ \cdot & \cdot & \cdot & & \cdot & \cdot \\ \cdot & \cdot & \cdot & & & \cdot \\ \cdot & \cdot & \cdot & & \cdot & \cdot \\ 0 & 0 & 0 & \ldots & 0 & 1 \\ -a_1 & -a_2 & -a_3 & \ldots & -a_{n-1} & -a_n \end{bmatrix}, \quad f = \begin{bmatrix} 0 \\ 0 \\ \cdot \\ \cdot \\ \cdot \\ 0 \\ 1 \end{bmatrix}. \tag{8}$$

The D-operator representation of the plant:

$$\prod_{i=1}^{n} (D - \alpha_i) x_1 = u \tag{9}$$

is also used widely in the sequel wherever it is more convenient.

Note that the cost functional (4) differs from the more conventional cost functional (1). Under the conditions (5), Wonham and Johnson (1964) have shown that there is no loss of generality in taking Q in (1) to be diag $(q_1, \ldots, q_n)$.

---

† Primes denote differentiation with respect to time t.

‡ I.e. x_0 may lie anywhere in the region of state space from which admissible trajectories to the origin exist.

Also it is shown, in Appendix I below, that for every cost functional (1), there exists an equivalent cost functional (4) which yields the same optimal control law, provided only that:

$$\left.\begin{array}{l} q_1 > 0 \\ \text{and the equation} \\ \sum_{i=1}^{n} q_i z^{i-1} = 0 \end{array}\right\} \quad (10)$$

has no positive real roots.

Thus it is too restrictive to require, as Wonham and Johnson have done, that Q be both diagonal and non-negative definite. The appropriate restrictions on Q are given by (10). In particular, some q_i could be negative without violating condition (10).

In the D-operator notation:

$$\langle c, x \rangle = c_p \prod_{i=1}^{p-1} (D - \beta_i) x_1. \quad (11)$$

It is shown in Appendix I that there is no loss of generality in choosing c such that $\text{Re}(\beta_i) < 0$. We shall also assume that:

$$\beta_i \neq \alpha_j \quad \text{all} \quad i, j. \quad (12)$$

It is shown in § 12 that if (12) is not satisfied, the problem can be reduced to one for which it is, so that it does not introduce a loss of generality.

The β_i can be regarded as the eigenvalues of the 'ideal model' corresponding to the performance index (4), which attains its absolute minimum value zero when applied to this model. When applied to any other plant, the resulting optimal controller causes the plant to best approximate this ideal model.

3. Pontryagin's equations

The present paper deals mainly with the case $\sigma = 0$, and for this case the Hamiltonian is:

$$H = \langle \phi, Ax + fu \rangle - \langle c, x \rangle^2. \quad (13)$$

From the maximum principle, the optimal control u is given by:

$$H(\phi, x, u^*) = \max_u H(\phi, x, u)$$

or

$$u^*(t) = \text{sign}\,[\phi_n(t)], \quad \phi_n(t) \neq 0. \quad (14)$$

The sign function on the right-hand side of (14) is undefined for zero argument and, therefore, singular solutions are possible for which $\phi_n(t)$ vanishes over a positive interval of time.

The canonical equations, satisfied by both singular and normal optimal controls, are

$$x_i' = [\partial H / \partial \phi_i]_{u=u^*},$$

$$\phi_i' = [-\partial H / \partial x_i]_{u=u^*}.$$

Written in full these are

$$\left.\begin{aligned} x_1{}' &= x_2, \\ &\;\;\vdots \\ x_{p-1}{}' &= x_p, \\ &\;\;\vdots \\ x_n{}' &= u - a_1 x_1 - \ldots - a_n x_n \end{aligned}\right\} \tag{15}$$

and

$$\left.\begin{aligned} \phi_1{}' &= 2c_1 \langle c, x \rangle + a_1 \phi_n, \\ &\;\;\vdots \\ \phi_p{}' &= -\phi_{p-1} + 2c_p \langle c, x \rangle + a_p \phi_n, \\ &\;\;\vdots \\ \phi_n{}' &= -\phi_{n-1} + a_n \phi_n. \end{aligned}\right\} \tag{16}$$

Also, since the settling time T is free, we have the transversality condition :

$$H = 0, \quad 0 \leqslant t \leqslant T. \tag{17}$$

4. Candidates for the optimal control

Equations (14)–(17) provide only necessary conditions for optimality. All trajectories which satisfy these necessary conditions yield candidates for the optimal control. The existence of at least one such candidate for the problems considered in this paper, follows from an existence theorem of Lee and Markus (1961). If there is more than one candidate trajectory, then the candidate(s) with the least cost is the optimal control. The next step, therefore, is the search for these candidates.

It has been shown by Rohrer and Sobral (1966) that all singular solutions of eqns. (14)–(17) lie in certain singular sub-spaces. Only one of these sub-spaces is, in general, comprised entirely of singular sub-arcs with convergent modes. All singular sub-arcs lying in this sub-space converge to the origin, while all sub-arcs not lying in this sub-space diverge away from the origin. Thus the final portion of an optimal trajectory, in general, must consist of a normal bang-bang arc impinging on this sub-space and a singular sub-arc lying in this sub-space. In special cases, the normal arc itself might pass through the origin and such trajectories do not, therefore, contain this terminal singular sub-arc.

A complete candidate trajectory would, in general, be comprised of singular sub-arcs in one or more of the singular sub-spaces which are joined together by normal bang-bang arcs. However, it is not plausible that singular sub-arcs containing divergent modes could be part of an optimal trajectory. For this reason, only candidates consisting of just one normal arc starting from the initial plant state plus a singular sub-arc converging to the origin will be considered.

It follows that only the singular sub-space containing the convergent sub-arcs need be determined and this will be done in the next section.

5. Singular sub-arcs

On a singular sub-arc:

$$\phi_n(t)=0,$$

whence, from the last $n-p$ eqns. (16):

$$\phi_p=\phi_{p+1}=\ldots=\phi_n=0, \tag{18}$$

while the pth equation becomes:

$$\phi_{p-1}=2c_p\langle c,x\rangle. \tag{19}$$

Using (19), the first $p-1$ eqns. (16) become:

$$\begin{bmatrix}\phi_1\\ \phi_2\\ \cdot\\ \cdot\\ \cdot\\ \phi_{p-1}\end{bmatrix}'=\begin{bmatrix}0 & 0 & \ldots & 0 & c_1/c_p\\ -1 & 0 & \ldots & 0 & c_2/c_p\\ \cdot & \cdot & & \cdot & \cdot\\ \cdot & \cdot & & \cdot & \cdot\\ \cdot & \cdot & & \cdot & \cdot\\ 0 & 0 & \ldots & -1 & c_{p-1}/c_p\end{bmatrix}\begin{bmatrix}\phi_1\\ \phi_2\\ \cdot\\ \cdot\\ \cdot\\ \phi_{p-1}\end{bmatrix}. \tag{20}$$

Equations (15), (20) are solved in Appendix II. It is shown there that on a convergent singular sub-arc:

$$\phi_i=0,\quad i=1,\ldots,n \tag{21}$$

and

$$\langle c,x\rangle=0. \tag{22}$$

Repeated differentiation of (22) $n-p$ times with respect to time yields:

$$\left.\begin{aligned}\langle(0\ \ c_1\ \ \ldots\ \ c_{p-1}\ \ c_p\ \ 0\ \ \ldots\ \ 0),x\rangle&=\langle c,Ax\rangle=0,\\ &\ \ \vdots\\ \langle(0\ \ 0\ \ \ldots\ \ 0\ \ c_1\ \ \ldots\ \ c_{p-1}\ \ c_p),x\rangle&=\langle c,A^{n-p}x\rangle=0.\end{aligned}\right\} \tag{23}$$

It follows that all singular sub-arcs lie in the $(p-1)$-dimensional linear sub-space K:

$$\{x|\langle c,A^ix\rangle=0,\quad i=0,\ldots,n-p\}. \tag{24}$$

One more differentiation of (22) yields:

$$\langle c,A^{n-p+1}x\rangle+\langle c,A^{n-p}fu\rangle=0,$$

i.e.

$$u=-\langle c,A^{n-p+1}x\rangle/\langle c,A^{n-p}f\rangle. \tag{25}$$

For motion in K:

$$\begin{aligned}x'&=(A-fcA^{n-p+1}/\langle c,A^{n-p}f\rangle)x\\ &=\hat{A}x.\end{aligned} \tag{26}$$

The solution of this homogeneous differential equation is:

$$x = \exp(\hat{A}t)x_{in} \tag{27}$$

for a starting point $x = x_{in}$.

Let the sub-set J of the linear sub-space K be defined as the set of all points invariant under the control law (25) with $|u(t)| \leqslant 1$.

For motion in J, from (25):

$$u = -\langle c, A^{n-p+1} \exp(\hat{A}t)x_{in} \rangle / \langle c, A^{n-p}f \rangle. \tag{28}$$

Therefore, a point $P = x_{in} \in K$ lies in J if and only if:

$$|\langle c, A^{n-p+1} \exp(\hat{A}t)x_{in} \rangle / \langle c, A^{n-p}f \rangle| \leqslant 1, \quad \text{all} \quad t \geqslant 0. \tag{29}$$

Equation (29) can be used to show that J is a closed convex sub-set of K. More explicitly, J is the sub-set of K bounded by the two parallel hyperplanes:

$$\langle c, A^{n-p+1}x \rangle = \pm \langle c, A^{n-p}f \rangle \tag{30}$$

and the singular sub-arcs tangential to them. Also, J contains the origin.

6. Normal optimal arcs

Necessary conditions have been given by Johnson (1965) for the existence of normal reverse time arcs starting from states in a singular sub-space S_s. However, these have been based on the assumption that a small neighbourhood of S_s can be chosen in which the switching function $\phi_n(t) \neq 0$. This is true, in general, only if $p = n$, i.e. when the singular sub-space is of dimension $n-1$. For $p < n$, the switching function usually contains an infinite number of zeros in any finite neighbourhood of S_s.

The existence of normal arcs impinging on the singular sub-set J can, however, be demonstrated by other means. It can be verified by substitution that singular sub-arcs in J do not add to the value of the cost functional (4). Therefore the normal arcs would be obtained as the optimal trajectories for the related problem of minimizing (4) for the plant (3) for the same initial plant state but with the final plant state being given by:

$$x(T) \in J \quad (T \text{ is free}), \tag{31}$$

instead of by (5).

The existence of optimal trajectories for the latter problem follows from a general existence theorem due to Lee and Markus (1961). However, it is to be noted that this only proves that normal arcs impinge on a non-empty sub-set of J and not necessarily on all points of J.

'Candidate' normal arcs can be obtained by the integration of the canonical eqns. (14)–(17) in reverse time, starting from states in J with the boundary value of the adjoint variable given by:

$$\phi(0) = 0, \tag{32}$$

from the continuity of the adjoint variable along an optimal trajectory.

For $p = n$, the validity of (32) can be shown from the transversality conditions at the starting point in J for the related problem, the adjoint variable being identical for the two problems.

For a trajectory terminating on an interior point of J, i.e. not on boundary (29), using (24) the transversality condition yields:

$$\phi(0)=\gamma c, \tag{33}$$

where γ is an arbitrary constant.

Also, since the settling time is free, we have from the transversality condition (17):

$$\langle \phi(0), Ax+fu(0)\rangle=0 \tag{34}$$

and using (33), we obtain:

$$\langle \gamma c, Ax+fu(0)\rangle=0 \tag{35}$$

and $u(0)=\pm 1$.

Now $\langle c, Ax\pm f\rangle=0$ only on the linear boundaries of J. Therefore for trajectories terminating at all interior points of J, eqn. (35) yields:

$$\gamma=0,$$

whence

$$\phi(0)=0. \tag{36}$$

The backward tracing of normal arcs is straightforward for $p=n$. In this case all normal arcs have a finite number of switches.

Difficulties arise when $p<n$ for the reason that the switching function $\phi_n(t)$ has an infinite number of zeros in any finite neighbourhood of the sub-set J. It is found, in general, that backward tracing with the exact initial values of the canonical variables yields only the switchless optimal trajectories (if any). The switching surface could, however, be computed approximately by starting the integration with perturbed initial values of the canonical variables. Even this is not straightforward for $n\geqslant 3$, due to the presence in the canonical equations of rapidly diverging monotonic modes which tend to swamp the oscillatory modes which define the switching surface. This has been reported by Grensted and Fuller (1965) for the problem of minimizing integral square error for a three integrator plant, this being an example of the case $n=3$, $p=1$.

Two examples, with $p=n$, are worked out in § 10.

7. Plants with zeros

In this section, the analysis is extended to include plants with zeros. Consider the proper†, linear, stationary plant

$$x'=Ax+fu, \tag{37a}$$

$$y=bx, \tag{37b}$$

where A is a $n\times n$ constant matrix, b,f are constant n vectors and y,u are scalars.

If the plant (37) is controllable, then, as before, A and f may be assumed to have the canonical forms (8). In general,

$$b=(b_0\quad b_1\quad \dots\quad b_m\quad 0\quad \dots\quad 0), \tag{38}$$

with

$$b_m\neq 0.$$

† A proper plant is defined as a plant whose transfer function has a numerator of lower order than the denominator.

The plant (37) can also be described by the transfer function:

$$M(s)/N(s), \tag{39}$$

between output y and input u, where

$$\left.\begin{aligned} M(s) &= b_m s^m + \ldots + b_1 s + b_0 = b_m \prod_{i=1}^{m} (s-\gamma_i), \\ N(s) &= s^n + \ldots + a_2 s + a_1 = \prod_{i=1}^{n} (s-\alpha_i). \end{aligned}\right\} \tag{40}$$

If, in addition, the plant (37) is completely observable, then $M(s)$ and $N(s)$ have no common factor (see, e.g., Zadeh and Desoer 1963).

8. The problem for plants with zeros

The problem studied is as follows: find the scalar control input $u(t)$ which minimizes the functional†:

$$\int_0^T \left[\left(\sum_{i=1}^{n-m} p_i y^{(i-1)} \right)^2 + \sigma u^2 \right] dt, \tag{41}$$

subject to (37) and the terminal conditions:

$$\left.\begin{aligned} x(0) &= x_0 \quad (x_0 \text{ is unrestricted}), \\ y(t) &= 0 \quad (t \geqslant T\,; T \text{ is free}). \end{aligned}\right\} \tag{42}$$

The control input u is subject to the saturation constraint:

$$|u| \leqslant 1. \tag{43}$$

It has been shown by Sivan (1965) that:

$$y^{(i)} = \langle b, A^i x \rangle, \quad i = 1, \ldots, n-m-1. \tag{44}$$

Thus y and its first $n-m-1$ time derivatives are all linear functions of the state variables and, therefore, continuous time functions. The cost functional (41) is the logical extension of the cost functional (4) and it includes only these derivatives of y in order that it be meaningful.

Equation (44) shows that the plant output is zero for all plant states in the m-dimensional linear sub-space F:

$$\{x \,|\, \langle b, A^i x \rangle = 0, \quad i = 0, \ldots, n-m-1\}. \tag{45}$$

However, for an observable system, a finite though definite control input is required to keep the representative point in F at every point of F except the origin. As the magnitude of the control input is bounded by (43), this is possible at all times $t \geqslant T$ only for points initially in a certain non-trivial sub-set G of F. G is the largest sub-set of F that is invariant under an allowable control.

For $\sigma = 0$, the cost function (41) becomes zero at the time T at which the plant state reaches the sub-set G. It could therefore be said that the control task is completed at time T, although a finite control input is still needed to keep the output at zero. Also this input is a simple linear function of the state variables. This view is also implicit in the work reported by Hutchinson (1963) and Lee (1962).

---

† $y^{(i)}$ denotes $d^i y/dt^i$.

However, for $\sigma > 0$, the cost function becomes equal to zero only when the control input reduces to zero, i.e. when the plant state arrives at the origin. In this case, therefore, the control task is obviously not complete when the plant output alone has been reduced to zero, and the optimal control input is identical to that for the system without zeros described by the same input-state eqn. (37 *a*).

9. The target sub-set

We shall now derive the equations defining the sub-set G. For motion in G, i.e. for $t \geqslant T$, we have from (37 *b*), (40) and (42):

$$b_m \prod_{i=1}^{m} (D-\gamma_i)x_1 = 0. \tag{46}$$

Let the multiplicity of the zero γ_i be m_i ($i = 1, \ldots, \eta$). Integration of eqn. (46) gives:

$$x_1 = \sum_{i=1}^{\eta} \sum_{j=1}^{m_i} V_{ij} t^{j-1} \exp(\gamma_i t) \tag{47}$$

and using (37 *a*) and (40), we have:

$$\begin{aligned} u &= \prod_{k=1}^{n} (D-\alpha_k)x_1 \\ &= \sum_{i=1}^{\eta} \exp(\gamma_i t) \prod_{k=1}^{n} (D-\alpha_k+\gamma_i)\left(\sum_{j=1}^{m_i} V_{ij} t^{j-1}\right). \end{aligned} \tag{48}$$

Inspection of eqn. (48) shows that the constraint (43) can be satisfied for all $t \geqslant T$, if and only if:

$$\left.\begin{aligned} V_{ij} &= 0, \quad j = 1, \ldots, m_i \quad \text{if} \quad \operatorname{Re}(\gamma_i) > 0, \\ V_{ij} &= 0, \quad j = 2, \ldots, m_i \quad \text{if} \quad \operatorname{Re}(\gamma_i) = 0. \end{aligned}\right\} \tag{49}$$

In general, we would have:

$$\left.\begin{aligned} \operatorname{Re}(\gamma_i) &< 0, \quad i = 1, \ldots, \mu, \\ \operatorname{Re}(\gamma_i) &= 0, \quad i = \mu+1, \ldots, \theta \end{aligned}\right\} \tag{50}$$

and, therefore,

$$\left.\begin{aligned} V_{ij} &= 0, \quad j = 1, \ldots, m_i, \quad i = \theta+1, \ldots, \eta, \\ V_{ij} &= 0, \quad j = 2, \ldots, m_i, \quad i = \mu+1, \ldots, \theta. \end{aligned}\right\} \tag{51}$$

Substituting (51) in (47), we obtain that, for $t \geqslant T$, x_1 satisfies the differential equation:

$$\prod_{i=1}^{\mu} (D-\gamma_i)^{m_i} \prod_{i=\mu+1}^{\theta} (D-\gamma_i)x_1 = 0. \tag{52}$$

Using (37 *a*), this can be written as:

$$\langle \hat{b}, x \rangle = 0. \tag{53}$$

Using the same arguments as before, eqn. (53) can be used to show that, for $t \neq T$, the plant state must lie in the linear sub-space:

$$\{x \mid \langle \hat{b}, A^{i-1}x \rangle = 0, \quad i = 1, \ldots, n-g\} \tag{54}$$

of dimension

$$g = \theta - \mu + \sum_{i=1}^{\mu} m_i,$$

and that G is a closed, convex sub-set of the linear sub-space (54) bounded by the two parallel hyperplanes:

$$\langle \hat{b}, A^{n-g}x \rangle = \pm \langle \hat{b}, A^{n-g-1}f \rangle \tag{55}$$

and the trajectories in (54) tangential to them.

Now, in general, in the cost function (41):

$$\left. \begin{array}{l} p_k \neq 0, \\ p = 0, \quad i = k+1, \ldots, n-m. \end{array} \right\} \tag{56}$$

From (44) and (56), the cost function (41) can be written in terms of the state variables as:

$$\left[\sum_{i=1}^{n-m} p_i y^{(i-1)} \right]^2 + \sigma u^2 = \left[\sum_{i=1}^{k} p_i \langle b, A^{i-1}x \rangle \right]^2 + \sigma u^2 \tag{57}$$

$$\triangle \langle c, x \rangle^2 + \sigma u^2,$$

where

$$\left. \begin{array}{l} c_p \neq 0, \\ c_i = 0, \quad i = p+1, \ldots, n \end{array} \right\} \tag{58}$$

and

$$p = m + k.$$

10. The optimal control

The problem, for $\sigma = 0$, can now be restated as follows: for the plant (37), find the control input u subject to the constraint (43) which minimizes the cost functional:

$$\int_0^T \langle c, x \rangle^2 \, dt, \tag{59}$$

with

$$\left. \begin{array}{ll} x(0) = x_0 & (x_0 \text{ is unrestricted}), \\ x(T) \in G & (T \text{ is free}). \end{array} \right\} \tag{60}$$

This problem differs from that considered in § 2 only in the constraint on the final plant state. It will be recalled that the optimal trajectory for that problem consisted of a normal arc impinging on the singular sub-set J, plus a singular sub-arc in J.

The singular sub-arc in J does not add to the value of the cost functional (41). Also, it can be verified by substitution of (54) in (24) that† $G \subseteq K$, and from the definitions of G and J it follows that $G \subseteq J$.

It is therefore clear that the portion of the above optimal trajectory from the initial plant state to the point where it impinges on G is the optimal trajectory

† $G \subseteq K$ denotes that G is a sub-set of K.

for the present problem. Thus, the optimal trajectory for the plant with zeros consists, in general, of a normal bang-bang arc from the initial plant state which impinges on the singular sub-set J, followed by a singular sub-arc in J which in turn impinges on the target sub-set G.

It can also be shown that if the normal arc does not impinge directly on G, then the singular sub-arc in J meets G only at the origin and the settling time is infinite. If the normal arc does impinge on G, there is, obviously, no singular sub-arc and the settling time is finite.

In particular, when the performance criterion is integral square error and the plant has only LHP zeros, G coincides with J and the settling time is finite for all initial plant states.

11. Examples

The results obtained in the previous sections will now be illustrated by examples of second and third-order plants.

Example 1

Minimize

$$\int_0^T (2x_1 + 3x_2 + x_3)^2 \, dt, \tag{61}$$

for the three integrator plant:

$$x' = \begin{bmatrix} 0 & 1 & 0 \\ 0 & 0 & 1 \\ 0 & 0 & 0 \end{bmatrix} x + \begin{bmatrix} 0 \\ 0 \\ 1 \end{bmatrix} u \tag{62}$$

for any initial plant state, with the final plant state $x(T) = 0$.

Using (24), we find that the singular sub-space K is the plane:

$$2x_1 + 3x_2 + x_3 = 0 \tag{63}$$

and from (30), the sub-set J of K is bounded by the planes:

$$2x_2 + 3x_3 = \pm 1 \tag{64}$$

and the singular trajectories in K tangential to them. The projection of J on the x_2x_3 plane is shown in fig. 1.

Fig. 1

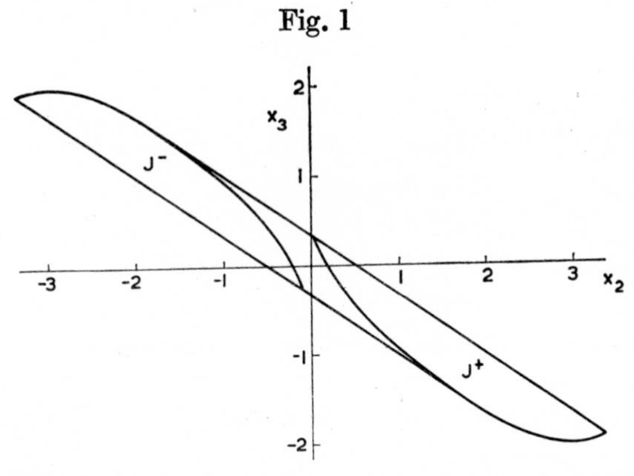

Projection of the singular sub-set on the x_2x_3 plane.

From (25), the singular control in J is given by:

$$u_s^* = -2x_2 - 3x_3. \tag{65}$$

By integration of the canonical eqns. (14)–(17) in negative (i.e. reverse) time with the boundary values:

$$\left.\begin{aligned} x(0) &= [-(3\alpha+\beta)/2, \alpha, \beta] \\ \phi(0) &= 0, \end{aligned}\right\} \tag{66}$$

and

where (α, β) lies within the projection of J on the x_2x_3 plane, we obtain the normal optimal control in the neighbourhood of J, until the first switch in reverse time, as:

$$u_n^*(t) = \operatorname{sign}\left[\{2\alpha + 3\beta + u(-0)\}\frac{t^2}{2} - (6\alpha + 7\beta)\frac{t^3}{6} + \{4\alpha - 5u(-0)\}\frac{t^4}{24} + \beta\frac{t^5}{30} + u(-0)\frac{t^6}{180}\right], \tag{67}$$

where $u(-0)$ is the value of the control input just before the normal arc impinges on the sub-set J.

A detailed analysis of $\phi_3(t)$, the argument of the sign function in (67), shows that normal arcs impinging on a sub-set J^+ with $u(-0)=1$ (see fig. 1) and on a sub-set J^- with $u(-0) = -1$ have no switches at all. For all other normal arcs, the first switch in reverse time occurs at the first zero of $\phi_3(t)$ for $t < 0$. Further switching points could be computed by piecewise integration of the canonical equations, preferably using an analogue or digital computer.

The optimal switching surface could be built up by plotting the coordinates of the successive switching points in reverse time while varying the starting point in J. This is quite straightforward, but the complete numerical results are omitted here for reasons of space. Figure 2 shows the plot of plant output x_1 and control input u versus time for a particular optimal trajectory. For comparison, the corresponding values for the time optimal trajectory from the same initial plant state are also plotted.

The above example is typical of the case $p = n$.

Example 2

Minimize integral square error for the plant:

$$(1-s)/s^2, \tag{68}$$

the desired plant output being zero.

The plant (68) can be described by the input–output-state equations:

$$\begin{bmatrix} x_1 \\ x_2 \end{bmatrix}' = \begin{bmatrix} 0 & 1 \\ 0 & 0 \end{bmatrix} \begin{bmatrix} x_1 \\ x_2 \end{bmatrix} + \begin{bmatrix} 0 \\ 1 \end{bmatrix} u, \tag{69}$$

$$y = x_1 - x_2. \tag{70}$$

The cost functional in terms of the state variables is:

$$\int_0^T (x_1 - x_2)^2 \, dt \tag{71}$$

and the settling time T is free.

Fig. 2

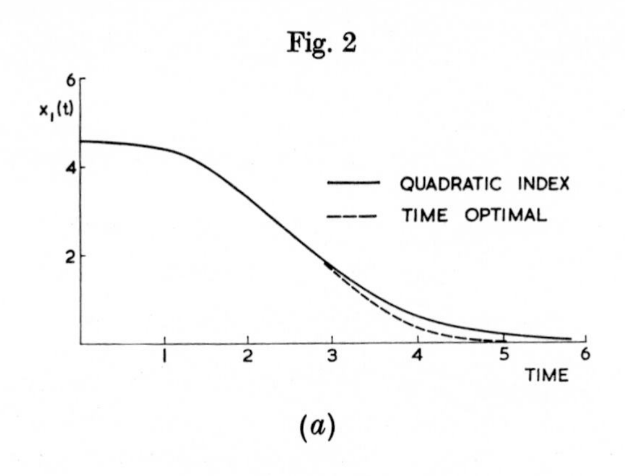

(*a*)

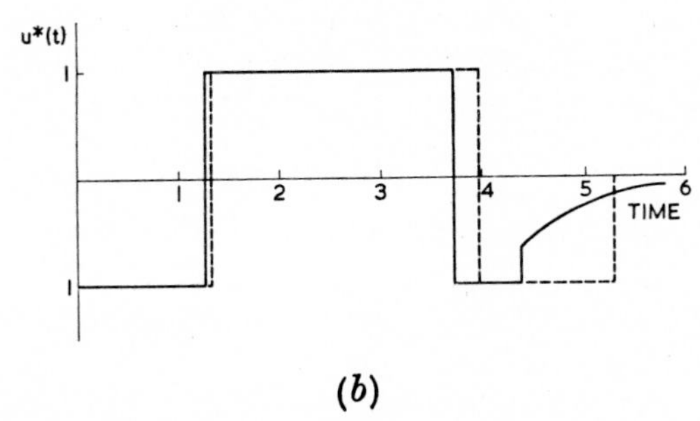

(*b*)

Typical plots of (*a*) $x_1(t)$ and (*b*) $u^*(t)$ for Example 1.

Using result (R 2) of Appendix I, we find that the cost functional (71) is equivalent to the cost functional:

$$\int_0^T (x_1 + x_2)^2 \, dt. \tag{72}$$

From (24), the linear sub-space K, which contains the convergent singular sub-arcs, is, therefore, the straight line:

$$x_1 + x_2 = 0 \tag{73}$$

and from (30), the singular sub-set J is the segment:

$$|x_2| \leqslant 1$$

of the straight line (73).

Using eqns. (52)–(55), we find that the target sub-set G is the origin of coordinates.

From (25), the singular control in J is given by:

$$u_s^* = -x_2. \tag{74}$$

To obtain the normal optimal arc, we integrate the canonical eqns. (14)–(17) in reverse time starting from the point:

$$x(0) = (\alpha, -\alpha), \quad |\alpha| \leqslant 1$$

in J, with the boundary value of the adjoint variable given by (32) as:

$$\phi(0) = 0$$

and we obtain the normal optimal control until the first switch in reverse time as:

$$u_n^* = \text{sign}\,[\phi_2(t)]$$
$$= \text{sign}\,[\{u(-0)-\alpha\}t^2/2+\alpha t^3/6-u(-0)t^4/24], \tag{75}$$

where $u(-0)$ is the value of u when the arc impinges on J. Thus, the first switch in reverse time occurs at time T_1 given by:

$$T_1 = 2\alpha u(-0) - 2[\alpha^2 - 3\alpha u(-0)+3]^{1/2}. \tag{76}$$

Further switching times could be obtained, as for the previous example, by piecewise integration of the canonical equations. The complete optimal trajectory consists of a normal arc outside J followed by a singular sub-arc in J.

The switching curve, for the first switch in reverse time, is shown in fig. 3 together with some typical optimal trajectories. Figure 4 shows the variation

Fig. 3

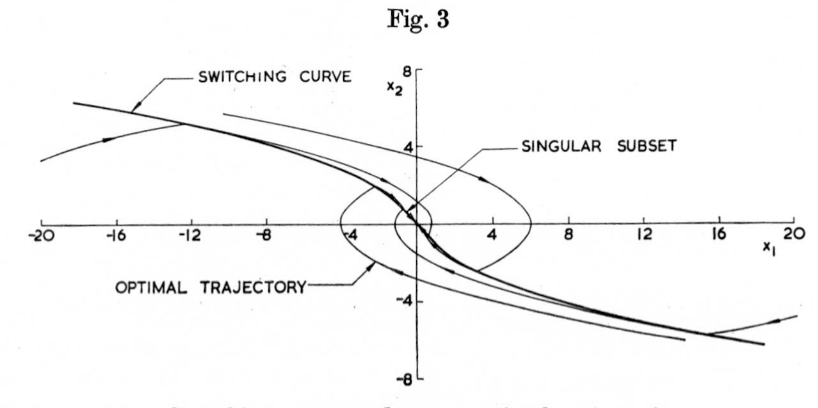

Switching curve and some optimal trajectories.

of the plant output y and the control input u for a particular optimal trajectory, and, for comparison, also for the time optimal trajectory from the same initial plant state.

The optimal controller for the present problem is the same as that obtained by Wonham and Johnson (1964) for the related problem of minimizing

$$\int_0^T (x_1^2 + x_2^2)\,dt$$

for the plant without zeros described by the same state eqn. (69).

12. Reduction of the order of the problem

Wonham and Johnson (1964) give sufficient conditions under which, for $\sigma=0$, the singular sub-set J coincides with the entire linear sub-space K. They consider only the case for which the dimension of K is $n-1$ for an nth-order plant, and, therefore, under these conditions the optimal control input is the same as that which minimizes integral square error for a related first-order plant. In the same paper, they also give conditions for equivalence with a second-order plant.

Fig. 4

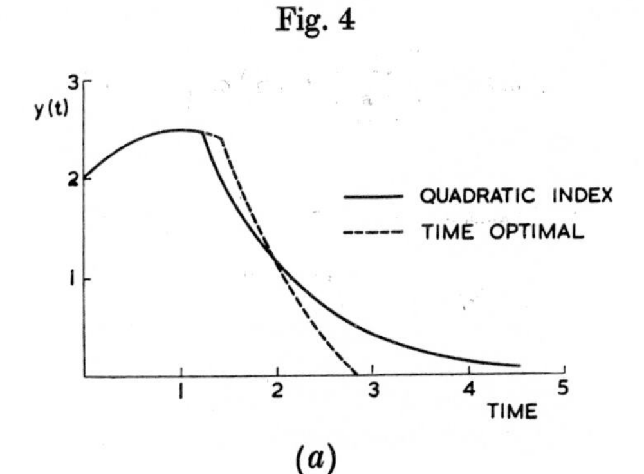

(*a*)

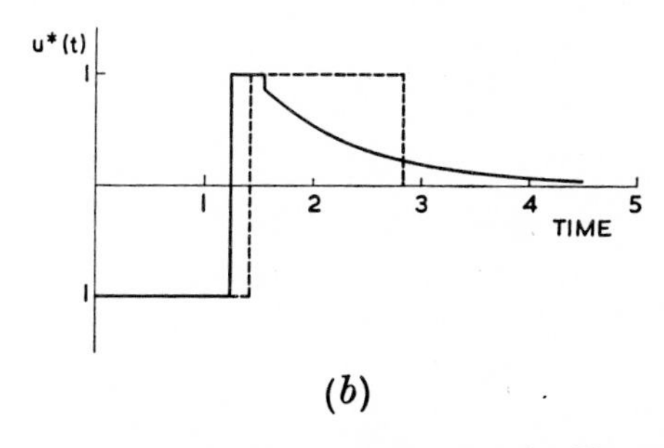

(*b*)

Typical plots of (*a*) $y(t)$ and (*b*) $u^*(t)$ for Example 2.

In another paper (Wonham and Johnson 1965), they give a similar result for $\sigma > 0$. Essentially the same problem is considered by Rekasius and Hsia (1964) who give conditions on the matrix Q in (1) such that linear switching is optimal, i.e. such that the plant is equivalent, for optimal control purposes, to one of first order.

A more general result, which is proved in Appendix III, is contained in the following theorem. This gives necessary and sufficient conditions for the reduction of the optimal control problem for an nth-order plant to that for an rth-order plant ($r < n$).

Theorem

If $u^*(t)$ minimizes:

$$\int_0^T L(x, u)\, dt \tag{77}$$

for the nth-order plant S:

$$x' = Ax + fu, \tag{78}$$

having eigenvalues:

$$\alpha_i, \quad i = 1, \ldots, n,$$

subject to the conditions:

$$\left.\begin{aligned} x(0) &= x_0 \quad (x_0 \text{ is unrestricted}), \\ x(T) &= 0 \quad (T \text{ is free}) \\ \text{and} \qquad |u| &\leqslant 1 \end{aligned}\right\} \tag{79}$$

and where

$$L(x,u)=\left[\prod_{i=1}^{p-1}(D-\beta_i)x_1\right]^2+\sigma u^2, \tag{80}$$

is chosen such that:

$$\beta_i=\alpha_{p+r-i}, \quad i=p-n+r,\ldots,p-1, \tag{81}$$

$$\beta_i \neq \alpha_j \quad \begin{cases} i=1,\ldots,p-n+r-1, \\ j=1,\ldots,r, \end{cases} \tag{82}$$

then

$$u^*(t) \quad \text{also minimizes} \quad \int_0^{\bar{T}} \bar{L}(\bar{x},u)\,dt \tag{83}$$

for the rth-order plant $\bar{S}$:

$$\prod_{i=1}(D-\alpha_i)\bar{x}_1=u, \tag{84}$$

subject to the conditions:

$$\left.\begin{aligned} &\bar{x}_i(0)=\langle d, A^{i-1}x_0\rangle, \\ &\bar{x}(\bar{T})=0 \quad (\bar{T} \text{ is free}) \\ &|u|\leqslant 1, \end{aligned}\right\} \tag{85}$$

and

where

$$\bar{L}(\bar{x},u)=\left[\prod_{i=1}^{p-n+r-1}(D-\beta_i)\bar{x}_1\right]^2+\sigma u^2 \tag{86}$$

and the vector d is defined by:

$$\langle d,x\rangle=\prod_{i=r+1}^{n}(D-\alpha_i)x_1, \tag{87}$$

provided the α_i satisfy the conditions:

complex α_i, $i=1,\ldots,r$ occur in conjugate pairs, (88)

$\mathrm{Re}\,(\alpha_i)<0$, $i=r+1,\ldots,n$. (89)

Equations (81), (82), (88) and (89) constitute a set of necessary conditions for this reduction of the order of the problem to be possible.

The theorem is illustrated by the block diagram in fig. 5. The optimal control input $u^*(t)$ depends only on the output $\bar{x}_1$ of the sub-plant $\bar{S}$, and, consequently, the poles of the sub-plant $\hat{S}$ are not changed by the optimal

Fig. 5

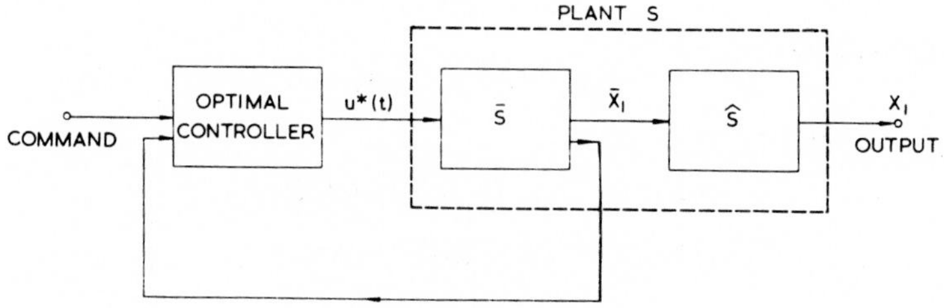

The system configuration.

controller. Condition (88) ensures that the sub-plant $\bar{S}$ is real, while condition (89) is necessary for the open-loop stability of the sub-plant $\hat{S}$, thus ensuring the stability of the entire control system.

The theorems given by Wonham and Johnson (1964, 1965) and Rekasius and Hsia (1964) on the conditions for the optimality of linear switching follow from the special case $r=1$ of the above theorem. $\hat{S}$ is then the first-order plant:

$$(D-\alpha_1)\bar{x}_1=u \tag{90}$$

and, from (86), the cost function which makes linear switching optimal is:

$$[(D-\alpha_2)\dots(D-\alpha_n)x_1]^2+\sigma u^2. \tag{91}$$

The necessary conditions (88), (89) for this case become:

$$\left.\begin{array}{l}\alpha_1 \text{ is real,}\\ \operatorname{Re}(\alpha_i)<0, \quad i=2,\dots,n.\end{array}\right\} \tag{92}$$

Their results on the conditions for the optimality of linear switching are, therefore, only partially correct.

In theorem (1) of Wonham and Johnson (1964), it is required that $n-1$ of the α_i be distinct. However, this condition is not necessary as can be seen from the proof of the above theorem in Appendix III. Similarly, the requirement in theorem (2) of the same paper that $n-2$ of the α_i be distinct is also unnecessary.

In the sequel to their theorem (1), they state that linear optimal switching may not always be possible even if conditions (92) are satisfied. This appears to stem from the requirement that $q_i \geqslant 0$ (all i) be satisfied by the elements of the diagonal matrix Q. This requirement was shown to be too restrictive in § 2 above.

The only necessary condition given by Rekasius and Hsia (1964) for the existence of a performance index yielding a linear switching law is that the plant must have at least one negative real open-loop pole. However, an inspection of the conditions (92) shows that this condition is not necessary.

13. Further examples

We shall conclude with two examples illustrating the theorem of § 12. The first is taken from Rekasius and Hsia (1964).

Example 3

For the linear plant:

$$\begin{bmatrix} x_1 \\ x_2 \\ x_3 \end{bmatrix}' = \begin{bmatrix} 0 & 1 & 0 \\ 0 & 0 & 1 \\ -1{\cdot}25 & -2{\cdot}25 & -2 \end{bmatrix} \begin{bmatrix} x_1 \\ x_2 \\ x_3 \end{bmatrix} + \begin{bmatrix} 0 \\ 0 \\ 1 \end{bmatrix} u, \tag{93}$$

find the quadratic performance indices which yield linear optimal switching laws for (i) $\sigma=0$, and (ii) $\sigma>0$.

The eigenvalues of the plant (93) are:

$$\left.\begin{array}{l}\alpha_1=-1,\\ \alpha_{2,3}=-0{\cdot}5\pm j.\end{array}\right\} \tag{94}$$

Since there is only one real pole, there is only one first-order plant to which the plant (93) can be made equivalent and this is:

$$(D+1)\bar{x}_1 = u. \tag{95}$$

The corresponding quadratic cost function is, from (80) and (81):

$$\begin{aligned} L(x,u) &= [(D-\alpha_2)(D-\alpha_3)x_1]^2 + \sigma u^2 \\ &= (1{\cdot}25x_1 + x_2 + x_3)^2 + \sigma u^2. \end{aligned} \tag{96}$$

(i) for $\sigma = 0$:

$$L(x,u) = (1{\cdot}25x_1 + x_2 + x_3)^2 \tag{97}$$

and the problem reduces to minimizing:

$$\int_0^T \bar{x}_1^2 \, dt$$

for the plant (95), and we have also:

$$\begin{aligned} \bar{x}_1 &= (D-\alpha_2)(D-\alpha_3)x_1 \\ &= 1{\cdot}25x_1 + x_2 + x_3. \end{aligned} \tag{98}$$

The optimal control law for the plant (93) is, therefore, obtained from well-known results for the plant (95) as:

$$\left.\begin{aligned} u^* &= -\operatorname{sign}(1{\cdot}25x_1 + x_2 + x_3), & 1{\cdot}25x_1 + x_2 + x_3 &\neq 0, \\ u^* &= 0, & 1{\cdot}25x_1 + x_2 + x_3 &= 0. \end{aligned}\right\} \tag{99}$$

(ii) Any $\sigma > 0$ in the cost function (96) yields a saturating linear optimal control law. The particular quadratic index derived by Rekasius and Hsia is obtained by setting $\sigma = 0{\cdot}8$, in which case:

$$L(x,u) = (1{\cdot}25x_1 + x_2 + x_3)^2 + 0{\cdot}8u^2 \tag{100}$$

and the problem reduces to minimizing:

$$\int_0^T (\bar{x}_1^2 + 0{\cdot}8u^2) \, dt, \tag{101}$$

for the plant (95). This readily yields the optimal control for the plant (93) as:

$$u^* = \operatorname{sat}[(1{\cdot}25x_1 + x_2 + x_3)/2], \tag{102}$$

where the sat function is defined by:

$$\left.\begin{aligned} \operatorname{sat}(z) &= z, & |z| &\leq 1, \\ \operatorname{sat}(z) &= \operatorname{sign}(z), & |z| &> 1. \end{aligned}\right\} \tag{103}$$

Using Appendix I, it can be shown easily that the cost functions obtained by Rekasius and Hsia are equivalent to the cost functions obtained here with much less labour.

Example 4

Minimize

$$\int_0^T (4x_1^2 + x_3^2) \, dt \tag{104}$$

for the fourth-order plant:

$$\begin{bmatrix} x_1 \\ x_2 \\ x_3 \\ x_4 \end{bmatrix}' = \begin{bmatrix} 0 & 1 & 0 & 0 \\ 0 & 0 & 1 & 0 \\ 0 & 0 & 0 & 1 \\ 0 & 0 & -2 & -2 \end{bmatrix} \begin{bmatrix} x_1 \\ x_2 \\ x_3 \\ x_4 \end{bmatrix} + \begin{bmatrix} 0 \\ 0 \\ 0 \\ 1 \end{bmatrix} u. \tag{105}$$

The plant (105), which we shall denote by S, has transfer function:

$$1/s^2(s^2+2s+2) \tag{106}$$

and, using the results of Appendix I, it can be shown that the cost function (104) is equivalent to the cost function:

$$L(x,u)=(2x_1+2x_2+\dot{x}_3)^2. \tag{107}$$

The plant S is partitioned such that the sub-plant $\hat{S}$ has transfer function:

$$1/(s^2+2s+2). \tag{108}$$

Now using the theorem of § 12, the problem reduces to minimizing:

$$\int_0^T \bar{x}_1{}^2\,dt \tag{109}$$

for the sub-plant $\bar{S}$ having transfer function:

$$1/s^2. \tag{110}$$

The optimal control for this problem is known to be (Fuller 1960):

$$u^* = -\operatorname{sign}(\bar{x}_1+0{\cdot}4446\bar{x}_2|\bar{x}_2|), \tag{111}$$

except if $x_1=x_2=0$, in which case:

$$u^*=0.$$

We also have:

$$\bar{x}_1=(D^2+2D+2)x_1$$

$$=2x_1+2x_2+x_3 \tag{112}$$

and

$$\bar{x}_2=2x_2+2x_3+x_4. \tag{113}$$

Thus the optimal control for the plant (105) is given by:

$$u^* = -\operatorname{sign}[(2x_1+2x_2+x_3)+0{\cdot}4446(2x_2+2x_3+x_4)|2x_2+2x_3+x_4|], \tag{114}$$

except if $2x_1+2x_2+x_3=2x_2+2x_3+x_4=0$, in which case:

$$u^*=0. \tag{115}$$

Acknowledgments

The author wishes to thank Professor J. F. Coales and the Control Group at Cambridge University for providing the facilities for this research and Dr. A. T. Fuller for many fruitful discussions. He is grateful to the Ministry of Education, Ceylon, and to the University of Ceylon, Peradeniya, for financial support.

Appendix I

In this Appendix we prove, for the plant (3) with the terminal constraints (5), the two results stated below.

(R 1) For every cost function:

$$\langle x, Qx \rangle + \sigma u^2, \tag{116}$$

where $Q = \operatorname{diag}(q_1, \ldots, q_n)$, there exists a cost function:

$$\langle c, x \rangle^2 + \sigma u^2, \tag{117}$$

which is equivalent to (116) in that it yields the identical optimal control law, provided that:

$$q_1 > 0 \tag{118}$$

and the equation:

$$\sum_{i=1}^{n} q_i z^{i-1} = 0 \tag{119}$$

has no positive real roots.

(R 2) If

$$\langle c, x \rangle = c_p \prod_{i=1}^{p-1} (D - \beta_i) x_1,$$

then the cost function (117) is equivalent to the cost function:

$$\left[c_p \prod_{i=1}^{p-1} \{D + \beta_i \operatorname{sign}(\operatorname{Re}(\beta_i))\} x_1 \right]^2 + \sigma u^2. \tag{120}$$

Proof:

Wonham and Johnson (1964) have shown that, under the conditions of the present problem, there is no loss of generality in taking the matrix Q in (116) to be diagonal. Following their method, integration by parts yields:

$$\int_0^T (\langle c, x \rangle^2 + \sigma u^2)\, dt = \int_0^T (\langle x, Qx \rangle + \sigma u^2)\, dt + \langle x_0, Bx_0 \rangle, \tag{121}$$

where the matrix B depends only on the vector c, and where

$$q_i = c_i^2 + 2 \sum_{j=1}^{i-1} (-1)^{i+j} c_j c_{2i-j}, \quad i = 1, \ldots, n. \tag{122}$$

Undefined c_i in (122) are to be taken as zero.

The quadratic form $\langle x_0, Bx_0 \rangle$ is a function of only the initial plant state x_0 and is, therefore, independent of the control input u. Thus, if c is any vector whose elements satisfy (122), then the controller which minimizes (117) would also minimize (116).

Now in general we would have:

$$\left. \begin{aligned} q_p &\neq 0, \\ q_i &= 0, \quad i = p+1, \ldots, n. \end{aligned} \right\} \tag{123}$$

Consider the following equation in β:

$$\sum_{i=1}^{p} q_i \beta^{i-1} (-\beta)^{i-1} = 0. \tag{124}$$

The roots of this equation occur in pairs $\pm \beta_i$, and all complex roots occur in conjugate pairs. If eqn. (119) has no positive real roots, then none of the

β_i are purely imaginary. Also, if (118) is satisfied, by renaming the pairs $\pm\beta_i$ where necessary, it is possible to ensure that:

$$\langle c, x\rangle = q_p^{1/2} \prod_{i=1}^{p-1} (D-\beta_i)x_1 \tag{125}$$

is real, and we then have the identity:

$$\left(\sum_{i=1}^{p} c_i\beta^{i-1}\right)\left\{\sum_{i=1}^{p} c_i(-\beta)^{i-1}\right\} \equiv \sum_{i=1}^{p} q_i\beta^{i-1}(-\beta)^{i-1}. \tag{126}$$

By comparing coefficients on both sides of (126), it can be verified that the c_i satisfy (122) and, therefore, the vector c defined by (125) yields a cost function (117) that is equivalent to the cost function (116). This proves result (R 1).

In practice we would want to control the state variable x_1, or, a combination of x_1 and some of the other state variables, and condition (118) would, therefore, be satisfied. Also, (125) shows that the β_i are the eigenvalues of the 'ideal model' corresponding to the cost function (117). It is therefore unlikely that any of the β_i would be purely imaginary as this would mean that the ideal model has undamped oscillatory modes. Thus the condition that eqn. (119) should have no positive real roots would also be satisfied in practice.

The solution c given by (125) is not unique as more than one arrangement of the pairs $\pm\beta_i$ might yield a real vector c. All such solutions define cost functions (117) which are equivalent to each other and to (116). In particular, they are all equivalent to the cost function (120), which, by inspection, can be verified as being real. This is result (R 2).

This result indicates that there is no loss of generality in taking $\text{Re}(\beta_i) < 0$, all i. If we make this restriction on the vector c, there is a one–one relationship between the cost functions (116) and (117).

Appendix II

Solution of the canonical eqns. (15) *and* (20) *for the singular sub-arcs which converge to the origin*

The characteristic equation for the system of eqns. (20) is:

$$c_1 - c_2\lambda + \ldots + (-1)^{p-1}c_p\lambda^{p-1} = 0. \tag{127}$$

By comparing (127) with (11), the roots of (127) are found to be:

$$\lambda_i = -\beta_i, \quad i = 1, \ldots, p-1. \tag{128}$$

In general, not all the β_i are distinct. Let the multiplicity of β_i be r_i, $i = 1, \ldots, \eta$.

The general solution of eqns. (20) for ϕ_{p-1} is then:

$$\phi_{p-1} = \sum_{i=1}^{\eta} \sum_{j=1}^{r_i} \mu_{ij} t^{j-1} \exp(-\beta_i t) \tag{129}$$

and from (19) and (129) we have therefore:

$$2c_p\langle c, x\rangle = \sum_{i=1}^{\eta} \sum_{j=1}^{r_i} \mu_{ij} t^{j-1} \exp(-\beta_i t). \tag{130}$$

We now assume, without loss of generality (see Appendix I), that:

$$\text{Re}(\beta_i) < 0, \quad \text{all} \quad i. \tag{131}$$

We are seeking only the singular sub-arcs which converge as $t \to \infty$, and from (130) and (131) it follows that on such sub-arcs:

$$\mu_{ij} = 0, \quad \text{all} \quad i, j. \tag{132}$$

From (130) and (132), therefore, on convergent sub-arcs:

$$\langle c, x \rangle = 0, \tag{133}$$

whence, from eqns. (20), (129) and (132):

$$\phi_i = 0, \quad i = 1, \ldots, p-1. \tag{134}$$

Appendix III

Proof of the theorem of § 12

The plant S is partitioned into two sub-plants $\bar{S}$ and $\hat{S}$ (see fig. 5) defined respectively by the equations:

$$\prod_{i=1}^{r} (D - \alpha_i)\bar{x}_1 = u \tag{135}$$

and

$$\prod_{i=r+1}^{n} (D - \alpha_i)x_1 = \bar{x}_1. \tag{136}$$

From (81) and (136) we have:

$$\bar{x}_1 = \prod_{i=p+r-n}^{p-1} (D - \beta_i)x_1 \tag{137}$$

and from (80) and (137), therefore,

$$L(x, u) = \left[\prod_{i=1}^{p-n+r-1} (D - \beta_i)\bar{x}_1 \right]^2 + \sigma u^2, \tag{138}$$

i.e.

$$L(x, u) = \bar{L}(\bar{x}, u). \tag{139}$$

Also, using (87), (136) can be written:

$$\bar{x}_1 = \langle d, x \rangle \tag{140}$$

and, by repeated differentiation of (140) with respect to time, we obtain the phase variables of the sub-plant $\bar{S}$ as:

$$\bar{x}_i = \langle d, A^{i-1}x \rangle, \quad i = 1, \ldots, r. \tag{141}$$

Thus, every point in the state (phase) space of $\bar{S}$ corresponds to an $(n-r)$-dimensional linear sub-space, defined by (141), in the state space of the plant S. In particular, the origin $\bar{x} = 0$ corresponds to the linear sub-space D:

$$\{x \mid \langle d, A^{i-1}x \rangle = 0, \quad i = 1, \ldots, r\}. \tag{142}$$

The settling time $\bar{T}$ of the sub-plant $\bar{S}$ is, therefore, the time at which the state of the plant S enters the sub-space D. Thus $\bar{T} \leqslant T$, and the optimal control input for $x \in D$ is given by:

$$u^*(t) = 0, \quad t \geqslant \bar{T}. \tag{143}$$

It is obvious from the system configuration that the sub-plant $\hat{S}$ must be open-loop stable, and we therefore have the necessary condition for the validity of the cost function $L(x, u)$:

$$\text{Re}(\alpha_i) < 0, \quad i = r+1, \ldots, n. \tag{144}$$

It then follows immediately from eqns. (139) and (141) that the optimal control input $u^*(t)$ which minimizes $L(x, u)$ for the plant S under the conditions (79) also minimizes $\bar{L}(\bar{x}, u)$ for the plant $\bar{S}$ with the conditions (85), thus proving the theorem.

It is clear that the control input $u^*(t)$ is also optimal for all initial states of the plant S which lie in the $(n-r)$-dimensional linear sub-space:

$$\{x \mid \langle d, A^{i-1}x \rangle = \langle d, A^{i-1}x_0 \rangle, \quad i = 1, \dots, r\}. \tag{145}$$

The results obtained above can be explained by the fact that under conditions (81) and (82), the plant S contains $n-r$ poles of the 'ideal model' corresponding to the cost function $L(x, u)$. These modes are not altered by the optimal controller and they are all contained in the sub-plant $\hat{S}$ of the optimal control system.

By a suitable permutation of the α_j and β_i, it is always possible to satisfy at least one of eqns. (81), and (82). If (81) cannot be satisfied for any such permutation, then the order of the sub-plant $\hat{S}$ is less than $n-r$ and, therefore, the problem cannot be reduced to that for a plant of an order as low as r. Thus (81) is a necessary condition for the theorem to be true.

On the other hand, if (82) cannot be satisfied, then the order of the sub-plant $\hat{S}$ is greater than $n-r$, and the problem reduces to that for a plant of order less than r. Thus (82) is a necessary condition in this sense.

References

Fuller, A. T., 1960, *Proc. IFAC Congress, Moscow* (London: Butterworths, 1961), **1**, 510.
Grensted, P. E. W., and Fuller, A. T., 1965, *Int. J. Control*, **2**, 33.
Hutchinson, C. E., 1963, Technical Report No. 6311–1, SEL, Stanford University.
Johnson, C. D., 1965, *Advances in Control Systems*, Vol. 2, edited by C. T. Leondes (New York: Academic Press), p. 209.
Kalman, R. E., 1964, *J. bas. Engng*, **86**, 51.
Lee, E. B., 1962, *Proc. ASK Symposium on Optimization*, Dayton, Ohio.
Lee, E. B., and Markus, L., 1961, *Archs ration. Mech. Analysis*, **8**, 36.
Letov, A. M., 1960, *Automn remote Control*, **21**, 389; 1961, *Ibid.*, **22**, 363.
Rekasius, Z. V., and Hsia, T. S., 1964, *I.E.E.E. Trans. autom. Control*, **9**, 370.
Rohrer, R. A., and Sobral, Jr., M., 1966, *J. bas. Engng*, **88**, 323.
Sivan, R., 1965, *I.E.E.E. Trans. autom. Control*, **10**, 193.
Wonham, W. M., and Johnson, C. D., 1964, *J. bas. Engng*, **86**, 106; 1965, *Ibid.*, **87**, 81.
Zadeh, L. A., and Desoer, C. A., 1963, *Linear Systems Theory* (New York: McGraw-Hill).

Optimal nonlinear control of systems with pure delay†

By A. T. Fuller
Engineering Department, Cambridge University

[Received February 28, 1968]

Abstract

When the plant contains a pure delay and the control signal is subject to a saturation constraint, the optimal controller and optimal performance can often be calculated, provided the corresponding results on the delay-free case are known. Several examples of the technique are given in the present paper. Some sub-optimal controllers are also discussed.

1. Introduction

Contemporary optimal nonlinear control theories such as dynamic programming and the maximum principle offer considerable difficulties when applied directly to systems of more than say third order. For high-order systems the calculation and implementation of the optimal controller is usually prohibitively complex. In practice however, most plants are of high order and there remains a need to adapt optimal nonlinear control theory to such plants.

One approach to this problem is as follows. The high-order plant is approximated by a low-order system together with a pure delay (i.e. a distance–velocity lag, or transport lag). Optimal control theory can then be applied to the approximate plant, even though it is of infinite order, as will be expounded in the present paper.

It is well known that the small lags in a plant have a combined effect similar to that of a pure delay. Thus the approximation of a high-order plant by a low-order system and a pure delay can be quite good (Küpfmüller 1928, Oldenbourg and Sartorius 1944). Indeed it may yield a more realistic mathematical model than a model of finite order, since in practice all plants have distributed parameters.

In the present paper attention will be centred on the case when the plant has only one control input and is represented by a pure delay followed by a linear time-invariant system. We shall say that the plant consists of a delay followed by a *sub-plant* (see fig. 1). The problem of optimal control of such a plant has partly been solved by previous workers. However, their solutions have not emerged clearly, because the writers were treating pure delay as a side-issue and were concentrating on other aspects such as sampled-data theory, or because they were writing before state space theory was well established, or because they did not distinguish between optimal and sub-optimal techniques. It will be easier to understand the contributions of these writers after we have first developed the theory for ourselves. Consequently a review of previous work will be postponed to the Appendix. For the present let us note that the early contributors were Bass (1956 a, b, c),

† Communicated by the Author.

Smith (1957 a, b, 1958, 1959), Kalman and Bertram (1957, 1958), Butkovski and Domanitski (1958), Varshavski (1958, 1960), Tsypkin (1959 a, b), Merriam (1959).

In fig. 1 the input v to the sub-plant is a delayed version of the controller output u. Thus we are in effect also dealing with the case when the control signal is subject to delay. Several proposed methods of control involve this type of delay. For example, predictive control†, and also on-line solution of the two-point boundary problem, require the controller to perform elaborate computations, possibly involving appreciable time-delay. Our results will perhaps help in the assessment of these methods, i.e. in the decision whether the increased delay outweighs the increased precision of computing.

Fig. 1

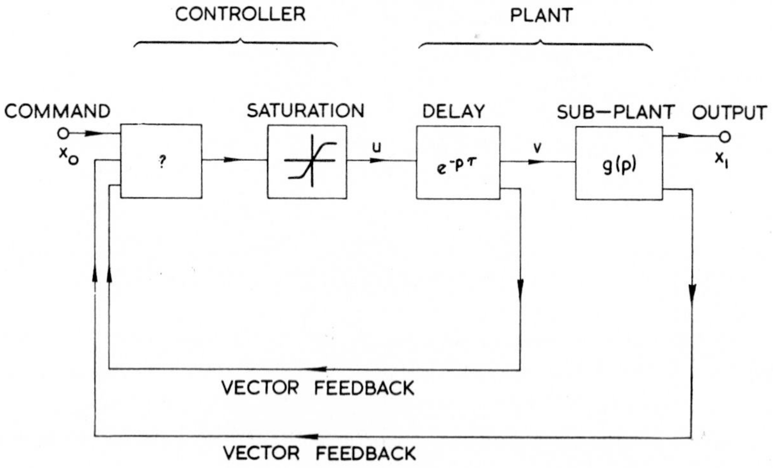

Control system with saturation and delay.

When the pure delay enters into the plant in a more general way the problem of optimal nonlinear control is much more difficult, and closed-form solutions of the type obtained in the present paper seem difficult to come by. The theory tends to concern itself with questions of the existence and uniqueness of solutions of the maximum principle or dynamic programming algorithms; see e.g. Oguztöreli (1966) for an introduction and further references.

2. Pure delay

In fig. 1 the pure delay is represented symbolically by its transfer function $e^{-p\tau}$. Let us now define the pure delay more explicitly.

The pure delay has scalar input $u(t)$ and scalar output $v(t)$, defined for $0 \leqslant t < \infty$, where t is time and $t=0$ is an arbitrary time origin. The functions $u(t)$ and $v(t)$ satisfy the relation

$$v(t)=u(t-\tau) \quad (\tau \leqslant t < \infty) \tag{2.1}$$

† The term predictive control is used in various senses. The method meant above is that developed especially by Coales and his collaborators (e.g. Adey *et al.* 1963, 1966), in which the controller contains a plant model simulated on a fast time scale. The best of a set of possible future control schedules is found by a rapid search process on the plant model, and is used to generate the present value of the real-time-scale control.

where the constant τ is the value of the delay. Relation (2.1) does not determine $v(t)$ for $0 \leqslant t < \tau$ and we may write, for this interval,

$$v(t) = \phi(t-\tau) \quad (0 \leqslant t < \tau) \tag{2.2}$$

where ϕ is arbitrary.

Let us define the state of the delay†. It is helpful, to begin with, to think of the delay as due to a magnetic tape passing at unit speed from a writing head with signal $u(t)$ to a reading head with signal $v(t)$. At $t=0$ the initial record on the tape segment between the heads is represented by the function $\phi(z-\tau)$ $(0 \leqslant z < \tau)$ where z represents distance along the tape measured from the reading head. Intuitively, then, $\phi(z-\tau)$ $(0 \leqslant z < \tau)$ represents the initial state of the delay. More generally, at a given instant t the state of the delay is represented by the record then on the tape segment between the heads, and this is the function $w(z,t)$ where, for $\tau \leqslant t < \infty$

$$w(z,t) = u(t-\tau+z) \quad (0 \leqslant z < \tau) \tag{2.3 a}$$

and, for $0 \leqslant t < \tau$, (see fig. 2)

$$w(z,t) = \begin{cases} u(t-\tau+z) & (\tau - t \leqslant z < \tau) \\ \phi(t-\tau+z) & (0 \leqslant z < \tau - t) \end{cases} \tag{2.3 b}$$

Other physical realizations of pure delays (e.g. delay lines, conveyor belts) have the same properties. We are therefore justified in *defining* a pure delay as having the properties of satisfying eqns. (2.1) and (2.2) and possessing the state (2.3). It only remains for us to verify that (2.3) meets the essential requirement of a state; namely the future of the output (of the delay in this case) must be uniquely determined by the present of the state and the present and future of the input. (This definition of state is similar to that given by Fuller (1960 a), but is modified to take account of the present of the input.)

Fig. 2

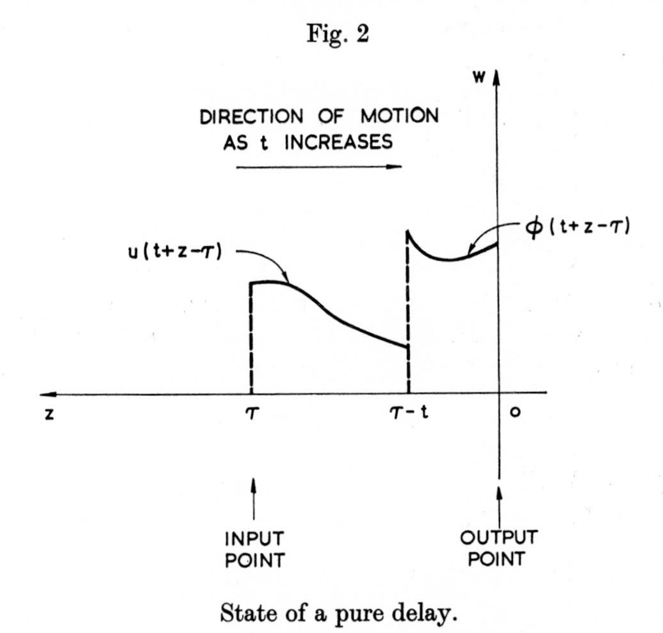

State of a pure delay.

† The state of a pure delay has also been discussed by Reeve (1968).

When t represents the present, and z varies from 0 up to τ, the right side of (2.3) describes the output up to the next τ units of time. The output subsequent to this is determined uniquely by the present and future of u, in view of (2.1). Hence (2.3) does meet the requirements of a state.

Thus the state of a pure delay has an infinity of coordinates, corresponding to the values of the function $w(z,t)$ when z varies from 0 up to τ.

When discussing a general vector variable $\theta(t)$ with a finite number of components $\theta_1(t), \theta_2(t), \ldots, \theta_m(t)$, it is customary often to drop the argument and write simply θ. In the same way we shall often drop the second argument from $w(z,t)$ and write simply

$$w(z,t) = w(z) \tag{2.4}$$

In what follows $w(z)$ will be used to represent either the state of the delay at a general instant t, or the state of the delay at the particular instant $t=0$, according to context.

Figure 2 depicts the state of the pure delay. The graph moves to the right as time increases, being continuously generated at the input point $z=\tau$, and continuously removed at the output point $z=0$.

3. The system

The control system is shown in fig. 1. The sub-plant satisfies the vector–matrix equation

$$\frac{dx}{dt} = Ax + Bv \tag{3.1}$$

where $x(t)$ is the vector $(x_1, x_2, \ldots, x_n)$, A and B are constant matrices, and $v(t)$ is a scalar variable. Thus $x_1, x_2, \ldots, x_n$ are the state coordinates of the sub-plant. x_1 is also the scalar output of the sub-plant. v is the input to the sub-plant.

The pure delay has scalar input $u(t)$ and output $v(t)$, satisfying (2.1) and (2.2). The state of the delay is $w(z,t)$, given by (2.3).

The command signal is shown in fig. 1 as the scalar variable x_0. The system error is $e(t)$ where

$$e = x_1 - x_0 \tag{3.2}$$

and the overall purpose of the control is to reduce the magnitude of e as time increases. We shall often have in mind the optimization of step response. For the simple examples to be considered this optimization problem is essentially the same as when the command signal is zero and the initial plant output $x_1(0)$ is non-zero (see e.g. Fuller 1960 b). Thus to simplify the argument we shall take:

$$x_0(t) = 0 \quad (0 \leqslant t < \infty) \tag{3.3}$$

but we shall still call the system output a step response, when appropriate.

The controller output is $u(t)$ and is subject to the saturation constraint

$$|u(t)| \leqslant 1 \quad (0 \leqslant t < \infty) \tag{3.4}$$

The controller must be physically realizable, i.e. must operate only on the present and past of available signals.

Two types of performance index will be adopted. The first type is

$$I = \int_0^\infty q(x_1, x_2, \dots, x_n)\, dt \tag{3.5}$$

i.e. the state of the sub-plant is penalized, but not the state of the delay. In particular, when

$$q(x_1, x_2, \dots, x_n) = x_1{}^2 \tag{3.6}$$

the performance index is integral-square-error. The second type of performance index is settling time, i.e. the time required for the sub-plant state and the delay state to become simultaneously zero.

The problem is to design the controller so that for any given initial state of the plant the performance index adopted is minimized.

When the delay is absent, it is known that the optimal controller output is simply a function of the plant state (see e.g. Fuller 1960 b). Analogously, when the delay is present, it is to be expected that the optimal controller output will be a function of the sub-plant state and a functional of the delay state. Such will turn out to be the case. Despite the apparent complexity of this result, closed-form solutions for the optimal controller will be obtained in simple examples.

4. Minimization of integral performance indices

In this section the performance index to be minimized is the integral type given by (3.5).

Consider first the case when the pure delay is absent. It is well known that in this case the optimal control is a function of the state coordinates of the plant† (see e.g. Fuller 1960 b)

$$u(t) = v(t) = C(x_1(t), x_2(t), \dots, x_n(t)) \quad (0 \leqslant t < \infty) \tag{4.1}$$

Also the minimal performance index is then a function of the initial values of the state coordinates

$$I = J(x_1(0), x_2(0), \dots, x_n(0)) \tag{4.2}$$

Next consider the case when the pure delay is present. The choice of the value of u at any given time t is equivalent to the choice of the value of v at time $t+\tau$, since

$$u(t) = v(t+\tau) \tag{4.3}$$

From (4.3) this choice of v is subject to the same saturation constraint as u. The best choice of v, the input to the sub-plant, at time $t+\tau$ is, in view of (4.1),

$$v(t+\tau) = C(x_1(t+\tau), x_2(t+\tau), \dots, x_n(t+\tau)) \tag{4.4}$$

From (4.3) and (4.4) the best choice of $u(t)$ is

$$u(t) = C(x_1(t+\tau), x_2(t+\tau), \dots, x_n(t+\tau)) \tag{4.5}$$

---

† It is assumed implicitly that q, A, and B are such that the performance index exists, either for all starting points x or for some sub-set of such points. This assumption must be satisfied for the problem to be meaningful. Similarly obvious existence conditions are assumed elsewhere in the paper, without being formulated explicitly.

Thus the optimal control is a function of the values of the sub-plant state coordinates at time τ ahead of the present. This function is the same function as in the delay-free case. We write (4.5) more compactly as

$$u(t) = C(x(t+\tau)) \tag{4.6}$$

It remains for us to express the future values $x_1(t+\tau), x_2(t+\tau), \ldots, x_n(t+\tau)$ of the sub-plant state coordinates in terms of their present values $x_1(t), x_2(t), \ldots, x_n(t)$ and of the present state $w(z,t)$ of the delay. This calculation is a standard application of matrix theory which goes back to Peano (1887, 1888), see e.g. Gantmacher (1954, Chap. 5, § 5) or Bellman (1960, Chap. 10, § 11). The result is that at time $t+\tau$, the response of system (3.1) to its input $v(t+s)$ $(0 \leqslant s < \tau)$ and initial conditions $x(t) = (x_1(t), x_2(t), \ldots, x_n(t))$, is†

$$x(t+\tau) = e^{A\tau}x(t) + e^{A\tau}\int_0^\tau e^{-As}Bv(t+s)\,ds \tag{4.7}$$

Also $w(z,t)$ describes the input to the sub-plant during the interval t to $t+\tau$

$$v(t+z) = w(z,t) \quad (0 \leqslant z < \tau) \tag{4.8}$$

From (4.7) and (4.8)

$$x(t+\tau) = e^{A\tau}x(t) + e^{A\tau}\int_0^\tau e^{-Az}Bw(z,t)\,dz \tag{4.9}$$

Substituting (4.9) in (4.6) we obtain

$$u(t) = C\left(e^{A\tau}x(t) + e^{A\tau}\int_0^\tau e^{-Az}Bw(z,t)\,dz\right) \tag{4.10}$$

which expresses the present value of the optimal control in terms of the present state of the sub-plant and the present state of the delay. In (4.10) let us drop the argument t, writing

$$w(z,t) = w(z) \tag{4.11}$$

Then the optimal controller is

$$u = C\left(e^{A\tau}x + e^{A\tau}\int_0^\tau e^{-Az}Bw(z)\,dz\right) \tag{4.12}$$

It is sometimes helpful in interpreting the argument of C in (4.12) to write the argument in terms of the impulse responses of the sub-plant (3.1). Suppose the function $h_{ij}(t)$ $(i = 1, 2, \ldots, n;\ j = 1, 2, \ldots, n)$ represents the response of coordinate x_i when a unit impulse forcing function applied at $t=0$ is added to the right side of the jth equation of the homogeneous version of system (3.1), i.e. the jth component of the vector equation

$$\frac{dx}{dt} = Ax \tag{4.13}$$

Suppose that the matrix of these impulse responses is

$$H(t) = [h_{ij}(t)] \tag{4.14}$$

† One may check (4.7) immediately, by verifying that if τ is written for t in (3.1), (4.7) satisfies (3.1), and by verifying that when $\tau = 0$ (4.7) yields the correct initial conditions $x(t)$.

Then the exponential matrix $e^{A\tau}$ which appears in (4.12) (and is known as the matrizant or as the transition matrix) is simply†

$$e^{A\tau}=H(\tau) \quad (\tau\geqslant 0) \tag{4.15}$$

(Kalman and Bertram 1960, Zadeh and Desoer 1963, Chap. 5). Consequently the optimal controller (4.12) can be written

$$u=C\left(H(\tau)x+\int_0^\tau H(\tau-z)Bw(z)\,dz\right) \tag{4.17}$$

Next let us calculate the minimal value of the performance index when the delay is present. We write (3.5) compactly as

$$I=\int_0^\infty q(x)\,dt \tag{4.18}$$

Thus

$$I=I_1+I_2 \tag{4.19}$$

where

$$I_1=\int_0^\tau q(x(t))\,dt \tag{4.20}$$

and

$$I_2=\int_\tau^\infty q(x(t))\,dt \tag{4.21}$$

I_1 is the contribution to the performance index during the initial period when changes of control have no effect on the sub-plant. I_1 is thus completely determined by the initial state of the plant. In fact during the initial interval, of duration τ, x is given by

$$x(t)=e^{At}x(0)+e^{At}\int_0^t e^{-As}Bv(s)\,ds \quad (0\leqslant t<\tau) \tag{4.22}$$

From (4.8) (with $t=0$) and (4.22)

$$x(t)=e^{At}x(0)+e^{At}\int_0^t e^{-Az}Bw(z,0)\,dz \quad (0\leqslant t<\tau) \tag{4.23}$$

From (4.20) and (4.23)

$$I_1=\int_0^\tau q\left(e^{At}x(0)+e^{At}\int_0^t e^{-Az}Bw(z,0)\,dz\right)dt \tag{4.24}$$

To calculate I_2 we note that the response of the sub-plant for $t\geqslant\tau$ is the same (apart from a time translation) as that for a delay-free plant, under optimal control and starting with initial conditions $x(\tau)$. Consequently the contribution to the performance index is then, in view of (4.2),

$$I_2=J(x_1(\tau),x_2(\tau),\ldots,x_n(\tau)) \tag{4.25}$$

Let us write J compactly as

$$J(x_1,x_2,\ldots,x_n)=J(x) \tag{4.26}$$

Then (4.25) is

$$I_2=J(x(\tau)) \tag{4.27}$$

---

† Note that (4.15) does not hold for $\tau<0$, since then $H=0$. For $\tau<0$, we replace (4.15) by

$$e^{A\tau}=(e^{A(-\tau)})^{-1}=H^{-1}(-\tau) \quad (\tau<0) \tag{4.16}$$

From (4.23) and (4.27)

$$I_2 = J\left(e^{A\tau}x(0) + e^{A\tau}\int_0^\tau e^{-Az}Bw(z,0)\,dz\right) \tag{4.28}$$

From (4.19), (4.24) and (4.28)

$$I = \int_0^\tau q\left(e^{At}x(0) + e^{At}\int_0^t e^{-Az}Bw(z,0)\,dz\right)dt$$
$$+ J\left(e^{A\tau}x(0) + e^{A\tau}\int_0^\tau e^{-Az}Bw(z,0)\,dz\right) \tag{4.29}$$

In (4.29) let us drop the argument 0, taking it for granted that the plant state referred to is the initial one, and write simply

$$I = \int_0^\tau q\left(e^{At}x + e^{At}\int_0^t e^{-Az}Bw(z)\,dz\right)dt$$
$$+ J\left(e^{A\tau}x + e^{A\tau}\int_0^\tau e^{-Az}Bw(z)\,dz\right) \tag{4.30}$$

(4.30) expresses the minimal performance index in terms of the initial state of the plant.

Written in terms of impulse responses (see (4.14) and (4.15)), (4.30) is

$$I = \int_0^\tau q\left(H(t)x + \int_0^t H(t-z)Bw(z)\,dz\right)dt$$
$$+ J\left(H(\tau)x + \int_0^\tau H(\tau-z)Bw(z)\,dz\right) \tag{4.31}$$

5. Example. First-order sub-plant and integral-square-error

Let the sub-plant be a single integrator, satisfying the scalar equation

$$\frac{dx}{dt} = bv \quad (b > 0) \tag{5.1}$$

where

$$x = x_1 \tag{5.2}$$

is the system error. Thus the complete plant transfer function is

$$\frac{b}{pe^{p\tau}} \tag{5.3}$$

Let the performance index be integral-square-error

$$I = \int_0^\infty x^2\,dt \tag{5.4}$$

The function (4.1) which describes the optimal controller when the delay is absent is (Fuller 1960 c)

$$C(x) = \begin{cases} -\operatorname{sgn} x & (x \neq 0) \\ 0 & (x = 0) \end{cases} \tag{5.5}$$

Also in the delay-free case, the function (4.2) which describes the minimal value of the performance index for starting point x is (Fuller 1960 c)

$$J(x) = \frac{1}{3b}|x|^3 \tag{5.6}$$

Hence, when the delay is present, application of result (4.12), or equivalently result (4.17), shows that the optimal controller is

$$u = -\operatorname{sgn}\left(x + b\int_0^\tau w(z)\,dz\right) \tag{5.7}$$

with sgn 0 interpreted as 0. Further, application of result (4.30), or equivalently result (4.31), shows that the minimal integral-square-error is then

$$I = \int_0^\tau \left(x + b\int_0^t w(z)\,dz\right)^2 dt + \frac{1}{3b}\left|x + b\int_0^\tau w(z)\,dz\right|^3 \tag{5.8}$$

In particular, for step response, i.e. when the delay state is initially zero, so that in (5.8)

$$w(z) = 0 \quad (0 \leqslant z < \tau) \tag{5.9}$$

the minimal integral-square-error is

$$I = x^2\tau + \frac{1}{3b}|x|^3 \tag{5.10}$$

Result (5.10) is readily verified. Thus during the initial interval of length τ after a step command signal the control has no effect on the sub-plant and the error remains fixed at x. Hence this interval yields a contribution $x^2\tau$ to the performance index. For $t \geqslant \tau$ the input $v(t)$ to the sub-plant reduces the error to zero at maximum speed, yielding a contribution which is the same as (5.6), namely $|x|^3/(3b)$, to the performance index.

6. Example. Second-order sub-plant and integral-square-error

Let the sub-plant consist of two integrators, satisfying

$$\begin{cases} \dfrac{dx_1}{dt} = x_2 & (6.1) \\ \dfrac{dx_2}{dt} = bv \quad (b > 0) & (6.2) \end{cases}$$

so that the transfer function of the plant from input u to output x_1 is

$$\frac{b}{p^2 e^{p\tau}} \tag{6.3}$$

Let the performance index be integral-square-error

$$I = \int_0^\infty x_1^2\,dt \tag{6.4}$$

When the delay is absent the solution to the optimization problem is known. The optimal controller is

$$C(x_1, x_2) = -\operatorname{sgn}\left(x_1 + \frac{k_1}{b}x_2|x_2|\right) \tag{6.5}$$

for points not on the switching curve, which is

$$x_1 + \frac{k_1}{b}x_2|x_2| = 0 \tag{6.6}$$

Here the constant k_1 is

$$k_1 = \left(\frac{\sqrt{33} - 1}{24}\right)^{1/2} \simeq 0{\cdot}444623560 \tag{6.7}$$

For points on the switching curve (6.6) and not at the origin any permissible control value is optimal†, e.g. we can put

$$C(x_1, x_2)=0 \tag{6.8}$$

At the origin the optimal control is

$$C(x_1, x_2)=0 \tag{6.9}$$

The minimal integral-square-error is (in the absence of delay)

$$J(x_1, x_2)=\frac{1}{b}x_1^2x_2+\frac{2}{3b^2}x_1x_2^3+\frac{2}{15b^3}x_2^5+\frac{k_2}{b^{1/2}}\left(x_1+\frac{1}{2b}x_2^2\right)^{5/2} \tag{6.10}$$

for points to the right of or on the switching curve, i.e. where

$$x_1+\frac{k_1}{b}x_2|x_2|\geqslant 0 \tag{6.11}$$

and

$$J(x_1, x_2)=-\frac{1}{b}x_1^2x_2+\frac{2}{3b^2}x_1x_2^3-\frac{2}{15b^3}x_2^5+\frac{k_2}{b^{1/2}}\left(-x_1+\frac{1}{2b}x_2^2\right)^{5/2} \tag{6.12}$$

for points to the left of or on the switching curve. Here the constant k_2 is

$$k_2=\frac{1}{20}(222+2\sqrt{33})^{1/2}\simeq 0{\cdot}764017548 \tag{6.13}$$

For the above results (6.5) to (6.13), see e.g. Fuller (1960 c, 1963, 1964 a, b, 1966).

When the delay is present, the optimal controller is‡, from (4.12) and (6.1) to (6.9)

$$u=C\left\{\begin{bmatrix}1 & \tau\\ 0 & 1\end{bmatrix}\begin{bmatrix}x_1\\ x_2\end{bmatrix}+\begin{bmatrix}1 & \tau\\ 0 & 1\end{bmatrix}\int_0^\tau\begin{bmatrix}1 & -z\\ 0 & 1\end{bmatrix}\begin{bmatrix}0\\ b\end{bmatrix}w(z)\,dz\right\} \tag{6.16}$$

$$=C\left\{\begin{bmatrix}x_1+\tau x_2+b\int_0^\tau(\tau-z)w(z)\,dz\\ x_2+b\int_0^\tau w(z)\,dz\end{bmatrix}\right\} \tag{6.17}$$

$$=-\operatorname{sgn}\left\{x_1+\tau x_2+b\int_0^\tau(\tau-z)w(z)\,dz+\frac{k_1}{b}\left(x_2+b\int_0^\tau w(z)\,dz\right)\left|x_2+b\int_0^\tau w(z)\,dz\right|\right\} \tag{6.18}$$

---

† Because for a given trajectory to the origin, the time spent on the switching curve and not at the origin has zero measure.

‡ For system (6.1, 6.2) the matrix A is

$$A=\begin{bmatrix}0 & 1\\ 0 & 0\end{bmatrix} \tag{6.14}$$

It follows that A^2 is zero; hence so are $A^3, A^4, \ldots$. Therefore

$$e^{A\tau}=E+A\tau=\begin{bmatrix}1 & \tau\\ 0 & 1\end{bmatrix} \tag{6.15}$$

E being the unit matrix.

Alternatively (6.15) can be obtained immediately by evaluating the impulse responses of the sub-plant (see (4.14) and (4.15)).

with sgn 0 treated as 0. Also, from (4.30) and (6.1) to (6.13), the minimal integral-square-error is then

$$I = \int_0^\tau \left(x_1 + t x_2 + b \int_0^\tau (t - z) w(z)\, dz \right)^2 dt$$
$$+ J\left(x_1 + \tau x_2 + b \int_0^\tau (\tau - z) w(z)\, dz,\ x_2 + b \int_0^\tau w(z)\, dz \right) \quad (6.19)$$

where $J(\alpha_1, \alpha_2)$ is the function defined by (6.10) to (6.12).

In particular, for step response, i.e. when in (6.19) x_2 and $w(z)$ are zero,

$$I = {x_1}^2 \tau + \frac{k_2}{b^{1/2}} |x_1|^{5/2} \quad (6.20)$$

7. Minimization of settling time

In this section the performance index to be minimized is settling time, i.e. the time for the sub-plant state and the delay state to become simultaneously zero. (After the plant has settled, the control u is to be maintained zero to keep the plant state at zero.)

Consider first the case when the pure delay is absent. It is well known that in this case the optimal control is a function of the state coordinates of the plant (see e.g. Fuller 1960 b)

$$u(t) = v(t) = C(x_1(t), x_2(t), \ldots, x_n(t)) \quad (0 \leqslant t < \infty) \quad (7.1)$$

with

$$C(0, 0, \ldots, 0) = 0 \quad (7.2)$$

Also the minimal settling time is then a function of the initial values of the state coordinates

$$I = J(x_1(0), x_2(0), \ldots, x_n(0)) = J(x(0)) \quad (7.3)$$

Next let us treat the case when the pure delay is present. Let us consider, to begin with, the problem of minimizing the settling time T of the sub-plant coordinates, regardless of the values of the delay state at time T.

Suppose

$$T \geqslant \tau \quad (7.4)$$

The argument of § 4 then applies, and gives the solution

$$u(t) = C(x_1(t+\tau), x_2(t+\tau), \ldots, x_n(t+\tau)) \quad (7.5)$$

for

$$t \leqslant T - \tau \quad (7.6)$$

Let us also choose

$$u(t) = 0 \quad (T - \tau < t < \infty) \quad (7.7)$$

Then the input to the delay will be zero during the interval of length τ preceding the instant T. Hence the delay state will be zero at T, i.e. the delay and sub-plant settle simultaneously, and indeed optimally.

On the other hand, suppose

$$T < \tau \tag{7.8}$$

i.e. suppose the initial state of the delay is such that the sub-plant settles at $t = T < \tau$ and remains settled during $T < t < \tau$. Then

$$v(t) = 0 \quad (T \leqslant t < \tau) \tag{7.9}$$

i.e. at T, part of the delay state is zero

$$w(z, T) = 0 \quad (0 \leqslant z < \tau - T) \tag{7.10}$$

Let us choose

$$u(t) = 0 \quad (0 \leqslant t < T) \tag{7.11}$$

Then at T another part of the delay state is zero (see (2.3))

$$w(z, T) = 0 \quad (\tau - T \leqslant z < \tau) \tag{7.12}$$

From (7.10) and (7.12), at T the complete delay state is zero

$$w(z, T) = 0 \quad (0 \leqslant z < \tau) \tag{7.13}$$

Thus with control (7.11) the complete plant settles at time T; and if we choose

$$u(t) = 0 \quad (0 \leqslant t < \infty) \tag{7.14}$$

the plant remains settled thereafter. (7.14) is the optimal control, since from (7.8) no control can make the sub-plant settle before $t = T$.

In view of (7.2), if controller (7.5) is used for all $t \geqslant 0$, control schedules (7.7) or (7.14) are generated automatically. We have thus proved that controller (7.5) minimizes the settling time not only of the sub-plant but also of the complete plant.

Writing (7.5) as

$$u(t) = C(x(t + \tau)) \tag{7.15}$$

and substituting from (4.9), (4.11) and (4.15), we express the optimal controller as

$$u = C\left(e^{A\tau}x + e^{A\tau}\int_0^\tau e^{-Az}Bw(z)\,dz\right) \tag{7.16}$$

or equivalently as

$$u = C\left(H(\tau)x + \int_0^\tau H(\tau - z)Bw(z)\,dz\right) \tag{7.17}$$

Thus the optimal controller is of the same nature as that for an integral performance index, namely (4.12) or equivalently (4.17). (The functions C are usually different for the two performance indices.)

Let us calculate the minimal value of the settling time when the delay is present. Suppose first that (7.4) applies, so that the initial plant state is such that the plant does not settle before $t=\tau$; i.e. suppose $x(\tau)\neq 0$:

$$e^{A\tau}x(0)+e^{A\tau}\int_0^\tau e^{-Az}Bw(z,0)\,dz\neq 0 \tag{7.18}$$

i.e. (pre-multiplying by $e^{-A\tau}$)

$$x(0)+\int_0^\tau e^{-Az}Bw(z,0)\,dz\neq 0 \tag{7.19}$$

or equivalently

$$x(0)+\int_0^\tau H^{-1}(z)Bw(z,0)\,dz\neq 0 \tag{7.20}$$

Then the settling time equals the delay interval τ plus the settling time of the sub-plant starting with initial state $x(\tau)$, i.e.

$$I=\tau+J(x(\tau)) \tag{7.21}$$

Thus, from (4.23),

$$I=\tau+J\left(e^{A\tau}x+e^{A\tau}\int_0^\tau e^{-Az}Bw(z)\,dz\right) \tag{7.22}$$

or equivalently

$$I=\tau+J\left(H(\tau)x+\int_0^\tau H(\tau-z)Bw(z)\,dz\right) \tag{7.23}$$

Suppose next that (7.8) applies, so that the initial plant state is such that the plant does settle before $t=\tau$. In this case the sub-plant state is zero from the settling time $t=T$ up to $t=\tau$

$$e^{At}x(0)+e^{At}\int_0^\tau e^{-Az}Bw(z,0)\,dz=0 \quad (T\leqslant t<\tau) \tag{7.24}$$

i.e.

$$x(0)+\int_0^s e^{-Az}Bw(z,0)\,dz=0 \quad (T\leqslant s<\tau) \tag{7.25}$$

i.e. the initial plant state satisfies

$$\left.\begin{aligned} x(0)+\int_0^T e^{-Az}Bw(z,0)\,dz=0 \\ \text{and} \qquad w(z,0)=0 \quad (T\leqslant z<\tau)\end{aligned}\right\} \tag{7.26}$$

or equivalently (see (4.16))

$$\left.\begin{aligned} x(0)+\int_0^T H^{-1}(z)Bw(z,0)\,dz=0 \\ \text{and} \qquad w(z,0)=0 \quad (T\leqslant z<\tau)\end{aligned}\right\} \tag{7.27}$$

More precisely, T is the *minimum* value of z for which (7.26), or equivalently (7.27), holds. Let us write this minimum value as $T(x(0),w(z,0))$, i.e. as $T(x,w(z))$. Then

$$I=T(x,w(z)) \tag{7.28}$$

8. Example. First-order sub-plant and settling time

Let the sub-plant be a single integrator, satisfying the scalar equation

$$\frac{dx}{dt} = bv \quad (b > 0) \tag{8.1}$$

so that the complete plant transfer function is

$$\frac{b}{pe^{p\tau}} \tag{8.2}$$

Let the performance index be settling time.

The function (7.1) which describes the optimal control when the delay is absent is (Fuller 1960 c)

$$C(x) = \begin{cases} -\operatorname{sgn} x & (x \neq 0) \\ 0 & (x = 0) \end{cases} \tag{8.3}$$

Also in the delay-free case, the function (7.3) which describes the minimal settling time for starting point x is (Fuller 1960 c)

$$J(x) = \frac{|x|}{b} \tag{8.4}$$

Hence, when the delay is present, application of result (7.16) or equivalently result (7.17) shows that the optimal controller is

$$u = -\operatorname{sgn}\left(x + b\int_0^\tau w(z)\,dz\right) \tag{8.5}$$

with sgn 0 interpreted as 0.

Also, when the delay is present, application of result (7.22), or equivalently (7.23), shows that the minimal settling time is

$$I = \tau + \frac{1}{b}\left|x + b\int_0^\tau u(z)\,dz\right| \tag{8.6}$$

if the initial plant state satisfies

$$x + b\int_0^\tau w(z)\,dz \neq 0 \tag{8.7}$$

On the other hand if the initial plant state satisfies

$$\left.\begin{aligned} x + b\int_0^T w(z)dz = 0 \\ w(z) = 0 \quad (T \leqslant z < \tau) \end{aligned}\right\} \tag{8.8}$$

$T(x, w(z))$ being the minimum value of T for which (8.8) holds, then result (7.28) shows that

$$I = T(x, w(z)) \tag{8.9}$$

In particular, for step response, i.e. when initially $w(z)$ $(0 \leqslant z < \tau)$ is zero, the minimal settling time is, from (8.6),

$$I = \tau + \frac{|x|}{b} \tag{8.10}$$

9. Example. Second-order sub-plant and settling time

Let the sub-plant consist of two integrators, satisfying

$$\begin{cases} \dfrac{dx_1}{dt} = x_2 & (9.1) \\ \dfrac{dx_2}{dt} = bv \quad (b > 0) & (9.2) \end{cases}$$

so that the transfer function of the plant from input u to output x_1 is

$$\frac{b}{p^2 e^{p\tau}} \tag{9.3}$$

Let the performance index be settling time.

When the delay is absent the optimal controller is, as is well known,

$$C(x_1, x_2) = -\operatorname{sgn}\left(x_1 + \frac{1}{2b} x_2 |x_2|\right) \tag{9.4}$$

for points not on the switching curve, which is

$$x_1 + \frac{1}{2b} x_2 |x_2| = 0 \tag{9.5}$$

For points on the switching curve (9.5), and not at the origin,

$$C(x_1, x_2) = -\operatorname{sgn} x_2 \tag{9.6}$$

and at the origin

$$C(x_1, x_2) = 0 \tag{9.7}$$

Results (9.4) to (9.7) go back to Feldbaum (1949) and Bushaw (1953); nowadays they are derived by means of the maximum principle (Pontryagin *et al.* 1961, Chap. 1, § 5). The minimal settling time is (in the absence of delay)

$$J(x_1, x_2) = \frac{1}{b} x_2 + \frac{2}{b^{1/2}}\left(x_1 + \frac{1}{2b} x_2^2\right)^{1/2} \tag{9.8}$$

for points to the right of or on the switching curve, i.e. where

$$x_1 + \frac{1}{2b} x_2 |x_2| \geqslant 0 \tag{9.9}$$

and

$$J(x_1, x_2) = -\frac{1}{b} x_2 + \frac{2}{b^{1/2}}\left(-x_1 + \frac{1}{2b} x_2^2\right)^{1/2} \tag{9.10}$$

for points to the left of or on the switching curve. For results (9.8) to (9.10) (which are for the delay-free case) see Sun Tszyan (1960) or Fuller (1963).

When the delay is present the optimal controller is, from (7.16) or (7.17) and (9.1) to (9.7),

$$u = -\operatorname{sgn}\left\{x_1 + \tau x_2 + b\int_0^\tau (\tau - z) w(z)\, dz + \frac{1}{2b}\left(x_2 + b\int_0^\tau w(z)\, dz\right)\left|x_2 + b\int_0^\tau w(z) dz\right|\right\} \tag{9.11}$$

where, if the argument of the sgn function is zero, u must be interpreted as -1, 0 or $+1$ according as

$$x_2 + b\int_0^\tau w(z)\, dz \tag{9.12}$$

is positive, zero or negative.

Also, when the delay is present, application of result (7.22) or equivalently (7.23) shows that the minimal settling time is

$$I = \tau + J\left(x_1 + \tau x_2 + b\int_0^\tau (\tau - z)w(z)\,dz, \quad x_2 + b\int_0^\tau w(z)\,dz\right) \tag{9.13}$$

where $J(\alpha_1, \alpha_2)$ is the function defined by (9.8) to (9.10), provided the initial plant state satisfies either

$$x_1 - b\int_0^\tau zw(z)\,dz \neq 0 \tag{9.14}$$

or

$$x_2 + b\int_0^\tau w(z)\,dz \neq 0 \tag{9.15}$$

(see (7.19) or (7.20)). On the other hand if neither (9.14) nor (9.15) is satisfied, then the initial plant state satisfies

$$\left.\begin{aligned} x_1 - b\int_0^T zw(z)\,dz &= 0 \\ x_2 + b\int_0^T w(z)\,dz &= 0 \\ w(z) = 0 \quad (T \leqslant z < \tau) \end{aligned}\right\} \tag{9.16}$$

for some $T < \tau$. If $T(x_1, x_2, w(z))$ is the minimum value of T for which (9.16) holds, result (7.28) shows that

$$I = T(x_1, x_2, w(z)) \tag{9.17}$$

In particular, for step response, i.e. when initially x_2 and $w(z)$ $(0 \leqslant z < \tau)$ are zero, the minimal settling time is, from (9.13),

$$I = \tau + 2\frac{|x_1|^{1/2}}{b^{1/2}} \tag{9.18}$$

10. Sub-optimal control. First technique

To construct the optimal control systems described above it is necessary to measure continually the present state of the pure delay, or to measure continually certain functionals of the present state of the pure delay. In simple cases this can conceivably be done. For example, if the delay is due to a conveyor belt, the functionals

$$\int_0^\tau w(z)\,dz \quad \text{and} \quad \int_0^\tau zw(z)\,dz \tag{10.1}$$

appearing in controllers (5.7), (6.18), (8.5) and (9.11) might be found by measuring continually the weight of the conveyor belt and contents, and its moment about one end. Often however the measurement of the delay state is not feasible, and it is necessary to adopt sub-optimal control.

One method of sub-optimal control is to incorporate in the controller a model of the delay, the input to the model delay being the controller output $u(t)$. The controller measures the state of the model delay in place of the state of the plant delay. During the first interval of length τ the states of the model delay and the plant delay are, in general, different; so that the control in this interval is non-optimal. However for $t > \tau$ the model delay state and the plant delay state are the same, and thus the control is, in a sense, optimal after the first delay interval. Overall, the control is sub-optimal.

This method belongs to the well known technique of simulating part of the plant in the controller in order to obtain approximate measurements of state coordinates which are inaccessible in the plant. The state coordinates of the model have been termed *copy coordinates* (Fuller 1960 b).

The problems involved in forming the functionals of the model delay state are still severe though not so difficult as for the plant delay. Usually the functionals would be approximated by sums of weighted measurements at discrete intervals along the distributed parameter of the model. The model delay itself might be only an approximation to a pure delay, obtained by sampled-data techniques in a digital computer, or lumped-parameter techniques† in an analogue computer.

11. Sub-optimal control. Second technique

The sub-optimal control technique of the previous section in effect replaces the (vector) term

$$\int_0^\tau H(\tau - z)Bw(z,t)\,dz \tag{11.1}$$

which appears in the optimal controller (see (4.17) and (7.17)) by the (vector) term

$$y(t) = \int_0^\tau H(\tau - z)Bu(t - \tau + z)\,dz \tag{11.2}$$

As stated in the previous section, one method of approximating (11.2) is to take a sum of weighted measurements at discrete intervals along a model delay line. Another method is as follows.

(11.2) is a linear functional of the past of the control signal u, and can thus be generated directly by passing u through a certain linear filter. Let us calculate the transfer function $f(p)$ of this filter. (11.2) may be written

$$y(t) = \int_0^\tau H(\alpha)Bu(t-\alpha)\,d\alpha \tag{11.3}$$

$$= \int_0^\infty H(\alpha)Bu(t-\alpha)\,d\alpha - \int_\tau^\infty H(\alpha)Bu(t-\alpha)\,d\alpha \tag{11.4}$$

The first integral on the right of (11.4) is the convolution of u with the impulse response of the sub-plant (i.e. the response of the vector x to an impulse applied to v). Thus the first term on the right of (11.4) is generated by passing the signal u through a filter with transfer function

$$g(p) \tag{11.5}$$

where $g(p)$ is the vector transfer function of the sub-plant (i.e. the transfer function from scalar input v to vector output x).

---

† For the synthesis of lumped-parameter approximations to pure delays, see e.g. Blake (1951), Štorch (1954), Kuh (1957), Smith (1959), and Reeve (1968).

The second term on the right of (11.4) may be written

$$-\int_0^\infty H(\tau+\beta)Bu(t-\tau-\beta)\,d\beta \tag{11.6}$$

in which, from (4.15),

$$H(\tau+\beta)=e^{A(\tau+\beta)}=e^{A\tau}e^{A\beta}=H(\tau)H(\beta) \tag{11.7}$$

Hence (11.6) is

$$-H(\tau)\int_0^\infty H(\beta)\,Bu(t-\tau-\beta)\,d\beta \tag{11.8}$$

The integral in (11.8) is the convolution of u, delayed by τ, with the impulse response of the sub-plant. Thus the second term on the right of (11.4) is generated by passing the signal u through a filter with transfer function

$$-H(\tau)e^{-p\tau}g(p) \tag{11.9}$$

Adding (11.5) and (11.9), we find the desired vector transfer function

$$f(p)=g(p)-H(\tau)g(p)e^{-p\tau} \tag{11.10}$$

i.e. the term $y(t)$ in the sub-optimal controller can be generated by feeding the control signal $u(t)$ through a filter with (vector) transfer function (11.10).

To synthesize $f(p)$ we can construct a model of the sub-plant, yielding $g(p)$, and put this in parallel with a path consisting of a pure delay $e^{-p\tau}$, followed by a model of the sub-plant, $g(p)$, followed by amplifiers giving the constant (matrix) gain $(-H(\tau))$. Thus it is still necessary to construct a model pure delay, as in § 10; however it is no longer necessary to tap off signals along the distributed parameter of the model delay.

Fig. 3

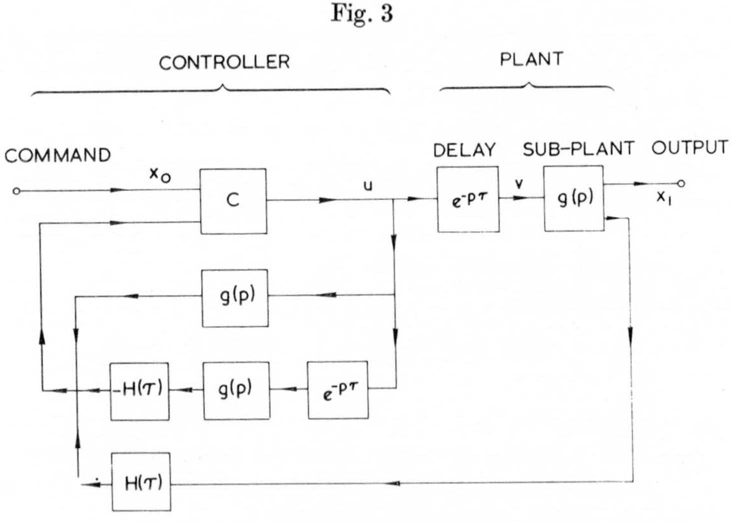

A sub-optimal control system.

The sub-optimal control system is as shown in fig. 3. One can avoid having in the controller two models of the sub-plant by interchanging the order of the delay model and the sub-plant model. However the delay model then operates on a vector input (if $n>1$) and must thus be synthesized by means of n scalar delays.

12. Sub-optimal control. Third technique

In the absence of the delay a sub-optimal method of control is to replace the function of the state coordinates which gives optimal control by a simpler function of these state coordinates. For example, one might use a linear combination of state coordinates, or a linear combination of nonlinear functions of single state coordinates (see e.g. Fuller 1967).

This technique can be adapted to the case when the delay is present. The optimal control is then a complicated function of the state coordinates of the sub-plant measured τ units of time ahead of the present. The sub-optimal control replaces this function by a simpler function of the same arguments.

More generally, suppose the sub-optimal control in the absence of delay is any functional of the past and present of the state coordinates. For example the sub-optimal controller might be a traditional PID (proportional + integral + derivative) controller operating on the system error, $e(t)$. Then a sub-optimal control in the presence of delay is the same functional of the state coordinates, but with these state coordinates translated in time by an amount τ. For example the sub-optimal controller might be a PID controller operating on $e(t+\tau)$. If the sub-optimal control in the absence of delay is stable, the corresponding control in the presence of delay is stable, since it gives the same system output, apart from a time translation.

13. Sub-optimal control. Fourth technique

Suppose the plant does not have pure delay, but consists of multiple small lags

$$\frac{1}{(1+p\tau_1)(1+p\tau_2)\ldots(1+p\tau_r)} \tag{13.1}$$

followed by a sub-plant of low order. As discussed in § 1 we may approximate (13.1) by a pure delay

$$e^{-p(\tau_1+\tau_2+\ldots+\tau_r)} = e^{-p\tau} \tag{13.2}$$

and then calculate an optimal controller by the preceding theory. If the state coordinates of the sub-plant are $x_1, x_2, \ldots, x_n$ the resulting controller is represented by a nonlinear function of n arguments. Each argument is a linear combination of $x_1, x_2, \ldots, x_n$ and a linear functional of the state of the delay, so that the controller is

$$\begin{aligned} u = C\Bigg(& c_{11}x_1 + c_{12}x_2 + \ldots + c_{1n}x_n + \int_0^\tau l_1(\tau - z)w(z)\,dz, \\ & c_{21}x_1 + c_{22}x_2 + \ldots + c_{2n}x_n + \int_0^\tau l_2(\tau - z)w(z)\,dz, \\ & \ldots, \\ & c_{n1}x_1 + c_{n2}x_2 + \ldots + c_{nn}x_n + \int_0^\tau l_n(\tau - z)w(z)\,dz\Bigg) \end{aligned} \tag{13.3}$$

where l_i represents the impulse response from v to x_i, and the c_{ij} are known constants.

On reverting to the original plant with lags (13.1), we may obtain a sub-optimal controller by replacing the linear functionals in (13.3) by linear combinations of the state coordinates $z_1, z_2, \ldots, z_r$ of the small lags (13.1). The resulting sub-optimal controller is

$$\begin{aligned} u = C(&c_{11}x_1 + \ldots + c_{1n}x_n + d_{11}z_1 + \ldots + d_{1r}z_r, \\ &c_{21}x_1 + \ldots + c_{2n}x_n + d_{21}z_1 + \ldots + d_{2r}z_r, \\ &\ldots, \\ &c_{n1}x_1 + \ldots + c_{nn}x_n + d_{r1}z_1 + \ldots + d_{rr}z_r) \end{aligned} \quad (13.4)$$

(13.4) shows how to modify a controller design based on the optimal controller for a low-order sub-plant, so as to take into account the multiple small lags in the complete plant. Namely, the state coordinates of the small lags are added *linearly* to the state coordinates of the sub-plant, in the control law. In practice the state coordinates of the small lags might well be replaced by copy coordinates of a model of (13.1) in the controller, as in § 10. Also the nonlinear function C might well be replaced by a simpler function of the same arguments, as in § 12.

As an example, consider the settling time problem for a plant with transfer function

$$\frac{1}{(1+p\tau)} \cdot \frac{b}{p^2} \quad (13.5)$$

By comparison with (9.11), a sub-optimal controller is

$$u = -\operatorname{sgn}\left\{x_1 + \tau x_2 + d_1 x_3 + \frac{1}{2b}(x_2 + d_2 x_3)|x_2 + d_2 x_3|\right\} \quad (13.6)$$

where x_3 is the state coordinate of the lag, i.e.

$$x_3 = \frac{1}{b}\frac{d^2 x_1}{dt^2} \quad (13.7)$$

The coefficients d_1 and d_2 can be found by evaluating the response, after an interval τ, of the sub-plant state to a unit initial value of x_3. One finds

$$\begin{cases} d_1 = e^{-1} b \tau^2 & (13.8) \\ d_2 = (1 - e^{-1}) b \tau & (13.9) \end{cases}$$

The switching function in (13.6) is a linear combination of nonlinear functions of linear combinations of state variables, and is thus relatively easy to synthesize. In contrast, the exactly optimal controller for plant (13.5), although calculable, is difficult to synthesize.

Sub-optimal controllers which involve relay control often result in sliding motion in the vicinity of the state origin (see e.g. Fuller 1967). In such cases the settling time may be, strictly speaking, infinite. However the controller can be near-optimal, in the sense of giving quick settling to within a small neighbourhood of the origin.

14. Generalizations

The theory given above can be generalized in various directions. The performance index can include the control u in its integrand, i.e. (3.5) can be generalized to

$$I = \int_0^\infty q(x_1, x_2, \ldots, x_n, u)\, dt \quad (14.1)$$

The control constraint (3.4) can be replaced by, or augmented by, an integral constraint:

$$\int_0^\infty Q(u)\,dt = L \tag{14.2}$$

Multi-variable systems, where the control u is a vector $(u_1, u_2, \ldots)$, can be treated, provided each control signal u_i is subject to the same delay τ before it reaches the input v_i to the sub-plant. If each control input u_i is subject to a different delay τ_i, the problem is more difficult, but results seem to be obtainable for the settling time performance index. A sub-optimal technique here is to insert extra delays in the faster control signal channels, so as to equalize the net delay in all channels.

Sampled-data systems can be dealt with, i.e. a discrete-time version of the above theory can be formulated.

15. Conclusions

When the plant contains a pure delay the optimal controller and minimal performance index may be calculated in certain cases. The results are useful for suggesting and assessing sub-optimal techniques. In formulating the optimal control law it is essential to take proper account of the state of the pure delay. In the literature, neglect of this point has sometimes resulted in the distinction between optimal control and sub-optimal control becoming blurred.

Appendix. Historical notes

In this Appendix a review will be given of such papers on the problem outlined in § 3 as are known to the present writer. It will be seen that several writers have realized that basically the controller must generate a function of a future state of the sub-plant. However they did not take complete account of the state of the pure delay, and thus arrived at controllers which are sub-optimal rather than optimal.

Bass (1956 a, b, c) had the idea of making the controller generate a function of the predicted state of the sub-plant, i.e. the state at time τ ahead of the present. His work was in connection with sub-optimal relay control systems. However his method of determining the predicted state of the sub-plant was imperfect. We have shown that it is necessary to construct a functional of the present delay state as in (4.9) or a functional of the control u during the past interval of length τ as in § 10. Instead, Bass used a function of the control u at time τ units ago, i.e. a function of $u(t-\tau)$. This technique is equivalent to measuring only one of the infinity of coordinates which make up the delay state. It only yields correct results whenever the present delay state is uniform along its length; i.e. provided the relay has not switched during the past interval of length τ. It is thus to be expected that Bass' technique results in an unstable relay control system†; and this had indeed been found to be so in a case studied by Korolev (1961). Nevertheless Bass' method can be considered as using an approximate measurement of the delay state, and

† It is assumed here that the relay is ideal, i.e. without dead zone. Relay control systems can often be stabilized by use of a large dead zone, but this involves poor steady state performance.

probably results in a smaller amplitude limit cycle than if no attempt is made to take account of the delay state.

Petrov and Rutkovski (1956), Mitsumaki (1959, 1960), Balakirev (1962) and Novoseltsev (1964) attempted to obtain optimal control of relay systems with pure delay by using switching curves and surfaces in the sub-plant state space. The switching curves and surfaces were similar to those in the absence of delay but distorted so as to yield earlier switching. This method leads to ambiguous requirements on the control due to the fact that the state of the delay is ignored; these ambiguities become particularly embarrassing near the origin of the sub-plant state space.

Smith (1957a, b, 1958, 1959) found controllers somewhat similar to those in §§ 11 and 12 for linear (i.e. non-saturating) systems with pure delay. His methods were based on formal block diagram manipulations, which took account of the response to step changes of command signal, but ignored the initial states of both the delay and the sub-plant. Consequently his controllers are sub-optimal. In his book (Smith 1958, § 17–9) he applied his methods to controllers with saturation, obtaining a system somewhat similar to that in fig. 3. In Smith's controller the blocks with transfer functions $H(\tau)$ and $-H(\tau)$ (see fig. 3) are missing; the reason being that he omitted to take account of initial conditions in the sub-plant. Thus his controller is less optimal than the sub-optimal controller of § 11. Buckley (1960) has discussed the practical application of Smith's controller.

Kalman and Bertram (1957, 1958) considered optimal linear control of linear sampled-data systems, and then extended their analysis to the case where the plant contains a pure delay. Their controller is sub-optimal in so far as it does not directly measure the state of the delay, and is the analogue for sampled-data non-saturating systems of the controller in § 11. Further work on the problem of Kalman and Bertram was done by Kurzweil (1963) and Koepcke (1965).

Butkovski and Domanitski (1958) considered the optimal settling time problem for the system of § 9. Their system is somewhat like that of fig. 3. However it omits the two lower feedback paths and instead arranges for appropriate initial conditions to be inserted in the block $g(p)$ in the upper feedback path. In so far as their system does not directly measure the state of the delay, it is sub-optimal. Also their system has only initial measurement of sub-plant state, and is thus less practical than that of fig. 3.

Varshavski (1958, 1960) considered, in effect, the optimal settling time problem for a plant containing a second-order sub-plant and a pure delay. He obtained a sub-optimal controller equivalent to that of fig. 3, and also replaced the nonlinear function generator C by a linear function as in § 12. Varshavski's methods were described in the book by Lerner (1961).

Tsypkin (1959a, b) treated non-optimal nonlinear sampled-data systems, with pure delay in the plant. He obtained a sub-optimal controller of the same nature as that discussed in § 12. His controller contained a sampled-data model of the pure delay, similar to that discussed in § 10.

Merriam (1959) applied dynamic programming to the problem of minimizing a quadratic integral performance index for a plant with delay and a first-order time-varying sub-plant. He obtained a linear controller which was sub-optimal to the extent that only an indirect measurement of the delay state was used, as in § 10. In his book (Merriam 1964, § 7–10) he generalized his treatment to the case of a sub-plant of nth order. For related work, also based on dynamic programming, see Krasovski (1963).

Sampled-data versions of Merriam's analysis have been given by Koepcke (1965) and Sawaragi and Inoue (1967).

Stratonovich and Shmalgauzen (1962) considered the optimal saturating control problem for a stationary stochastic plant with delay. They obtained a controller which generates a function of a future state of the sub-plant, similar to the deterministic controller of the present paper.

References

Adey, A. J., Coales, J. F., and Stiles, J. A., 1963, *Proc. I.F.A.C. Congress, Basle. Theory* (London: Butterworths, 1961), p. 41.

Adey, A. J., Billingsley, J., and Coales, J. F., 1966, *Proc. I.F.A.C. Congress, London* (London: Butterworths, 1967), p. 40F.1.

Balakirev, V. S., 1962, *Avtomat. Telemekh.*, **23**, 1014.

Bass, R. W., 1956 a, *Q. appl. Math.*, **14**, 415; 1956 b, *Jet Propul.*, **26**, 644; 1956 c, *Proc. Symp. Nonlinear Circuit Analysis* (Polytechnic Inst. Brooklyn), p. 163.

Bellman, R., 1960, *Introduction to Matrix Analysis* (New York: McGraw-Hill).

Blake, D. V., 1951, *Automatic and Manual Control* (*Proc. Cranfield Conf.*), edited by A. Tustin (London: Butterworths, 1952), p. 539.

Buckley, P. S., 1960, *Proc. I.F.A.C. Congress, Moscow* (London: Butterworths, 1961), **1**, 33.

Bushaw, D. W., 1953, *Rep. exp. Tow. Tank, Stevens Inst. Technol.*, No. 469.

Butkovski, A. G., and Domanitski, S. M., 1958, *Teoriya i Primenenie Diskretnich Sistem* (Proc. Conf. on Theory and Application of Discrete Systems, Akad. Nauk SSSR, September 1958), p. 27.

Feldbaum, A. A., 1949, *Avtomat. Telemekh.*, **10**, 249.

Fuller, A. T., 1960 a, *J. Electron. Contr.*, **8**, 381; 1960 b, *Ibid.*, **8**, 465; 1960 c, *Proc. I.F.A.C. Congress, Moscow* (London: Butterworths, 1961), **1**, 510; 1963, *J. Electron. Contr.*, **15**, 63; 1964 a, *Ibid.*, **17**, 283; 1964 b, *Ibid.*, **17**, 301; 1966, *Int. J. Control*, **3**, 359; 1967, *Ibid.*, **5**, 197.

Gantmacher, F. R., 1954, *Teoriya Matrits* (Moscow: Fizmatgiz). English translation: *Theory of Matrices* (New York: Chelsea, 1959).

Kalman, R. E., and Bertram, J. E., 1957, *Proc. Computers in Control Systems Conf., Atlantic City* (Amer. Inst. Elect. Engrs, special publication T–101), p. 130; 1958, *Trans. Am. Inst. elect. Engrs* II, **77**, 602; 1960, *J. bas. Engng*, **82**, 371.

Koepcke, R. W., 1965, *J. bas. Engng*, **87**, 74.

Korolev, N. A., 1961, *Avtomat. Telemekh.*, **22**, 605.

Krasovski, N. N., 1963, *Proc. I.F.A.C. Congress, Basle. Theory* (London: Butterworths, 1964), p. 327.

Kuh, E. S., 1957, *I.R.E. natn. Conv. Rec.*, **5**, 160.

Küpfmüller, K., 1928, *Elekt. Nachr. Tech.*, **5**, 451.

Kurzweil, F., 1963, *I.E.E.E. Trans. autom. Control*, **8**, 27.

Lerner, A. Ya., 1961, *Printsipi Postroeniya Bistrodeistvushikh Sledyashikh Sistem i Regulyatorov* (Moscow: Gosenergoizdat). German translation: *Schnelligkeitsoptimale Regelungen* (Munich: Oldenbourg, 1963).

Merriam, C. W., 1959, *Trans. Am. Inst. elect. Engrs* II, **78**, 506; 1964, *Optimization Theory and the Design of Feedback Control Systems* (New York: McGraw-Hill).

Mitsumaki, T., 1959, *Bull. Jap. Soc. mech. Engrs*, **2**, 348; 1960, *Proc. I.F.A.C. Congress, Moscow* (London: Butterworths, 1961), **1**, 520.

Novoseltsev, V. N., 1964, *Avtomat. Telemekh.*, **25**, 1545.

Oguztöreli, M. N., 1966, *Time-lag Control Systems* (New York: Academic Press).

Oldenbourg, R. C., and Sartorius, H., 1944, *Dynamik Selbsttätiger Regelungen* (Munich: Oldenbourg). English translation: *The Dynamics of Automatic Controls* (New York: Amer. Soc. Mech. Engrs, 1948).

PEANO, G., 1887, *Atti Accad. Sci., Torino*, **22,** 437; 1888, *Math. Annln*, **32,** 450.
PETROV, V. V., and RUTKOVSKI, V. I., 1956, *Izv. Akad. Nauk SSSR, Otd. Tech. Nauk*, No. 4, 16.
PONTRYAGIN, L. S., BOLTYANSKI, V. G., GAMRELIDZE, R. V., and MISHCHENKO, E. F., 1961, *Mathematical Theory of Optimal Processes* (in Russian, Moscow: Fizmatgiz). (English translation, New York: Wiley, 1962.)
REEVE, P. J., 1968, *Int. J. Control*, **8,** 53.
SAWARAGI, Y., and INOUE, K., 1967, *Rep. Engng Res. Inst. Kyoto Univ.*, No. 131.
SMITH, O. J. M., 1957 a, *I.R.E. natn. Conv. Rec.*, **5,** 8; 1957 b, *Chem. Engng Prog.*, **53,** 217; 1958, *Feedback Control Systems* (New York: McGraw-Hill); 1959, *I.S.A. Jl*, **6,** 28.
STORCH, L., 1954, *Proc. Inst. Radio Engrs*, **42,** 1666.
STRATONOVICH, R. L., and SHMALGAUZEN, V. N., 1962, *Izv. Akad. Nauk SSSR, O.T.N., Energ. Avt.*, No. 5, 131.
SUN TSZYAN, 1960, *Izv. Akad. Nauk SSSR, O.T.N., Energ. Avt.*, No. 5, 96.
TSYPKIN, YA. Z., 1959 a, *Dokl. Akad. Nauk SSSR*, **124,** 812; 1959 b, *Regelungstechnik*, **7,** 196.
VARSHAVSKI, O. G., 1958, *Teoriya i Primenenie Diskretnich Sistem* (Proc. Conf. on Theory and Application of Discrete Systems, Akad. Nauk SSSR, September 1958), 36; 1960, *Priborostroenie, Leningr.*, No. 8, 4.
ZADEH, L. A., and DESOER, C. A., 1963, *Linear System Theory* (New York: McGraw-Hill).

Part III.

Optimal nonlinear stochastic control

K

INTRODUCTION TO PART III

In Part III a number of simple nonlinear stochastic control systems are optimized. Although the systems are of low order, having state spaces of dimension at most two, the results are of preliminary interest. Furthermore the results can be extended to higher-order systems by the approximate technique of Paper 10 (this technique goes over to the stochastic case without difficulty).

More work needs to be done on low-order nonlinear stochastic optimization. One approach would be to devise efficient approximate procedures for the solution of the stochastic version of Bellman's equation, which version is introduced by Paper 15.

Papers 11–15 deal with the case when the state is completely accessible to measurement. Paper 16 discusses the case when the state measurements are contaminated by noise. Although the latter discussion is concerned with linear systems, the results can be generalized to apply to saturating systems (see Paper 18).

Paper 11

Reprinted from JOURNAL OF ELECTRONICS AND CONTROL, Vol. 9, No. 1, pp. 65–80, July, 1960.

Paper 12

Reprinted from JOURNAL OF ELECTRONICS AND CONTROL, Vol. 10, No. 1, pp. 61–80, January, 1961.

Paper 13

Reprinted from JOURNAL OF ELECTRONICS AND CONTROL, Vol. 10, No. 2, pp. 157–164, February, 1961.

Paper 14

Reprinted from JOURNAL OF ELECTRONICS AND CONTROL, Vol. 11, No. 6, pp. 463–479, December, 1961.

Paper 15

Reprinted from JOURNAL OF ELECTRONICS AND CONTROL, Vol. 10, No. 6, pp. 473–488, June, 1961.

Paper 16

Reprinted from JOURNAL OF ELECTRONICS AND CONTROL, Vol. 13, No. 3, pp. 263–279, September, 1962.

Optimization of Non-linear Control Systems with Random Inputs†

By A. T. Fuller
Department of Engineering, University of Cambridge

[Received April 26, 1960]

Abstract

It is shown that for a wide class of saturating control systems with random inputs, the optimum controller is an instantaneous non-linear function of the input phase coordinates and the output phase coordinates. If the system is a relay control system, the optimum controller is represented by a switching surface in the corresponding phase space. In special cases the optimum controller can be simplified by the use of error phase coordinates. These results are applicable when the random input is a generalized Markov process, a Gaussian process, or a Gaussian signal plus a Gaussian noise.

The synthesis of the optimum switching surface by a self-optimizing technique is discussed.

§ 1. Introduction

In two previous papers (Fuller 1960 a, b) the writer investigated the nature of the optimum controller for non-linear control systems with deterministic transient inputs. It was shown that, in many cases, the optimum controller simply generates an instantaneous non-linear function of the input phase coordinates and of the output phase coordinates. In the present paper the nature of the optimum controller when the inputs are random is investigated. It is found that again the optimum controller is often represented by an instantaneous non-linear function of certain 'input phase coordinates' (when these are suitably defined) and of the output phase coordinates. Readers who are familiar with the transient input case dealt with by Fuller (1960 a, b) will find the argument of the present paper fairly straightforward.

§ 2. Statement of the Problem

The block diagram of the system to be considered is shown in fig. 1. The justification for adopting this block diagram was given in a previous paper (Fuller 1960 a).

2.1. *Main Process*

The main process (i.e. the plant) is assumed to be a specified *determinate* process with phase coordinates $Y_1, Y_2, \dots Y_n$. As defined previously (Fuller 1960 a), this means that if the present values of $Y_1, Y_2, \dots Y_n$ are known, and if the complete future of the input L to the main process is known, then the future of the output Y of the main process is completely determined. Questions of the physical meaning and measurement of the phase coordinates have already been discussed (Fuller 1960 a, b).

† Communicated by J. F. Coales.

2.2. *Amplifier*

The amplifier, which drives the main process, is assumed to be completely specified by its given input–output characteristic,

$$L=g(C), \quad \ldots \ldots \ldots (1)$$

i.e. the amplifier is free from time-lags, hysteresis, etc. $g(C)$ is assumed to be a single-valued monotonic function with a finite positive maximum and a finite negative minimum. This last assumption is made for definiteness and is intended to imply that the optimum systems we shall discuss are not trivial in the sense of being perfect. However, in many cases (e.g. when the main process itself saturates) this assumption is unnecessary for the validity of the ensuing argument.

Fig. 1

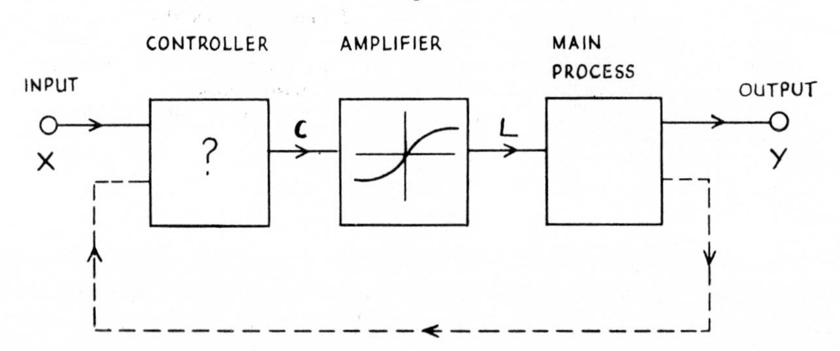

Block diagram of saturating control system. Note: the dashed line **represents** possibly multiple connections from the main process to the controller.

2.3. *Controller*

The controller, which drives the amplifier, is assumed to be physically realizable (i.e. operates only on the present and past of available signals) and is otherwise unrestricted. It is to be designed so as to minimize one of the performance criteria discussed later.

2.4. *Input*

The nature of the random input will be defined and discussed in §§ 3, 4 and 5.

2.5. *Performance Criterion*

The error E of the control system is defined by

$$E=X-Y \quad \ldots \ldots \ldots (2)$$

where X is the system input and Y is the system output. When the input is random, the case of most interest is that when E is a stationary random function of time t. In this case a general performance criterion (Fuller 1959 b) is M_ν' where

$$M_\nu' \equiv \lim_{T\to\infty} \frac{1}{2T}\int_{-T}^{T} |E|^\nu\, dt, \quad (\nu \geqslant 0) \quad \ldots \ldots (3)$$

$$\equiv \overline{|E|^\nu}, \quad (\nu \geqslant 0). \quad \ldots \ldots \ldots (4)$$

Thus M_2' is the well-known performance criterion mean-square-error. M_0' is the fractional duration of non-zero error (Fuller 1959b). A more general criterion is

$$\overline{q(E)} \equiv \lim_{T\to\infty} \frac{1}{2T}\int_{-T}^{T} q(E)\,dt \quad . \quad . \quad . \quad (5)$$

where $q(E)$ is any single-valued function for which the time-average exists. For most of the paper we shall adopt criterion (5).

2.6. *The Problem*

The control problem we shall consider is to find a controller which minimizes the ensemble-average of some given performance criterion as defined in § 2.5. (It is assumed that the criterion chosen can exist when a suitable controller is incorporated in the system.) More general formulations of the problem will be discussed in § 11.

§ 3. Classification of Random Signals

Most of the expositions of stochastic processes written for control engineers are centred on Wiener's (1949) theory of optimum linear filters. Emphasis is placed on the first and second probability distributions and their properties. When, however, we remove the restriction of linearity and seek an *absolutely* optimum filter or controller, it becomes necessary to look more closely into the nature of random signals. Consequently in this and the next two sections we shall classify and discuss random signals generally. Our approach is similar to that of the textbook of Blanc-Lapierre and Fortet (1953).

By a random signal $x(t)$ is meant any member of an ensemble of signals $x_1(t)$, $x_2(t)$, ..., each of which has a probability (in a mathematical sense) of occurring. $x(t)$ may be (i) analytic, (ii) piece-wise analytic, (iii) completely non-analytic, or (iv) otherwise.

(i) Perhaps the simplest stochastic process is the *analytic random signal.* Each member of the ensemble is an analytic function of t in the range $t_0 \leqslant t < \infty$, i.e.

$$x(t) = k_0 + k_1 t + k_2 t^2 + \ldots, \quad (t_0 \leqslant t < \infty), \quad . \quad . \quad . \quad (6)$$

where one or more of the k's are random variables, and t_0 is some fixed time origin. If $x(t)$, $x'(t)$, ... are known at any instant $t > t_0$ the future of x is completely determined. The theory of control optimization for such an input becomes the same as that for 'transient' inputs; this has already been dealt with (Fuller 1960b) and will not be mentioned further in the present paper. One might say that analytic random signals are stochastically trivial.

(ii) Next in order of complexity is the case when each member of the ensemble is not analytic but is composed of successive parts each of which is analytic in its interval. Each such interval may be finite or infinite, and discontinuities in $x(t)$ or its time-derivatives occur at the isolated points where the intervals join. In any interval bounded by a pair of neighbouring discontinuities, $x(t)$ may be represented by a power series:

$$x(t) = k_{i0} + k_{i1}(t - t_i) + k_{i2}(t - t_i)^2 + \dots, \quad (t_i \leqslant t < t_{i+1}, \quad i = \dots -1, 0, 1, 2, \dots) \qquad (7)$$

where one or more of the k_{ij} and the t_i are random variables. We call such processes *piece-wise analytic random signals.*

(iii) Let us suppose that, in a piece-wise analytic signal, the frequency with which discontinuous changes occur is increased without limit, the change-points t_i being distributed so that some appear in every finite interval. We call such processes *completely non-analytic random signals.* The simplest example is a Gaussian process, which will be discussed in the next section.

The above classification is intended to give the reader a preliminary orientation. In the next two sections we shall be more specific.

§ 4. Generation of Random Signals

In practice random signals are generated by mechanisms which are partly determinate. These mechanisms have certain effects (outputs) which are determined by random causes (inputs). Usually the outputs have time lags behind the inputs, i.e. operate on the pasts of the inputs, or roughly speaking, are weighted time-averages of the inputs. Thus the outputs are less random than the inputs. These inputs are outputs of other mechanisms which themselves have inputs which are still more random. Proceeding backwards along the chains of cause-and-effect, we thus eventually arrive at purely random signals. By a *purely random signal* we mean a signal of which the value at any instant t is a random variable which is statistically independent of t and of the value at any other instant.

The concept of a purely random signal is a straightforward generalization of the concept of 'white noise' with which engineers are familiar†. For white noise, the values at any two different instants have zero *correlation,* but are not necessarily statistically independent.

Often, in order to arrive at non-trivial stochastic processes, we must consider the causative purely random signals as limiting cases of non-purely random signals, in which the amplitudes are allowed to increase indefinitely. This makes it difficult to define a precise mathematical model of such purely random signals. In fact mathematicians prefer to consider what are in effect integrals of the latter, or 'processes with independent increments' (see Blanc-Lapierre and Fortet (1953)). However, there is little doubt that in due course white noise and its generalizations will be given mathematical respectability, just as has been the case with Dirac's delta-function.

There are two types of purely random signal which are particularly important. The first we term a *Poisson purely random signal,* $\zeta(t)$; this

† See, e.g., Bode and Shannon (1950) for a discussion of white noise in the context of Wiener's theory.

is described as follows. $\zeta(t)$ is constant for all t except at isolated instants when $\zeta(t)$ has impulse functions. The latter are Dirac delta-functions, i.e. have infinite height, infinitesimal width, and finite value (area). Each delta-function has the same value, and the mean value of $\zeta(t)$ is zero. The probability of an impulse appearing in any infinitesimal interval Δt is $\mu\Delta t$ where μ is a constant. A well-known consequence (see e.g. Feller (1950)) is that the probability of there being K impulses in a given finite interval T is

$$P(K) = \frac{(\mu T)^K}{K!} \exp(-\mu T), \qquad \dots\dots (8)$$

i.e. K has a Poisson distribution. This is why we termed $\zeta(t)$ a Poisson purely random signal.

The second main type of purely random signal we shall term a *Gaussian purely random signal*, $\xi(t)$. It may be regarded as a limiting case of the Poisson purely random signal, in which the impulse value is made proportional to $\mu^{-1/2}$, and μ is made infinitely great. If $\xi(t)$ is fed into a linear smoothing filter, the output is a Gaussian process (see Blanc-Lapierre and Fortet (1953)), i.e. a signal of which the joint probability distribution at any set of instants is Gaussian. This is why we termed $\xi(t)$ a Gaussian purely random signal.

A great variety of random signals can be obtained by feeding $\zeta(t)$ or $\xi(t)$ into a linear or non-linear filter, in fact all the simple types which are commonly discussed in the context of control systems. We have already mentioned that Gaussian processes are generated by passing $\xi(t)$ through a linear filter. In particular, if the filter is simply an integrator, the output is termed *Brownian motion* (see Wax (1954)). If $\zeta(t)$ is fed into a linear smoothing filter, the output is termed *shot-effect* (Rice (1944, 1945)). If $\zeta(t)$ is fed into a bistable flip-flop, the output is a *random telegraph signal* (Rice (1944, 1945)).

The impulses of $\zeta(t)$ often give rise to the discontinuities in piece-wise analytic signals of class (ii). The infinitesimal impulses of $\xi(t)$ often give rise to completely non-analytic signals of class (iii).

§ 5. Phase State of Random Signals

We are now in a position to define the following general model: a random signal is to be regarded as the output of a determinate system which has as inputs a set of purely random signals. A *determinate* system (Fuller (1960 a)) is a system of which the future output is completely determined by the futures of the inputs and by the present values of a set of variables termed phase coordinates.

The state of the random signal at a given instant t_0 is completely characterized by the phase state of the determinate part of the model. To see this, note that the signal at any future instant t_1 is completely determined by (i) the phase state of the determinate system at t_0, (ii) the futures of the purely random inputs (i.e. for $t_0 \leqslant t < t_1$). But the futures of the purely random inputs are, by definition, statistically independent of their past values. Therefore the probability distribution of the signal at t_1, conditional on all the present and past values of the phase coordinates, depends statistically only on (i). We call the phase state of the determinate system in the model the *phase state of the random signal.*

Suppose the phase state of the random signal $x(t)$ has m coordinates $x_1, x_2, \ldots x_m$. Then it follows from the above definitions that the probability†

$$P(x_1{}^1, x_2{}^1, \ldots x_m{}^1;\ t_1 \,|\, x_1{}^0, x_2{}^0, \ldots x_m{}^0;\ t_0) \qquad (9)$$

that the signal will be in state $(x_1{}^1, x_2{}^1, \ldots x_m{}^1)$ at t_1, given that it was in state $(x_1{}^0, x_2{}^0, \ldots x_m{}^0)$ at t_0 $(t_0 < t_1)$, is independent of the states of the signal before t_0. This property defines the signal as an m-dimensional *Markov process.* In the present paper we shall be concerned only with such Markov processes. Note that it is difficult to conceive of random signals which are not Markov processes in this sense if it is widely interpreted, i.e. if infinitely many phase coordinates are allowed. (Thus we could take as the phase coordinates of $x(t)$ the values of $x(t)$ at all past instants of time.) However, our theory is of practical interest only when the number of phase coordinates is not large.

§ 6. The Optimum Controller

Since the controller is physically realizable, its output $C(t)$ can in general be any non-linear or linear operation, however complicated, on the pasts of all available signals; i.e. $C(t)$ is a general *functional* of these signals. We shall now show that when the controller is optimum in the sense of § 2, $C(t)$ simplifies to an 'instantaneous' *function* of the present values of the input phase coordinates and of the output phase coordinates‡.

At any instant t_0, the future of the error $E(t)$ for $t > t_0$ is completely determined by

(i) the future of the input $X(t)$,
(ii) the initial values of the main process phase coordinates,
(iii) the known constants of the input and of the main process,
(iv) the input to the main process, $L(t)$ for $t \geqslant t_0$.

The future of $X(t)$ is statistically dependent only on the present values $X_1(t_0), X_2(t_0), \ldots X_m(t_0)$ of the phase coordinates of the input. Therefore the future of $E(t)$ is statistically dependent only on

(i) $X_1(t_0), X_2(t_0), \ldots X_m(t_0)$,
(ii) $Y_1(t_0), Y_2(t_0), \ldots Y_n(t_0)$,
(iii) known constants,
(iv) $L(t)$ for $t \geqslant t_0$.

At time t_0 items (i), (ii) and (iii) constitute all the relevant known data affecting the future contributions to $\overline{q(E)}$, and item (iv) is to be chosen on the basis of these data. Therefore the best value of $L(t_0)$, in any well-defined

† In expression (9) the superscripts are for labelling purposes only.
‡ A more precise demonstration of this result is given in the Appendix.

sense, must be a function of arguments (i), (ii) and (iii) and of t_0:

$$L_{\text{opt}}(t_0) = F[X_1(t_0), X_2(t_0), \dots X_m(t_0), Y_1(t_0), Y_2(t_0), \dots Y_n(t_0), t_0]. \quad (10)$$

Here F is some unknown function of arguments (i), (ii), (iii) and t_0. We have not written arguments (iii) explicitly in (10) as they are not needed for present purposes. Because the purely random signals which generate the input $X(t)$ are statistically stationary, F in (10) cannot depend explicitly on the last argument, t_0. Hence we can write

$$L_{\text{opt}}(t_0) = G[X_1(t_0), \dots X_m(t_0), Y_1(t_0), \dots Y_n(t_0)]. \quad . \quad . \quad (11)$$

Since this equation holds for all t_0, we can drop the suffix and write

$$L_{\text{opt}}(t) = G[X_1(t), \dots X_m(t), Y_1(t), \dots Y_n(t)], \quad . \quad . \quad (12)$$

or, more briefly,

$$L_{\text{opt}} = G[X_1, X_2, \dots X_m, Y_1, Y_2, \dots Y_n]. \quad . \quad . \quad . \quad (13)$$

This is the required result, stating that the optimum control input to the main process is an instantaneous non-linear function of the input phase coordinates and of the output phase coordinates.

§ 7. Relay Control Systems

Consider an ideal relay control system, in which

$$L = \operatorname{sgn} C. \quad . \quad . \quad . \quad . \quad . \quad . \quad . \quad . \quad . \quad (14)$$

For such a system, L_{opt} as given by eqn. (13) is a discontinuous function of the phase coordinates $X_1, X_2, \dots X_m, Y_1, Y_2, \dots Y_n$. For a given value of each of the latter coordinates, L_{opt} is either $+1$ or -1. If L_{opt} is $+1$, the corresponding point in the phase space is termed a P-point. Similarly if L_{opt} is -1, the point is termed an N-point. The whole phase space consists of regions of P-points and regions of N-points. The boundary between these regions is a *switching surface.* Thus the concept of a switching surface which has been developed for relay control systems with transient inputs (see e.g. Fuller (1960 b)) applies equally well when the input is random. Questions of the realization and interpretation of switching surfaces have been discussed in the previous paper (Fuller (1960 b)).

§ 8. Error Coordinates

In the previous paper (Fuller. 1960 b) it was shown that, when the input is transient, it is theoretically advantageous to replace some of the input coordinates by error coordinates, if both the 'input generating system' and the main process include pure integrators. We shall now show briefly that the argument still applies when the input is random.

Suppose the control system input $X \equiv X_1$ is generated by integrating r times a random signal X_{r+1} which itself has phase coordinates $X_{r+1}, X_{r+2}, \dots X_m$. Suppose also that the main process output $Y \equiv Y_1$

is obtained by integrating r times a signal Y_{r+1} which itself has phase coordinates $Y_{r+1}, Y_{r+2}, \ldots Y_n$. Define $X_2, X_3, \ldots X_r$ and $Y_2, Y_3, \ldots Y_r$ by the relations

$$\frac{dX_i}{dt} = X_{i+1}, \quad (i=1, 2, \ldots r) \qquad (15)$$

and

$$\frac{dY_i}{dt} = Y_{i+1}, \quad (i=1, 2, \ldots r). \qquad (16)$$

Define $E_1, E_2, \ldots E_r$ by

$$E_i \equiv X_i - Y_i, \quad (i=1, 2, \ldots r). \qquad (17)$$

Then

$$E_1 = E, \ E_2 = E', \ \ldots \ E_r = E^{(r-1)}. \qquad (18)$$

Subtract (16) from (15) and substitute from (17). The result is

$$\frac{dE_i}{dt} = E_{i+1}, \quad (i=1, 2, \ldots r-1) \qquad (19)$$

and

$$\frac{dE_r}{dt} = X_{r+1} - Y_{r+1}. \qquad (20)$$

Equations (19) and (20) are in canonical form (Fuller 1960 a) with forcing function $X_{r+1} - Y_{r+1}$, and therefore have phase coordinates $E_1, E_2, \ldots E_r$ together with the phase coordinates of $X_{r+1}(t)$ and $Y_{r+1}(t)$. We can therefore adopt as phase coordinates for the determination of E_1, i.e. of E:

$$E_1, E_2, \ldots E_r, X_{r+1}, X_{r+2}, \ldots X_m, Y_{r+1}, Y_{r+2}, \ldots Y_n, \qquad (21)$$

which may be written, in view of eqns. (18),

$$E, E', \ldots E^{(r-1)}, X_{r+1}, X_{r+2}, \ldots X_m, Y_{r+1}, Y_{r+2}, \ldots Y_n. \qquad (22)$$

Application of the argument of § 6 now shows that the optimum controller must generate an instantaneous function of coordinates (22). These coordinates are $(m+n-r)$ in number, whereas the combined input and output coordinates are $(m+n)$ in number. Thus the use of error coordinates results in a simpler controller, when the input generator and the main process each include pure integrators.

§ 9. Examples

9.1. *Poisson–Markov Inputs*

One of the simplest of the piece-wise analytic random signals described in § 3 is the *Poisson–Markov* step process, in which the points of discontinuity have a Poisson distribution along the time-axis, and $x(t)$ is constant between discontinuities, the value of the constant being a random variable which depends statistically only on the value in the previous continuous interval. This process could be generated by a Poisson purely random signal, the impulses of which trigger a sampling device which samples a signal compounded of the present values of $x(t)$ and another purely random signal. Clearly the only phase coordinate of $x(t)$ is x itself,

Thus if this signal is the input $X(t)$ to the control system, the optimum controller must generate simply an instantaneous function of the form $C(X, Y_1, Y_2, \ldots Y_n)$.

9.2. *Random Telegraph Input*

As a particular case of the inputs of § 9.1, let the input $X(t)$ be a *random telegraph signal* (Rice 1944, 1945), in which $X(t)$ takes alternately the values $+\sigma$ and $-\sigma$ the points of changeover having a Poisson distribution along the time-axis with mean rate μ per unit time. Let the system be an ideal relay control system with a main process consisting of a single integrator, i.e. having a transfer function $1/a_1 p$. Then the optimum controller is represented by a switching surface in (X, Y) phase space:

$$C\,(X, Y) = 0. \qquad (23)$$

Further, starting from result (23) it is possible to show (Fuller 1959 a) that the optimum switching function, if the criterion of performance is mean-square error, is

$$C(X, Y) = \frac{(1+6\mu a_1 \sigma)^{1/3} - 1}{2\mu a_1 \sigma} X - Y. \qquad (24)$$

9.3. *Shot-effect Inputs*

If the input to a linear filter is a Poisson purely random signal, the output $x(t)$ is said to be a shot-effect. Let us consider the case when the filter has a transfer function

$$\frac{1}{c_0 + c_1 p + \ldots + c_m p^m}. \qquad (25)$$

The phase coordinates of this filter are (Fuller 1960 a) $x(t), x'(t), \ldots x^{(m-1)}(t)$. Hence if $x(t)$ is made the input $X(t)$ to the control system, the optimum controller generates a function of the form

$$C(X, X', \ldots X^{(m-1)}, Y_1, Y_2, \ldots Y_n).$$

9.4. *Gaussian Inputs*

Consider a Gaussian signal $x(t)$ with a known spectral density of the form

$$\Phi(\omega) = \frac{1}{\alpha_0 + \alpha_1 \omega^2 + \ldots + \alpha_m \omega^{2m}}. \qquad (26)$$

This signal may be regarded as generated by passing a Gaussian purely random signal through a stable filter with a transfer function

$$H(p) = \frac{1}{c_0 + c_1 p + \ldots + c_m p^m}, \qquad (27)$$

where the coefficients are determined, as is well known, by the relation

$$|H(j\omega)|^2 = \Phi(\omega). \qquad (28)$$

The phase coordinates of filter (27) are, as mentioned in § 9.3, $x(t)$, $x'(t)$, $\ldots x^{(m-1)}(t)$. Hence if the control system input $X(t)$ is Gaussian with a spectral density of form (26), the optimum controller generates a function of the form $C(X, X', \ldots X^{(m-1)}, Y_1, Y_2, \ldots Y_n)$.

Next consider the more general case of a Gaussian signal $x(t)$ with a known spectral density of the form

$$\Phi(\omega) = \frac{\beta_0 + \beta_1\omega^2 + \ldots + \beta_k\omega^{2k}}{\alpha_0 + \alpha_1\omega^2 + \ldots + \alpha_m\omega^{2m}}, \quad (k < m). \qquad (29)$$

This signal may be regarded as generated by passing a Gaussian purely random signal through a stable minimum-phase† filter with a transfer function

$$H(p) = \frac{d_0 + d_1 p + \ldots + d_k p^k}{c_0 + c_1 p + \ldots + c_m p^m}, \qquad (30)$$

where the coefficients are determined by the relation

$$|H(j\omega)|^2 = \Phi(\omega). \qquad (31)$$

It has been shown (Fuller 1960 a) that the phase coordinates

$$z(t), z'(t), \ldots z^{(m-1)}(t)$$

of filter (30) are represented by

$$\frac{x}{d_0 + d_1 D + \ldots + d_k D^k}, \frac{Dx}{d_0 + d_1 D + \ldots + d_k D^k}, \ldots \frac{D^{m-1}x}{d_0 + d_1 D + \ldots + d_k D^k}. \qquad (32)$$

This symbolism means the following. The first phase coordinate, z, is a particular solution of the differential equation

$$(d_0 + d_1 D + \ldots + d_k D^k)z = x. \qquad (33)$$

The other phase coordinates are simply derivatives of z.

Other particular solutions of eqn. (33), obtained by inserting different initial conditions, become equal to $z(t)$ when the initial transient has had time to die away. Thus we can copy $z(t)$ by feeding $x(t)$ into a filter with a transfer function

$$\frac{1}{d_0 + d_1 p + \ldots + d_k p^k}. \qquad (34)$$

The output of this filter is termed a *copy coordinate*, z_c, (Fuller 1960 a). The complete set of copy coordinates‡ is $z_c, z_c', \ldots z_c^{(m-1)}$. It follows that the optimum controller, which is represented by the function

$$C(z, z', \ldots z^{(m-1)}, Y_1, Y_2, \ldots Y_n), \qquad (35)$$

can also be written

$$C(z_c, z_c', \ldots z_c^{(m-1)}, Y_1, Y_2, \ldots Y_n), \qquad (36)$$

---

† i.e. such that the zeros of the transfer function are all in the left half plane.

‡ It is interesting to note that the optimum prediction of a future value of x (in Wiener's sense) can be shown to be a linear combination of these copy coordinates (Fuller 1959 a).

or symbolically:

$$C\left(\frac{X}{d_0+d_1p+\ldots+d_kp^k}, \frac{pX}{d_0+d_1p+\ldots+d_kp^k}, \right.$$
$$\left.\ldots\frac{p^{m-1}X}{d_0+d_1p+\ldots+d_kp^k}, Y_1, Y_2, \ldots Y_n\right). \quad \ldots \quad (37)$$

Usually the coordinates z, z', $\ldots z^{(m-1)}$ are not directly available for measurement, and then it is appropriate to adopt controller (37). The block diagram of the complete system with this controller is shown in fig. 2.

Fig. 2

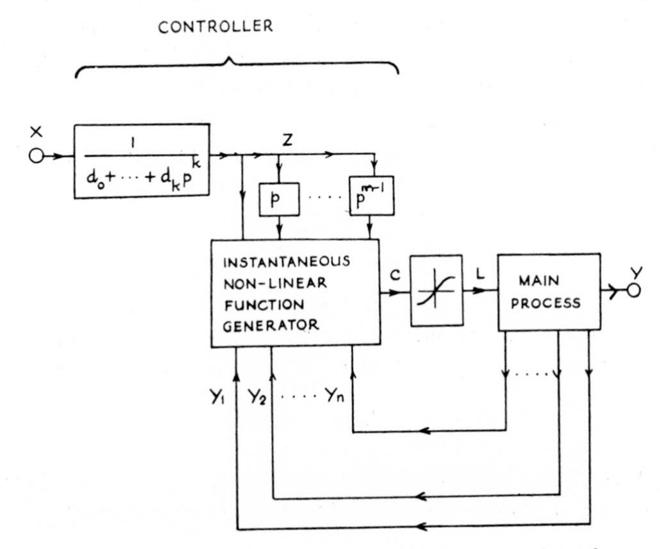

Saturating control system with a Gaussian input X and with an optimum controller.

9.5. *Brownian Motion Input*

As a particular case of the inputs of § 9.4, let the input $X(t)$ be Brownian motion, generated by feeding a Gaussian purely random signal into a filter having a transfer function

$$H(p)=\frac{1}{c_1p}. \quad \ldots\ldots \quad (38)$$

Let the main process have a transfer function $1/a_1p$. Then the results of § 9.4 show that the optimum controller generates a signal of the form $C(X, Y)$. However, using the error coordinates of § 8, this form can be simplified to $C(E)$.

Furthermore, starting from the last result, it can be shown (Fuller 1959 a) that for optimum performance, relay operation of the saturation characteristic is called for, and that the optimum switching function is simply

$$C(E)=E \quad \ldots\ldots \quad (39)$$

for any performance criterion of type (4).

9.6. *Noisy Gaussian Inputs*

When a process $X(t)$ consists of signal $X_S(t)$ and noise $X_N(t)$

$$X = X_S + X_N, \qquad (40)$$

in practice the phase coordinates of X_S cannot be measured. Nevertheless it may be possible to show that $X(t)$ is statistically equivalent to a process of which the copy coordinates can be measured. Since X is the only available input, it is plausible that the optimum controller† must then generate an instantaneous non-linear function of these copy coordinates and of the main process phase coordinates. We shall show how the relevant copy coordinates may be obtained when both signal and noise are Gaussian; for more general cases this problem is unsolved.

Since X is the sum of two Gaussian processes, it is itself Gaussian, and is completely characterized by its spectral density $\Phi_{XX}(\omega)$. It is statistically equivalent to any other Gaussian process having the same spectral density. In particular X is equivalent to a process generated by smoothing a Gaussian purely random signal with a filter having a transfer function $H(p)$ where

$$|H(j\omega)|^2 = \Phi_{XX}(\omega) \qquad (42)$$

$$= \Phi_{SS}(\omega) + \Phi_{SN}(\omega) + \Phi_{NS}(\omega) + \Phi_{NN}(\omega), \qquad (43)$$

where Φ_{SS}, Φ_{NN}, Φ_{NS}, Φ_{SN} are the spectral densities and cross spectral densities of X_S and X_N. Assuming that all the Φ's are rational functions of ω^2, we can solve eqn. (43) to yield $H(p)$ in stable minimum-phase form:

$$H(p) = \frac{d_0 + d_1 p + \ldots + d_k p^k}{c_0 + c_1 p + \ldots + c_m p^m}. \qquad (44)$$

As discussed in § 9.4, the copy coordinates for this filter are obtained by passing its output through filters having transfer functions‡

$$\frac{1}{d_0 + d_1 p + \ldots + d_k p^k}, \quad \frac{p}{d_0 + d_1 p + \ldots + d_k p^k}, \quad \ldots \quad \frac{p^{m-1}}{d_0 + d_1 p + \ldots + d_k p^k}. \qquad (45)$$

Hence the optimum controller is represented symbolically by the function

$$C\left(\frac{X}{d_0 + d_1 p + \ldots + d_k p^k}, \quad \frac{pX}{d_0 + d_1 p + \ldots + d_k p^k}, \right.$$

$$\left. \ldots \frac{p^{m-1}X}{d_0 + d_1 p + \ldots + d_k p^k}, \quad Y_1, Y_2, \ldots Y_n\right). \qquad (46)$$

The block diagram for the optimum control system is the same in form as in fig. 2 for the non-noisy Gaussian input (though of course the d's, k's, m's and the instantaneous function generators are different in the two cases).

---

† In § 9.6 the criterion of performance is no longer $\overline{q(E)}$ but is $\overline{q(E_S)}$ where

$$E_S \equiv X_S - Y. \qquad (41)$$

‡ It is interesting to note that the optimum Wiener filter for the signal X_S can be expressed as a linear combination of transfer functions (45) (Fuller 1959 a).

§ 10. Practical Techniques

We have shown that the optimum controller is an instantaneous non-linear function of the phase coordinates. The problem reduces to finding this function. If the number of coordinates is not large (say < 4), it may be feasible to find the optimum function empirically on an analogue computer. However, if the number of coordinates is large, or if great accuracy is required, it is appropriate to find the function by self-optimizing techniques. One such technique is as follows.

Let us simplify the problem by assuming that the system is a relay control system, so that we have to find a switching surface in the phase space. The phase space is quantized, i.e. divided into cells, as shown for instance in fig. 3 for a two-dimensional phase space. To each cell is assigned a polarity, either $+$ or $-$. The boundary between $+$ cells and $-$ cells

Fig. 3

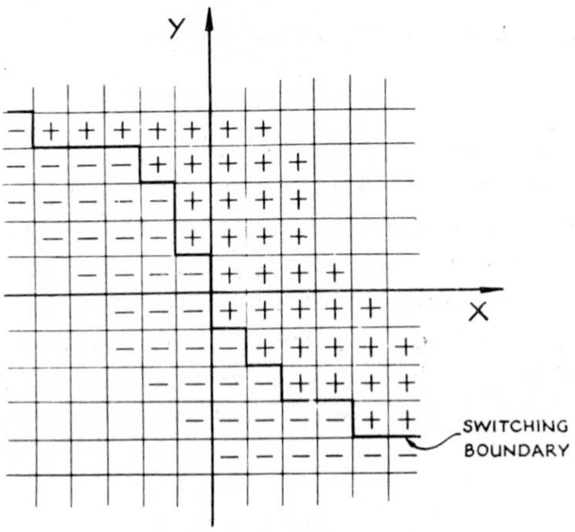

Quantized phase space.

is the switching boundary. The polarities of all the cells are stored in a digital 'memory' (e.g. a magnetic matrix of ferrite cores), which can be quite simple: each element of the memory is a two-state device which corresponds to one of the cells, and its state depends on whether the cell is $+$ or $-$. Throughout operation, the controller measures the phase state of the control system input and output, reads the sign of the corresponding memory cell, and orders the relay to switch positively if the cell is $+$, and to switch negatively if the cell is $-$. In self-optimization, the boundary between $+$ cells and $-$ cells is altered one cell at a time and the effect on system performance is then measured. If there is an improvement, the switching boundary is further changed in the same direction. If there is a deterioration the boundary is changed in the reverse direction. If there is no change in performance, the boundary is changed at a different point and the cycle repeated. This process can be expected to lead to the optimum in most cases.

If mathematical or numerical formulae for the optimum function are known, the function can be synthesized by means of a digital computer, or more cheaply by means of a multi-pole diode–resistor network. A synthesis procedure for the latter networks has been given by Stern (1956).

§ 11. Generalizations

The results in §§ 6 and 7 can be readily generalized in various directions.

If there are several controllable inputs L_i to the main process, each will require a separate controller, and the principles of § 6 will apply to each controller output C_i. For such a *multi-loop* system, when optimum, each C_i will be an instantaneous function of the phase coordinates of X and the phase coordinates of the main process.

If there are several unwanted random disturbances, i.e. *load variations*, acting on the main process, each C_i will be an instantaneous function of the phase coordinates of X, the phase coordinates of the main process, and the phase coordinates of each disturbance.

§ 12. Conclusions

We have seen that the optimum controller simply generates an instantaneous function of the phase coordinates even when the input is random. This result enables the optimum non-linear controller to be completely calculated in very simple cases (see §§ 9.2 and 9.5), or to be found by a straightforward self-optimizing technique. When the optimum performances of a range of typical control systems have been found by these techniques, it will be possible to assess the validity of the various known approximate methods of optimization.

APPENDIX

Nature of the Optimum Controller

In this appendix we show that the optimum controller generates simply an instantaneous function of the phase coordinates. The argument given here is more precise than that given in § 6, however complete rigour is not attempted.

To simplify the notation, let us denote by $\mathbf{V}(t)$ the vector with components which are the phase coordinates $X_1(t), X_2(t), \ldots X_m(t), Y_1(t), Y_2(t), \ldots Y_n(t)$. Also let us denote by $\boldsymbol{\psi}(t)$ the vector with components which are the purely random signals which generate $X(t)$. Then the definition of the phase coordinates implies that, at time t,

$$q = U\left[\mathbf{V}(t_0), (t-t_0), \begin{pmatrix}\boldsymbol{\psi}(u) \text{ for all} \\ t_0 \leqslant u < t\end{pmatrix}, \begin{pmatrix}L(u) \text{ for all} \\ t_0 \leqslant u < t\end{pmatrix}\right], \quad (t \geqslant t_0), \quad . \quad . \quad (47)$$

i.e. q is some unique functional of the pasts of $\boldsymbol{\psi}$ and L, which functional depends only on the arguments written explicitly in (47) (and on known constants).

Suppose

$$L(t) = L_{\text{opt}}(t) \quad \text{for all} \quad t < t_0. \qquad (48)$$

Then the problem reduces to minimizing

$$Q_0 \equiv \mathscr{E}\left\{ \lim_{T\to\infty} \frac{1}{2T} \int_{t_0}^{T} U\, dt \right\} \qquad (49)$$

where $\mathscr{E}\{\ \}$ represents the expectation operator. We assume that it is permissible to change the order of operations. Then

$$Q_0 = \lim_{T\to\infty} \frac{1}{2T} \int_{t_0}^{T} \mathscr{E}\{U\}\, dt. \qquad (50)$$

Thus at time $t = t_0$ the problem is to choose $L(t_0)$ so that

$$Q_0 = \lim_{T\to\infty} \frac{1}{2T} \int_{t_0}^{T} \mathscr{E}\left\{ U \left[\mathbf{V}(t_0), (t-t_0), \begin{pmatrix} \psi(u) \text{ for all} \\ t_0 \leqslant u < t \end{pmatrix}, \begin{pmatrix} L(u) \text{ for all} \\ t_0 \leqslant u < t \end{pmatrix} \right] \right\} dt \quad (51)$$

is minimized with respect to $L(t_0)$. At time $t = t_0$ the only quantities appearing in expression (51) which are known† are $\mathbf{V}(t_0)$ and t_0. Therefore $L_{\text{opt}}(t_0)$ must be chosen on the basis of these data, i.e.

$$L_{\text{opt}}(t_0) = F[\mathbf{V}(t_0), t_0]. \qquad (52)$$

We shall show that the second argument in the right side of (52) is redundant.

Similarly at time $t = t_1$ the problem is to choose $L(t_1)$ so that

$$Q_1 = \lim_{T\to\infty} \frac{1}{2T} \int_{t_1}^{T} \mathscr{E}\left\{ U \left[\mathbf{V}(t_1), (t-t_1), \begin{pmatrix} \psi(u) \text{ for all} \\ t_1 \leqslant u < t \end{pmatrix}, \begin{pmatrix} L(u) \text{ for all} \\ t_1 \leqslant u < t \end{pmatrix} \right] \right\} dt \quad (53)$$

is minimized, yielding

$$L_{\text{opt}}(t_1) = F[\mathbf{V}(t_1), t_1]. \qquad (54)$$

In expression (53) put

$$\tau \equiv t - t_1 + t_0, \qquad (55)$$

$$L\{u\} \equiv L(u + t_1 - t_0), \qquad (56)$$

$$\psi\{u\} \equiv \psi(u + t_1 - t_0), \qquad (57)$$

$$S \equiv T - t_1 + t_0. \qquad (58)$$

Then

$$Q_1 = \lim_{S\to\infty} \frac{2S}{2(S + t_1 - t_0)} \cdot \frac{1}{2S} \int_{t_0}^{S} \mathscr{E}\left\{ U \left[\mathbf{V}(t_1), (\tau - t_0), \begin{pmatrix} \psi(u) \text{ for all} \\ t_1 \leqslant u < \tau + t_1 - t_0 \end{pmatrix}, \begin{pmatrix} L(u) \text{ for all} \\ t_1 \leqslant u < \tau + t_1 - t_0 \end{pmatrix} \right] \right\} d\tau \quad (59)$$

$$= \lim_{S\to\infty} \frac{1}{2S} \int_{t_0}^{S} \mathscr{E}\left\{ U \left[\mathbf{V}(t_1), (\tau - t_0), \begin{pmatrix} \psi\{u\} \text{ for all} \\ t_0 \leqslant u < \tau \end{pmatrix}, \begin{pmatrix} L\{u\} \text{ for all} \\ t_0 \leqslant u < \tau \end{pmatrix} \right] \right\} d\tau. \quad (60)$$

---

† We assume that $\psi(t_0)$ cannot be measured at $t = t_0$, because $\psi(t_0)$ is impulsive in character (see § 4) and can therefore only be measured by its *subsequent* effect on X.

Now $\psi\{u\}$ is an independent random variable with a probability distribution which is independent of u, by definition. Therefore expression (60) is unchanged if $\psi\{u\}$ is replaced by $\psi\{u+k\}$ where k is any constant. Let us choose $k=t_0-t_1$. Then

$$Q_1=\lim_{S\to\infty}\frac{1}{2S}\int_{t_0}^{S}\mathscr{E}\left\{U\left[\mathbf{V}(t_1),\ (\tau-t_0),\ \begin{pmatrix}\psi(u)\text{ for all}\\ t_0\leqslant u<\tau\end{pmatrix},\ \begin{pmatrix}L\{u\}\text{ for all}\\ t_0\leqslant u<\tau\end{pmatrix}\right]\right\}d\tau. \quad . \quad . \quad . \quad (61)$$

The problem now is to choose $L\{t_0\}$ so that this expression is minimized. Comparison of expressions (51) and (61) shows that the problem is the same in the two cases, except for trivial changes of notation. Hence the solution is the same as in eqn. (52) with these changes of notation, i.e.

$$L_{\text{opt}}\{t_0\}=F[\mathbf{V}(t_1),\ t_0]. \quad . \quad . \quad . \quad . \quad . \quad (62)$$

From eqns. (54), (56) and (62)

$$F[\mathbf{V}(t_1),\ t_1]=F[\mathbf{V}(t_1),\ t_0]. \quad . \quad . \quad . \quad . \quad . \quad (63)$$

Since t_0 and t_1 are arbitrary, eqn. (63) shows that the second argument of the function F is redundant:

$$F[\mathbf{V}(t_1),\ t_1]=G[\mathbf{V}(t_1)]. \quad . \quad . \quad . \quad . \quad . \quad (64)$$

From eqns. (54) and (64)

$$L_{\text{opt}}(t_1)=G[\mathbf{V}(t_1)]. \quad . \quad . \quad . \quad . \quad . \quad . \quad (65)$$

Since t_1 is arbitrary, we can drop the suffix and write

$$L_{\text{opt}}(t)=G[\mathbf{V}(t)], \quad . \quad . \quad . \quad . \quad . \quad . \quad (66)$$

or, more briefly

$$L_{\text{opt}}=G[\mathbf{V}], \quad . \quad . \quad . \quad . \quad . \quad . \quad . \quad (67)$$

i.e.

$$L_{\text{opt}}=G[X_1,\ X_2,\ \ldots X_m,\ Y_1,\ Y_2,\ \ldots Y_n], \quad . \quad . \quad . \quad (68)$$

which is the required result.

References

Blanc-Lapierre, A., and Fortet, R., 1953, *Théorie des Fonctions Aléatoires* (Paris: Masson).

Bode, H. W., and Shannon, C. E., 1950, *Proc. Inst. Radio Engrs, N.Y.*, **38,** 417.

Feller, W., 1950, *An Introduction to Probability Theory and its Applications* (New York: Wiley).

Fuller, A. T., 1959 a, Ph.D. Thesis, Cambridge University; 1959 b, *J. Electron. Contr.*, **7**, 456; 1960 a, *J. Electron. Contr.*, **8**, 381; 1960 b, *Ibid.*, **8**, 5.

Rice, S. O., 1944, *Bell Syst. Tech. J.*, **23,** 282; 1945, *Ibid.*, **24,** 46.

Stern, T. E., 1956, *Proc. Symp. Nonlin. Circuit Analysis* (Polytechnic Institute of Brooklyn), p. 315.

Wax, N., 1954, *Selected Papers on Noise and Stochastic Processes* (New York: Dover Publications).

Wiener, N., 1949, *Extrapolation, Interpolation and Smoothing of Stationary Time Series* (New York: Wiley).

Optimization of a Non-linear Control System with a Random Telegraph Signal Input†

By A. T. Fuller
Department of Engineering, Cambridge University

[Received November 8, 1960]

ABSTRACT

Controllers which minimize various performance criteria, such as mean-square-error, are calculated for a first-order relay control system with a random square wave input. This input is the ' random telegraph signal '.

§ 1. Introduction

In previous papers (Fuller 1959 b, 1960 a, b, c, d) the writer has discussed the general nature of optimum controllers for relay control systems (and saturating systems) when the input is random. In the present paper a particular example, for which exact results are obtainable, will be evaluated. Apart from their intrinsic interest, these results are useful for testing the accuracy of other more general but only approximate methods of optimization.

Fig. 1

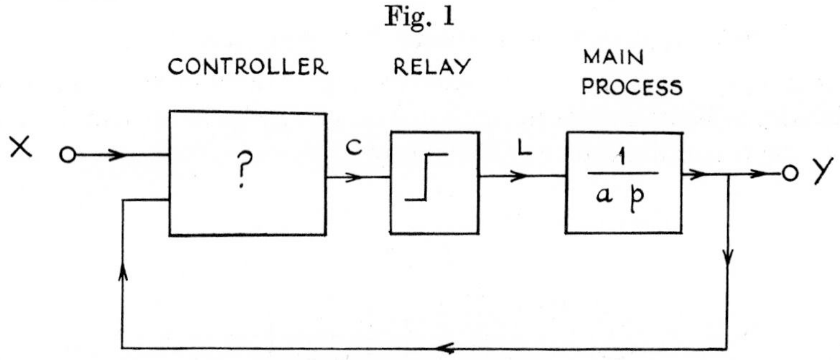

Relay control system.

The system to be considered is the ideal relay control system shown in fig. 1. The main process (i.e. the plant) is a simple integrator, having a transfer function $1/ap$, $(a>0)$. The relay is assumed perfect, giving a relay output L where

$$L = \operatorname{sgn} C, \qquad (1)$$

C being the input to the relay. The controller may perform any physically realizable operation on the input and on the main process and is to be designed so as to minimize the time-average of $|E|^{\nu}$. Here ν is a non-negative number, and E is the error, which is defined by

$$E = X - Y, \qquad (2)$$

† Communicated by J. F. Coales.

X being the system input and Y the system output. The performance criteria $\overline{|E|^{\nu}}$ have been discussed in a previous paper (Fuller 1959 b).

The input is the random square wave shown in fig. 2, and is the 'random telegraph signal' (Kenrick 1929, Rice 1944). $X(t)$ takes alternately the values $+\sigma$ and $-\sigma$, the points of changeover having a Poisson distribution along the t-axis (t being time), with average rate of occurrence μ per unit time.

Fig. 2

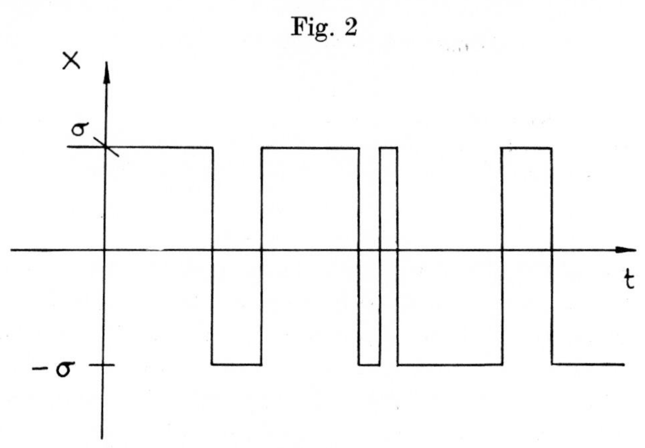

Random square wave.

§ 2. Analysis of a Simple Non-optimum System

It is a useful preliminary to analyse the system with a particularly simple controller as shown in fig. 3. The controller subtracts Y from X, so that the controller output C is simply

$$C=E. \quad \quad (3)$$

Fig. 3

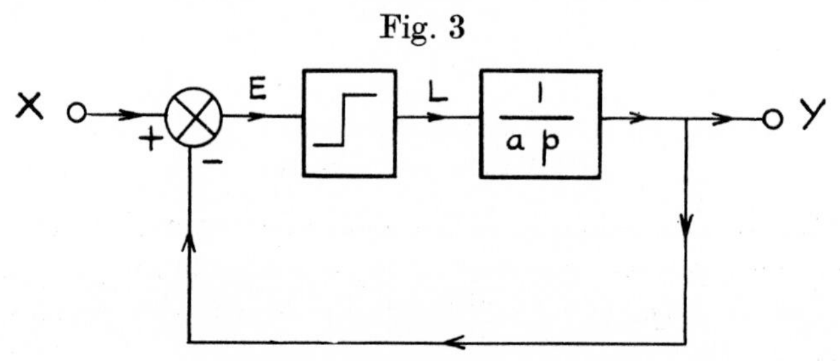

System with a simple controller.

We shall calculate the joint probability density function of the system input and output, and hence find the probability density function of the error. Throughout we shall implicitly invoke the ergodic hypothesis, that is, we shall equate time-averages with ensemble-averages.

Consider an ensemble of inputs, each with its own control system. At time t let the conditional probability density function of Y, given that $X=+\sigma$ ($\sigma>0$), be $D_{P}\{Y, \sigma\}$; and let the conditional probability density function of Y, given that $X = -\sigma$, be $D_N\{Y, \sigma\}$. Then the joint probability density function of X and Y is $D_J(X, Y)$ where

$$D_J(X, Y) = \tfrac{1}{2}D_P\{Y, \sigma\}\delta(X-\sigma) + \tfrac{1}{2}D_N\{Y, \sigma\}\delta(X+\sigma) \quad . \quad . \quad . \quad (4)$$

where $\delta(X \pm \sigma)$ represents a delta-function at $X = \mp\sigma$.

The time-variation of Y is as illustrated in fig. 4. Y moves towards X, and whenever Y reaches X, Y stays† with X until the next change of X. Thus for a finite fraction of the time Y remains at $+\sigma$ or at $-\sigma$ and elsewhere changes at a rate $\pm 1/a$. It follows that D_P has a delta-function at $Y = +\sigma$ and is finite elsewhere, and that D_N has a delta-function at $Y = -\sigma$ and is finite elsewhere.

Fig. 4

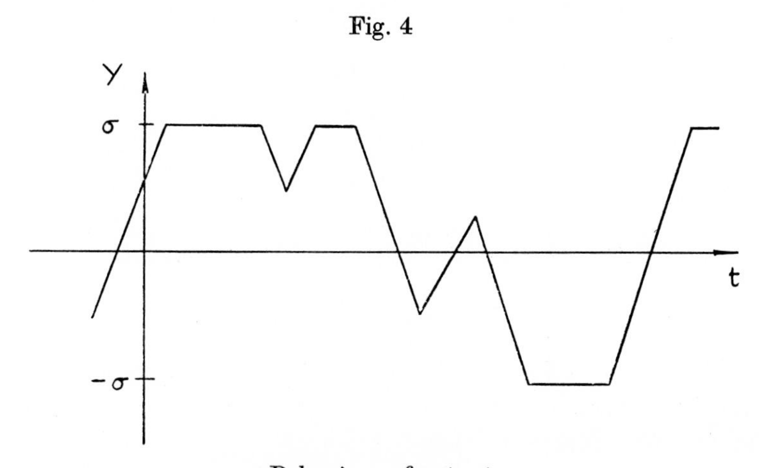

Behaviour of output.

It is shown‡ in Appendix I that the finite parts of D_P and D_N are rectangular (as in fig. 5), and that D_P and D_N are given by

$$D_P = \begin{cases} \dfrac{\mu a + \delta(Y-\sigma)}{1+2\mu a\sigma}, & |Y| \leqslant \sigma \\ 0, & |Y| > \sigma \end{cases} \quad . \quad . \quad . \quad . \quad . \quad (5)$$

and

$$D_N = \begin{cases} \dfrac{\mu a + \delta(Y+\sigma)}{1+2\mu a\sigma}, & |Y| \leqslant \sigma, \\ 0, & |Y| > \sigma. \end{cases} \quad . \quad . \quad . \quad . \quad . \quad (6)$$

---

† To see this, insert a small pure delay in the main process, and then let the delay become infinitesimal. Alternatively, insert an infinitesimal dead-zone in the relay.

‡ The method of Appendix I is first to guess D_P and D_N and then to show that an ensemble starting with the guessed distributions remains stationary as time increases. A more satisfactory and more general method is to set up and solve partial differential equations for D_P and D_N. The writer's colleague, W. M. Wonham, has proposed and is currently investigating the latter method.

Fig. 5

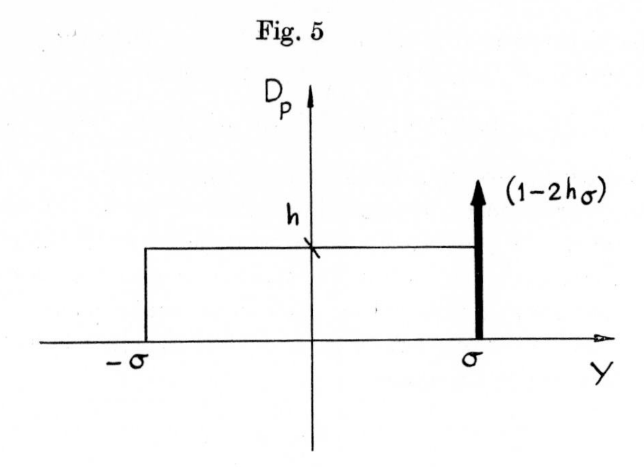

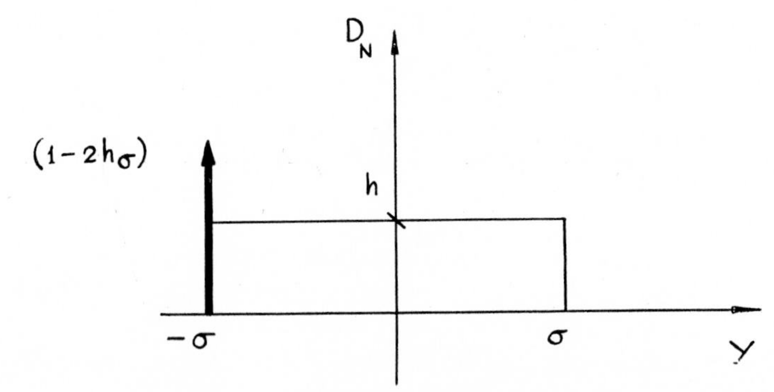

Form of D_P and D_N. Note: ↑ (A) represents a delta-function of magnitude A.

From eqns. (2), (4), (5) and (6) it follows (see Appendix I) that the probability density function G of the error E is

$$G(E) = \begin{cases} \dfrac{\frac{1}{2}\mu a + \delta(E)}{1+2\mu a\sigma}, & |E| \leqslant 2\sigma, \\ 0, & |E| > 2\sigma. \end{cases} \qquad (7)$$

This density function is drawn in fig. 6.

From eqn. (7) and from the relation

$$\overline{|E|^{\nu}} = \int_{-\infty}^{\infty} |E|^{\nu} G(E)\, dE. \qquad (8)$$

it follows that the performance criterion is given by

$$\overline{|E|^{\nu}} = \frac{2^{\nu+1}\mu a\sigma^{\nu+1}}{(\nu+1)(1+2\mu a\sigma)}. \qquad (9)$$

In particular, the mean-square-error satisfies

$$\frac{\overline{E^2}}{\sigma^2} = \frac{8\mu a\sigma}{3(1+2\mu a\sigma)}; \qquad (10)$$

Fig. 6

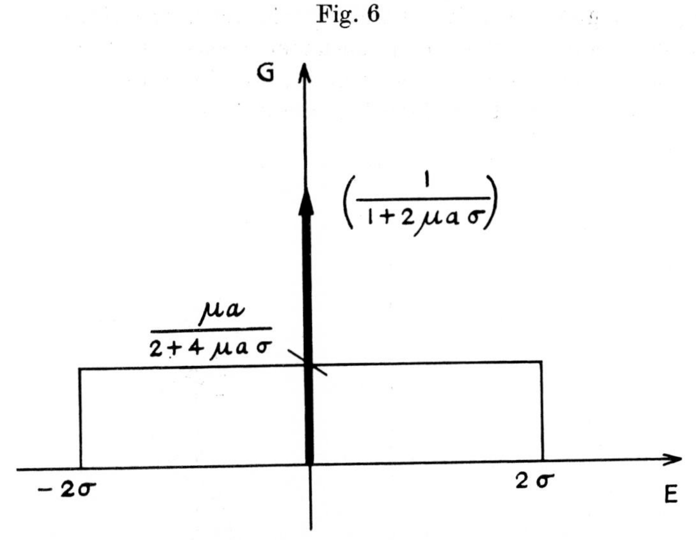

Probability density function of error for the system of fig. 3.

and the fractional time during which the error is non-zero, which is $\overline{|E|^0}$ (see Fuller 1959 b), is

$$\overline{|E|^0} = \frac{2\mu a\sigma}{1+2\mu a\sigma}. \qquad (11)$$

§ 3. Nature of the Optimum Controller

We now turn from the special system of § 2 to a consideration of the optimum system. It follows from general principles (Fuller 1960 c) that the optimum controller for the system described in § 1 and for a given performance criterion is represented by some switching curve† in the (X, Y) phase-plane. Because the input is limited to the two values $\pm\sigma$, the representative point travels only along the vertical lines $X = \pm\sigma$. (However, the trajectories are easier to visualize if we assume that the input changes are not quite perfectly sudden. Then when the input changes, the representative point may be considered as travelling horizontally from one value of X to the other.) The switching curve divides these lines into segments of positive switching and segments of negative switching, as indicated in fig. 7. Thus the controller is characterized by the points of intersection of the switching curve with the lines $X = \pm\sigma$. We call these points *I points*, and it will turn out that only two I points need be taken into account.

If the segment of the line $X = \pm\sigma$ just above a given I point gives positive switching, then when near the I point (and on $X = \pm\sigma$), the representative point moves away from it; and we call this I point a *point of repulsion.*

† On one side of a switching curve the controller output is positive, thus making $L = +1$, and on the other side the controller output is negative, thus making $L = -1$.

If the segment gives negative switching, the representative point moves towards the I point, and we call the latter a *point of attraction*. The I points consist of points of repulsion and attraction alternating along the line $X=+\sigma$ and alternating along the line $X=-\sigma$.

Fig. 7

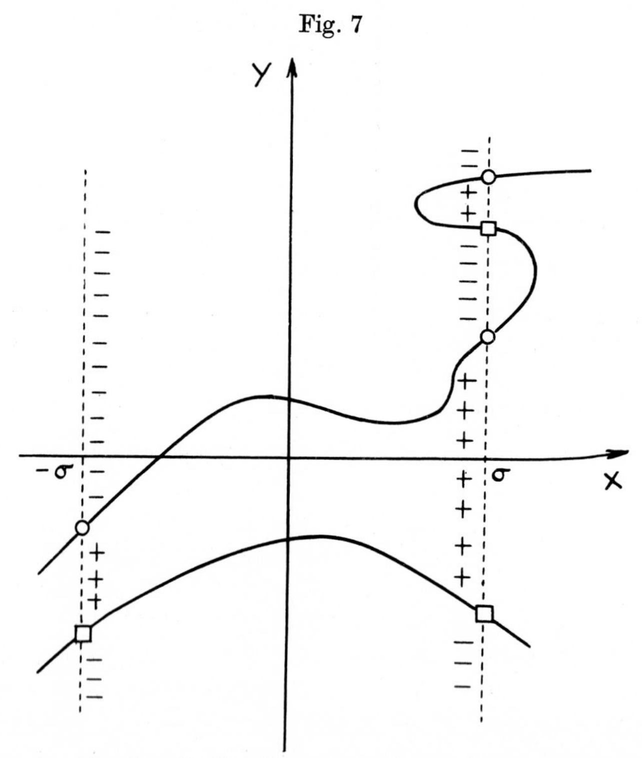

Switching curve in the (X, Y) plane. (○ points of attraction; □ points of repulsion.)

It is shown in Appendix II that, for an optimum system, it is sufficient for the controller to have only two I points, namely a point of attraction on the line $X=+\sigma$ and a point of attraction on the line $X=-\sigma$. We designate these points as (σ, α) and $(-\sigma, \beta)$ and the problem reduces to finding the best values of the ordinates α and β. Any switching curve is then optimum which cuts $X=\pm\sigma$ only in two points of attraction which coincide with (σ, α_{opt}) and $(-\sigma, \beta_{opt})$.

§ 4. Probability Density Function of the Error

As a step towards finding the best values of α and β we first find the expression for the probability density function of the error E when the controller has general values of α and β. This can be done readily by adapting the results of § 2 for the simple non-optimum system (i.e. for the case $\alpha=\sigma$, $\beta=-\sigma$), as in Appendix III. The latter appendix shows that the probability density function Q of error E is given by

$$Q(E)=\left\{\begin{array}{c}\dfrac{\mu a+\delta(E-\sigma+\alpha)+\delta(E+\sigma+\beta)}{2[1+\mu a(\alpha-\beta)]} \\ \text{in } \sigma-\alpha\leqslant E\leqslant\sigma-\beta \\ \text{and } -\sigma-\alpha\leqslant E\leqslant -\sigma-\beta \\ 0,\text{ elsewhere,}\end{array}\right\} \quad . \quad . \quad . \quad (12)$$

provided $$\alpha>\beta. \quad . \quad . \quad . \quad . \quad . \quad . \quad . \quad . \quad . \quad . \quad (13)$$

Appendix III also shows that inequality (13) must hold when the system is optimum (we are taking $\sigma>0$). Density function (12) is drawn in fig. 8.

Fig. 8

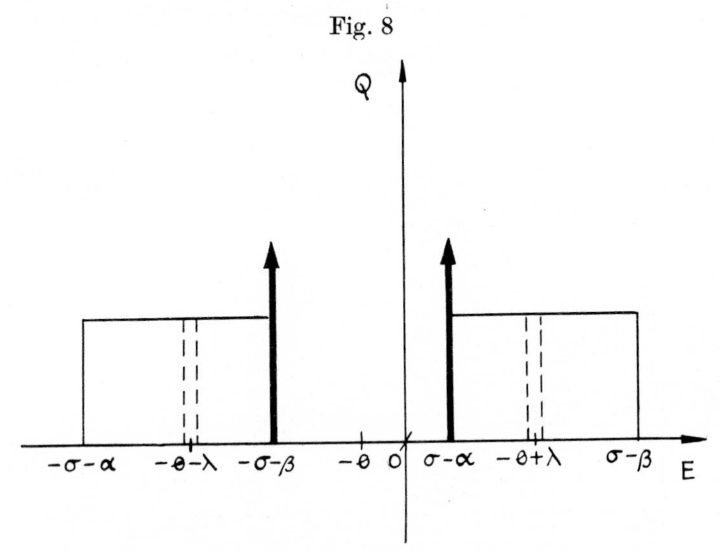

Probability density function of error for a general system.

§ 5. Minimization of $\overline{|E|^{\nu}}$

The minimization with respect to α and β of the performance criteria $\overline{|E|^{\nu}}$, when E has the probability density function (12), is moderately straightforward and is carried out in Appendix IV. The results are as follows, for various values of ν. These results are expressed in terms of the dimensionless parameter J, where

$$J\equiv 2\mu a\sigma. \quad . \quad . \quad . \quad . \quad . \quad . \quad . \quad . \quad . \quad (14)$$

$\underline{\nu=0}$

If $J\leqslant 1$,

$$\left.\begin{array}{r}\alpha_{\text{opt}}=-\beta_{\text{opt}}=\sigma, \\ \overline{|E|^{0}}_{\min}=\dfrac{J}{1+J}.\end{array}\right\} \quad . \quad . \quad . \quad . \quad . \quad . \quad (15)$$

If $J \geqslant 1$,

$$\left.\begin{aligned} \alpha_{\text{opt}} &= \beta_{\text{opt}} = -\sigma, \\ \overline{|E|^0}_{\min} &= \tfrac{1}{2}. \end{aligned}\right\} \qquad (16)$$

$\underline{\nu = 1}$

$$\left.\begin{aligned} \alpha_{\text{opt}} &= -\beta_{\text{opt}} = \sigma, \\ \frac{\overline{|E|}_{\min}}{\sigma} &= \frac{J}{1+J}. \end{aligned}\right\} \qquad (17)$$

$\underline{\nu = 2}$

$$\left.\begin{aligned} \frac{\alpha_{\text{opt}}}{\sigma} &= -\frac{\beta_{\text{opt}}}{\sigma} = \frac{3}{1+(1+3J)^{1/3}+(1+3J)^{2/3}}, \\ \frac{\overline{E^2}_{\min}}{\sigma^2} &= \frac{(1+3J)^{2/3}-1-2J+J^2}{J^2}. \end{aligned}\right\} \qquad (18)$$

$\underline{0 \leqslant \nu \leqslant 1}$

If $J \leqslant 1$,

$$\left.\begin{aligned} \alpha_{\text{opt}} &= -\beta_{\text{opt}} = \sigma, \\ \frac{\overline{|E|^\nu}_{\min}}{\sigma^\nu} &= \frac{2^\nu}{1+\nu} \cdot \frac{J}{1+J}. \end{aligned}\right\} \qquad (19)$$

If $J \geqslant 1$,

$$\beta_{\text{opt}} = -\sigma, \qquad (20)$$

and, with this value of β,

$$\frac{\overline{|E|^\nu}}{\sigma^\nu} = \frac{\left(1-\dfrac{\alpha}{\sigma}\right)^\nu + \dfrac{J}{2(1+\nu)}\left[2^{1+\nu} + \left(1+\dfrac{\alpha}{\sigma}\right)^{1+\nu} - \left(1-\dfrac{\alpha}{\sigma}\right)^{1+\nu}\right]}{2+J\left(1+\dfrac{\alpha}{\sigma}\right)}. \qquad (21)$$

For given values of ν and J, expression (21) can be readily plotted against α/σ over the relevant range, which is shown in Appendix II to be

$$-1 \leqslant \alpha/\sigma \leqslant 1, \qquad (22)$$

and the minimum $\overline{|E|^\nu}$ and corresponding optimum value of α found.

$\underline{\nu \geqslant 1}$

It is optimum to make the controller symmetric, i.e. to put

$$\beta = -\alpha, \qquad (23)$$

and then

$$\frac{\overline{|E|^\nu}}{\sigma^\nu} = \frac{\left(1-\dfrac{\alpha}{\sigma}\right)^\nu + \dfrac{J}{2(\nu+1)}\left[\left(1+\dfrac{\alpha}{\sigma}\right)^{\nu+1} - \left(1-\dfrac{\alpha}{\sigma}\right)^{\nu+1}\right]}{1+J\dfrac{\alpha}{\sigma}}. \qquad (24)$$

For given values of ν and J, expression (24) can be readily plotted against α/σ over the relevant range, which is shown in Appendix IV to be

$$0 \leqslant \alpha/\sigma \leqslant 1, \qquad (25)$$

and the minimum $\overline{|E|^{\nu}}$ and the corresponding optimum value of α found.

§ 6. Switching Curves

As pointed out in § 3, any switching curve is optimum if it cuts the lines $X = \pm\sigma$ only at the points $(\sigma, \alpha_{\text{opt}})$ and $(-\sigma, \beta_{\text{opt}})$, and in such a way that both points are points of attraction. The simplest optimum controller is thus represented by the straight line through the two points, namely

$$\tfrac{1}{2}(\alpha_{\text{opt}} - \beta_{\text{opt}})\sigma^{-1}X - Y + \tfrac{1}{2}(\alpha_{\text{opt}} + \beta_{\text{opt}}) = 0. \qquad (26)$$

If we represent the controller by the relation $C(X, Y)$ between its output C and its inputs X and Y, then the switching curve (26) is realized by the simple linear controller

$$C(X, Y) = \tfrac{1}{2}(\alpha_{\text{opt}} - \beta_{\text{opt}})\sigma^{-1}X - Y + \tfrac{1}{2}(\alpha_{\text{opt}} + \beta_{\text{opt}}) \; ; \qquad (27)$$

e.g. for $\nu = 2$ we obtain from eqns. (18) and (27)

$$C(X, Y) = \left[\frac{3}{1 + (1 + 3J)^{1/3} + (1 + 3J)^{2/3}}\right] X - Y. \qquad (28)$$

However, it is advantageous to take as the switching curve the loci† of the points $(\sigma, \alpha_{\text{opt}})$ and $(-\sigma, \beta_{\text{opt}})$ when σ is varied. The controller is then optimum, not merely for a single value of σ, but for all input amplitudes σ. The switching curves obtained in this way, for $\nu = 0$, 1 and 2, are as follows:

For $\nu = 0$:

$$X \operatorname{sgn}(1 - 2\mu a X) - Y = 0. \qquad (29)$$

For $\nu = 1$:

$$X - Y = 0, \qquad (30)$$

so that in this case the simple linear controller of § 2 is optimum.

For $\nu = 2$:

$$X - Y - 2\mu a Y|Y| - \tfrac{4}{3}\mu^2 a^2 Y^3 = 0. \qquad (31)$$

These switching curves are shown in fig. 9. Notice that near the origin of the phase-plane, the optimum switching curve is either approximately or exactly $X - Y = 0$, which is the optimum switching curve for simple step-inputs (Fuller 1960 d). It follows that when the potential performance of the system is good, the approximation of using a design based on step-inputs is justified (e.g. if the minimized r.m.s. error is less than one third of the r.m.s. input, comparison of eqns. (10) and (18) shows that the step-input design yields an r.m.s. error which is less than 3% greater than the minimum).

---

† These loci are obtained: for $X \geqslant 0$, by writing X for σ and Y for α_{opt} in the equation for α_{opt}; and for $X \leqslant 0$ by writing $-X$ for σ and Y for β_{opt} in the equation for β_{opt}.

Fig. 9

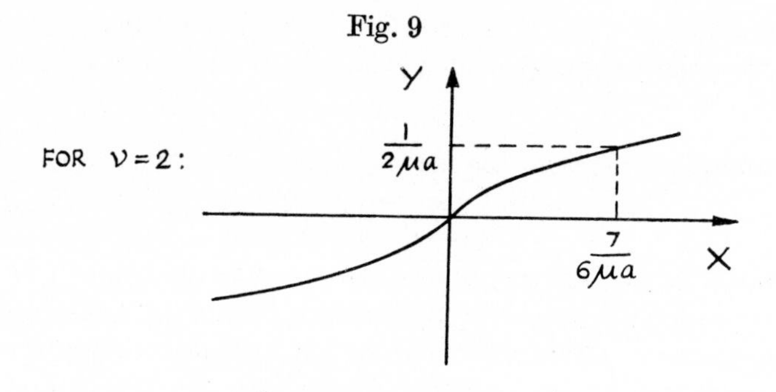

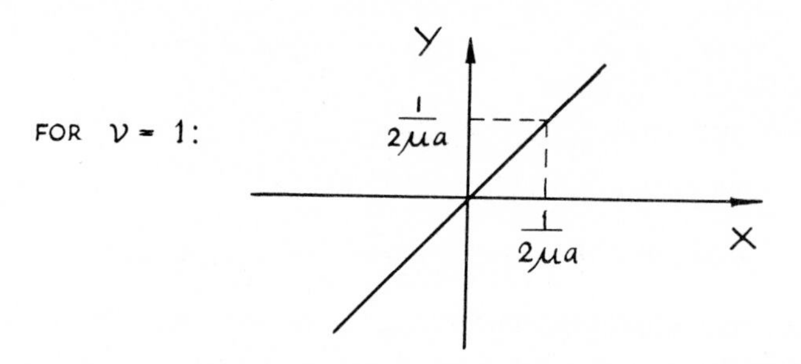

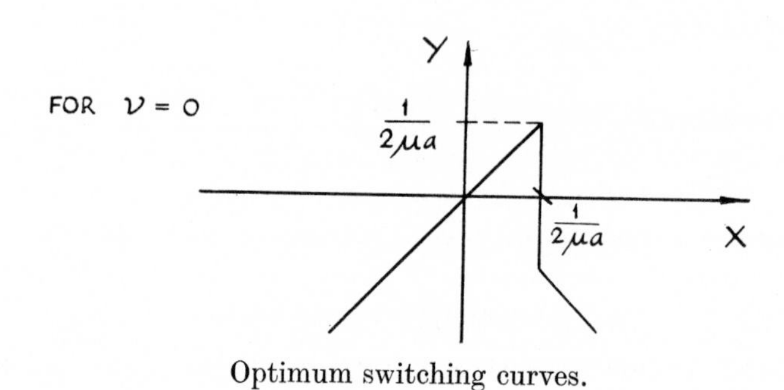

Optimum switching curves.

§ 7. Conclusions

The present paper shows that in a simple case it is possible to calculate the absolutely optimum controller for a nonlinear system with a random input. Another such case is given in the writer's thesis (Fuller 1959 a). This encourages the hope that other simple systems may also be mathematically tractable, and that sufficient results may be found to enable the designer to assess the various approximate methods of optimization.

APPENDIX I

Probability Density Functions for a Special System†

In this appendix probability density functions associated with the simple non-optimum system of § 2 are evaluated. We shall assume that the finite parts of D_P and D_N are rectangular, as in fig. 5, and then justify the assumption by showing that if the ensemble starts with some such distribution, the latter is maintained as time increases.

At any given moment, a given member of the ensemble has as input either $X = +\sigma$ or $X = -\sigma$. When this member has $X = +\sigma$, we call it a *P-system*. When it has $X = -\sigma$, we call it an *N-system*. Half the ensemble consists of P-systems, and half consists of N-systems. As time increases, some of the P-systems change to N-systems, and some of the N-systems change to P-systems. We assume that, at $t=0$, the P-systems are distributed with density function

$$D_P = \begin{Bmatrix} h+(1-2h\sigma)\delta(Y-\sigma), & |Y| \leqslant \sigma \\ 0, & |Y| > \sigma, \end{Bmatrix} \quad . \quad . \quad . \quad (32)$$

and the N-systems are distributed with density function

$$D_N = \begin{Bmatrix} h+(1-2h\sigma)\delta(Y+\sigma), & |Y| \leqslant \sigma \\ 0, & |Y| > \sigma, \end{Bmatrix} \quad . \quad . \quad . \quad (33)$$

when we take

$$\sigma > 0 \quad . \quad . \quad . \quad . \quad . \quad . \quad . \quad . \quad . \quad . \quad (34)$$

(h is the height of the rectangular part of the density function and the coefficient $(1-2h\sigma)$ of the delta-function is chosen to make the total probability equal to unity).

Consider those P-systems which, at $t=0$, occupy an infinitesimal region R of distribution (32), defined by

$$-\sigma < Y_1 \leqslant Y \leqslant Y_1 + \Delta Y < \sigma. \quad . \quad . \quad . \quad . \quad . \quad (35)$$

As time increases, P-systems cross the ordinate through Y_1, into R, at rate h/a per unit time. Similarly P-systems cross the ordinate through $Y_1 + \Delta Y$, out of R, at rate h/a per unit time. Therefore the motion of the P-systems towards $Y=\sigma$ does not change the number of P-systems in R. However, some of the P-systems in R change to N-systems in R, as t increases. Because the changes have a Poisson distribution along the time-axis, future changes are independent of past changes, i.e. each P-system has the same probability

$$\mu \Delta t \quad . \quad . \quad . \quad . \quad . \quad . \quad . \quad . \quad . \quad . \quad (36)$$

of changing to an N-system in an infinitesimal time interval Δt. Therefore the P-systems in R change to N-systems in R at rate $h\Delta Y\mu$ per unit time. But this loss is made good by the change of N-systems in R to P-systems in

† The concept used here of a constant rate of conversion of P-systems into N-systems and of N-systems into P-systems was suggested by W. M. Wonham (1960). This concept results in a shorter and clearer argument than that originally given in the writer's thesis (Fuller 1959 a). A general exposition of the concept will be given in Wonham's thesis, to be submitted in Dec. 1960.

R, which also occurs at rate $\mu h \Delta Y$ per unit time. Hence as time increases, the number of P-systems in R remains constant. Similarly the number of N-systems in R remains constant. This is true for all regions R except possibly those at the edges of the distribution. Hence the distribution of the ensemble remains constant, except possibly at the edges.

It remains to investigate the edges of the distribution. Consider those P-systems which, at $t=0$, occupy an infinitesimal region S defined by

$$-\sigma \leqslant Y \leqslant -\sigma+\Delta Y. \qquad (37)$$

As time increases, N-systems in the delta function of distribution (33) convert to P-systems at rate $(1-2h\sigma)\mu$ per unit time, and enter region S. Also P-systems cross the ordinate through $Y=-\sigma+\Delta Y$ at rate h/a per unit time. Further, the loss of P-systems in S due to conversion to N-systems is exactly cancelled by the conversion to P-systems of N-systems in the region

$$-\sigma < Y \leqslant -\sigma+\Delta Y. \qquad (38)$$

Therefore the net rate of gain of P-systems in S is

$$(1-2h\sigma)\mu - h/a, \qquad (39)$$

and is thus zero if

$$h=\frac{\mu a}{1+2\mu a\sigma}. \qquad (40)$$

The delta-function at $Y=-\sigma$ gains N-systems from region (38) at rate h/a, and loses N-systems by conversion to P-systems, at rate $(1-2h\sigma)\mu$. Hence the net rate of change of the magnitude of the delta-function is zero if h satisfies expression (40). Therefore the distribution of the ensemble remains constant in the region (37), and by symmetry the same is also true of the right-hand edge of the distribution.

Substituting (40) in (32) and (33), we find that the probability density functions of the P-systems and N-systems are

$$D_{\mathrm{P}}=\left\{\begin{array}{ll}\dfrac{\mu a+\delta(Y-\sigma)}{1+2\mu a\sigma}, & |Y|\leqslant\sigma \\ 0, & |Y|>\sigma\end{array}\right\} \qquad (41)$$

and

$$D_{\mathrm{N}}=\left\{\begin{array}{ll}\dfrac{\mu a+\delta(Y+\sigma)}{1+2\mu a\sigma}, & |Y|\leqslant\sigma \\ 0, & |Y|>\sigma\end{array}\right\}. \qquad (42)$$

To calculate the probability density function $G(E)$ from results (41) and (42), we argue as follows. When $X=+\sigma$,

$$E=\sigma-Y \qquad (43)$$

where Y is distributed as in (41). Hence

$$G(E|X=+\sigma)=\left\{\begin{array}{ll}\dfrac{\mu a+\delta(E)}{1+2\mu a\sigma}, & 0\leqslant E\leqslant 2\sigma \\ 0, & \text{elsewhere.}\end{array}\right\} \qquad (44)$$

Similarly,

$$G(E|X=-\sigma)=\begin{cases}\dfrac{\mu a+\delta(E)}{1+2\mu a\sigma}, & -2\sigma\leqslant E\leqslant 0,\\ 0, & \text{elsewhere.}\end{cases} \quad . \quad . \quad (45)$$

Also $$G(E)=\tfrac{1}{2}G(E|X=+\sigma)+\tfrac{1}{2}G(E|X=-\sigma). \quad . \quad . \quad . \quad (46)$$

From eqns. (44), (45) and (46)

$$G(E)=\begin{cases}\dfrac{\frac{1}{2}\mu a+\delta(E)}{1+2\mu a\sigma}, & |E|\leqslant 2\sigma,\\ 0, & |E|>2\sigma.\end{cases} \quad . \quad . \quad . \quad . \quad (47)$$

APPENDIX II

Phase Plane Trajectories

In this appendix it is shown that the optimum controller is characterized by two points of attraction in the (X, Y) phase plane. The argument was begun in § 3 and continues as follows.

Sooner or later the representative point will arrive at some point of attraction, say at U_P on the line $X=\sigma$ (because sooner or later there is a very long time-interval before X changes). It will stay there until X changes, then move to U_N (which is the point on the line $X=-\sigma$ and opposite U_P), then move towards the nearest point of attraction, at V_N (see fig. 10). If the input does not change too soon, the representative point will reach V_N, then when the input does change, move to V_P. We now consider two cases: (*a*) no point of repulsion between U_P and V_P, (*b*) one or more points of repulsion between U_P and V_P.

Case (*a*)

After reaching V_P the representative point will start to travel back to U_P. Thereafter, it can never pass U_P, because U_P is a point of attraction, and similarly it can never pass V_N. In fact, the representative point will wander randomly along the segments $V_P\,U_P$ and $U_N\,V_N$, crossing from one to the other whenever the input changes. We call the trajectory $U_PU_NV_NV_P$ the *bounding cycle* of the motion. U_P and V_N are the only controller parameters which affect system performance.

Case (*b*)

After reaching V_P the representative point will start to travel towards the nearest point of attraction W_P. If W_P lies in the direction of U_P, then sooner or later W_P will be reached. Thereafter, the situation is the same as in case (*a*), except that the bounding cycle will be $W_PW_NV_NV_P$ instead of $U_PU_NV_NV_P$. On the other hand, if W_P lies in the opposite direction to that of U_P, the representative point will start to travel further away from U_P. The trajectory from V_N onwards is now qualitatively the same as that with which we started at U_P. Therefore the representative point will either move within a bounding cycle $V_NV_PW_PW_N$, or will recede still further from U_P. Repeating the argument indefinitely, we see that either a bounding cycle will be reached†, or the representative point will tend continually to recede from the origin. The latter type of performance would, of course, be useless and certainly not optimum.

Thus in all cases the optimum controller is determined only by two points of attraction, one on the line $X = +\sigma$ and the other on the line $X = -\sigma$. We designate these points as (σ, α) and $(-\sigma, \beta)$, and *the problem reduces to finding the best values of the ordinates* α *and* β.

Fig. 10

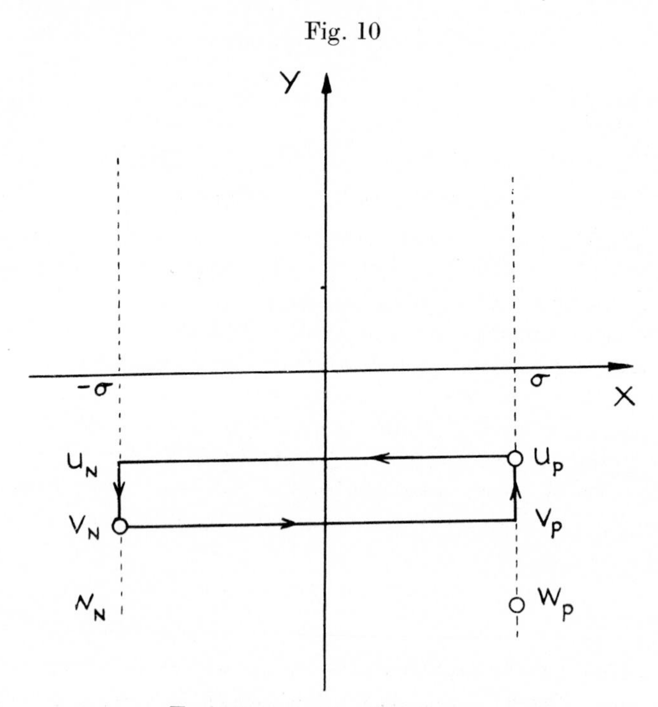

Trajectory in the (X, Y) plane.

Finally, note that for optimum performance Y should stay between or at the two input values $\pm\sigma$ (once it reaches this range), i.e. α and β must satisfy

$$\left.\begin{aligned} |\alpha| &\leqslant |\sigma|, \\ \text{and} \quad |\beta| &\leqslant |\sigma|. \end{aligned}\right\} \qquad (48)$$

To prove this, consider a system (i) in which (48) does not hold, say because $\alpha > |\sigma|$. Then $Y > |\sigma|$ during certain intervals T_j of time t. Next consider a system (ii) which has the same input and output as system (i), except that during T_j, Y is held at $|\sigma|$. Then the errors of the two systems are the same, except that during T_j, system (ii) has error of smaller magnitude than that of system (i). System (i) is therefore worse than system (ii), and cannot be optimum.

† If U_P, V_P, W_P, ... form an infinite sequence with limit point Z_P, then ultimately the trajectory is effectively along the horizontal line Z_PZ_N, which is equivalent to a bounding cycle of zero height.

APPENDIX III

Probability Density Functions for General Systems

First Generalization

In Appendix I it is assumed that σ is positive. If σ is made negative, the system operation is obviously unaffected, and the magnitude of h, the height of the rectangular density function, is unchanged, i.e.

$$h=\frac{\mu a}{1-2\mu a\sigma}, \quad \sigma<0. \qquad (49)$$

Thus the generalization of eqns. (41) and (42) for any sign of σ is

$$D_{\mathrm{P}}\{Y,\sigma\}=\left\{\begin{matrix} \dfrac{\mu a+\delta(Y-\sigma)}{1+2\mu a|\sigma|}, & |Y|\leqslant|\sigma|, \\ 0, & |Y|>|\sigma|, \end{matrix}\right\} \qquad (50)$$

and

$$D_{\mathrm{N}}\{Y,\sigma\}=\left\{\begin{matrix} \dfrac{\mu a+\delta(Y+\sigma)}{1+2\mu a|\sigma|}, & |Y|\leqslant|\sigma|, \\ 0, & |Y|>|\sigma|. \end{matrix}\right\} \qquad (51)$$

Further, if any constant ρ is added to the square-wave input, so that X takes alternately the values $\rho+\sigma$ and $\rho-\sigma$, the error distribution will be unchanged (since only *changes* in X affect the error after Y has caught up with X for the first time). The output distribution will be as before, except that it will be displaced a distance ρ along the Y-axis. The joint input–output distribution will have the density function

$$D_{\mathrm{J}}(X,Y)=\tfrac{1}{2}D_{\mathrm{P}}\{(Y-\rho),\sigma\}\delta(X-\rho-\sigma)+\tfrac{1}{2}D_{\mathrm{N}}\{(Y-\rho),\sigma\}\delta(X-\rho+\sigma) \qquad (52)$$

where D_{P} and D_{N} are as in (50) and (51).

Second Generalization

Let us denote as system 1 the system with the simple controller $C=E$ (which was analysed in § 2), and let us denote as system 2 the system with the controller with the parameters α and β (which we wish to optimize).

For system 2, Y always moves towards or remains at α when $X=+\sigma$, and moves towards or remains at β when $X=-\sigma$. Y thus varies in precisely the same way as does the output of system 1 when the input of the latter is a random square wave taking alternately the values $X=\alpha$ and $X=\beta$. But the joint density function of the latter system is obtained by substituting $\rho=\frac{1}{2}(\alpha+\beta)$ and $\sigma=\frac{1}{2}(\alpha-\beta)$ in eqn. (52), and is thus

$$D_{\mathrm{J1}}(X,Y)\equiv\tfrac{1}{2}D_{\mathrm{P}}\{(Y-\theta),\phi\}\delta(X-\alpha)+\tfrac{1}{2}D_{\mathrm{N}}\{(Y-\theta),\phi\}\delta(X-\beta) \qquad (53)$$

where

$$\theta\equiv\tfrac{1}{2}(\alpha+\beta),\quad \phi\equiv\tfrac{1}{2}(\alpha-\beta). \qquad (54)$$

Therefore the joint probability density function of system 2 is

$$D_{\mathrm{J2}}(X,Y)\equiv\tfrac{1}{2}D_{\mathrm{P}}\{(Y-\theta),\phi\}\delta(X-\sigma)+\tfrac{1}{2}D_{\mathrm{N}}\{(Y-\theta),\phi\}\delta(X+\sigma). \qquad (55)$$

If the probability density function of the error of system 2 is $Q(E)$, it follows from eqns. (2) and (55) that

$$Q(E|X=+\sigma)=D_{\mathrm{P}}\{(-E+\sigma-\theta),\,\phi\} \quad . \quad . \quad . \quad . \quad (56)$$

and

$$Q(E|X=-\sigma)=D_{\mathrm{N}}\{(-E-\sigma-\theta),\,\phi\}. \quad . \quad . \quad . \quad (57)$$

Adding (56) and (57) we obtain

$$Q(E)=\tfrac{1}{2}D_{\mathrm{P}}\{(-E+\sigma-\theta),\,\phi\}+\tfrac{1}{2}D_{\mathrm{N}}\{(-E-\sigma-\theta),\,\phi\}. \quad . \quad (58)$$

From eqns. (50), (51) and (58), the probability density function of the error is

$$Q(E)=\left\{\begin{array}{r}\dfrac{\mu a+\delta(E-\sigma+\theta+\phi)+\delta(E+\sigma+\theta-\phi)}{2(1+2\mu a|\phi|)}\\ \text{in } \sigma-\theta-|\phi|\leqslant E\leqslant \sigma-\theta+|\phi|,\\ \text{and } -\sigma-\theta-|\phi|\leqslant E\leqslant -\sigma-\theta+|\phi|,\\ 0,\ \text{elsewhere.}\end{array}\right\} \quad . \quad . \quad (59)$$

If $\alpha>\beta$, we obtain from eqns. (54) and (59)

$$Q(E)=\left\{\begin{array}{r}\dfrac{\mu a+\delta(E-\sigma+\alpha)+\delta(E+\sigma+\beta)}{2[1+\mu a(\alpha-\beta)]},\\ \text{in } \sigma-\alpha\leqslant E\leqslant \sigma-\beta,\\ \text{and } -\sigma-\alpha\leqslant E\leqslant -\sigma-\beta,\\ 0,\ \text{elsewhere,}\end{array}\right\} \quad . \quad . \quad (60)$$

which is shown in fig. 8.

If $\alpha<\beta$ we obtain similarly,

$$Q(E)=\left\{\begin{array}{r}\dfrac{\mu a+\delta(E-\sigma+\alpha)+\delta(E+\sigma+\beta)}{2[1+\mu a(\beta-\alpha)]},\\ \text{in } \sigma-\beta\leqslant E\leqslant \sigma-\alpha,\\ \text{and } -\sigma-\beta\leqslant E\leqslant -\sigma-\alpha,\\ 0,\ \text{elsewhere.}\end{array}\right\} \quad . \quad . \quad (61)$$

Let us take

$$\sigma>0, \quad . \quad . \quad . \quad . \quad . \quad . \quad . \quad . \quad . \quad . \quad (62)$$

as we may now again do without loss of generality. If we consider $Q(E)$ first when α and β are two fixed values such that $\alpha>\beta$, and then when the two values are interchanged so that $\alpha<\beta$, comparison of eqns. (60) and (61) shows that the only effect of the interchange is to move the two delta-functions further away from the origin. Therefore the case with $\alpha>\beta$ is better than the case with $\alpha<\beta$, and we need consider only the case $\alpha>\beta$ in our optimization.

APPENDIX IV

Optimum Controller Parameters

Elimination of One Parameter

The problem is to find the values of the parameters α and β, where $\alpha > \beta$, which minimize $\overline{|E|^\nu}$ when E has the density function (12). To simplify the algebra, we shall first use the symmetry of the density function to eliminate one parameter.

The density function (12) is symmetric with respect to its mean, which is $-\theta$. Two small elements, each of infinitesimal area Δ, at distances $\pm\lambda$ from the mean (as shown dotted in fig. 8), make a contribution m to $\overline{|E|^\nu}$, where

$$m = (-\theta+\lambda)^\nu\Delta + (\theta+\lambda)^\nu\Delta \quad . \quad . \quad . \quad . \quad . \quad . \quad . \quad (63)$$

Let θ vary and keep ϕ and λ fixed. It follows from eqn. (59) that the only change in the distribution Q is a translation along the E-axis. The effect of this translation on contribution m may be evaluated as follows. From eqn. (63)

$$\frac{dm}{d\theta} = \nu[-(-\theta+\lambda)^{\nu-1} + (\theta+\lambda)^{\nu-1}]\Delta \quad . \quad . \quad . \quad . \quad (64)$$

$$= 0 \quad \text{when} \quad \theta = 0. \quad . \quad . \quad . \quad . \quad . \quad . \quad . \quad . \quad . \quad (65)$$

Also

$$\frac{d^2m}{d\theta^2} = \nu(\nu-1)[(-\theta+\lambda)^{\nu-2} + (\theta+\lambda)^{\nu-2}]\Delta. \quad . \quad . \quad . \quad (66)$$

In expression (66) the quantities $(-\theta+\lambda)$ and $(\theta+\lambda)$ are positive. (This may be deduced from fig. 8, bearing in mind that α and β are restricted to the ranges (48) and σ is positive.)
Therefore

$$\frac{d^2m}{d\theta^2} \gtrless 0 \quad \text{according as } \nu \gtrless 1. \quad . \quad . \quad . \quad . \quad (67)$$

If $\nu > 1$ it follows from relations (65) and (67) that there is a minimum of m at $\theta = 0$ and this is the only minimum. If $\nu < 1$, it follows from relation (67) that m is minimized by giving θ either its largest or its least value consistent with restrictions (48); this is obtained either by increasing θ until $\alpha = \sigma$ or by decreasing θ until $\beta = -\sigma$. (The case $\nu = 1$ may be treated as a limiting case of $\nu < 1$ or of $\nu > 1$ indifferently.)

The above results hold for all the values of λ and therefore for $\overline{|E|^\nu}$ taken as a whole. Hence, if $\nu \geqslant 1$, $\theta_{\text{opt}} = 0$; and so one of the parameters α and β may be eliminated by making

$$\beta = -\alpha. \quad . \quad . \quad . \quad . \quad . \quad . \quad . \quad . \quad . \quad (68)$$

If $\nu \leqslant 1$, one parameter may be eliminated by making either $\beta = -\sigma$ or $\alpha = \sigma$, say for definiteness

$$\beta = -\sigma. \quad . \quad . \quad . \quad . \quad . \quad . \quad . \quad . \quad . \quad (69)$$

Note from eqn. (68) that when $\nu \geqslant 1$ the controller gives symmetric operation. On the other hand, when $\nu < 1$, it turns out that *asymmetric* operation can be optimum in some cases.

Minimization with Respect to the Remaining Parameter

In this section we treat separately the cases $\nu \geqslant 1$, $\nu = 1$, $\nu = 2$, $\nu \leqslant 1$, $\nu = 0$.

$\underline{\nu \geqslant 1}$

From eqns. (60) and (68)

$$Q(E) = \begin{cases} \dfrac{\mu a + \delta(E - \sigma + \alpha) + \delta(E + \sigma - \alpha)}{2(1 + 2\mu a \alpha)}, \\ \quad \text{in } \sigma - \alpha \leqslant E \leqslant \sigma + \alpha, \\ \quad \text{and } -\sigma - \alpha \leqslant E \leqslant -\sigma + \alpha, \\ 0, \quad \text{elsewhere.} \end{cases} \quad . \quad . \quad (70)$$

Hence

$$\overline{|E|^\nu} = \frac{\left[\int_{\sigma - \alpha}^{\sigma + \alpha} + \int_{-\sigma - \alpha}^{-\sigma + \alpha}\right] \mu a |E|^\nu dE + 2|\sigma - \alpha|^\nu}{2(1 + 2\mu a \alpha)}. \quad . \quad . \quad (71)$$

Since

$$0 \leqslant \alpha \leqslant \sigma, \quad . \quad . \quad . \quad . \quad . \quad . \quad . \quad . \quad (72)$$

which follows from relations (13), (48), (62), (68), the evaluation of (71) is straightforward and yields result (24).

$\underline{\nu = 1}$

The substitution $\nu = 1$ in eqn. (24) gives

$$\frac{\overline{|E|}}{\sigma} = 1 - \left(\frac{\sigma}{\alpha} + J\right)^{-1}. \quad . \quad . \quad . \quad . \quad . \quad (73)$$

Thus $\overline{|E|}$ decreases as α increases in range (72).

Therefore

$$\alpha_{\text{opt}} = \sigma, \quad (\text{hence } \beta_{\text{opt}} = -\sigma), \quad . \quad . \quad . \quad . \quad . \quad (74)$$

yielding

$$\frac{\overline{|E|}_{\min}}{\sigma} = \frac{J}{1 + J}. \quad . \quad . \quad . \quad . \quad . \quad . \quad . \quad . \quad (75)$$

$\underline{\nu = 2}$

The substitution $\nu = 2$ in eqn. (24) gives

$$\frac{\overline{E^2}}{\sigma^2} = \frac{1 + (J - 2)\left(\dfrac{\alpha}{\sigma}\right) + \left(\dfrac{\alpha}{\sigma}\right)^2 + \dfrac{J}{3}\left(\dfrac{\alpha}{\sigma}\right)^3}{1 + J\left(\dfrac{\alpha}{\sigma}\right)}. \quad . \quad . \quad . \quad (76)$$

It follows that $d\overline{E^2}/d\alpha = 0$ when

$$\left(\frac{\alpha}{\sigma}\right)^3 + \frac{3}{J}\left(\frac{\alpha}{\sigma}\right)^2 + \frac{3}{J^2}\left(\frac{\alpha}{\sigma}\right) - \frac{3}{J^2} = 0, \quad \ldots \quad (77)$$

of which equation the only real root is

$$\left.\begin{aligned} \left(\frac{\alpha}{\sigma}\right)_{\text{opt}} &= \frac{(1+3J)^{1/3}-1}{J} \\ &= \frac{3}{1+(1+3J)^{1/3}+(1+3J)^{2/3}} = -(\beta/\sigma)_{\text{opt}}. \end{aligned}\right\} \quad \ldots \quad (78)$$

From eqns. (76) and (78)

$$\frac{\overline{E^2}_{\min}}{\sigma^2} = \frac{(1+3J)^{2/3}-1-2J+J^2}{J^2}. \quad \ldots \quad (79)$$

$\underline{\nu \leqslant 1}$

From eqns. (60) and (69)

$$Q(E) = \left\{\begin{array}{c} \dfrac{\mu a + \delta(E-\sigma+\alpha) + \delta(E)}{2[1+\mu a(\sigma+\alpha)]} \\ \text{in } \sigma-\alpha \leqslant E \leqslant 2\sigma, \\ \text{and } -\sigma-\alpha \leqslant E \leqslant 0, \\ 0, \text{ elsewhere.} \end{array}\right\} \quad \ldots \quad (80)$$

From relations (48), (62) and (80), result (21) may be readily deduced. The proof of results (19) is as follows. Let b be defined as

$$b \equiv \frac{\sigma-\alpha}{2\sigma}, \quad \ldots \quad (81)$$

so that

$$0 \leqslant b \leqslant 1. \quad \ldots \quad (82)$$

From relations (21), (81) and (82):

$$\frac{\overline{|E|^\nu}}{\sigma^\nu} = \frac{b^\nu(1+\nu)+J[1+(1-b)^{1+\nu}-b^{1+\nu}]}{2^{1-\nu}(1+\nu)[1+J(1-b)]} \quad \ldots \quad (83)$$

$$\geqslant \frac{b(1+\nu)+J[1+\{1-(1+\nu)b\}-b]}{2^{1-\nu}(1+\nu)[1+J(1-b)]} \quad \ldots \quad (84)$$

$$\equiv \left[\frac{\overline{|E|^\nu}}{\sigma^\nu}\right]_{b=0} + \frac{b(1-J)(1+\nu+\nu J)}{2^{1-\nu}(1+v)(1+J)[1+J(1-b)]}. \quad (85)$$

$$\geqslant \left[\frac{\overline{|E|^\nu}}{\sigma^\nu}\right]_{b=0} \quad \text{if} \quad J \leqslant 1. \quad \ldots \quad (86)$$

Hence if $J \leqslant 1$, the best value of b is $b=0$, i.e.

$$\alpha_{\text{opt}} = \sigma, \quad \ldots \quad (87)$$

and results (19) follow.

$\underline{\nu=0}$

Substituting $\nu=0$ in eqn. (21) we find

$$\overline{|E|^0}=\frac{F(\alpha/\sigma)+J(1+\alpha/\sigma)}{2+J(1+\alpha/\sigma)}, \qquad (88)$$

where

$$F(\alpha/\sigma)=\begin{cases}1, & \alpha/\sigma<1,\\ 0, & \alpha/\sigma=1.\end{cases} \qquad (89)$$

Hence $\overline{|E|^0}$ increases as α increases in the range $-\sigma\leqslant\alpha<\sigma$, then decreases discontinuously, at $\alpha=\sigma$. Therefore there are local minimum values at $\alpha=-\sigma$ and $\alpha=+\sigma$, these minima being

$$\overline{|E|^0}=\tfrac{1}{2} \quad \text{and} \quad \overline{|E|^0}=\frac{J}{1+J} \text{ respectively.} \qquad (90)$$

From (90), the absolute minimum is at $\alpha=-\sigma$ or $\alpha=+\sigma$ according as $J>1$ or $J<1$, yielding results (15) and (16).

References

Fuller, A. T., 1959 a, Ph.D. Thesis, Cambridge; 1959 b, *J. Electron. Contr.*, **7,** 456; 1960 a, *Ibid.*, **8,** 381; 1960 b, *Ibid.*, **8,** 465; 1960 c, *Ibid.*, **9,** 65; 1960 d, *Proc. I.F.A.C. Congress, Moscow* (London: Butterworths).

Kenrick, G. W., 1929, *Phil. Mag.*, **7,** 176.

Rice, S. O., 1944, *Bell Syst. tech. J.*, **23,** 282; 1945, *Ibid.*, **24,** 46.

Wonham, W. M., 1960, *I.R.E. Trans.*, **IT–6** (in the press).

Optimization of a Saturating Control System with Brownian Motion Input†

By A. T. Fuller
Department of Engineering, University of Cambridge

[Received November 8, 1960]

ABSTRACT

A non-linear control system with a random input is optimized theoretically. The main process (i.e. the plant) consists of an integrator and is preceded by a saturating amplifier. The random input is Brownian Motion (i.e. an integrated Gaussian white noise). It is found that, for various performance criteria, relay operation of the saturation characteristic is called for, the relay switching function being simply the system error.

§ 1. Introduction

In previous papers (Fuller 1959, 1960 a, b, c, d) the writer discussed the nature of the optimum controller for saturating control systems with random and non-random inputs. One case of complete optimization of a non-linear system with a random input has already been treated (Fuller 1961), and the object of the present paper is to present another such case. Apart from their intrinsic interest, the results are useful for testing the accuracy of various approximate methods for optimizing control systems with random inputs.

§ 2. The System

The system to be considered is shown in fig. 1. The main process (i.e. the plant) is a simple integrator with transfer function

$$\frac{1}{ap}, \quad a>0, \qquad (1)$$

Fig. 1

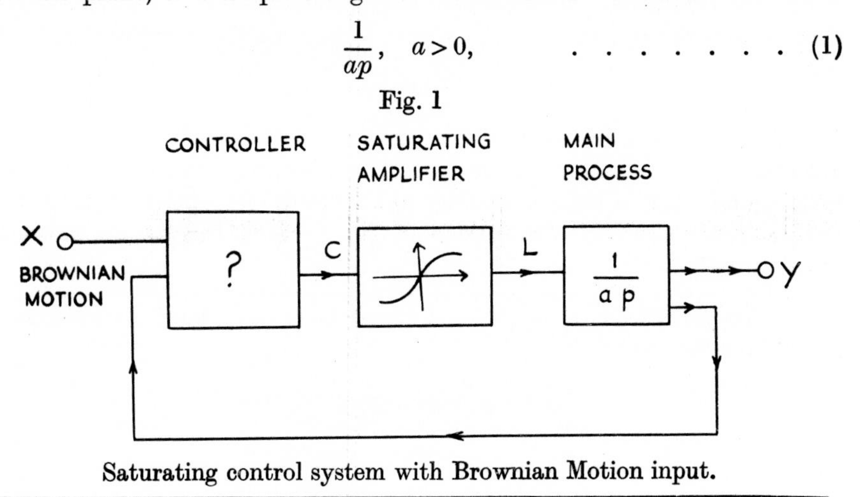

Saturating control system with Brownian Motion input.

† Communicated by J. F. Coales.

from the main process input L to the control system output Y. The main process is driven by an amplifier which saturates. The amplifier is assumed to be free from hysteresis, time-lag, etc. and is completely specified by its given input–output characteristic which is monotonic increasing and is asymptotic to positive and negative saturation levels of equal magnitude. For convenience, the saturation level is normalized to unity. The control system input X is Brownian Motion, i.e. a Gaussian process with spectral density† of the form

$$\frac{\eta}{\omega^2} \qquad (2)$$

where η is a constant. X may be regarded as being generated by integrating a Gaussian purely random signal‡ $\xi(t)$, so that

$$\frac{dX}{dt} = \xi(t) \qquad (3)$$

where t is time. The controller output C drives the amplifier, and is restricted to be a physically realizable signal, i.e. must operate only on the past and present of available signals. The problem is to design the controller so that the time-average of

$$|E|^{\nu}, \quad \nu > 0, \qquad (4)$$

is minimized, where E is the system error,

$$E = X - Y, \qquad (5)$$

and ν is a constant. The performance criteria $\overline{|E|^{\nu}}$ have been discussed in a previous paper (Fuller 1959).

It follows from general principles (Fuller 1960 c) that the optimum L is an instantaneous non-linear or linear function of E. The problem is thus to find the optimum $L(E)$, where L satisfies the constraint

$$|L| \leqslant 1. \qquad (6)$$

The block diagram for this problem is shown in fig. 2, and we shall confine attention to the system shown there. (When the optimum $L(E)$ has been found, it is a simple matter to calculate the optimum controller characteristic $C(E)$ for the system of fig. 1, using the known amplifier characteristic $L(C)$.)

---

† In the present paper we use as a formal basis the following conventional definition of spectral density $\Phi(\omega)$:

$$\Phi(\omega) = \int_{-\infty}^{\infty} \phi(\tau) \exp(-j\omega\tau)\, d\tau$$

where $\phi(\tau)$ is the autocorrelation. (Many different conventions are used in the literature.)

‡ $\xi(t)$ is a type of ' white noise ' (Fuller 1960 c).

§ 3. Partial Differential Equations for Probability Density Functions

Consider a general system (not necessarily a control system) represented by the canonical differential equations

$$\frac{dz_i}{dt}=f_i(z_1,\ z_2,\dots z_n)+\xi_i,\quad i=1,\ 2,\dots n \quad . \quad . \quad . \quad (7)$$

where the ξ_i are Gaussian purely random signals, and the z_i are phase coordinates. Let

$$p(z_1^{\,1}, z_2^{\,1}, \dots z_n^{\,1}, t_1 | z_1^{\,0}, z_2^{\,0}, \dots z_n^{\,0}, t_0)\, dz_1\, dz_2 \dots dz_n \quad . \quad . \quad (8)$$

be the conditional probability that the system will be between phase states $(z_1^{\,1}, z_2^{\,1}, \dots z_n^{\,1})$ and $(z_1^{\,1}+dz_1, z_2^{\,1}+dz_2, \dots, z_n^{\,1}+dz_n)$ at time t_1 given that the system was in state $(z_1^{\,0}, z_2^{\,0}, \dots z_n^{\,0})$ at time $t_0 (t_0 < t_1)$. Then according to Barrett (1960), p satisfies the partial differential equation (generalized Fokker–Planck equation)

$$\frac{\partial p}{\partial t_1} = -\sum_{i=1}^{n} \frac{\partial}{\partial z_i^{\,1}}(f_i p)+\tfrac{1}{2}\sum_{i,j=1}^{n} \eta_{ij}\frac{\partial^2 p}{\partial z_i^{\,1}\partial z_j^{\,1}} \quad . \quad . \quad . \quad . \quad (9)$$

Fig. 2

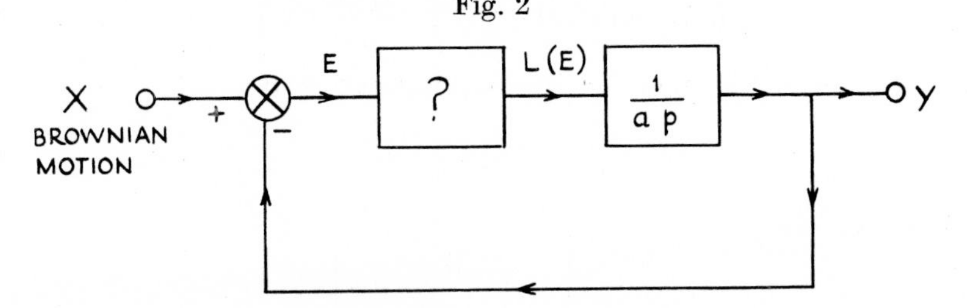

Simplified block diagram.

where η_{ij} is the (specified) cross-spectral-density between the white noises ξ_i and ξ_j. (Thus η_{ii} is the 'power per unit bandwidth' of ξ_i.) We assume that, in cases of practical interest, p settles down to some fixed probability density function independent of initial conditions when t_1 becomes large, so that (9) then yields

$$0 = -\sum_{i=1}^{n} \frac{\partial}{\partial z_i}(f_i p) + \tfrac{1}{2}\sum_{i,j=1}^{n} \eta_{ij}\frac{\partial^2 p}{\partial z_i \partial z_j} \quad . \quad . \quad . \quad . \quad (10)$$

as the partial differential equation of the stationary probability density function of the phase state.

§ 4. Analysis of a General System†

The control system of fig. 2 obeys the following differential equation:

$$\frac{dY}{dt} = \frac{1}{a}L(E). \quad . \quad . \quad . \quad . \quad . \quad . \quad . \quad . \quad . \quad (11)$$

† The analysis in § 4 is due to J. F. Barrett (1958).

L'

From eqns. (3), (5) and (11)

$$\frac{dE}{dt} = -\frac{1}{a}L(E) + \xi. \quad \ldots \ldots \quad (12)$$

Equation (12) is in canonical form (7) (with only one phase coordinate, E). Hence, applying eqn. (10), we find that the probability density function of the error E satisfies

$$\frac{d}{dE}\left[\frac{L(E)p(E)}{a}\right] + \frac{\eta}{2}\frac{d^2p(E)}{dE^2} = 0. \quad \ldots \quad (13)$$

(This is an ordinary rather than a partial differential equation because there is only one phase coordinate). Integrating eqn. (13) once, we obtain

$$\frac{L(E)p(E)}{a} + \frac{\eta}{2}\frac{dp(E)}{dE} = \text{constant} \quad \ldots \quad (14)$$

$$= 0, \quad \ldots \ldots \quad (15)$$

since $p(E) \to 0$ and $dp(E)/dE \to 0$ for large E. Integrating eqn. (15) we have

$$p(E) = \text{const.} \exp\left[-\int_0^E \frac{2L(u)}{\eta a} du\right]. \quad \ldots \quad (16)$$

Here the constant is fixed by the requirement that

$$\int_{-\infty}^{\infty} p(E)\, dE = 1. \quad \ldots \ldots \quad (17)$$

Hence

$$p(E) = \frac{\exp\left[-\int_0^E \frac{2L(u)}{\eta a} du\right]}{\int_{-\infty}^{\infty} \exp\left[-\int_0^E \frac{2L(u)}{\eta a} du\right] dE}, \quad \ldots \quad (18)$$

which is Barrett's (1958) result for the system of fig. 2.

§ 5. Optimization

The problem now reduces to finding the function $L(E)$ which, subject to constraint (6), minimizes $\overline{|E|^{\nu}}$, where E has probability density function (18). This problem is treated in the Appendix, where it is shown that the optimum L is simply

$$L_{\text{opt}} = \operatorname{sgn} E \quad \ldots \ldots \quad (19)$$

so that relay operation of the saturation characteristic is called for. (This operation may be achieved in the system of fig. 1 by inserting a high-gain amplifier before the saturating amplifier and making E the input to the high-gain amplifier.) In the optimum system the probability density function of the error is

$$p(E) = \frac{1}{\eta a}\exp\left(-\frac{2|E|}{\eta a}\right) \quad \ldots \ldots \quad (20)$$

and the value of the performance criterion is therefore

$$\overline{|E|^{\nu}}_{\min} = \Gamma(\nu+1)\left(\frac{\eta a}{2}\right)^{\nu}, \quad (\nu > 0). \quad . \quad . \quad . \quad (21)$$

In particular, the minimized mean-square-error is

$$\overline{E^2}_{\min} = \tfrac{1}{2}\eta^2 a^2. \quad . \quad . \quad . \quad . \quad . \quad . \quad . \quad . \quad . \quad (22)$$

It is interesting to note that the optimum controller is the same as that given by Fuller (1960 d) for a system with the same main process but with X a step-input instead of Brownian Motion.

§ 6. Conclusions

The present paper and a previous one (Fuller 1961) show that in at least two cases it is possible to optimize completely a non-linear control system with a random input.

APPENDIX

Elimination of Asymmetric Functions

We first show that we can restrict attention to symmetric functions $L(E)$, i.e. such that

$$L(E) = -L(-E). \quad . \quad . \quad . \quad . \quad . \quad . \quad . \quad . \quad (23)$$

Consider an asymmetric system α which has

$$L(E) = \begin{cases} F_1(E), & E < 0, \\ F_2(E), & E > 0. \end{cases} \quad . \quad . \quad . \quad . \quad . \quad (24)$$

For system α, define the numerator of the right side of eqn. (18) as $K_1(E)$ for $E < 0$ and $K_2(E)$ for $E > 0$. Then $\overline{|E|^{\nu}}$ for system α is

$$\overline{|E|^{\nu}_{\alpha}} = \frac{\int_{-\infty}^{0} (-E)^{\nu} K_1(E)\, dE + \int_{0}^{\infty} E^{\nu} K_2(E)\, dE}{\int_{-\infty}^{0} K_1(E)\, dE + \int_{0}^{\infty} K_2(E)\, dE}. \quad . \quad . \quad (25)$$

Define the four integrals in this expression as S_1, S_2, S_3, S_4, so that

$$\overline{|E|_{\alpha}^{\nu}} = \frac{S_1 + S_2}{S_3 + S_4}. \quad . \quad . \quad . \quad . \quad . \quad (26)$$

Owing to the asymmetry of $L(E)$, we can have $S_1/S_3 \neq S_2/S_4$. For definiteness, suppose that

$$\frac{S_1}{S_3} \geqslant \frac{S_2}{S_4}. \quad . \quad . \quad . \quad . \quad . \quad . \quad . \quad . \quad . \quad (27)$$

Therefore:

$$S_1S_4 \geqslant S_2S_3, \qquad (28)$$

$$S_1S_4 + S_2S_4 \geqslant S_2S_3 + S_2S_4, \qquad (29)$$

or

$$\frac{S_1+S_2}{S_3+S_4} \geqslant \frac{S_2}{S_4}. \qquad (30)$$

Now consider a symmetrical system β, which has

$$L(E) = \begin{Bmatrix} -F_2(-E), & E<0, \\ F_2(E), & E>0. \end{Bmatrix} \qquad (31)$$

It follows that $\overline{|E|^\nu}$ for system β is

$$\overline{|E|_\beta{}^\nu} = \frac{\int_{-\infty}^{0} (-E)^\nu K_2(-E)\,dE + \int_0^\infty E^\nu K_2(E)\,dE}{\int_{-\infty}^{0} K_2(-E)\,dE + \int_0^\infty K_2(E)\,dE} \qquad (32)$$

$$= \frac{2\int_0^\infty E^\nu K_2(E)\,dE}{2\int_0^\infty K_2(E)\,dE} \qquad (33)$$

$$= \frac{S_2}{S_4}. \qquad (34)$$

From eqns. (26) and (34) and inequality (30)

$$\overline{|E|_\alpha{}^\nu} \geqslant \overline{|E|_\beta{}^\nu}, \qquad (35)$$

i.e. the symmetrical system β is superior to, or at least as good as the asymmetrical system α. Therefore we need only consider symmetrical systems in our optimization.

Optimum Symmetric Function

Suppose then that the symmetrical system β has some characteristic $L \equiv L_\beta(E)$ such as shown in fig. 3, for $E>0$. Suppose also that another symmetrical system γ has the same characteristic for $E>0$, except that, in an infinitesimal region near $E=E_0$, L_γ is slightly greater than L_β (as shown dotted in fig. 3). We shall show that system γ is better than system β. Let the difference in the areas under L_γ and L_β for $E>0$ be Δ. Then we have for system γ,

$$\exp\left[-\int_0^E \frac{2L_\gamma(u)\,du}{\eta a}\right] = \begin{Bmatrix} \exp\left[-\int_0^E \frac{2L_\beta(u)\,du}{\eta a}\right], & 0<E<E_0 \\ \exp\left[-\int_0^E \frac{2L_\beta(u)\,du}{\eta a} - \frac{2\Delta}{\eta a}\right], & E_0<E \end{Bmatrix} \qquad (36)$$

$$= \begin{Bmatrix} K_2(E), & 0<E<E_0 \\ K_2(E)\exp(-A), & E_0<E \end{Bmatrix} \qquad (37)$$

where

$$A \equiv \frac{2\Delta}{\eta a}. \qquad (38)$$

Fig. 3

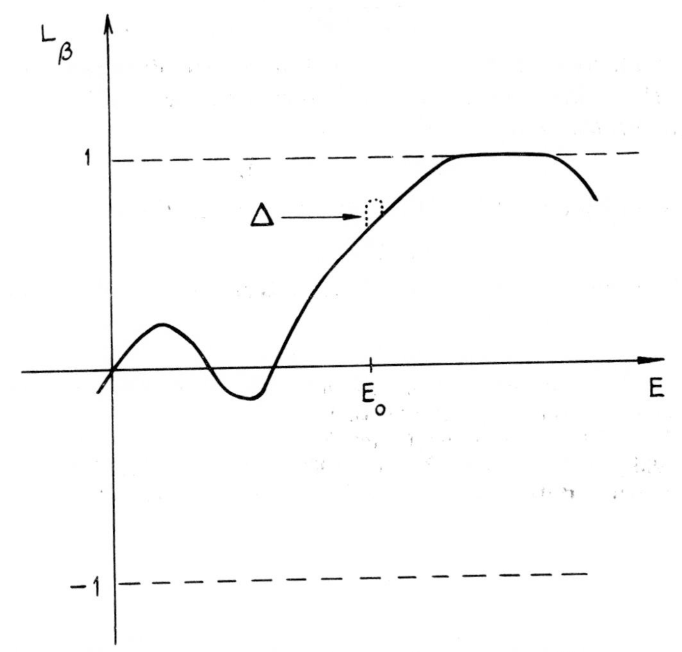

Combined characteristic of controller and amplifier.

It follows that (applying eqn. (18) and symmetry)

$$\overline{|E|_\gamma{}^\nu} = \frac{\int_0^\infty E^\nu K_2(E)\,dE - [1-\exp(-A)]\int_{E_0}^\infty E^\nu K_2(E)\,dE}{\int_0^\infty K_2(E)\,dE - [1-\exp(-A)]\int_{E_0}^\infty K_2(E)\,dE} \qquad (39)$$

Define

$$B \equiv 1-\exp(-A) > 0. \qquad (40)$$

Then eqn. (39) can be written

$$\overline{|E|_\gamma{}^\nu} = \frac{S_2 - BS_2{}^*}{S_4 - BS_4{}^*}, \text{ say.} \qquad (41)$$

Now S_2/S_4 is the νth generalized moment of the part of $p_\beta(E)$ from $E=0$ to $E=\infty$. Also $S_2{}^*/S_4{}^*$ is the νth moment of the same function but truncated from $E=0$ to $E=E_0$. The effect of such truncation on the left of any positive function is obviously to increase its moments. Therefore

$$\frac{S_2{}^*}{S_4{}^*} > \frac{S_2}{S_4}. \qquad (42)$$

Hence

$$-BS_2{}^*S_4 < -BS_2S_4{}^*, \qquad (43)$$

$$S_2S_4 - BS_2{}^*S_4 < S_2S_4 - BS_2S_4{}^*, \qquad (44)$$

or

$$\frac{S_2 - BS_2{}^*}{S_4 - BS_4{}^*} < \frac{S_2}{S_4}. \qquad (45)$$

From eqns. (34) and (41) and inequality (45)

$$\overline{|E|_{\gamma}^{\nu}} < \overline{|E|_{\beta}^{\nu}}, \qquad (46)$$

i.e. any small local increase in $L(E)$, (for $E>0$), improves the control system. If we keep on making such increases for all $E>0$, subject to constraint (6), the best system is reached when

$$L(E)=1, \quad (E>0). \qquad (47)$$

Using the symmetry relation (23), we thus obtain the result

$$L_{\text{opt}}(E)=\operatorname{sgn} E. \qquad (48)$$

Results (20) and (21) follow immediately from eqns. (18) and (48).

References

Barrett, J. F., 1958, Ph.D. Thesis, Cambridge; 1960, *Proc. I.F.A.C. Conf., Moscow* (London: Butterworths).

Fuller, A. T., 1959, *J. Electron. Contr.*, **7**, 456; 1960 a, *Ibid.*, **8**, 381; 1960 b, **8**, 465; 1960 c, *Ibid.*, **9**, 65; 1960 d, *Proc. I.F.A.C. Conf., Moscow* (London: Butterworths); 1961, *J. Electron. Contr.*, **10**, 61.

Saturating Servomechanism to Follow a Random Process†

By P. H. Govaerts
University of California

[Received September 16, 1961]

Abstract

Solutions to the problem of obtaining the optimal bounded input to a servomechanism which must follow with minimum error a discrete-time random process are given for first-order systems. When the random process is a sequence of independent random variables, the optimal solution, which minimizes a function of the error, is independent of the probability distribution of the random variables when the discount factor a is restricted to $0 \leqslant a \leqslant \frac{1}{2}$. In the case of Markov dependence and symmetric distribution functions, there is no restriction on the discount factor.

§ 1. Introduction

Much work has been done in the last ten years on the subject of optimizing control systems with deterministic bounded inputs. General results for a system described by an nth order linear differential equation with constant coefficients were indicated by Bellman *et al.* (1958). Boltyanskii *et al.* (1960) in their Maximum Principle, stated necessary conditions for optimality in the cases of linear and non-linear systems. The same results can be obtained from a dynamic programming approach, as pointed out by Desoer (1961); La Salle (1960) studied more extensively the time optimal control problem.

It is only very recently however that attempts have been made to extend that theory to the probabilistic case. For a finite state machine, Eaton and Zadeh (1961) establish the optimal policy for going to a certain state at minimal cost when the probabilities of transition from state to state depend on the control. Earlier, Newton (1952) and Booton (1953) had optimized saturating servomechanisms with random inputs. Their solutions however were based on linearization procedures whose accuracy is always questionable. Fuller (1961) has given exact solutions for an integrator, whose input is restricted to $+1, 0, -1$ and which must follow a Poisson process. In this paper, we consider a first-order system whose input is bounded in absolute value by γ and which must follow a discrete-time random process, described by a sequence of independent or of Markov dependent random variables. When the random process is a finite Markov chain, the optimization of general servomechanisms whose state vector has a finite

† Communicated by the Author.

number of components, each of them with finite range only, can be performed by an iteration procedure (see Govaerts 1961).

§ 2. Statement of the Problem. Summary of the Results

Let $x(t)$ be a discrete-time random process; $x(t)$ is a sequence of pulses of constant amplitude X_n in the interval of time $((n-1)T, nT]$, the changes occurring at nT, for all positive integers n; X_n are random variables. We assume that the value x_n taken by X_n in $((n-1)T, nT]$ can be observed at time $(n-1)T^+$ (see fig. 1).

Fig. 1

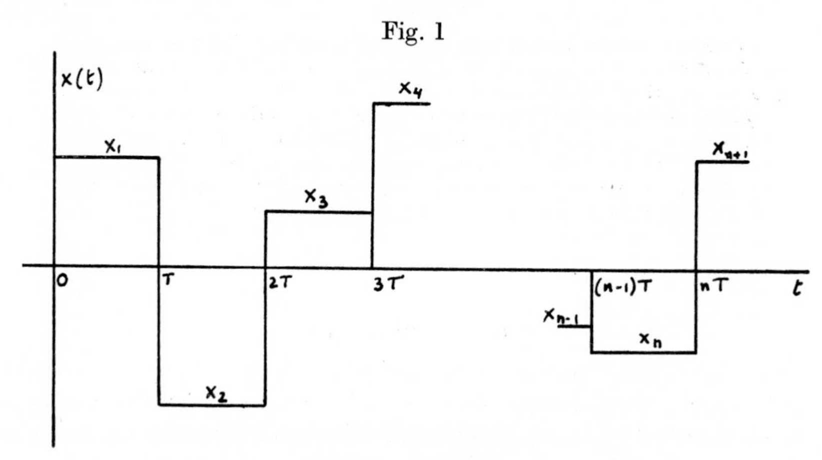

Random process $x(t)$.

The probability density function $p(X_{n+1}|X_n = x_n)$ is given for all n. It will be assumed independent of x_n in § 3 and of the form $p(|X_{n+1} - x_n|)$ in § 4.

Let the servomechanism be an integrator: the future output depends only on the present value of the output and of the present input; it does not depend on the way these present values have been reached. The results obtained can be extended to servomechanisms with transfer function $G(s) = 1/s + b$, b any constant.

The input $r(t)$ to the integrator assumes the constant value r_n in each interval $((n-1)T, nT]$ for all n; it is chosen, according to the optimal policy to be found, at $(n-1)T^+$ knowing x_n and the output $y((n-1)T) = y_{n-1}$ of the integrator. It is restricted to the interval $[-\gamma, +\gamma]$.

Let $y(t)$ be the output of the integrator and define the error $e(t)$ by $e(t) = x(t) - y(t)$. The cost function $h(e)$ is the absolute value of e (§ 3) or, more generally in § 4, is a function satisfying:

$h(0) = 0$,

$h(-e) = +h(e)$,

$h'(e)$ goes through zero at $e = 0$ and is piecewise continuous,

$h'(e)$ is monotone non-decreasing.

At $t=(i-1)T^+$, knowing x_i, y_{i-1}, the cost functional to be minimized with respect to r_i is the sum of the cost $h(x_i-y_i)$ at $t=iT$, and the expected value of the minimal cost functional for the following intervals, the importance of which is decreased by multiplication by the discount factor a $(0<a\leqslant 1)$.

In view of the stationary character of the sequence of random variables $\{X_n\}$, we may consider the problem at $t=0$ only; then, given x_1, y_0, the problem is to find r_1 optimal when the random process $x(t)$ extends on a number N (finite or not) of intervals.

The optimal r_1 expressed as a function of x_1 and y_0 is called the optimal policy for N intervals (fig. 2).

Fig. 2

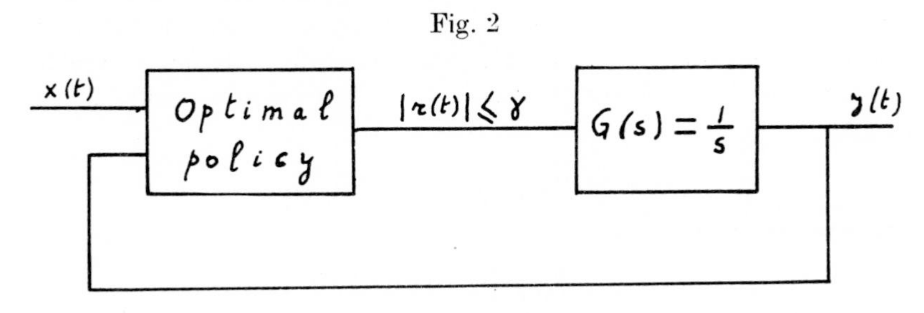

Optimal policy.

2.1. *Summary of the Results*

In § 3 a bound is established for the discount factor a in order that the optimal policy for N intervals be independent of N, and so be identical to that for $N=1$ which is

$$r_1=\gamma\,\text{sat}\left(\frac{x_1-y_0}{\gamma T}\right),$$

where

$$\text{sat}\,x=\begin{cases} x & |x|\leqslant 1,\\ \dfrac{x}{|x|} & |x|\geqslant 1.\end{cases}$$

In § 4 it is proved that policy is also the optimal one for symmetric Markov processes, whatever the discount factor $a\geqslant 0$.

§ 3. Independent Case

3.1. *Hypotheses*

We first assume that the random process $x(t)$ has duration NT. The random variables X_1, X_2, $\ldots X_N$ which are the amplitudes of $x(t)$ in the intervals $(0,T]$, $(T,2T]$, $\ldots((N-1)T,NT]$ are independent, defined by their distribution functions $F(X_1)$, $F(X_2)$, $\ldots F(X_n)$, and belong to L_1 (i.e. have finite first absolute moment). The cost function is the absolute value of the error.

The method used to obtain the optimal r_1 is to get an expression of the functional f_N to minimize as a function of r_1 and find the sign of its derivative with respect to r_1, for all r_1. Then, as r_1 is restricted in $[-\gamma, +\gamma]$, the optimal policy is:

(*a*) if the derivative is always positive in $[-\gamma, +\gamma]$: $r_1^0 = -\gamma$,

(*b*) if the derivative is always negative in $[-\gamma, +\gamma]$: $r_1^0 = +\gamma$,

(*c*) if the derivative goes through zero in $[-\gamma, +\gamma]$: $r_1^0 =$ a value where the derivative is zero.

Here and throughout the whole text the superscript 0 always indicates an optimal solution.

The functional f_N to be minimized is the sum of the cost of the first interval and the expected value (decreased by the discount factor a) of the cost f_{N-1}^0 encountered when following an optimal policy in the $N-1$ remaining intervals. These expressions will be evaluated below for some values of N.

3.2. *Optimal Policy for One Interval* ($N=1$)

The cost functional f_1 is:

$$f_1 = |e_1| = |x_1 - y_0 - r_1 T|. \qquad (1)$$

By differentiation of (1):

$$\frac{\partial |e_1|}{\partial r_1} = \frac{\partial}{\partial r_1}|x_1 - y_0 - r_1 T| = T \operatorname{sgn}\left(r_1 - \frac{x_1 - y_0}{T}\right).$$

The optimal policy for one interval is then:

$$r_1^0 = \begin{cases} +\gamma & \text{when } x_1 - y_0 \geqslant +\gamma T, \\ \dfrac{x_1 - y_0}{T} & \text{when } -\gamma T \leqslant x_1 - y_0 \leqslant \gamma T, \\ -\gamma & \text{when } x_1 - y_0 \leqslant -\gamma T, \end{cases}$$

or

$$r_1^0 = \gamma \operatorname{sat}\left(\frac{x_1 - y_0}{\gamma T}\right). \qquad (2)$$

The dependence of the optimal cost function $f_1^0(x_1, y_0)$ on the variables x_1 and y_0 is as follows:

$$f_1^0(x_1, y_0) = \underset{r_1 \in [-\gamma, +\gamma]}{\text{Min}} |x_1 - y_0 - r_1 T| = |x_1 - y_0 - r_1^0 T|; \qquad (3)$$

f_1^0 is zero in a strip of width $2\gamma T$ around the straight line $x_1 = y_0$ and is represented elsewhere by two half-planes at angle 45° on the (x_1, y_0) plane. (See fig. 3.)

Remark

$$\frac{\partial f_1^0(x_1, y_0)}{\partial y_0}$$

takes only the values -1, 0, $+1$. We denote this fact by:

$$\frac{\partial f_1^0}{\partial y_0} \in \{-1, 0, +1\}. \qquad (4)$$

Fig. 3

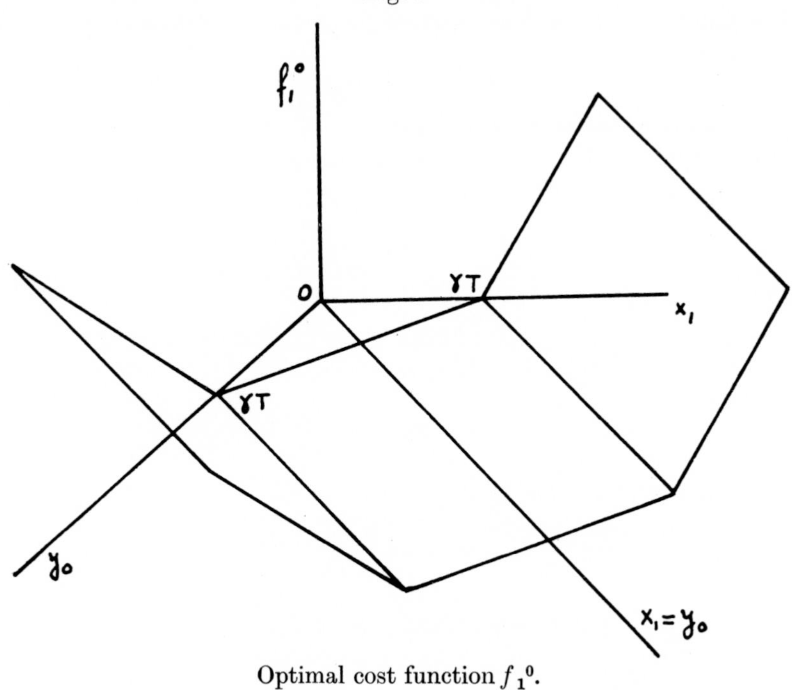

Optimal cost function f_1^0.

3.3. *Optimal Policy for Two Intervals* ($N=2$)

$$f_2^0(x_1, y_0) = \operatorname*{Min}_{r_1} f_2(x_1, y_0, r_1)$$

$$= \operatorname*{Min}_{r_1 \in [-\gamma, +\gamma]} \{|x_1 - y_0 - r_1 T| + a \mathop{E}_{x_2} f_1(x_2, y_1)\}, \quad . \quad . \quad (5)$$

where $y_1 = y_0 + r_1 T$.

Some reasons for introducing the discount factor a will be discussed in the N intervals case.

Let us evaluate

$$\frac{\partial f_2(x_1, y_0, r_1)}{\partial r_1}:$$

$$\frac{\partial |x_1 - y_0 - r_1 T|}{\partial r_1} = T \operatorname{sgn}\left(r_1 - \frac{x_1 - y_0}{T}\right),$$

$$\frac{\partial}{\partial r_1} f_1^0(x_2, y_1) = T \frac{\partial}{\partial y_1} f_1^0(x_2, y_1) = T \frac{\partial}{\partial y_0} f_1^0(x_2, y_0) = T\{-1, 0, +1\}$$

by a preceding remark (eqn. (4)).

Then

$$\left| a \mathop{E}_{x_2} \frac{\partial f_1^0(x_2, y_1)}{\partial r_1} \right| = \left| a T \mathop{E}_{x_2} \{-1, 0, +1\} \right| \leqslant aT$$

and the second term of $|\partial f_2 / \partial r_1|$ is always less than the first one if $a < 1$.

Therefore, when $a < 1$, $\partial f_2/\partial r_1$ changes its sign when the first term does, and, whenever $0 \leqslant a < 1$, the optimal policy for two intervals is:

$$e_1{}^0 = \gamma \operatorname{sat}\left(\frac{x_1 - y_0}{\gamma T}\right). \qquad (6)$$

It is the same policy as for one interval (eqn. (2)).

After applying $r_1{}^0$ at $t=0$, we get, at time $t=T$, the value $y_1 = y_0 + r_1{}^0 T$ and we apply in $(T, 2T]$ the optimal policy for one interval:

$$r_2{}^0 = \gamma \operatorname{sat}\left(\frac{x_2 - y_1}{\gamma T}\right).$$

Remark

$$\frac{\partial f_2{}^0(x_1, y_0)}{\partial y_0} = \frac{\partial f_1{}^0(x_1, y_0)}{\partial y_0} + aE_{x_2}\frac{\partial f_1{}^0(x_2, y_1)}{\partial y_0}$$

$$= \{-1, 0, +1\} + a \,.\, \frac{\partial y_1}{\partial y_0} E_{x_2} \frac{\partial f_1{}^0(x_2, y_1)}{\partial y_1};$$

but $\partial y_1/\partial y_0 = \{0, +1\}$ since $y_1 = y_0 + r_1{}^0 T$, hence

$$\left|\frac{\partial f_2{}^0(x_1, y_0)}{\partial y_0}\right| \leqslant 1 + a\left|\frac{\partial f_1{}^0(x_1, y_0)}{\partial y_0}\right| \leqslant 1 + a. \qquad (7)$$

3.4. *Optimal Policy for N Intervals*

$$f_N{}^0(x_1, y_0) = \underset{r_1 \in [-\gamma, +\gamma]}{\operatorname{Min}} f_N(x_1, y_0, r_1)$$

$$= \underset{r_1 \in [-\gamma, +\gamma]}{\operatorname{Min}} \{|x_1 - y_0 - r_1 T| + aE_{x_2} f_{N-1}{}^0(x_2, y_1)\}. \qquad (8)$$

The discount factor $a < 1$ decreases the influence of future errors. In fact, if $a=1$, the error e_1 and the expected values of all the errors e_j $(j = 2, \ldots N)$ are given the same importance. As their absolute values are always non-negative $f_N{}^0$ tends to $+\infty$ when N goes to $+\infty$, and the minimization does not make sense. One way of preventing it would be to use f_N/N, an 'average' cost function (the discrete time equivalent of

$$\frac{1}{T}\int_0^T |e(t)|\,dt\cdot).$$

But in this case, when $N \to \infty$, the first term (error of the first interval divided by N) disappears and the optimization with respect to r_1 loses its significance.

With the introduction of the discount factor a, which multiplies the successive future errors by the successive powers of a, f_N remains finite when $N \to \infty$. (See Appendix I.)

We have now:

$$\frac{\partial f_N}{\partial r_1} = \frac{\partial}{\partial r_1}|x_1 - y_0 - r_1 T| + a\frac{\partial}{\partial r_1} E_{x_2} f_{N-1}{}^0(x_2, y_1)$$

$$= \frac{\partial}{\partial r_1}|x_1 - y_0 - r_1 T| + aE_{x_2}\frac{\partial}{\partial r_1} f_{N-1}{}^0(x_2, y_1). \qquad (9)$$

Assuming

$$\left|\frac{\partial f_{N-1}{}^0(x_1, y_0)}{\partial y_0}\right| \leqslant c < \infty \quad (c = \text{constant}), \quad . \quad . \quad . \quad (10)$$

we get

$$\left|\frac{\partial f_N}{\partial r_1}\right| \leqslant |\{-T, +T\}| + ac|\underset{x_2}{E}\{-T, 0, +T\}|$$

and

$$r_1{}^0 = \gamma \operatorname{sat}\left(\frac{x_1 - y_0}{\gamma T}\right) \text{ if } aTc < T \text{ or } ac < 1. \quad . \quad . \quad (11)$$

When (10) is satisfied we have also:

$$\frac{\partial f_N{}^0(x_1, y_0)}{\partial y_0} = \{-1, 0, +1\} + a\frac{\partial y_1}{\partial y_0}\underset{x_2}{E}\frac{\partial f_{N-1}{}^0(x_2, y_1)}{\partial y_1};$$

but, as

$$\frac{\partial y_1}{\partial y_0} = \{0, +1\},$$

$$\left|\frac{\partial f_N{}^0(x_1, y_0)}{\partial y_0}\right| \leqslant 1 + a\left|\frac{\partial f_{N-1}{}^0(x_1, y_0)}{\partial y_0}\right| \leqslant 1 + ac, \quad . \quad . \quad . \quad (12)$$

which is a bound for the N intervals case, given the bound for $N-1$ intervals.

For $N = 2$, we had

$$\left|\frac{\partial f_2{}^0}{\partial y_0}\right| \leqslant 1 + a \quad (\text{eqn. (7)}),$$

then by induction on (12):

$$\left|\frac{\partial f_{N-1}{}^0(x_1, y_0)}{\partial y_0}\right| \leqslant 1 + a + \ldots + a^{N-2} = c \quad . \quad . \quad . \quad . \quad (13)$$

and $ac < 1$ is equivalent to $1 + a + \ldots + a^{N-1} < 2$.

Conclusion

If $1 + a + \ldots + a^{N-1} < 2$, the optimal policy for N intervals is the optimal policy for one interval; furthermore it is unique.

The uniqueness property derives immediately from the fact that $\partial f_N / \partial r_1$ goes through zero at only one value of r_1 or has a constant sign over the interval $-\gamma \leqslant r_1 \leqslant +\gamma$.

3.5. *Asymptotic Behaviour*

Let N go to ∞. Then $1 + a + \ldots + a^{N-1} \to 1/1 - a$. We have a unique optimal policy if $1/1 - a < 2$ or:

$$a < \tfrac{1}{2}. \quad . \quad . \quad . \quad . \quad . \quad . \quad . \quad . \quad . \quad (14)$$

As before the uniqueness may not hold if the equality sign is allowed in (14); but even for $a = \frac{1}{2}$:

$$r_1{}^0 = \gamma \operatorname{sat}\left(\frac{x_1 - y_0}{\gamma T}\right) \quad . \quad . \quad . \quad . \quad . \quad . \quad . \quad (15)$$

is still an optimal policy.

Conclusion

With $0 \leqslant a \leqslant \frac{1}{2}$, independent random variables with finite first absolute moment, and the absolute value as error criterion, an optimal policy is

$$r_1^0 = \gamma \operatorname{sat}\left(\frac{x_1 - y_0}{\gamma T}\right).$$

Remark

The bound $a = \frac{1}{2}$ and the independence of the optimal policy on the distribution functions $F(X_i)$ are due to the cost function adopted. With that cost function differentiating terms of the cost functional under the integral sign leaves us only with distribution functions, which, by definition, are always bounded by 1. From that we get the recurrence relation for the bound on the derivatives of f_N, and the bound for a.

3.6. *Extension to First-order Servomechanisms*

For servomechanisms with transfer function $G(s) = 1/s + b$ (b = any finite constant), the output y_1 at $t = T$, given y_0, is:

$$y_1 = y_0 + \int_0^T \exp[-b(T-\tau)] r(\tau)\, d\tau. \quad . \quad . \quad . \quad . \quad (16)$$

Then the region reachable from y_0 is $[y_0 - B, y_0 + B]$ where

$$B = \int_0^T \exp[-b(T-\tau)]\gamma\, d\tau = \frac{\gamma}{b}[1 - \exp(-bT)]. \quad . \quad . \quad (17)$$

Instead of finding $r_1 \in^0 [-\gamma, +\gamma]$, we may equivalently look for

$$y_1^0 \in [y_0 - B, y_0 + B].$$

The same arguments used in the case of the integrator apply here, provided that we perform the optimization on y_1. As $y_1 = y_0 + r_1 T$, differentiations with respect to y_1 or with respect to r_1 yield the same answer, except for the multiplicative constant T.

The result will be, as in the previous case:

$$y_1^0 = y_0 + B \operatorname{sat}\left(\frac{x_1 - y_0}{B}\right), \quad . \quad . \quad . \quad . \quad . \quad (18)$$

whose meaning is: try to go as near as possible to x_1, obeying the constraints.

The value r_1^0 is immediately obtained from y_1^0, by inverting (16).

§ 4. Symmetric Markov Process

The random variables $\{X_n\}$ are now Markov dependent (Feller 1959). The conditional probability density $p(X_{n+1}|X_1 \ldots X_n)$ does not depend on $X_1, \ldots X_{n-1}$ by definition of the Markov property. For the dependence on X_n, we assume:

$$p(X_{n+1}|X_n = x_n) = p(X_{n+1} - x_n) = p_{n+1}(\cdot), \quad . \quad . \quad . \quad (19)$$

so that X_{n+1} is the sum of the value x_n assumed by X_n in $((n-1)T, nT]$ and a random variable of probability density $p_{n+1}(\cdot)$.

With the assumptions made on $p(X_{n+1}|X_n)$ the Chapman-Kolmogorov equation relating $p(X_{n+1}|X_1)$ to $p(X_n|X_1)$ becomes:

$$\begin{aligned} p(X_{n+1}|X_1=x_1) &= \int_{-\infty}^{+\infty} p(X_{n+1}|X_n)p(X_n|X_1=x_1)\,dX_n \\ &= \int p(X_{n+1}-X_n)p(X_n|X_1)\,dX_n \\ &= p_{n+1}(\cdot)*p(X_n|X_1), \end{aligned}$$

where * stands for convolution, and, by iteration

$$p(X_{n+1}|X_1)=p_{n+1}(\cdot)*p_n(\cdot)*\ldots*p_2(\cdot). \quad \ldots \quad (20)$$

4.1. *A Property of the Process*

When $p_{n+1}(\cdot)=p(\cdot)$ is the same for any n, $p(X_{n+1}|X_1)$ is equal to $p(\cdot)$ convolved $n-1$ times with itself.

Let us consider n independent, identically distributed random variables $Y_1, Y_2, \ldots Y_n$ with probability density $p(Y)$. Then

$$S_n=\sum_{i=1}^{n} Y_i$$

has probability density $p(S_n)$ equal to $p(Y)$ convolved $n-1$ times with itself, and the distribution function of $X_{n+1}-x_1$ given $X_1=x_1$ is the same as that of S_n.

Assuming $p(\cdot)$ is centred at expectation and has finite first absolute moment, we can apply Kolmogorov's strong law of large numbers (Loève 1960):

$$P\{\bigcup_{k\geqslant n} |X_{k+1}-x_1|>k\epsilon\}\to 0 \quad \text{whatever } \epsilon>0. \quad \ldots \quad (21)$$

From (21) we conclude that, if at time $t=0^+$, we knew $X(t,\omega)$ for all ω's (where ω denotes as usual a point in the probability space), then it would be possible, for y_0 finite and $\gamma>\epsilon>0$, to catch $X(t,\omega)$ after a certain number of intervals with probability one.

If $p(\cdot)$ is identically zero outside the interval $[-\gamma T, +\gamma T]$, then it is possible to follow the process for all t after it has been reached once; the cost will be positive only during the 'transient' period, i.e. before $y(t)$ has reached $x(t)$. When $p(\cdot)$ does not vanish outside $[-\gamma T, +\gamma T]$, it will not be possible to follow $x(t)$ for all t, but, by (21), $x(t)$ will be reached again after a certain number of intervals with probability one. Unfortunately the future of $X(t,\omega)$ is not known at $t=0^+$, and such an optimal behaviour cannot be achieved.

In the rest of this section the conditional probability $p(X_{n+1}|X_n)$ will be symmetric around X_n, i.e.

$$p(X_{n+1}|X_n)=p(|X_{n+1}-X_n|)=p_{n+1}(\cdot), \quad \ldots \quad (22)$$

however, it may depend upon n.

From this, it results immediately $E(X_{n+1}|X_n)=X_n$ for all n, which implies that $\{X_n\}$ form a martingale (Loève 1960).

4.2. *Cost Criterion*

The cost criterion $h(e)$ obeys the following conditions:

(1) $h(0)=0$,

(2) $h(-e)=+h(e)$,

(3) $h'(e)$ goes through zero at $e=0$ and is piecewise continuous,

(4) $h'(e)$ is monotone non-decreasing.

Functions satisfying these requirements are, for example, the absolute value, and any power of degree greater than 1; these conditions imply $h(e)$ is a continuous convex function.

4.3. *Optimal Policy for Two Intervals*

Given the servomechanism, an integrator, x_1 and y_0, the optimal policy for one interval is a direct consequence of the properties of $h(e)$. As $e=x_1-y_1, f_1(x_1, y_1)=h(x_1-y_1)$ is a minimum when $h'(x_1-y_1)$ goes through zero at

$$y_1=x_1 \quad \text{and} \quad r_1{}^0=\gamma \operatorname{sat}\left(\frac{x_1-y_0}{\gamma T}\right).$$

An equivalent description, using $y_1{}^0$, is:

$$\begin{aligned} y_1{}^0&=y_0+\gamma T && \text{for} \quad x_1>y_0+\gamma T, \\ y_1{}^0&=x_1 && \text{for} \quad y_0-\gamma T \leqslant x_1 \leqslant y_0+\gamma T, \\ y_1{}^0&=y_0-\gamma T && \text{for} \quad x_1<y_0-\gamma T. \end{aligned}$$

The cost $f_1{}^0(x_1, y_1{}^0)$ is zero as long as $y_1{}^0=x_1$.

For two intervals:

$$\begin{aligned} f_2(x_1, y_1) &= h(x_1-y_1)+a\underset{x_2}{E} f_1{}^0(x_2, y_1) \\ &= h(x_1-y_1)+a\int_{y_1+\gamma T}^{+\infty} f_1(x_2, y_2=y_1+\gamma T)p(x_2|x_1)\,dx_2 \\ &\quad +a\int_{-\infty}^{y_1-\gamma T} f_1(x_2, y_2=y_1-\gamma T)p(x_2|x_1)\,dx_2. \end{aligned} \tag{23}$$

Hence

$$\begin{aligned} \frac{\partial f_2(x_1, y_1)}{\partial y_1} &= h'(y_1-x_1)+a\int_{y_1+T}^{\infty}\left(\frac{\partial f_1}{\partial y_2}(x_2, y_2)\right)_{y_2=y_1+T} p(x_2|x_1)\,dx_2 \\ &\quad +a\int_{-\infty}^{y_1-\gamma T}\left(\frac{\partial f_1}{\partial y_2}(x_2, y_2)\right)_{y_2=y_1-\gamma T} p(x_2|x_1)\,dx_2 \quad . \end{aligned} \tag{24}$$

as $f_1(x_2, y_2)=0$ when $x_2=y_2$.

Theorem 1

When (22) holds and $h(e)$ obeys the conditions of § 3.2, then $\partial f_2(x_1, y_1)/\partial y_1$ as a function of y_1 has the properties:

(1) it goes through zero for $y_1=x_1$ and is piecewise continuous,

(2) it is odd symmetric around x_1,

(3) it is monotone non-decreasing.

The same theorem applies in the N intervals case and will be proved there.

We illustrate it when $h(e)=|e|$. Then (24) becomes:

$$\frac{\partial f_2(x_1, y_1)}{\partial y_1} = \operatorname{sgn}(y_1 - x_1) - a\int_{y_1+\gamma T}^{+\infty} p(x_2|x_1)\, dx_2 + a\int_{-\infty}^{y_1-\gamma T} p(x_2|x_1)\, dx_2. \quad (25)$$

At $y_1 = x_1$ the first term goes through zero and the two integrals cancel each other.

As $p(X_2|X_1)$ is symmetric around X_1, the difference of the two integrals has opposite values for $y_1 = x_1 + b$ and $y_1 = x_1 - b$. If we derive both integrals in (25) with respect to y_1, we get

$$ap(x_2 = y_1 + \gamma T) + ap(x_2 = y_1 - \gamma T) \geqslant 0,$$

which ensures the monotone non-decreasing property.

Corollary

The optimal policy is

$$r_1^0 = \gamma \operatorname{sat}\left(\frac{x_1 - y_0}{\gamma T}\right).$$

Proof

From the properties of the derivative mentioned in the theorem, it results that the cost functional will be the smallest when y_1, restricted to $[y_0 - \gamma T, y_0 + \gamma T]$ is as near x_1 as possible. Expressed for r_1, that statement becomes the claimed optimal policy.

4.4. *Optimal Policy for N Intervals*

From $p(X_{n+1}|X_n) = p(X_{n+1} - X_n)$ and $h(e) = h(x - y)$, it follows the cost functional f_N depends only on $x_1 - y_0$ for each N.

We have:

$$f_N(x_1, y_1) = h(x_1 - y_1) + a \underset{x_2}{E}\, f_{N-1}^0(x_2, y_1), \quad (26)$$

$$f_{N-1}^0(x_2, y_1) = \underset{y_2}{\operatorname{Min}}\, f_{N-1}(x_2, y_2) \quad \text{where} \quad y_2 \in [y_1 - \gamma T, y_1 + \gamma T]. \quad (27)$$

Theorem 2

Let $f_{N-1}(x_2, y_2)$ be such that $\partial f_{N-1}(x_2, y_2)/\partial y_2$, as a function of y_2:

(1) goes through zero at $y_2 = x_2$ and is piecewise continuous,

(2) is odd symmetric around x_2,

(3) is monotone non-decreasing.

Then $\partial f_N(x_1, y_1)/\partial y_1$ has the same properties, with the subscripts changed from 2 to 1. The proof is given in Appendix II.

Corollary 1

The properties of $\partial f_N(x_1, y_1)/\partial y_1$ given in the theorem are true for all N.

Proof

It suffices to show it to be true for $N=2$. Then by the induction relation of the theorem, it will be established for all N. When $N=2$, $f_{N-1}(x_2, y_2)=h(x_2-y_2)$, for which the properties are true, by definition of the cost function h itself.

Corollary 2

The optimal policy, for any N, is:

$$r_1^0=\gamma \operatorname{sat}\left(\frac{x_1-y_0}{\gamma T}\right).$$

This results immediately from the properties of $\partial f_N(x_1, y_1)/\partial y_1$. It is independent of the 'discount' factor a which may take any value $0 \leqslant a < \infty$.

Conclusion

For probability densities $p(X_{n+1}|X_n)=p(|X_{n+1}-x_n|)$, for the cost functional equal to the sum of the cost of the first interval and the minimal expected cost of the following intervals multiplied by the discount factor a, and for the cost function $h(e)$ satisfying the four conditions of § 4.2, then it has been shown the optimal policy is (16) for any N and any a, i.e. is the policy minimizing the cost at the end of the first interval.

Remarks

(1) The exchange between the differentiation with respect to y_1 and the expectation taken with respect to x_2 is allowed only when the functions considered are integrable with respect to x_2. If, for example, $h(e)$ is bounded above by $|e|^\nu$, ν finite, then requiring the random variables $\{X_n\}$ to belong to L_ν will assure integrability.

(2) The extension to servos with transfer function $G(s)=1/s+b$ is immediate, as outlined in the previous part (§ 3.6). The optimal policy is

$$y_1^0=y_0+B \operatorname{sat}\left(\frac{x_1-y_0}{B}\right),$$

where

$$B=(\gamma/b)[1-\exp(-bT)].$$

(3) If we weaken the conditions on $h(e)$, by replacing the monotone non-decreasing condition of $h'(e)$ by $eh'(e)>0$ for $e\neq 0$, we have that $\partial f_N(x_1, y_1)/\partial y_1$:

(*a*) goes through zero at $y_1=x_1$,

(*b*) is odd symmetric around x_1,

(*c*) $(y_1-x_1)\cdot\dfrac{\partial f_N(x_1, y_1)}{\partial y_1}>0$, $y_1\neq x_1$.

Then the cost functional f_N will still be extremal for $r_1=\gamma \operatorname{sat}(x_1-y_0)/\gamma T$ but it may not correspond to a minimum. For such an example see Appendix III.

4.5. *Extension to Discrete Random Variables*

We had previously assumed the existence of probability density functions $p(X_{n+1}|X_n)$, i.e. that the distribution functions $F(X_{n+1}|X_n)$ were Lebesgue absolutely continuous.

If now the random variables X_{n+1} are discrete, their distribution functions are step functions and the set of discontinuity points is countable.

The symmetry of X_{n+1} around $X_n = x_n$ is still assumed to hold, what is written $F(X_{n+1} - x_n) = 1 - F(x_n - X_{n+1})$.

The same hypotheses are made on the cost function $h(e)$. Theorem 2 proved before is still true, as we did not use the absolute continuity of $F(X_{n+1}|X_n)$ but only its symmetry property. Then the policy given by (16) is still the optimal one.

The only change in the proof of the theorem is that $\partial f_{N-1}/\partial y_1$ may have discontinuities on a countable set instead of on a finite one; its set of discontinuity points will remain finite if the random variables $\{X_{n+1}\}$ $(n = 1, 2, \ldots N-1)$ take their whole measure on finite sets only.

§ 5. Conclusions

For two different classes of discrete-time random processes, and for first-order systems, conditions have been given so that the optimal policy depends only on the error at the end of the first interval and not on the future remaining intervals. The extension to servomechanisms of higher order would be of great interest; in that case solutions can be given when the range of the state vector is only a finite set, and the range of the random process is finite too.

Acknowledgments

The results reported here are extracted from a Ph.D. thesis submitted at the University of California, Berkeley.

The author wishes to thank Professor C. A. Desoer, Department of Electrical Engineering, for proposing the subject and giving many helpful suggestions during the course of the research. This work was sponsored in part by the U.S. Air Force under Contract AF 19(604)-5466.

APPENDIX I

For random variables with finite first absolute moment and when the discount factor a is less than 1, f_N remains finite when $N \to \infty$.

Proof

Let us first assume $y_0 = 0$. f_N is the cost resulting from applying a policy in the first interval and the optimal policy in the $N-1$ remaining ones.

It is certainly less than the cost obtained when, at each step, we maximize the error: from

$$|e(nT)| = |x_n - y_{n-1} - r_n{}^0T| \leqslant |x_n + n\gamma T \operatorname{sgn} X_n| \quad . \quad \text{(A 1)}$$

we obtain:

$$f_N \leqslant |x_1 - y_1| + \sum_{n=2}^{N} \underset{X_2 \ldots X_n}{E} |X_n + n\gamma T \operatorname{sgn} X_n| a^{n-1}$$

$$= |x_1 - y_1| + \sum_{n=2}^{N} \underset{X_n}{E} |X_n + n\gamma T \operatorname{sgn} X_n| a^{n-1}$$

by independence of the X_n's.

$$f_N \leqslant |x_1 - y_1| + \sum_{n=2}^{N} a^{n-1} E|X_n| + \sum_{n=2}^{N} n\gamma T a^{n-1} E|\operatorname{sgn} X_n|$$

$$= |x_1 - y_1| + \sum_{n=2}^{N} a^{n-1} E|X_n| + \gamma T \sum_{n=2}^{N} n a^{n-1}.$$

As the right member remains finite as $N \to \infty$, when a is strictly less than one, f_N itself is finite.

When $y_0 \neq 0$, instead of replacing $y_{n-1} + r_n{}^0T$ by $-n\gamma T \operatorname{sgn} X_n$ in (A 1), we replace it now by $-(|y_0| + n\gamma T) \operatorname{sgn} X_n$ and the same arguments apply.

As $f_{N-1}{}^0(x_2, y_1) \leqslant f_{N-1} < \infty$ for all finite y_1, $f_{N-1}{}^0(x_2, y_1)$ is integrable with respect to x_2, whatever y_1 finite; by assumption

$$\left| \frac{\partial f_{N-1}{}^0(x_2, y_1)}{\partial y_1} \right| \leqslant c < \infty,$$

(eqn. (3.10)), then we may differentiate under the integral sign (Loève 1960) and eqn. (3.9) is correct.

APPENDIX II

Proof of Theorem 2

From the assumptions on $f_{N-1}(x_2, y_2)$, in the second interval:

$$r_2{}^0 = \gamma \operatorname{sat} \left(\frac{x_2 - y_1}{\gamma T} \right)$$

and

$$f_{N-1}{}^0(x_2, y_2) = \begin{cases} f_{N-1}(x_2, y_2 = x_2), & y_1 - \gamma T \leqslant x_2 \leqslant y_1 + \gamma T, \\ f_{N-1}(x_2, y_2 = y_1 + \gamma T), & x_2 > y_1 + \gamma T, \\ f_{N-1}(x_2, y_2 = y_1 - \gamma T), & x_2 < y_1 - \gamma T. \end{cases}$$

Hence

$$\begin{aligned} f_N(x_1, y_1) = h(y_1 - x_1) &+ a f_{N-1}(x_2, y_2 = x_2) P[y_1 - \gamma T \leqslant x_2 \leqslant y_1 + \gamma T] \\ &+ a \int_{y_1 + \gamma T}^{+\infty} f_{N-1}(x_2, y_2 = y_1 + \gamma T) p(x_2 | x_1) \, dx_2 \\ &+ a \int_{-\infty}^{y_1 - \gamma T} f_{N-1}(x_2, y_2 = y_1 - \gamma T) p(x_2 | x_1) \, dx_2, \end{aligned} \quad \text{(A 2)}$$

and by differentiating and simplifying

$$\frac{\partial f_N(x_1, y_1)}{\partial y_1} = h'(y_1 - x_1) + a \int_{y_1+\gamma T}^{+\infty} \frac{\partial f_{N-1}}{\partial y_2}(x_2, y_2 = y_1 + \gamma T) p(x_2|x_1)\, dx_2$$

$$+ a \int_{-\infty}^{y_1-\gamma T} \frac{\partial f_{N-1}}{\partial y_2}(x_2, y_2 = y_1 - \gamma T) p(x_2|x_1)\, dx_2. \qquad \text{(A 3)}$$

Value at $y_1 = x_1$

From (A 3). with the change of variable $x_2 = u + x_1 + \gamma T$ in the first integral and $x_2 = x_1 - \gamma T - u$ in the second one, these two integrals become:

$$a \int_0^{\infty} \frac{\partial f_{N-1}}{\partial y_2}(u + x_1 + \gamma T, y_2 = x_1 + \gamma T) p(|u + \gamma T|)\, du$$

$$+ a \int_0^{\infty} \frac{\partial f_{N-1}}{\partial y_2}(x_1 - \gamma T - u, y_2 = x_1 - \gamma T) p(|-u - \gamma T|)\, du.$$

As $\partial f_{N-1}/\partial y_2(x_2, y_2)$ goes through zero at $x_2 = y_2$, i.e. at $u = 0$ and is odd symmetric, the functions under the integral signs have opposite values for all u, and the two integrals cancel each other.

Therefore, when $y_1 = x_1$ (A 3) reduces to $h'(y_1 - x_1)$ which goes through zero. As the integrals in (A 3) are continuous functions of y_1, if follows that $h'(y_1 - x_1)$ be piecewise continuous implies the same for $\partial f_N(x_1, y_1)/\partial y_1$.

Odd Symmetry

Let us denote

$$F = a \frac{\partial}{\partial y_1} \underset{x_2}{E} f_{N-1}{}^0(x_2, y_1). \qquad \text{(A 4)}$$

By the preceding section $F(y_1 = x_1) = 0$. F is odd symmetric around x_1 if $F(y_1 = x_1 + b) = -F(y_1 = x_1 - b)$:

$$F(y_1 = x_1 + b) = a \int_{-\infty}^{x_1+b-\gamma T} \frac{\partial f_{N-1}}{\partial y_2}(x_2, y_2 = x_1 + b - \gamma T) p(x_2|x_1)\, dx_2$$

$$+ a \int_{y_1+b+\gamma T}^{+\infty} \frac{\partial f_{N-1}}{\partial y_2}(x_2, y_2 = x_1 + b + \gamma T) p(x_2|x_1)\, dx_2, \qquad \text{(A 5)}$$

$$F(y_1 = x_1 - b) = a \int_{-\infty}^{x_1-b-\gamma T} \frac{\partial f_{N-1}}{\partial y_2}(x_2, y_2 = x_1 - b - \gamma T) p(x_2|x_1)\, dx_2$$

$$+ a \int_{x_1-b+\gamma T}^{+\infty} \frac{\partial f_{N-1}}{\partial y_2}(x_2, y_2 = x_1 - b + \gamma T) p(x_2|x_1)\, dx_2. \qquad \text{(A 6)}$$

In the first term of (A 5) let $x_2 = x_1 + b - \gamma T - u$, and in the second term of (A 6) let $x_2 = x_1 - b + \gamma T + u$. Then by the symmetry of $p(x_2|x_1)$ and the odd symmetry of $\partial f_{N-1}/\partial y_2(x_2, y_2)$, these two terms have opposite values. The same being true for the other terms of (A 5) and (A 6) we have:

$$F(y_1 = x_1 + b) = -F(y_1 = x_1 - b).$$

Monotone Non-decreasingness

$h'(y_1-x_1)$ being monotone non-decreasing by assumption, let us look at the integrals only:

$$\begin{aligned}\frac{\partial^2}{\partial y_1^2} f_{N-1}^0(x_2, y_1) = a\frac{\partial f_{N-1}}{\partial y_2}(x_2 = y_2 = y_1 - \gamma T)p(x_2 = y_1 - \gamma T) \\ - a\frac{\partial f_{N-1}}{\partial y_2}(x_2 = y_2 = y_1 + \gamma T)p(x_2 = y_1 + \gamma T) \\ + a\int_{-\infty}^{y_1-\gamma T} \frac{\partial^2 f_{N-1}}{\partial y_2^2}(x_2, y_2 = y_1 - \gamma T)p(x_2|x_1)\,dx_2 \\ + a\int_{y_1+\gamma T}^{+\infty} \frac{\partial^2 f_{N-1}}{\partial y_2^2}(x_2, y_2 = y_1 + \gamma T)p(x_2|x_1)\,dx_2. \qquad \text{(A 7)}\end{aligned}$$

The first two terms are zero, and as $\partial^2 f_{N-1}/\partial y_2^2 \geqslant 0$ by the monotone non-decreasing property of $\partial f_{N-1}/\partial y_2$, the second derivative of f_{N-1}^0 is greater or equal to zero, which proves the monotone non-decreasing property.

APPENDIX III

Weakened Assumptions for Theorem 2

Let the cost function $h(e)$ be:

$$h(e) = \begin{cases} 2|e| & -\tfrac{1}{2} \leqslant e \leqslant \tfrac{1}{2}, \\ 1 + \epsilon(|e| - \tfrac{1}{2}) & \text{elsewhere, with } \epsilon > 0. \end{cases}$$

Then

$$\begin{cases} eh'(e) > 0 & e \neq 0, \\ h'(e) & \text{odd symmetric,} \\ h'(e) & \text{goes through zero at } e = 0. \end{cases}$$

In the following example, we show that with these weakened assumptions, the policy (4.16) is no more optimal.

Let the initial values be: $x_1 = 3$, $y_0 = 1$, $N = 2$, $\gamma T = 1$, and the random variable X_2 defined by:

$$\begin{aligned} X_2 &= x_1 + 3 = 6, && \text{with probability } \tfrac{1}{2}, \\ X_2 &= x_1 - 3 = 0, && \text{with probability } \tfrac{1}{2}. \end{aligned}$$

For this two-interval process we have:

$$f_2(x_1, y_0) = h(3 - y_1) + (a/2)h(6 - y_2) + (a/2)h(-y_2).$$

We compare now the two policies:

(1) The 'optimal' policy (4.16), i.e. $r_1 = \gamma \operatorname{sat}(x_1 - y_0)/\gamma T$. Here we get: $r_1 = \gamma$, $y_1 = y_0 + \gamma T = 2$, and in the second interval:

$$\begin{aligned} r_2 &= +\gamma, \quad \text{if } X_2 = 6, \\ r_2 &= -\gamma, \quad \text{if } X_2 = 0. \end{aligned}$$

Hence

$$f_2(x_1, y_0) = h(3-2) + (a/2)h(6-3) + (a/2)h(-1) = (1 + \epsilon/2) + (a/2)(2 + 3\epsilon).$$

(2) From y_0, try to reach one of the possible values of X_2, here $X_2=0$. while keeping the error at $t=T$ as small as possible. Then:

$$r_1=0, \quad r_2=+\gamma, \quad \text{if} \quad X_2=6,$$
$$r_2=-\gamma, \quad \text{if} \quad X_2=0.$$

Hence

$$f_2(x_1, y_0)=h(3-1)+(a/2)h(6-2)+(a/2)h(0)=1+\frac{3\epsilon}{2}+\frac{a}{2}\left(1+\frac{7\epsilon}{2}\right).$$

The second policy is better than the first one if and only if its cost functional is smaller, i.e. if and only if:

$$1+\frac{3\epsilon}{2}+\frac{a}{2}\left(1+\frac{7\epsilon}{2}\right) \leqslant 1+\frac{\epsilon}{2}+\frac{a}{2}(2+3\epsilon)$$

or

$$a \geqslant \frac{2\epsilon}{1-\epsilon/2}.$$

As ϵ may be as small as we wish, policy (4.16) is no more the optimal policy whatever $a>0$.

References

Bellman, R., Glicksberg, I., and Gross, O., 1958, *Rand Rep.* R–313.

Boltyanskii, V. G., Gamkrelidze, R. V., and Pontryagin, L. S., 1960, *Bull. Akad. Sci. U.R.S.S.*, **24,** 3.

Booton, R. C., 1953, *Proc. Brooklyn Symposium on Non-linear Circuit Analysis*, **2,** 369.

Desoer, C. A., 1961, *J. Franklin Inst.*, **271,** 361.

Eaton, J., and Zadeh, L. A., 1961, " Optimal Pursuit Strategies in Discrete State Probabilistic Systems ". presented at the J.A.C.C., Boulder, Colorado, June.

Feller, W., 1959, *An Introduction to Probability Theory and its Applications*, 2nd edition (New York : John Wiley & Sons Inc.), p. 368.

Fuller, A. T., 1961, *J. Electron. Contr.*, **10,** 61.

Govaerts, P. H., 1961, *Ph.D. Thesis*, University of California.

LaSalle, J. P., 1960, *Theory of Nonlinear Oscillations*, **5** (Princeton University Press), p.1.

Loève, M., 1960, *Probability Theory*, 2nd edition (New York : D. van Nostrand) pp. 239, 389, 126.

Newton, G. C., 1952, *J. Franklin Inst.*, **254,** 281, 391.

Optimal Control of Continuous Time, Markov, Stochastic Systems†

By J. J. FLORENTIN
Electrical Engineering Department,
The Imperial College of Science and Technology, London

[Received June 8, 1961]

ABSTRACT

The control problem considered is that of a non-linear, time varying plant which is disturbed by, or must follow, some stochastic process. Further, the system of plant plus inputs must be adequately represented by a mathematical model whose probability function description satisfies the Markov condition. The control is to be chosen, within allowed limits, so as to optimize a given performance index.

Using concepts based on Bellman's Dynamic Programming, partial differential equations are set up whose solution yields the value of the performance index, and the optimal control settings over some prescribed time interval. A discussion shows that continuous observation of the system is a theoretical necessity, in contrast to the deterministic situation. Usually the observations will entail feedback.

For the special class of problems with linear plant dynamics, quadratic performance indices, and Gaussian random components, all with time varying parameters, the partial differential equations are reduced to a set of ordinary differential equations, well suited to machine solution. Because of its simplicity, this method supplants previous solutions to these special problems, based on complex variable and integral equation techniques.

LIST OF SYMBOLS

$A(x;t)$	Term in plant equation.
$\mathbf{B}(t)$	Cost of control matrix.
$\mathbf{D}(t)$	Control matrix.
E	Expectation.
$F(z)$	Distribution function of random impulses.
$G(\Delta x/x;t)$	Distribution function of increments in x at time t.
$K(x_0, t_0)$	Value of performance index when system starts from x_0, t_0.
k_0, $\mathbf{K}_1$, $\mathbf{K}_2$	Scalar, vector, and matrix in expansion of K.
$L(x, t, \theta)$	Integrand in path integral performance index.
m	Mean value of random process.
n	Order of system equations.
$P(x;t/x_0;t_0)$	Probability density of x at time t, conditional on $x=x_0$ at time t_0.
$\mathbf{Q}(t)$	Cost of state matrix.
t, T	Time.

M

† Communicated by the Author.

u	Dummy time variable.
$\mathbf{x}$	State vector.
z	Magnitude of random impulse.
λ	Mean rate of Poisson process.
θ	Control variable.
ξ	A completely random process.
σ_{rs}^2	Co-variance of process r and s.
$\sum$	Summation sign.
τ	Future time.
Φ	Performance function.
$*$	Distinguishes optimal value of quantity.

§ 1. Introduction

Methods for the optimal control of dynamic systems have been thoroughly investigated under deterministic conditions. When the plant to be controlled is subject to random inputs with known statistical parameters, the problem has been called 'stochastic' by Bellman (1956). The present paper is an extension of the approach pioneered by Bellman; further collateral reading is to be found in the book by Howard (1960).

The control situation considered is that shown in fig. 1. The equations of the plant are given, $\boldsymbol{\xi}$ represents a set of random inputs with known statistical distributions, $\mathbf{x}$ is the state vector; $\boldsymbol{\theta}$ is a set of variables which can be used to control the plant. A suitable performance index is given, and the controller must bring this to an extremum by applying a suitable sequence of settings to the control variables, $\boldsymbol{\theta}$. The present method enables the partial differential equation of optimal control to be set up, the solution of this equation leads to the decision rule which generates the required $\boldsymbol{\theta}$ sequence over some defined time period.

Fig. 1

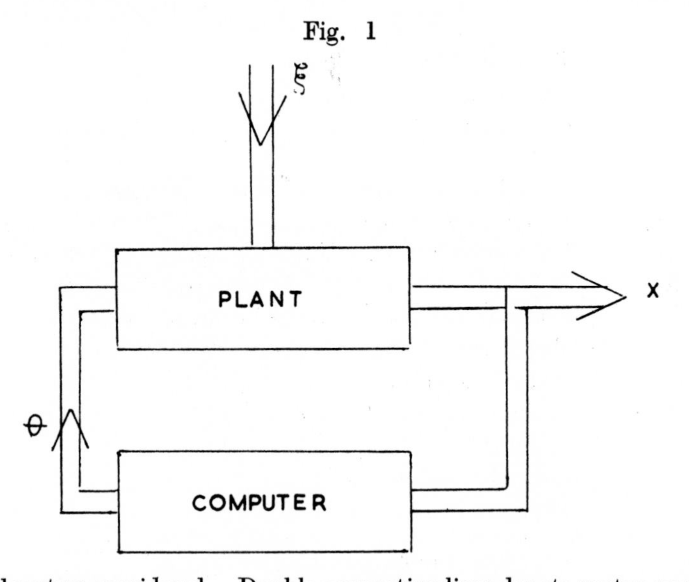

General system considered. Double connecting lines denote vector quantities.

This basic problem has been studied for 20 years, starting with the well-known work of Wiener; the present method goes beyond existing work by providing control equations for any system which, by correct choice of state vector components, may have a Markov description. In practice, nearly all systems of engineering interest may be adequately represented by mathematical models having Markov properties. These properties are not restrictive, and the plant and random process may be non-linear and non-stationary; further, the resulting optimal control may itself be a non-linear, non-stationary function of the system state vector components.

In many problems the present method leads to numerical formulae whose form is well suited to machine computation, examples of this are given. In those examples where it may be compared to previous techniques, it leads to numerical formulae which are substantially simpler.

An interesting point is that the equations which are set up imply continuous observation of the plant, together with a continuous up-dating of the actual control policy, usually this will mean feedback. This need for continuous observation marks a critical difference between deterministic and stochastic control.

§ 2. The Dynamic Transition Equations

The state vector description of the system will be used; this is a set of numbers which, at any single time instant, completely specify the system.

Initially, to simplify the presentation, terminal control only is considered. Because the system is subject to random inputs, the performance index will always be an averaged quantity, at this stage the average of some function Φ, of the final state, $\mathbf{x}(T)$. Thus, the performance index, K, is

$$K = E\{\Phi(\mathbf{x}(T))\}, \quad \text{(2.1)}$$

where E denotes expectation.

The correct description of a stochastic system is the joint probability function of the state vector components. If the system is Markov the probability density functions at different times are related by the Chapman–Kolmogorov equation, here used in retrospective form as

$$P(\mathbf{x}; T \mid \mathbf{x}_0; t_0) = \int P(\mathbf{x}; T \mid \mathbf{x}_0 + \Delta\mathbf{x}_0; t_0 + \Delta t_0)\, dG(\Delta\mathbf{x}_0 \mid \mathbf{x}_0; t_0), \quad \text{(2.2)}$$

where $t_0 < T$. $P(\mathbf{x}; T \mid \mathbf{x}_0; t_0)\, d\mathbf{x}$ is the probability of the system representative point being in a small region round $\mathbf{x}$ at time T, given that it was at $\mathbf{x}_0$ at time t_0. $G(\Delta\mathbf{x} \mid \mathbf{x}; t)$ is the distribution function of changes in system position up to, and including $\Delta\mathbf{x}$, when the system is at $\mathbf{x}$ at time t.

Multiplying both sides of eqn. (2.2) by $\Phi(\mathbf{x}(T))$, and integrating over $\mathbf{x}(T)$ gives

$$K(\mathbf{x}_0; t_0) = \int K(\mathbf{x}_0 + \Delta\mathbf{x}_0; t_0 + \Delta t_0)\, dG(\Delta\mathbf{x}_0 \mid \mathbf{x}_0; t_0). \quad \text{(2.3)}$$

Here the performance index has been made a function of the system initial state, and a dynamic transition equation for performance index has been obtained. This is a key relation in what follows.

In continuous dynamic systems attention may be concentrated on the effects of very small changes in starting time. First note that G is a distribution and integrates to unity, so that

$$K(\mathbf{x}_0; t_0) - K(\mathbf{x}_0; t_0 + \Delta t_0)$$
$$= \int [K(\mathbf{x}_0 + \Delta\mathbf{x}_0; t_0 + \Delta t_0) - K(\mathbf{x}_0; t_0 + \Delta t_0)]\, dG(\Delta\mathbf{x}_0 \,|\, \mathbf{x}_0; t_0). \quad (2.4)$$

On dividing by Δt_0, assuming a limit,

$$-\frac{\partial K}{\partial t_0} = \lim_{\Delta t_0 \to 0} \int \frac{[K(\mathbf{x}_0 + \Delta\mathbf{x}_0; t_0 + \Delta t_0) - K(\mathbf{x}_0; t_0 + \Delta t_0)]}{\Delta t_0}\, dG(\Delta\mathbf{x}_0 \,|\, \mathbf{x}_0; t_0). \quad (2.5)$$

If the limit on the right-hand side exists, a partial differential equation for the performance index transitions will have been obtained.

In eqn. (2.5) it is the function $G(\Delta\mathbf{x} \,|\, \mathbf{x}; t)$ which expresses the physical details of the system. The change in system state in a short time will be partly deterministic, depending on the time constants, non-linearities and other restoring forces and partly sudden jumps due to the action of the purely random components. If the purely random components are generated by processes having infinitely divisible increments, then a Markov system model can usually be constructed. Moreover, the assumption of processes with infinitely divisible increments limits the classes of distributions of the purely random portion to either Gaussian, or generalized Poisson. This further leads to a straightforward development of partial differential equations from eqn. (2.5). The assumption of infinitely divisible increments will be made here, for an extensive discussion of these processes, and their physical implications see Levy (1948).

In order to make the discussion less abstract a very simple process will be considered, and examples of partial differential equations, typical of the allowed distributions, will be derived. Consider the single dimensional process

$$\dot{x} = A(x; t) + \xi, \quad (2.6)$$

where ξ is a process having independent increments, i.e. completely random noise. Some explicit distributions for ξ will now be taken.

2.1. *Gaussian Distribution*

If ξ is non-stationary, non-linear, Gaussian, with mean $m(x; t)$ and variance $\sigma^2(x; t)$, then for Δt small, dropping the arguments for convenience:

$$dG(\Delta x \,|\, x; t) = \frac{1}{\sqrt{(2\pi\sigma^2 \Delta t)}} \exp \frac{-(\Delta x - A\Delta t - m\Delta t)^2}{2\sigma^2 \Delta t}. \quad (2.7)$$

Now Δx will be very small, so expand the terms in the square bracket in eqn. (2.5), obtaining terms such as

$$K + \Delta x_0 \frac{\partial K}{\partial x_0} + \Delta t_0 \frac{\partial K}{\partial t_0} + \frac{\Delta^2 x_0}{2!} \frac{\partial^2 K}{\partial x_0^2} + \Delta x_0 \Delta t_0 \frac{\partial^2 K}{\partial x_0 \partial t_0} + \frac{\Delta^2 t_0}{2!} \frac{\partial^2 K}{\partial t_0^2}.$$
$$- K - \Delta t_0 \frac{\partial K}{\partial t_0} - \frac{\Delta^2 t_0}{2!} \frac{\partial^2 K}{\partial t_0^2}. \quad (2.8)$$

Using the distribution of (2.7) and integrating the quantities $(\Delta x_0)^r$, also retaining terms of order Δt_0 only, (2.8) then becomes

$$(m\Delta t_0 + A\Delta t_0)\frac{\partial K}{\partial x_0} + \frac{\sigma^2}{2}\Delta t_0 \frac{\partial^2 K}{\partial x_0{}^2}; \quad \ldots \ldots \quad (2.9)$$

further, dividing by Δt_0, and taking the limit, yields the partial differential equation

$$-\frac{\partial K}{\partial t_0} = (m+A)\frac{\partial K}{\partial x_0} + \frac{\sigma^2}{2}\frac{\partial^2 K}{\partial x_0{}^2}. \quad \ldots \quad (2.10)$$

2.2. *Poisson Distribution*

As a first example of a Poisson distribution suppose ξ to be an impulse of magnitude z, which may occur in any short time interval Δt with probability $\lambda\Delta t$, then to first order in Δt

$$\begin{aligned} \Delta x_0 &= A\Delta t_0 + z \text{ with probability } \lambda\Delta t_0 \\ &= A\Delta t_0 \quad \text{ with probability } (1-\lambda\Delta t_0). \end{aligned}$$

The square bracketed terms in eqn. (2.5) become

$$K(x_0 + A\Delta t_0 + z)\lambda\Delta t_0 + K(x_0 + A\Delta t_0)(1-\lambda\Delta t_0) - K. \quad . \quad (2.11)$$

Because Δt_0 is very small this may be expanded in terms of $A\Delta t_0$, giving

$$\lambda\Delta t_0\left(K(x_0+z) + A\Delta t_0 \frac{\partial K}{\partial(x_0+z)}\right) + \left(K(x_0) + A\Delta t_0\frac{\partial K}{\partial x_0}\right)(1-\lambda\Delta t_0) - K. \quad \ldots \quad (2.12)$$

Dividing by Δt_0, and taking the limit, yields the partial differential equation

$$-\frac{\partial K}{\partial t_0} = A\frac{\partial K}{\partial x_0} + \lambda K(x_0+z) - \lambda K. \quad \ldots \quad (2.13)$$

If, further, the magnitude of the randomly occurring impulse is not fixed, but at each occurrence is independently chosen from a distribution $F(z)$, the equation must be modified to

$$-\frac{\partial K}{\partial t_0} = A\frac{\partial K}{\partial x_0} + \int \lambda K(x_0+z)\,dF(z) - \lambda K. \quad \ldots \quad (2.14)$$

2.3. *Discrete Level Poisson Process*

Suppose ξ is a discrete level process, that is, a multi-level step wave, with a denumerable set of levels ξ_i. The chance of a jump in level occurring in a short time Δt, when in state i, is $\lambda_i\Delta t$; and, given a jump has occurred, the conditional probability of going to state j is p_{ij}. The process $\{x\}$ is no longer Markov, but the joint process $\{x, \xi\}$ is Markov. Consider the set of initial states (x_0, ξ_i) with associated performance indices $K(x_0, \xi_i) = K_i(x_0)$.

Following a very similar argument to that in § 2.2 leads to the following set of partial differential equations

$$-\frac{\partial K_i(x_0)}{\partial t_0} = A_i\frac{\partial K_i}{\partial x_0} + \sum_{j\neq i}\lambda_i p_{ij} K_j - \lambda_i K_i, \quad \ldots \quad (2.15)$$

where $i = 1, 2, \ldots$ and $A_i = A(x_0; t_0) + \xi_i$.

2.4. *Multi-dimensional Processes*

Consider now the multi-dimensional process, with state equations

$$\dot{x}_i = A_i(x_1, x_2, \ldots x_n) + \xi_i, \quad i = 1, 2, \ldots n, \quad \ldots \quad (2.16)$$

where ξ is a multi-dimensional process.

The above partial differential equations may readily be generalized; to quote one example, the Gaussian eqn. (2.10) becomes

$$-\frac{\partial K}{\partial t_0} = \sum (m_i + A_i) \frac{\partial K}{\partial x_{0,i}} + \sum\sum \frac{\sigma_{ij}^2}{2} \frac{\partial^2 K}{\partial x_i \partial x_j}, \quad \ldots \quad (2.17)$$

where σ_{ij}^2 is the covariance of the ξ_i, ξ_j components.

The above equations are similar in form to retrospective probability equations. Equations for probabilities in Markov control systems have been given by Barrett (1960), and forward equations for the step-wave case have been given by Wonham (1960).

§ 3. Scheme of Optimal Control

Control is exercised in a stochastic system when the transition probabilities can be adjusted, within limits, by the controller. This is indicated by introducing the control variable, $\boldsymbol{\theta}$, into the transition probability; in application $\boldsymbol{\theta}$ would be an adjustable set point, a relay setting, or generally the variable by which the controller adjusts the plant. The averaged value of the performance index when the system starts from $\mathbf{x}_0$, t_0 is now

$$K(\mathbf{x}_0, t_0) = \int K(\mathbf{x}_0 + \Delta\mathbf{x}_0; t_0 + \Delta t_0)\, dG(\Delta\mathbf{x}_0 \,|\, \mathbf{x}_0; t_0; \boldsymbol{\theta}). \quad . \quad (3.1)$$

The possibility now exists of choosing $\boldsymbol{\theta}$ as a function of time, over some prescribed interval, so that K is minimized (maximized). When the minimum (maximum) of K has been found, subject to the constraints on $\boldsymbol{\theta}$, it will be called optimal, and where necessary distinguished as K^*.

Taking up the iterative structure and computational scheme of Bellman's Dynamic Programming, a required relation would be

$$K^*(\mathbf{x}_0, t_0) = \min_{\boldsymbol{\theta}} (\max) E_{\Delta\mathbf{x}_0} \{K^*(\mathbf{x}_0 + \Delta\mathbf{x}_0; t_0 + \Delta t_0)\}. \quad . \quad (3.2)$$

Then computation will be done entirely in terms of the optimal, starred, quantity. It has sometimes been argued that eqn. (3.2) follows by the same reasoning as is involved in deterministic dynamic programming. This is *not* correct, and as it happens the correct mathematical interpretation of eqn. (3.2) supplies a valuable insight into the physical arrangement of control. In the deterministic situation starting with a known $\mathbf{x}_0$, t_0 and a final value criterion at time T, it is possible to find the required $\boldsymbol{\theta}(t)$ over the entire interval $t_0 < t < T$. This is also possible in the stochastic situation, and will be discussed in detail in a forthcoming paper by Katz. However, this is not the arrangement indicated by eqn. (3.2), in order to satisfy the alternating sequence of minimizing (maximizing) and averaging it is necessary that $\boldsymbol{\theta}$ be chosen as a number in the period $[t_0, t_0 + \Delta t_0]$, but only as a decision rule in $[t_0 + \Delta t_0, T]$. Thus the actual value of $\boldsymbol{\theta}$ in the second and succeeding intervals will only be known when the particular realization of the random process in the preceding intervals has been observed. To implement eqn. (3.2) it is necessary to observe the sequence of states, as the process unfolds, and calculate the required $\boldsymbol{\theta}$ according to which of the possible random trajectories is actually being followed. This can be done either by observing the random inputs and computing the resulting changes in system state, or by directly observing the changes of system state, in which case feedback is being used.

With the reservations made above, note that eqn. (3.2) is simply the dynamic transition equation of the performance index with the addition of an extremum condition. Thus all the equations of § 2, when an appropriate $\boldsymbol{\theta}$ is incorporated may be minimized (maximized) to obtain the equations of optimal control.

§ 4. Equations of Optimal Control

Usually control will be exercised through the dynamics of system, by adding a further forcing term, or varying a parameter in the plant equation, although the present theory is of such generality that it includes the possibility of control by varying the parameters of the noise. In order to illustrate an equation of optimal control consider the simple example of a first order linear system where control is effected by a forcing function

$$\dot{x} = A(x\,;\,t) + \theta(t) + \xi. \qquad (4.1)$$

The actual equation of optimal control depends on the distribution function of the random term, for instance, for ξ Gaussian, the equation required is (2.10) with an added extremum condition, here taken as a minimum, that is

$$-\frac{\partial K^*}{\partial t_0} = \min_{\theta}\left[(A+\theta)\frac{\partial K^*}{\partial x_0} + \frac{\sigma^2}{2}\frac{\partial^2 K^*}{\partial {x_0}^2}\right]. \qquad (4.2)$$

In eqn. (4.2) the minimizing is to be carried out first, resulting in the optimal θ being expressed as a function of the current state and derivatives of K. After substituting for θ^*, the equation is to be solved with appropriate boundary conditions derived by reference to the performance index. Usually the performance index will be a simple function of the terminal conditions, so that the value of K starting at time T, i.e. with no time left to go, at a specified x, will be known, e.g., if the performance index is $E\{x^2(T)\}$ then

$$K^*(x', T) = (x')^2. \qquad (4.3)$$

Other boundary conditions, e.g. as $x \to \infty$, will usually be apparent.

In many systems the performance index is some function summed up over the entire trajectory, rather than a simple end point condition, in this case for a given starting $\mathbf{x}_0$:

$$K^*(\mathbf{x}_0\,;\,t_0) = \min_{\boldsymbol{\theta}}\,(\max)\,E\left\{\int_{t_0}^{T} L(\mathbf{x}, u, \boldsymbol{\theta})\,du\right\}. \qquad (4.4)$$

On putting $K = x_{n+1}$, there results $\dot{x}_{n+1} = L(\mathbf{x}, t, \boldsymbol{\theta})$, which is just an extra dynamic equation, and the optimizing is now equivalent to minimizing (maximizing) $x_{n+1}(T)$, so that the problem is equivalent to the previous one. This equivalence is, of course, well known in the deterministic problem, and has been pointed out by Rozonoer (1959). When it comes to writing out the control equations x_{n+1} need not actually appear. A way of seeing this is to note that $K^*(\mathbf{x}_0, t_0)$ may also be written

$$K^*(\mathbf{x}_0, t_0) = \min_{\boldsymbol{\theta}} (\max) E_{\Delta \mathbf{x}_0} \left\{ \int_{t_0}^{t_0 + \Delta t_0} L(\mathbf{x}, u, \boldsymbol{\theta})\, du + K^*(\mathbf{x}_0 + \Delta \mathbf{x}_0, t_0 + \Delta t_0) \right\} \quad . \; . \; . \quad (4.5)$$

To first order in Δt_0 the averaged value of the integral is $L \Delta t_0$, so that

$$K^*(\mathbf{x}_0, t_0) = \min_{\boldsymbol{\theta}} (\max) \left[L(\mathbf{x}_0, t_0, \boldsymbol{\theta})\, \Delta t_0 + E_{\Delta \mathbf{x}_0} \{ K^*(\mathbf{x}_0 + \Delta \mathbf{x}_0, t_0 + \Delta t_0) \} \right]. \quad (4.6)$$

On expanding $K^*(\mathbf{x}_0 + \Delta \mathbf{x}_0, t_0 + \Delta t_0)$ and averaging, the terms $K^*(\mathbf{x}_0, t_0)$ cancel. On further dividing through by Δt_0 a partial differential equation will be obtained, similar in form to the previous equations, but containing the function L, of course no term in x_{n+1} appears.

To illustrate this type of path integral performance index, and the resulting optimal control equation take the example of equation (4.1) with a performance index

$$K^*(\mathbf{x}_0, t_0) = \min_{\theta} E \left\{ \int_{t_0}^{T} (x^2 + \theta^2)\, du \right\}. \quad . \; . \; . \; . \quad (4.7)$$

The resulting equation of optimal control is

$$-\frac{\partial K^*}{\partial t_0} = \min_{\theta} \left[x_0{}^2 + \theta^2 + (A + \theta) \frac{\partial K^*}{\partial x_0} + \frac{\sigma^2}{2} \frac{\partial^2 K^*}{\partial x_0{}^2} \right]. \quad . \quad (4.8)$$

What has really been indicated is that

$$\frac{\partial K^*}{\partial x_{n+1}} = 1.$$

§ 5. Formulation and Solution of Control Problems

The above theory gives a basis for the systematic solution of stochastic control problems. For brevity, the steps involved in setting up, and solving the control problem are given in order:

(i) Write down the dynamic equations of the plant, including the random, and control, terms.

(ii) From (i) find the transition probabilities of the system; then using the performance index specification, write down the transition equation for the performance index.

(iii) Find the $\boldsymbol{\theta}$ in the allowed range, as a function of the state, and performance index derivatives, which minimizes, (maximizes) the control equation.

(iv) Substitute for $\boldsymbol{\theta}^*$, and solve the resulting equation for optimal performance index, using boundary conditions obtained by considering the nature of the performance index.

(v) Use the optimal performance index solution to find $\boldsymbol{\theta}^*$ as a function of current state, and a set of time varying parameters.

(vi) The resultant explicit expression for $\boldsymbol{\theta}^*$ is a full mathematical specification of the controller operation. The practical instrumentation of the controller is a realization of the mathematical specification, this will not be considered further.

The difficult step is (iv), that is solving an equation, which may be integro-partial differential, and non-linear. There is, however, one class of problems whose solution is straightforward; these are problems with quadratic performance indices, linear dynamics, and additive Gaussian noise. The reader may immediately object, saying that these can be solved by very well-known, earlier, techniques based on the work of Wiener. However, in practice, problems involving non-stationary parameters, and a finite optimizing time interval, were almost completely intractable by the earlier methods. In the present method such problems are elementary and lead immediately to practical computing techniques. To explain the technique two very simple examples will be given.

5.1. *Servo Following a Stochastic Command Signal*

The system is shown in fig. 2. The plant is an integrator with equation

$$\dot{x}_1 = \theta. \qquad (5.1)$$

The command signal is assumed to be equivalent to white noise of zero mean, and unit variance, passed through a unit time constant, and has equation

$$\dot{x}_2 = -x_2 + \xi. \qquad (5.2)$$

Fig. 2

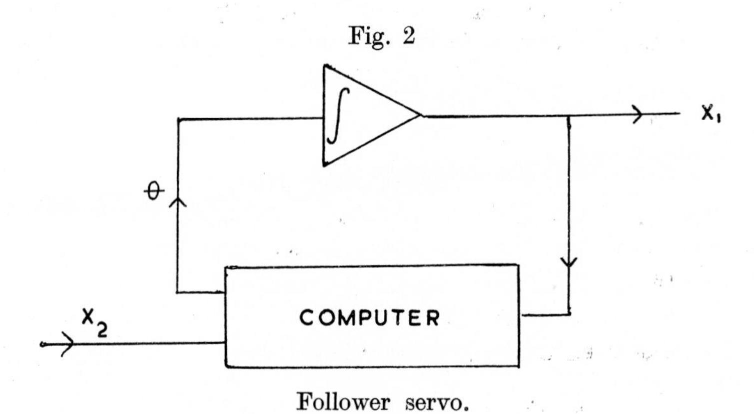

Follower servo.

Equal weight is given to following error, and cost of control, and the performance index is

$$K^*(\mathbf{x}_0, t_0) = \min E\left\{\int_{t_0}^{T} [(x_1 - x_2)^2 + \theta^2]\, du\right\}. \qquad (5.3)$$

For convenience put $\tau = T - t_0 =$ time left to go. The resulting equation of optimal control is

$$\frac{\partial K^*}{\partial \tau} = \min_{\theta}\left[(x_1 - x_2)^2 + \theta^2 + \theta\frac{\partial K^*}{\partial x_1} - x_2\frac{\partial K^*}{\partial x_2} + \frac{\sigma^2}{2}\frac{\partial^2 K^*}{\partial {x_2}^2}\right]. \qquad (5.4)$$

Differentiating with respect to θ, the minimum is found for

$$\theta^* = -\tfrac{1}{2}\frac{\partial K^*}{\partial x_1}. \qquad (5.5)$$

M1

On substituting for θ^* the equation to be solved is

$$\frac{\partial K^*}{\partial \tau}=(x_1-x_2)^2-\tfrac{1}{4}\left(\frac{\partial K^*}{\partial x_1}\right)^2-x_2\frac{\partial K^*}{\partial x_2}+\frac{\sigma^2}{2}\frac{\partial K^*}{\partial {x_2}^2}. \quad (5.6)$$

No special conditions have to be met at the terminus, and the boundary conditions are

$$K^*(\mathbf{x},0)=0, \quad \text{all } \mathbf{x}; \quad K^*(\mathbf{x},\tau)\to\infty \quad \text{as} \quad \|\mathbf{x}\|\to\infty. \quad (5.7)$$

Try fitting a series solution, this is similar to the approach used in the deterministic case by Merriam III (1960):

$$K^*(\mathbf{x},\tau)=k_0(\tau)+\sum k_i(\tau)x_i+\sum\sum k_{ij}(\tau)x_ix_j+\sum\sum\sum k_{ijm}(\tau)x_ix_j\; x_m, \quad (5.8)$$

where the k-coefficients depend only on time.

On substituting it will be found that only the constant and quadratic terms are required, and that one can choose $k_{ij}=k_{ji}$. If the matrix of the quadratic coefficients is $\mathbf{K}_2$, the resultant coefficient equations are

$$\left.\begin{aligned} &\dot{k}_0=\sigma^2k_{22},\\ &\dot{\mathbf{K}}_2=\begin{bmatrix}1, & -1\\ -1, & 1\end{bmatrix}-\mathbf{K}_2\begin{bmatrix}1, 0\\ 0, 0\end{bmatrix}\mathbf{K}_2-\begin{bmatrix}0, 0\\ 0, 1\end{bmatrix}\mathbf{K}_2-\mathbf{K}_2\begin{bmatrix}0, 0\\ 0, 1\end{bmatrix}, \end{aligned}\right\} \quad (5.9)$$

with boundary conditions, $k(0)=0$. These may easily be solved on an analogue, or digital, computer. The author actually used a digital machine, and the results are shown in fig. 3. The explicit expression for θ^* is

$$\theta^*=-k_{11}x_1-k_{12}x_2, \quad (5.10)$$

see fig. 4 for controller.

5.2. *Regulator with Noise*

The system is shown in fig. 5; the plant is again an integrator, and the disturbances have the same statistical structure as before, the dynamic equations are

$$\dot{x}_1=x_2+\theta, \quad \dot{x}_2=-x_2+\xi. \quad (5.11)$$

Wishing to keep x_1 small, the performance index is

$$K^*(x_0,t_0)=\min_{\theta} E\left\{\int_{t_0}^{T}(x_1{}^2+\theta^2)\,du\right\}. \quad (5.12)$$

It is assumed that the noise can be observed, so that the controller may have a section to provide a copy coordinate of x_2, see Fuller (1960). The equation of optimal control is

$$\frac{\partial K^*}{\partial \tau}=\min_{\theta}\left[x_1{}^2+\theta^2+(\theta+x_2)\frac{\partial K^*}{\partial x_1}-x_2\frac{\partial K^*}{\partial x_2}+\frac{\sigma^2}{2}\frac{\partial K^*}{\partial {x_2}^2}\right]. \quad (5.13)$$

Differentiation gives the optimal control as

$$\theta^*=-\tfrac{1}{2}\frac{\partial K^*}{\partial x_1}, \quad (5.14)$$

and the equation to be solved is

$$\frac{\partial K^*}{\partial \tau}=x_1{}^2-\tfrac{1}{4}\left(\frac{\partial K^*}{\partial x_1}\right)^2+x_2\left(\frac{\partial K^*}{\partial x_1}-\frac{\partial K^*}{\partial x_2}\right)+\frac{\sigma^2}{2}\frac{\partial^2 K^*}{\partial {x_2}^2}. \quad (5.15)$$

Fig. 3

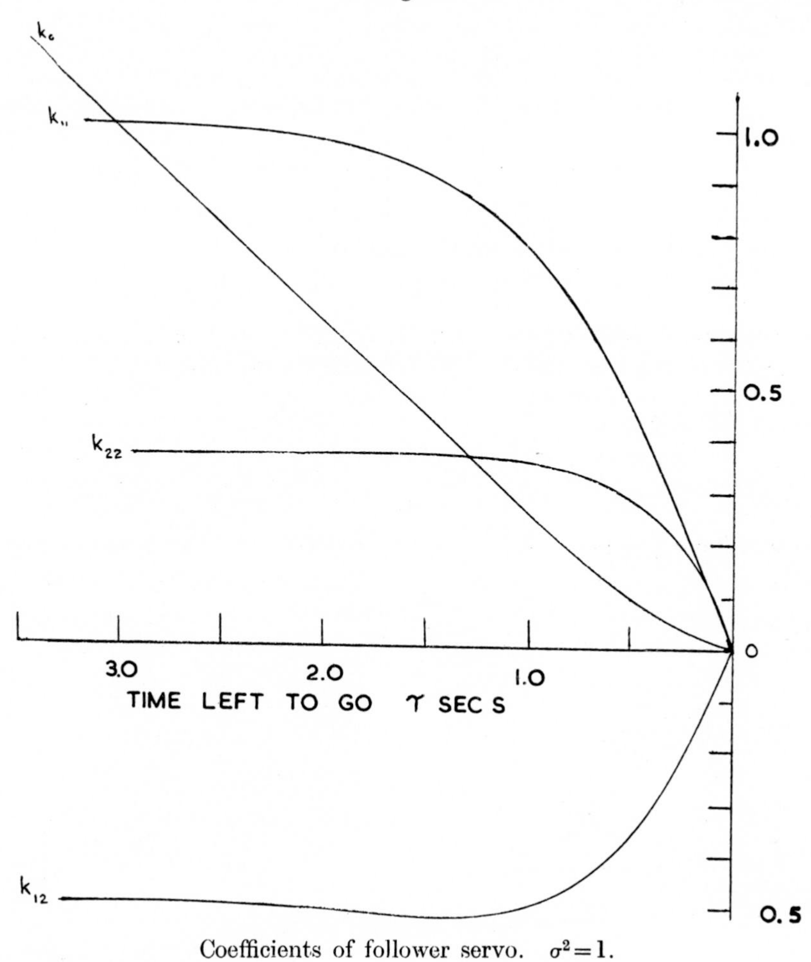

Coefficients of follower servo. $\sigma^2=1$.

Fig. 4

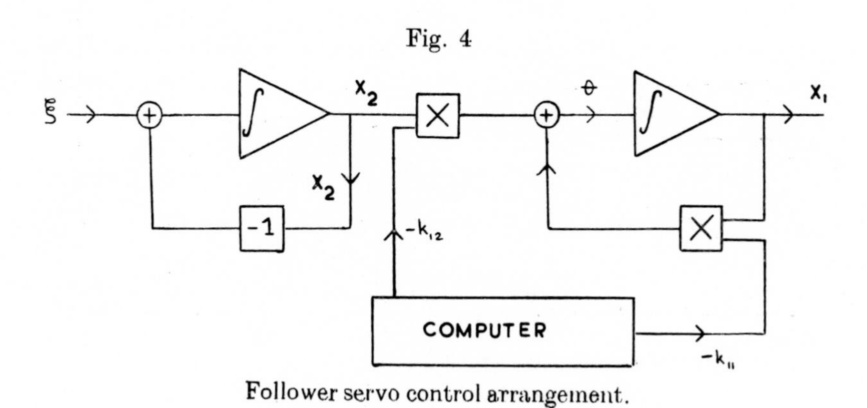

Follower servo control arrangement.

Fig. 5

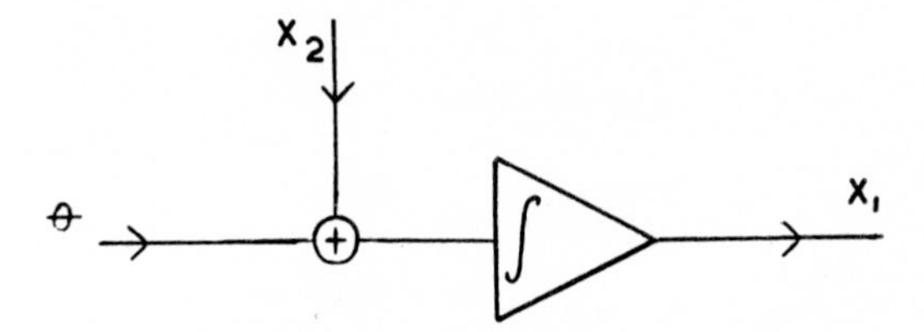

Simple regulator.

The solution is again of the form of eqn. (5.8), and only constant and quadratic terms are required. The coefficient equations are

$$\left.\begin{aligned} \dot{k}_0 &= \sigma^2 k_{22}, \\ \dot{\mathbf{K}}_2 &= \begin{bmatrix} 1, 0 \\ 1, 0 \end{bmatrix} - \mathbf{K}_2 \begin{bmatrix} 1, 0 \\ 0, 0 \end{bmatrix} \mathbf{K}_2 + \begin{bmatrix} 0, & 0 \\ 1, & -1 \end{bmatrix} \mathbf{K}_2 + \mathbf{K}_2 \begin{bmatrix} 0, & 1 \\ 0, & -1 \end{bmatrix}. \end{aligned}\right\} \quad (5.16)$$

The solution is graphed in fig. 6, and the controller arrangement is shown in fig. 7.

Fig. 6

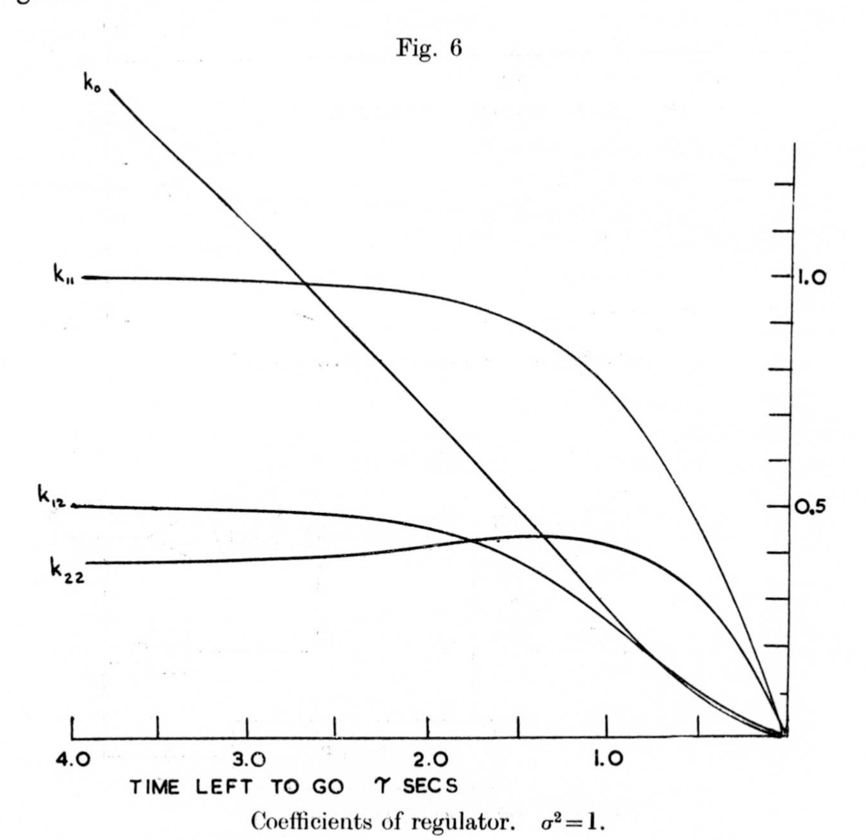

Coefficients of regulator. $\sigma^2 = 1$.

Fig. 7

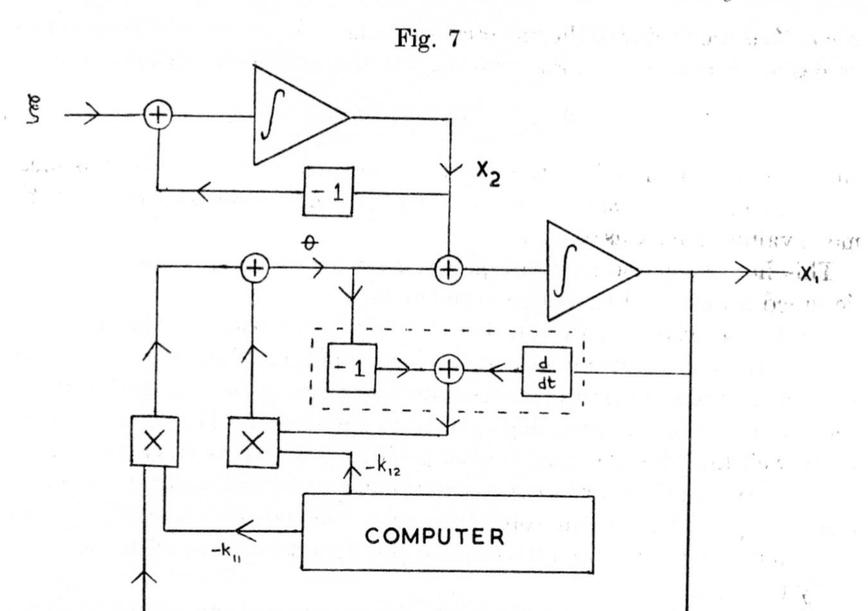

Regulator control arrangement. The block within dotted lines produces a copy coordinate of x_2.

5.3. *Comments on the Examples*

These comments are intended to draw out general points in the method.

(i) It will be seen that the dispersion (or variance) of the random process does not affect the decision, θ^*, but does affect the value of the performance index. In the servo follower example the deterioration in the performance index is related to the essential error variance of prediction of the command signal.

It is of interest to see what effect the mean value of the random process has on the decision. Take the second, noisy regulator, example as an illustration. Suppose ξ, has mean m and unit variance, then $\xi = m + \xi'$ where ξ' has zero mean, and unit variance.

The new equation of optimal control is

$$\frac{\partial K^*}{\partial \tau} = \min_{\theta}\left[x_1{}^2 + \theta^2 + (\theta + x_2)\frac{\partial K^*}{\partial x_1} + (-x_2 + m)\frac{\partial K^*}{\partial x_2} + \frac{\sigma^2}{2}\frac{\partial^2 K^*}{\partial x_2{}^2}\right]. \qquad \ldots \quad (5.17)$$

On carrying through the minimizing and series solution, it will be found that the linear term in the series is required, and the coefficient equations are

$$\left.\begin{aligned}
\dot{k}_0 &= -\tfrac{1}{4}k_1{}^2 + mk_2 + \sigma^2 k_{22},\\
\dot{\mathbf{K}}_1{}^T &= -[k_1 k_{11}, k_1 k_{12}] + [0, k_1 - k_2] + 2m\,[k_{21}, k_{22}]\\
\dot{\mathbf{K}}_2 &= \begin{bmatrix}1, 0\\ 0, 0\end{bmatrix} - \mathbf{K}_2\begin{bmatrix}1, 0\\ 0, 0\end{bmatrix}\mathbf{K}_2 + \begin{bmatrix}0, & 0\\ 1, & -1\end{bmatrix}\mathbf{K}_2 + \mathbf{K}_2\begin{bmatrix}0, & 1\\ 0, & -1\end{bmatrix},
\end{aligned}\right\} \qquad (5.18)$$

where $\mathbf{K}_1$ is the vector of the linear coefficients As $T \rightarrow \infty$ the assymptotic values are $k_1 = m$, $k_{11} = 1$, $k_{12} = 0{\cdot}5$, $k_{22} = 0{\cdot}375$, giving the optimal control

$$\theta^* = -0{\cdot}5m - x_1 - 0{\cdot}5x_2. \qquad (5.19)$$

But x_2 is compounded of the mean value plus a term, with zero mean x_2' say, hence $\theta^* = -m - x_1 - 0{\cdot}5x_2'$; thus the controller subtracts out the mean value, exactly as expected.

This has the practical consequence that an optimal controller can be designed using only knowledge of the mean.

(ii) For a system with suitable stability properties, as the final time $T \rightarrow \infty$, the control becomes a stationary function of state. The control coefficients (except k_0) can then be found by putting the time derivatives to zero, and solving the resulting quadratic equations. This is the solution of the well-known stationary Wiener problem; it involves no manipulation of complex variables, or integral equations, and indeed makes the problem elementary. The present solutions are connected with a similar simplification of the time series filtering and prediction problem by Kalman and Bucy (1961).

(iii) In more general problems the plant parameters, Gaussian distribution parameters and the performance index coefficients will be time varying. It is possible to give a compact general form for the solution as follows:

The plant equations are

$$\mathbf{x}' = \mathbf{A}(t)\mathbf{x} + \mathbf{D}(t)\boldsymbol{\theta} + \boldsymbol{\xi}, \qquad (5.20)$$

where $\boldsymbol{\xi}$ is a multi-dimensional white noise process with mean value vector $\mathbf{m}$, and variance/co-variance matrix $\boldsymbol{\sigma}$, $\mathbf{D}(t)$ is an $n \times n$ matrix which allows more general control action.

The performance index is

$$K^*(\mathbf{x}_0, t_0) = \min_{\boldsymbol{\theta}} E\left\{\int_{t_0}^{T} [\mathbf{x}^T\mathbf{Q}(t)\mathbf{x} + \boldsymbol{\theta}^T\mathbf{B}(t)\boldsymbol{\theta}]\, du\right\}, \qquad (5.21)$$

where $\mathbf{Q}(t)$, $\mathbf{B}(t)$ are symmetric $n \times n$ matrices.

The partial differential equation of optimal control is

$$-\frac{\partial K^*}{\partial t_0} = \min_{\boldsymbol{\theta}}\left[\boldsymbol{\theta}^T\mathbf{B}\boldsymbol{\theta} + \mathbf{x}^T\mathbf{Q}\mathbf{x} + (\mathbf{A}\mathbf{x} + \mathbf{D}\boldsymbol{\theta} + \mathbf{m})^T\left[\frac{\partial K^*}{\partial \mathbf{x}}\right] + \tfrac{1}{2}\left[\frac{\partial}{\partial \mathbf{x}}\right]^T \boldsymbol{\sigma}\left[\frac{\partial}{\partial \mathbf{x}}\right]K^*\right]. \qquad (5.22)$$

Differentiation yields

$$\boldsymbol{\theta}^* = -\tfrac{1}{2}\,\mathbf{B}^{-1}\mathbf{D}^T\left[\frac{\partial K^*}{\partial \mathbf{x}}\right]. \qquad (5.23)$$

If $\mathbf{B}$ is singular, then $\mathbf{B}^{-1}$ is to be taken in the sense of the pseudo-inverse (Penrose 1955), which roughly means that a suitable set of linear equations can replace this expression.

The resulting equation is

$$-\frac{\partial K^*}{\partial t_0} = -\tfrac{1}{4}\left[\frac{\partial K^*}{\partial \mathbf{x}}\right]^T \mathbf{DB}^{-1}\mathbf{D}^T\left[\frac{\partial K^*}{\partial \mathbf{x}}\right] + \mathbf{x}^T\mathbf{Qx} + (\mathbf{Ax}+\mathbf{m})^T\left[\frac{\partial K^*}{\partial \mathbf{x}}\right] + \tfrac{1}{2} \times \left[\frac{\partial}{\partial \mathbf{x}}\right]^T \boldsymbol{\sigma} \left[\frac{\partial}{\partial \mathbf{x}}\right] K^*. \quad (5.24)$$

Substituting series (5.8) into this equation, and making extensive use of the properties of symmetric matrices, the coefficient equations are

$$\left.\begin{aligned} -k_0' &= -\tfrac{1}{4}\mathbf{K}_1{}^T\mathbf{DB}^{-1}\mathbf{D}^T\mathbf{K}_1 + \sum\sum \sigma_{rs}k_{rs} + \mathbf{m}^T\mathbf{K}_1, \\ -\mathbf{K}_1{}'^T &= -\mathbf{K}_1{}^T\mathbf{DB}^{-1}\mathbf{D}^T + \mathbf{K}_1{}^T\mathbf{A} + 2\mathbf{m}^T\mathbf{K}_2, \\ -\mathbf{K}_2{}' &= \mathbf{Q} - \mathbf{K}_2\mathbf{DB}^{-1}\mathbf{D}^T\mathbf{K}_2 + \mathbf{A}^T\mathbf{K}_2 + \mathbf{K}_2\mathbf{A}, \end{aligned}\right\} \quad (5.25)$$

with boundary conditions $k_0(T) = \mathbf{K}_1(T) = \mathbf{K}_2(T) = 0$.

For a system whose time-varying parameters are well behaved mathematically, the eqns. (5.25) show that only elementary computing is required. Thus the non-stationary, finite time interval problem is provided with a direct computing algorithm.

(iv) It has sometimes been suggested that the stochastic optimal control problem could be solved by substituting for the random terms a predicted value, and then solving the problem as if it were deterministic. In certain, suitable examples this quasi-stochastic technique will actually provide the correct control policy; what it cannot provide is an estimate of the deterioration of the performance index due to the inherent errors in prediction, the present method, of course, does this.

§ 6. Conclusions

Mathematical arguments leading to the partial differential equations of optimal control for systems whose stochastic descriptions satisfy the Markov condition have been presented. Examples of control equations for systems which are non-linear, non-stationary, and whose random components have Gaussian, or generalized Poisson distributions, have been given.

For problems with quadratic performance indices, linear plant dynamics and Gaussian random components, the partial differential equations of control have been reduced to a set of ordinary differential equations, which are well suited to solution on an automatic machine. This technique supplies a simple computing algorithm for many problems which previously had been solved by complex variable, or integral equation methods.

Acknowledgments

The author was greatly helped by the tireless discussion and questioning of Dr. Stanley Katz. He also benefited from discussion with Mr. John Pearson.

This work was supported by a contract from the National Physical Laboratory, to whom thanks are due for permission to publish it.

References

Barrett, J. F., 1960, *Proc. I.F.A.C. Congress, Moscow* (London: Butterworths).

Bellman, R., 1956, *Dynamic Programming* (Princeton: University Press).

Fuller, A. T., 1960, *J. Electron. Contr.*, **9,** 65.

Howard, R. A., 1960, *Dynamic Programming and Markov Processes* (The Technology Press and Wiley).

Kalman, R. E., and Bucy, R. S., 1961, *J. Amer. Soc. mech. Engrs*, Paper No. 60–JAC–12, March.

Levy, P., 1948, *Processus Stochastiques et Mouvement Brownien* (Paris: Gauthier-Villars).

Merriam, C. W., 1960, *Inform. Contr.*, **3,** 32.

Penrose, R., 1955, *Proc. Camb. phil. Soc.*, **51,** 406.

Rozonoer, L. I., 1959, *Automatika i Telemekhanika*, **20,** Nos. 10–12. English translation in *Automation and Remote Control.*

Wonham, W. M., 1960, *Trans. Inst. Radio Engrs, N.Y.*, **6,** 539–544.

Partial Observability and Optimal Control†

By J. J. Florentin

Electrical Engineering Department, Imperial College, London, S.W.7, England

[Received May 21, 1962]

Abstract

Optimal control theory usually assumes that the state vector components of the controlled object can be measured exactly. However, practical measuring instruments frequently give significant error; also it is often impossible to monitor each component of the state. It is pointed out that in many situations, including adaptive ones, it is still possible to compute an optimal control even with both observational restrictions.

Some simple examples are presented to illustrate the methods. These examples lead to a useful extension of the Certainty Equivalence Principle.

These methods give an approach to the fundamental problem of partial observability, which is to say what kinds of observational restrictions allow effective controllers to be devised. Some tentative remarks are made about this.

Table of Main Symbols

$\mathbf{A}$	System matrix.
$\mathbf{B}$	Cost of state matrix.
$\mathbf{D}$	Control action matrix.
E	Expectation operator.
$f(\mathbf{S}, m)$	Performance index with sufficient statistic $\mathbf{S}$, and m decisions to be made.
$F(c)$	Cumulative distribution function.
$\mathbf{H}$	Observation matrix.
$\mathbf{I}$	Unit matrix.
k	Indexes time backwards.
$\mathbf{K}$	Matrix operating on observations.
$\mathbf{L}$	Matrix of control cost.
$\mathbf{M}$	Mean value vector.
$\mathbf{N}$	Noise contribution to performance index.
$P(\mathbf{X}\|\mathscr{O})$	Probability of state $\mathbf{X}$ given observation $\mathscr{O}$.
$\mathbf{Q}$	Variance–covariance matrix of disturbances.
r	Indexes time forward.
$\mathbf{S}$	Sufficient statistic vector.
$\mathbf{U}$	Variance–covariance matrix of estimate.

† Communicated by the Author. This work was supported by a contract from the National Physical Laboratory.

v_1, v_2	Variance of disturbances.
$\mathbf{W}$	Matrix of coefficients in performance index.
$\mathbf{X}$	State vector.
$\mathbf{Y}$	Random element in estimated value of $\mathbf{X}$.
s	An observation.
$\mathcal{S}$	Observation vector.
Γ	Factor involving observation.
$\mathbf{\Delta}$	Coefficient of control.
η	A disturbance.
$\boldsymbol{\theta}$	Control vector.
$\boldsymbol{\xi}$	A disturbance.
$\boldsymbol{\sigma}$	Covariance matrix.

§ 1. Introduction

In current optimal systems theory it is assumed that all the components of the state vector of the controlled object can be measured exactly. In practice this is frequently impossible, first because the available measuring instruments may be imperfect and make errors, and secondly because not all variables are conveniently accessible for observation. This paper points out that even with both measurement restrictions, in many situations of practical interest the optimizing problem can still be formulated and solved by using statistical decision theory (Wald 1950). Actually to produce solutions, Bellman's (1961) method of Dynamic Programming will be used and is a key reference for this paper.

Since the principles are somewhat general, several examples are given to illustrate their use. An important aspect of the control computation is that the value of control depends on all the past observations, the compounding of past observations being a machine learning process. From the examples a useful extension of the Certainty Equivalence Principle (Simon 1956) can be deduced.

As well as having obvious practical applications these techniques also form an approach to the fundamental problems of observability and controllability. The need for such a theory and the important role of these concepts was first recognized by Kalman (1960, 1961), who also gave certain definitions and mathematical relationships. The material in the present paper leads naturally to an intuitive idea of partial observability for a wide class of systems. Roughly speaking, a system is partially observable when past knowledge and present observations can be combined so as to produce all the probability functions needed to solve properly a control decision problem of interest. At the present stage it does not appear possible to propose general definitions, but appropriate questions are raised. In particular, it is suggested that partial observability is a property of the system and the desired performance, rather than of the controlled object alone.

§ 2. Outline of Principles

To make the discussion more concrete consider a controlled object having a state vector $\mathbf{X}$, some of whose components can be observed, and to which control can be applied at unit time intervals $r = 0, 1, 2, \ldots$ only. The optimal control problem begins with a scalar performance index f, whose value depends on the realized value of $\mathbf{X}(r)$ over some defined time period, e.g.

$$f = \sum_{0}^{m} ||\mathbf{X}(r)||$$

or

$$f = ||\mathbf{X}(m)||,$$

with m a fixed time instant. Control can be applied at certain inputs, $\boldsymbol{\theta}(r)$ is the control vector at time r. The solution to the optimizing problem is a method of computing a sequence $\{\boldsymbol{\theta}(r)\}$ so as to maximize, or minimize, the performance index. If the future behaviour of the system is known only statistically, then the performance index is modified to an average over the possible future states. If the past and current states, and the parameters of the controlled object are known exactly, but the future inputs are known statistically, then the problem is conventionally called stochastic. If, further, certain parameters of the controlled object are known only statistically, then the problem is called adaptive. It is suggested that if, in addition, the current and past states are known only statistically, then the problem may be called incompletely observable. Incomplete observability thus occurs with inaccurate measuring instruments, or restricted access to measuring points.

In all the above categories the optimal value of control can be found by a feed-back procedure. An observation is made at each time instant, and then compounded with the previously determined probability functions to form a new probability description of the future behaviour of the controlled object. This new probability function is immediately used to find the current numerical value of the control. Such a combination of sequential observations and decisions has exactly the structure of Wald's statistical decision theory.

Actual computing operations with probability functions and the whole recorded set of past observations would be very laborious and, if possible, it is best to summarize all the information in the observations in a set of sufficient statistics. It is intuitively obvious that any decision made on the basis of the probability function can be made with the sufficient statistic alone, but a more formal justification is shown by Raiffa and Schlaifer (1961). Raiffa and Schlaifer also show the advantage of using the class of distribution functions having certain self-reproducing properties; this will be seen in the examples below. The reducing of observations into sets of statistics constitutes a machine learning procedure.

When the general principles briefly outlined above are applied to particular problems in control, the result is to pose a mathematical problem in which a numerical value of $\boldsymbol{\theta}(r)$ must be found for each r so as to maximize, or minimize, the given performance index. This mathematical problem may, or may not, have a solution; for instance in the mathematical problem it might be that $f=\infty$ regardless of the value of $\boldsymbol{\theta}$. Given a system with incomplete observability, if the mathematical decision problem has a solution for the prescribed performance index, then the system will be said to be partially observable with respect to that index.

§ 3. Examples

Rather than attempt an abstract demonstration of the principles, some fully worked examples will be given. Although very simple they illustrate the discussion of the next section.

Fig. 1

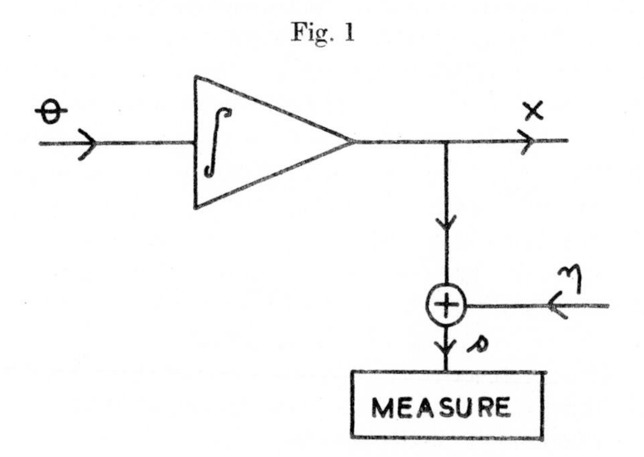

Simple regulator.

3.1. *Simple Regulator with Inaccurate Measurements*

The system is shown in fig. 1, and has equation

$$x(r+1)=x(r)+\theta(r). \qquad (1)$$

The observations $s(r)$ are corrupted by noise, so that

$$s(r)=x(r)+\eta(r), \qquad (2)$$

where $\{\eta(r)\}$ is an independent sequence of Gaussian random variables with zero mean, and variance v_1.

At time $r=0$ an observation $s(0)$ is made. This leads to an *a posteriori* density for $x(0)$ of

$$\frac{1}{\sqrt{(2\pi v_1)}} \exp - \frac{1}{2v_1}(x(0)-s(0))^2.$$

Note that the variable of interest occurs only in the exponential term. This part containing the variable of interest will be called the significant part, and below only this part of the probability functions will be quoted.

At time $r=1$ an observation $\mathfrak{z}(1)$ is made. This gives a likelihood for $x(1)$ of

$$\exp -\frac{1}{2v_1}(x(1)-\mathfrak{z}(1))^2.$$

This density is not, however, the best that can be obtained, for from eqns. (1) and (2) the *a priori* density for $x(1)$ has significant part

$$\exp -\frac{1}{2v_1}(x(1)-\mathfrak{z}(0)-\theta(0))^2,$$

so that the full *a posteriori* density of $x(1)$ has significant part

$$\exp -\frac{1}{2v_1}[(x(1)-\mathfrak{z}(0)-\theta(0))^2+(x(1)-\mathfrak{z}(1))^2],$$

which can be reduced to

$$\exp -\frac{1}{2}\cdot\frac{2}{v_1}\left(x(1)-\frac{\mathfrak{z}(1)+\mathfrak{z}(0)+\theta(0)}{2}\right)^2.$$

Thus, the *a posteriori* density of $x(1)$ is Gaussian, as was the density for $x(0)$; this is the self-reproducing property mentioned earlier. For the Gaussian distribution the mean $m(r)$ and the variance $u(r)$ form a sufficient statistic vector $\mathbf{S}(r)$, hence

$$\mathbf{S}(1)=[m(1),\, u(1)]=[\tfrac{1}{2}(\mathfrak{z}(1)+\mathfrak{z}(0)+\theta(0)),\, \tfrac{1}{2}v_1].$$

At stage 1 past knowledge and current observations were combined so as to exhaust all the potential information; this is the important machine learning process. The process can be continued so that, for instance,

$$\mathbf{S}(3)=[m(3), u(3)]=[\tfrac{1}{4}(\mathfrak{z}(0)+\theta(0)+\theta(1)+\theta(2)+\mathfrak{z}(1)+\theta(1)+\theta(2)+\mathfrak{z}(2) +\theta(2)+\mathfrak{z}(3)),\, \tfrac{1}{4}v_1].$$

It is clear that in this problem the learning is effectively a repetition of the observation of the initial state. On a stage-by-stage basis the compounding of observations is expressed by

$$\left.\begin{aligned} m(r+1)&=\frac{(r+1)}{(r+2)}m(r)+\frac{(r+1)}{(r+2)}\theta(r)+\frac{1}{(r+2)}\mathfrak{z}(r+1),\\ u(r+1)&=\frac{v_1}{(r+2)}\,. \end{aligned}\right\} \quad \ldots \quad (3)$$

Note that $\lim_{r\to\infty}[u(r)]=0.$

When making a decision at time r it is necessary to have distributions for $x(r+1)$ and $\mathbf{S}(r+1)$. It is useful to write

$$x(r+1)=m(r)+\theta(r)+y(r), \quad \ldots\ldots\ldots \quad (4)$$

where $y(r)$ expresses the uncertainty in the *a posteriori* density of $x(r)$, and is Gaussian with zero mean and variance $u(r)$. Now

$$\mathfrak{z}(r+1)=x(r+1)+\eta(r+1)=m(r)+y(r)+\theta(r)+\eta(r+1). \quad \ldots \quad (5)$$

This leads to the *a priori* stochastic equations for the sufficient statistics

$$\left.\begin{aligned} m(r+1) &= m(r)+\theta(r)+(y(r)+\eta(r+1))(1+v_1/u(r))^{-1}, \\ u(r+1) &= v_1(1+v_1/u(r))^{-1}. \end{aligned}\right\} \quad . \quad . \quad (6)$$

To illustrate the computation of control a particular performance index will be taken:

$$\left.\begin{aligned} f(m) &= \min E\left\{\sum_{1}^{m}(\theta^2(k)+x^2(k-1))\right\}, \\ f(0) &= 0. \end{aligned}\right\} \quad . \quad . \quad . \quad . \quad (7)$$

It should be noted that k will now be taken to index time backwards, with terminus at $k=0$. This change from forward to backward time indexing will be done in each example. The author regrets any inconvenience to the reader; this change preserves the distinction between information gathering—a function of the past—and control—a function of the future.

Noting that the performance index is a function of $\mathbf{S}$, Bellman's Principle of Optimality now gives the feed-back relation

$$f(\mathbf{S}(k), k) = \min_{\theta} E\{\theta^2(k)+x^2(k-1)+f(\mathbf{S}(k-1), k-1)\}. \qquad (8)$$

Computation is conveniently begun at $k=1$:

$$\begin{aligned} f(\mathbf{S}(1), 1) &= \min_{\theta} E\{\theta^2(1)+x^2(0)\} \\ &= \min_{\theta} E\{\theta^2(1)+(m(1)+y(1)+\theta(1))^2\} \\ &= \min_{\theta} [2\theta^2(1)+m^2(1)+2m(1)\theta(1)+u(1)]. \quad . \quad . \quad . \quad . \quad (9) \end{aligned}$$

By differentiation the optimal control and performance index are found to be

$$\theta^*(1) = -\tfrac{1}{2}m(1), \quad f(\mathbf{S}(1), 1) = \tfrac{1}{2}m^2(1)+u(1).$$

For $k=2$:

$$\begin{aligned} f(\mathbf{S}(2), 2) &= \min_{\theta} E\{\theta^2(2)+x^2(1)+f(\mathbf{S}(1), 1)\} \\ &= \min_{\theta} E\{\theta^2(2)+(m(2)+y(2)+\theta(2))^2+\tfrac{1}{2}m^2(1)+u(1)\}. \quad (10) \end{aligned}$$

On substituting

$$m(1) = m(2)+\theta(2)+y(2)+\eta(1),$$

$$u(1) = \tfrac{1}{2}v_1,$$

followed by averaging and minimizing, it will be found that

$$\theta^*(2) = -\tfrac{3}{5}m(2),$$

$$f(\mathbf{S}(2), 2) = \tfrac{3}{5}m^2(2)+u(2)+\frac{\tfrac{1}{2}(u(2)+v_1)}{(1+v_1/u(2))^2}+u(1).$$

All subsequent stages are calculated in the same way, yielding

$$\left.\begin{aligned} &\theta^*(k) = -c(k)m(k), \\ &f(\mathbf{S}(k), k) = c(k)m^2(k) + [u(k) + c(k-1)(u(k)+v_1)(1+v_1/u(k))^{-2}] \\ &\qquad + [\quad] \ldots \quad + u(1), \\ &c(k) = \frac{1+c(k-1)}{2+c(k-1)}, \quad c(0) = 0. \end{aligned}\right\} \quad (11)$$

To implement control observations are made, and $\mathbf{S}(k)$ calculated, using eqns. (6); then the control is found from eqns. (11). It will be noted that the form of control is the same as would be found for the system with the disturbance source removed and with m replacing x. The performance index has the same substitution plus added terms containing the variance of the disturbance.

Fig. 2

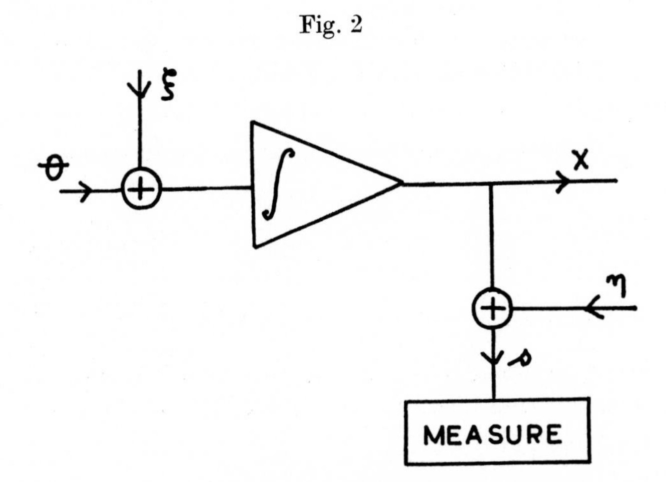

Noisy regulator.

3.2. *Noisy Regulator with Inaccurate Measurements*

The system is shown in fig. 2, and has the equation

$$x(r+1) = x(r) + \theta(r) + \xi(r), \quad \ldots \ldots \quad (12)$$

where $\{\xi(r)\}$ is an independent Gaussian sequence with zero mean and variance v_2. The observations are corrupted by noise as in the previous example.

Suppose that at time $r=0$ the information density of $x(0)$ is given as Gaussian with mean $m(0)$, and variance $u(0)$; then $x(0)$ can be written

$$x(0) = m(0) + y(0), \quad \ldots \ldots \quad (13)$$

where $y(0)$ is Gaussian with zero mean and variance $u(0)$. The *a priori* density for $x(1)$ is conveniently written as a stochastic equation:

$$x(1) = x(0) + \theta(0) + \xi(0) = m(0) + y(0) + \theta(0) + \xi(0). \quad \ldots \quad (14)$$

At time $r=1$ an observation $s(1)$ is made leading to the likelihood for $x(1)$

$$\exp -\frac{1}{2v_1}(x(1)-s(1))^2.$$

Compounding the *a priori* density and likelihood leads to an *a posteriori* density with significant part

$$\exp -\tfrac{1}{2}[(x(1)-s(1))^2 v_1^{-1}+(x(1)-m(0)-\theta(0))^2(v_2+u(0))^{-1}].$$

This leads to relations for the sufficient statistics at time 1 in terms of the current observation and past statistics:

$$\left.\begin{aligned} m(1)&=\left[s(1)v_1^{-1}+\frac{m(0)+\theta(0)}{v_2+u(0)}\right]\left[v_1^{-1}+\frac{1}{v_2+u(0)}\right]^{-1}, \\ u(1)&=\left[v_1^{-1}+\frac{1}{v_2+u(0)}\right]^{-1}. \end{aligned}\right\} \quad . \quad . \quad (15)$$

Again it is necessary to find distributions for $x(r+1)$ and $\mathbf{S}(r+1)$ in order to make a decision at time r. Noting that

$$s(r+1)=x(r+1)+\eta(r+1)=m(r)+\theta(r)+y(r)+\xi(r)+\eta(r+1), \quad (16)$$

it is easily verified that the stochastic equations for the mean and variance are

$$\left.\begin{aligned} m(r+1)&=m(r)+\theta(r)+(y(r)+\xi(r)+\eta(r+1))\left(1+\frac{v_1}{v_2+u(r)}\right)^{-1}, \\ u(r+1)&=\left[v_1^{-1}+\frac{1}{v_2+u(r)}\right]^{-1}. \end{aligned}\right\} \quad (17)$$

The same performance index will be used as in the previous problem. For $k=1$ (remembering that time is now indexed backwards)

$$\begin{aligned} f(\mathbf{S}(1),1)&=\min_{\theta} E\{\theta^2(1)+x^2(0)\} \\ &=\min_{\theta} E\{\theta^2(1)+(m(1)+\theta(1)+y(1)+\xi(1))^2\} \\ &=\min\,[2\theta^2(1)+m^2(1)+2m(1)\theta(1)+u(1)+v_2]. \quad . \quad . \quad . \quad (18) \end{aligned}$$

Again it is found that

$$\theta^*(1)=-\tfrac{1}{2}m(1), \quad f(\mathbf{S}(1),1)=\tfrac{1}{2}m^2(1)+u(1)+v_2.$$

With $k=2$:

$$\begin{aligned} f(\mathbf{S}(2),2)&=\min_{\theta} E\{\theta^2(2)+x^2(1)+f(\mathbf{S}(1),1)\} \\ &=\min E\Bigg\{\theta^2(2)+(m(2)+\theta(2)+y(2)+\xi(2))^2 \\ &\qquad +\tfrac{1}{2}\left(m(2)+\theta(2)+\frac{y(2)+\xi(2)+\eta(1)}{1+v_1(v_2+u(2))^{-1}}\right)^2\Bigg\}, \end{aligned} \quad (19)$$

which leads to $\theta^*(2)=-\tfrac{3}{5}m(2)$,

$$f(\mathbf{S}(2),2)=\tfrac{3}{5}m^2(2)+u(2)+v_2+\frac{\tfrac{1}{2}(u(2)+v_1+v_2)}{[1+v_1(v_1+u(2))^{-1}]^2}.$$

It will be found that the control is exactly the same as in the previous example, but that the general performance index contains extra additive terms in v_2. Control is implemented as before, with the appropriate modification of eqn. (15) being used to calculate the sufficient statistic.

3.3. *The General Linear System with Quadratic Performance Indices*

The above results can be generalized to time-varying linear systems having

(i) Gaussian disturbances of known correlation,

(ii) measurement points such that the components of the state vector can only be measured in certain linear combinations, there being fewer observation points than state vector components,

(iii) observations corrupted by correlated Gaussian noise,

(iv) a quadratic path sum performance index,

(v) dynamic equation coefficients satisfying certain modifications of Kalman's observability and controllability criteria.

Fig. 3

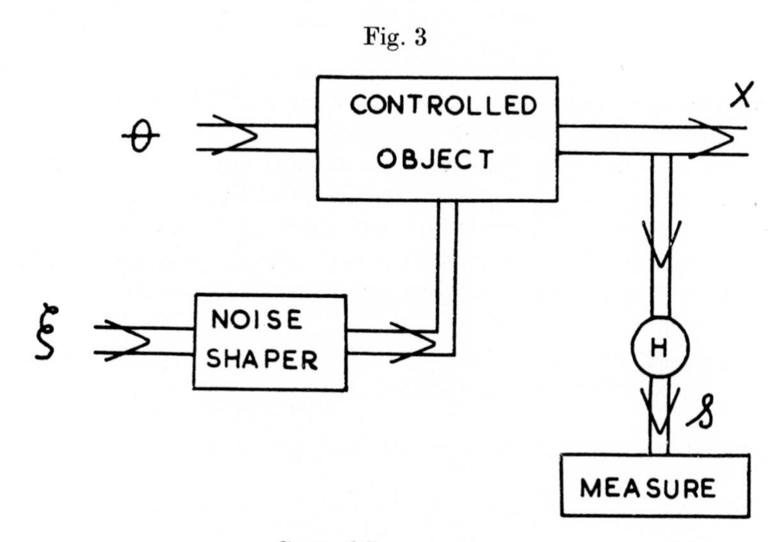

General linear system.

All the above properties can be summed up in the general dynamic and observational equations, see also fig. 3

$$\left.\begin{aligned}\mathbf{X}(r+1)&=\mathbf{A}(r)\mathbf{X}(r)+\mathbf{D}(r)\boldsymbol{\theta}(r)+\boldsymbol{\xi}(r),\\ \mathscr{S}(r)&=\mathbf{H}(r)\mathbf{X}(r),\end{aligned}\right\} \quad . \quad . \quad . \quad (20)$$

where $\mathbf{X}$, $\boldsymbol{\theta}$ are n-vectors, $\mathbf{A}$, $\mathbf{D}$ are $n\times n$ matrices, $\{\boldsymbol{\xi}(r)\}$ is a sequence of independent Gaussian n-vectors with zero mean and variance–covariance matrix $\mathbf{Q}(r)$. $\mathbf{H}$ is an $n\times p$ matrix with $p<n$, and $\mathscr{S}$ is a p-vector.

To illustrate the application of these equations consider a simple regulator with equation

$$x_1(r+1) = x_1(r) + \theta_1(r),$$

and suppose that the observations are corrupted by a first-order correlated noise x_2 such that

$$x_2(r+1) = \tfrac{1}{2}x_2(r) + \xi_2.$$

Then

$$\mathbf{A} = \begin{bmatrix} 1, & 0 \\ 0 & \frac{1}{2} \end{bmatrix}, \quad \mathbf{D} = \begin{bmatrix} 1, & 0 \\ 0 & 0 \end{bmatrix}, \quad \mathbf{Q} = \begin{bmatrix} 0, & 0 \\ 0, & v_2 \end{bmatrix}, \quad \mathbf{H} = [1, 1].$$

The notation will now be adopted that $\mathbf{X}$ is $N[\mathbf{M}, \mathbf{U}]$ means that $\mathbf{X}$ has a multi-variate Gaussian distribution with mean vector $\mathbf{M}$ and variance–covariance matrix $\mathbf{U}$. Occasionally $||\mathbf{Z}|\mathbf{B}||$ will mean $\mathbf{Z}^T\mathbf{B}\mathbf{Z}$ If the control problem actually has a solution it can be found as follows.

To begin the observation sequence assume that at time $r=0$ it has been given that $\mathbf{X}(0)$ is $N[\mathbf{M}(0), \mathbf{U}(0)]$ then the *a priori* equation for $\mathbf{X}(1)$ is

$$\mathbf{X}(1) = \mathbf{A}\mathbf{M}(0) + \mathbf{A}\mathbf{Y}(0) + \mathbf{D}\boldsymbol{\theta}(0) + \boldsymbol{\xi}(0), \quad . \; . \; . \quad (21)$$

where $\mathbf{Y}(0)$ is $N[0, \mathbf{U}(0)]$ and the explicit dependence of the coefficients on time has been omitted. Hence the *a priori* distribution for $\mathbf{X}(1)$ is $N[\mathbf{A}\mathbf{M}(0) + \mathbf{D}\boldsymbol{\theta}(0), \mathbf{A}\mathbf{U}(0)\mathbf{A}^T + \mathbf{Q}]$.

At time $r=1$ $\mathscr{S}(1)$ is observed. It is clear that $\mathscr{S}(1)$ and $\mathbf{X}(1)$ are jointly Gaussian; but in view of the relation $\mathscr{S}(1) = \mathbf{H}\mathbf{X}(1)$ the distribution is singular. In spite of this singularity it is still possible to find the conditional probability $P(\mathbf{X}(1)|\mathscr{S}(1))$; see, for instance, Raiffa and Schlaifer (1961). In particular, the conditional mean of $\mathbf{X}(1)$ given $\mathscr{S}(1)$ is found from

$$\mathbf{M}(1) = E\{\mathbf{X}(1)\} + \mathrm{Cov}\{\mathbf{X}(1)\mathscr{S}^T(1)\}[\mathrm{Cov}\{\mathscr{S}(1)\mathscr{S}^T(1)\}]^{-1}[\mathscr{S}(1) - E\{\mathscr{S}(1)\}]. \quad . \; . \; . \quad (22)$$

Noting that $\mathrm{Cov}\{\mathscr{S}(1)\mathscr{S}^T(1)\} = \mathbf{H}\,\mathrm{Cov}\{\mathbf{X}(1)\mathbf{X}^T(1)\}\mathbf{H}^T$,

$$\mathrm{Cov}\{\mathscr{S}(1)\mathbf{X}^T(1)\} = \mathbf{H}\,\mathrm{Cov}\{\mathbf{X}(1)\mathbf{X}^T(1)\},$$

it is found that

$$\mathbf{M}(1) = [\mathbf{I} - \mathbf{K}(1)\mathbf{H}][\mathbf{A}\mathbf{M}(0) + \mathbf{D}\boldsymbol{\theta}(0)] + \mathbf{K}(1)\mathscr{S}(1), \quad . \; . \; . \quad (23)$$

where

$$\mathbf{K}(1) = [\mathbf{A}\mathbf{U}(0)\mathbf{A}^T + \mathbf{Q}]\mathbf{H}^T[\mathbf{H}(\mathbf{A}\mathbf{U}(0)\mathbf{A}^T + \mathbf{Q})\mathbf{H}^T]^{-1}. \quad . \; . \quad (24)$$

The covariance of $\mathbf{X}(1)$ given $\mathscr{S}(1)$ is not dependent on the observation, and is found from the formula

$$\mathbf{U}(1) = \mathrm{Cov}\{\mathbf{X}(1)\mathbf{X}^T(1)\} - \mathrm{Cov}\{\mathbf{X}(1)\mathscr{S}^T(1)\}[\mathrm{Cov}\{\mathscr{S}(1)\mathscr{S}^T(1)\}]^{-1} \times \mathrm{Cov}\{\mathscr{S}(1)\mathbf{X}^T(1)\} \quad . \; . \; . \; . \; . \; . \; . \; . \; . \; . \quad (25)$$

$$= [\mathbf{I} - \mathbf{K}(1)\mathbf{H}][\mathbf{A}\mathbf{U}(0)\mathbf{A}^T + \mathbf{Q}]. \quad . \; . \; . \; . \; . \quad (26)$$

In order to compute the control an *a priori* distribution for the sufficient statistic is required. Taking stage $r=0$ as a typical example this distribution can be found by noting that

$$\mathscr{S}(1)=\mathbf{HX}(1)=\mathbf{H}[\mathbf{AM}(0)+\mathbf{D}\boldsymbol{\theta}(0)+\mathbf{AY}(0)+\boldsymbol{\xi}(0)], \quad . \quad . \quad (27)$$

when it follows from eqn. (23) that

$$\mathbf{M}(1)=\mathbf{AM}(0)+\mathbf{D}\boldsymbol{\theta}(0)+\mathbf{K}(1)\mathbf{H}[\mathbf{AY}(0)+\boldsymbol{\xi}(0)]. \quad . \quad . \quad . \quad (28)$$

Thus $\mathbf{M}(1)$ is $N[\mathbf{AM}(0)+\mathbf{D}\boldsymbol{\theta}(0),\ \mathbf{K}(1)\mathbf{H}[\mathbf{AU}(0)\mathbf{A}^T+\mathbf{Q}](\mathbf{K}(1)\mathbf{H})^T]$.

As noted above, $\mathbf{U}(1)$ is not a random variable at stage 0.

The performance index to be considered is a generalization of the previous ones:

$$f(\mathbf{S}(m),\ m)=\min E\left\{\sum_1^m(\boldsymbol{\theta}^T(k)\mathbf{L}\boldsymbol{\theta}(k)+\mathbf{X}^T(k-1)\mathbf{BX}(k-1))\right\},$$

where $\mathbf{L}$, $\mathbf{B}$ are symmetric, positive definite matrices whose elements are functions of time only.

To compute control the Principle of Optimality will be used as before. It should again be noted that k will index time backwards. For $k=1$:

$$\begin{aligned} f(\mathbf{S}(1),\ 1)&=\min_\theta E\{\boldsymbol{\theta}^T(1)\mathbf{L}\boldsymbol{\theta}(1)+\mathbf{X}^T(0)\mathbf{BX}(0)\} \\ &=\min_\theta E\{||\boldsymbol{\theta}(1)|\mathbf{L}||+||\mathbf{AM}(1)+\mathbf{D}\boldsymbol{\theta}(1)+\mathbf{Y}(1)+\boldsymbol{\xi}(1)|\mathbf{B}||\} \\ &=\min_\theta[||\boldsymbol{\theta}(1)|\mathbf{L}||+||\mathbf{AM}(1)+\mathbf{D}\boldsymbol{\theta}(1)|\mathbf{B}||]+\mathbf{N}(1), \quad . \quad . \quad . \quad (29) \end{aligned}$$

where $\mathbf{N}(1)$ represents the contribution due to disturbances, and does not involve $\boldsymbol{\theta}(1)$:

$$\mathbf{N}(1)=E\{||\mathbf{Y}(1)+\boldsymbol{\xi}(1)|\mathbf{B}||\}. \quad . \quad . \quad . \quad . \quad . \quad (30)$$

On differentiating it is found that

$$\boldsymbol{\theta}^*(1)=-[\mathbf{D}^T\mathbf{BD}+\mathbf{L}]^{-1}\mathbf{D}^T\mathbf{BAM}(1). \quad . \quad . \quad . \quad (31)$$

For convenience put

$$\boldsymbol{\theta}^*(1)=-\boldsymbol{\Delta}(1)\mathbf{M}(1). \quad . \quad . \quad . \quad . \quad . \quad . \quad . \quad . \quad (32)$$

Substituting for $\boldsymbol{\theta}^*(1)$ it is found that

$$f(\mathbf{S}(1),\ 1)=||\mathbf{M}(1)\boldsymbol{\Delta}(1)|\mathbf{L}||+||\mathbf{M}(1)(\mathbf{A}-\boldsymbol{\Delta}(1))|\mathbf{B}||+\mathbf{N}(1), \quad . \quad . \quad (33)$$

which can be written as

$$f(\mathbf{S}(1),\ 1)=||\mathbf{M}(1)|\mathbf{W}(1)||+\mathbf{N}(1). \quad . \quad . \quad . \quad . \quad (34)$$

For $k=2$:

$$\begin{aligned} f(\mathbf{S}(2),\ 2)&=\min_\theta E\{||\boldsymbol{\theta}(2)|\mathbf{L}||+||\mathbf{X}(1)|\mathbf{B}||+f(\mathbf{S}(1),\ 1)\} \\ &=\min_\theta[||\boldsymbol{\theta}(2)|\mathbf{L}||+||\mathbf{AM}(2)+\mathbf{D}\boldsymbol{\theta}(2)|\mathbf{B}+\mathbf{W}(1)||]+\mathbf{N}(2)+\mathbf{N}(1), \end{aligned} \quad . \quad . \quad . \quad (35)$$

where

$$\mathbf{N}(2)=E\{||\mathbf{Y}(2)+\boldsymbol{\xi}(2)|\mathbf{B}||+||\mathbf{K}(1)\mathbf{H}(\mathbf{AY}(2)+\boldsymbol{\xi}(2))|\mathbf{W}(1)||\}.$$

It is now obvious that stage $k=2$ has the same equations as stage $k=1$, but with $\mathbf{B}$ replaced by $[\mathbf{B}+\mathbf{W}(1)]$. The optimal control and performance index can now be written down by inspection. These relationships hold at every stage, and the general iteration may readily be written down:

$$\left.\begin{aligned}
\mathbf{\Delta}(k) &= -[\mathbf{D}^T(\mathbf{B}+\mathbf{W}(k-1))\mathbf{D}+\mathbf{L}]^{-1}\mathbf{D}^T(\mathbf{B}+\mathbf{W}(k-1))\mathbf{A}, \\
\mathbf{W}(k) &= ||\mathbf{\Delta}(k)|\mathbf{L}|| + ||\mathbf{A}-\mathbf{\Delta}(k)|\mathbf{B}+\mathbf{W}(k-1)||, \\
\boldsymbol{\theta}^*(k) &= -\mathbf{\Delta}(k)\mathbf{M}(k), \\
\mathbf{N}(k) &= E\{||\mathbf{Y}(k)+\boldsymbol{\xi}(k)|\mathbf{B}|| + ||\mathbf{K}(k-1)\mathbf{H}(\mathbf{A}\mathbf{Y}(k) \\
&\qquad +\boldsymbol{\xi}(k))|\mathbf{W}(k-1)||\}, \\
f(\mathbf{S}(k), k) &= ||\mathbf{M}(k)|\mathbf{W}(k)|| + \mathbf{N}(k) + \mathbf{N}(k-1) + \ldots + \mathbf{N}(1).
\end{aligned}\right\} \quad (36)$$

It is evident that the iteration for the optimal control alone is the same as for the deterministic case with all the disturbances removed, and $\mathbf{X}$ replaced by $\mathbf{M}$. The performance index has more terms than the deterministic case; see, for instance, Kalman *et al.* (1959) and Merriam (1960).

Control can be implemented by making observations, then computing the new mean and variance with eqns. (23) and (25), appropriately replacing indexes 0 and 1 by r and $(r+1)$. The value of optimal control is then computed from eqns. (36). This system of equations gives a solution if a satisfactory one exists; some factors affecting the nature of the solution are discussed in §4.

3.4. *A Partially Observable Adaptive System*

It is quite possible to control a system which has unknown parameters, and which therefore requires adaptive control; and at the same time the observations are corrupted by noise. As an example consider a simple regulator having an unknown contant in the dynamic equation

$$x(r+1)=x(r)+c+\theta(r)+\xi(r), \quad \ldots\ldots \quad (37)$$

where c is the unknown constant. The observations are

$$s(r)=x(r)+\eta(r). \quad \ldots\ldots\ldots \quad (38)$$

To find the estimation equations assume that at time $r=0$ a joint Gaussian density for $x(0)$ and c has been given; this has mean vector $[m_x(0), m_c(0)]$ and variance–covariance matrix $\mathbf{U}(0)$. Then the *a priori* stochastic equation for $x(1)$ can be written

$$x(1)=m_x(0)+m_c(0)+y_x(0)+y_c(0)+\theta(0)+\xi(0), \quad \ldots \quad (39)$$

$y_x(0)$ and $y_c(0)$ being random variables which are jointly Gaussian with zero mean and variance–covariance matrix $\mathbf{U}(0)$. From equation (39) it is possible to find the following *a priori* moments:

$$\operatorname{Var}\{x(1)\}=\sigma_1, \quad \operatorname{Cov}\{cx(1)\}=\sigma_2, \quad \operatorname{Cov}\{[x(1), c]^T[x(1), c]\}=\boldsymbol{\sigma}.$$

The mean and variance of the conditional distribution $P(x(1), c|\jmath(1))$ can now be found using the formulae of the last example. Noting that

$$\mathrm{Var}\{\jmath(1)\} = \sigma_1 + v_1,$$
$$\mathrm{Cov}\{[x(1), c]^T \jmath(1)\} = [\sigma_1, \sigma_2]^T,$$

it is found that the new mean vector after observing $\jmath(1)$ is

$$\begin{bmatrix} m_x(1) \\ m_c(1) \end{bmatrix} = \begin{bmatrix} m_x(0) + m_c(0) + \theta(0) \\ m_c(0) \end{bmatrix} + \Gamma \begin{bmatrix} \sigma_1 \\ \sigma_2 \end{bmatrix}, \quad . \quad . \quad . \quad (40)$$

where

$$\Gamma = \frac{\jmath(1) - m_x(0) - m_c(0) - \theta(0)}{\sigma_1 + v_1}.$$

The new variance–covariance matrix is

$$\mathbf{U}(1) = \boldsymbol{\sigma} - \frac{1}{\sigma_1 + v_1} \begin{bmatrix} \sigma_1 \\ \sigma_2 \end{bmatrix} [\sigma_1, \sigma_2]. \quad . \quad . \quad . \quad . \quad . \quad (41)$$

This form of estimation equation holds at every time instant.

In order to compute the control it is necessary to have an *a priori* stochastic equation for the next mean vector. This equation can be found by noting that

$$\jmath(1) = x(1) + \eta(1), \quad . \quad . \quad . \quad . \quad . \quad . \quad . \quad . \quad (42)$$

and then substituting for $x(1)$ from eqn. (39). This value for $\jmath(1)$ is now substituted in Γ whence eqn. (40) becomes the necessary *a priori* equation.

If a quadratic performance index is used, and the functional iteration carried out, it will be found that the expression for optimal control is the same as would be obtained for the equivalent deterministic system, but with x and c replaced by m_x and m_c. The method of this example extends directly to certain time-varying adaptive systems. These are systems where the time-varying element can be described by an additive term

$$\sum c_i \phi_i(r),$$

where the c_i are unknown constants, and the $\phi_i(r)$ are known functions of time.

To illustrate this consider the system of eqn. (37) modified to

$$x(r+1) = x(r) + c\phi(r) + \theta(r) + \xi(r). \quad . \quad . \quad . \quad . \quad (43)$$

Given the same observation equation and initial conditions as before the *a priori* equation for $x(1)$ now becomes

$$x(1) = m_x(0) + m_c(0)\phi(0) + y_x(0) + y_c(0)\phi(0) + \theta(0) + \xi(0). \qquad (44)$$

The required covariances are labelled as before, and the resulting equation for up-dating the mean is

$$\begin{bmatrix} m_x(1) \\ m_c(1) \end{bmatrix} = \begin{bmatrix} m_x(0) + m_c(0)\phi(0) + \theta(0) \\ m_c(0) \end{bmatrix} + \Gamma \begin{bmatrix} \sigma_1 \\ \sigma_2 \end{bmatrix}. \quad . \quad . \quad (45)$$

The revised factor Γ is now

$$\Gamma = \frac{\jmath(1) - m_x(0) - m_c(0)\phi(0) - \theta(0)}{\sigma_1 + v_1}. \tag{46}$$

All other equations carry through as before. On calculating the iteration for control according to a quadratic performance index it will again be found that the expression for optimal control is that for the equivalent deterministic case, with x and c replaced by m_x and m_c.

A point of interest in this example is that the observations lead to a joint distribution for the state and unknown parameter, so that a single observation can carry information on different types of unknowns.

3.5. *Some Miscellaneous Applications of Partial Observability*

There are other problems in control where the techniques described above give a unifying viewpoint. One is when using several instruments to monitor one measurement point. This has practical uses; for instance, one instrument is adjusted to register small values of the quantity being observed, and the other adjusted for large values. In order properly to combine the readings of the separate instruments the error probabilities of the different instruments should be compounded in the same way as in the above examples.

Another problem arises when there is a time delay in the system, or measuring arrangements; then the current state has to be estimated from measurements made in the past. If a probability function for the current state can be found, and then expressed as a set of sufficient statistics, then these sufficient statistics enter the control problem in exactly the same way as if the measurements have been made at the same time. For suitable systems the extended Certainty Equivalence Principle discussed below holds, and then the best least squares prediction is used to replace the state variables in the control expression originally obtained for deterministic control.

It should perhaps be mentioned that the estimation formulae of the examples hold whenever the system is time-varying linear, and the disturbances Gaussian. If a different performance index is used, then only the functional iteration needs to be changed. Also, if a linear system has bounded control, a relay for instance, then it is the minimizing in the functional iteration which is changed. In these cases only a computational solution appears feasible.

§ 4. Comments on the Examples

Some general points arising from the examples will now be discussed.

4.1. *Extension of the Certainty Equivalence Principle*

The examples are constructive proofs of the rule: for time-varying linear systems with Gaussian disturbances, quadratic performance indices, and where the state vector cannot be measured directly, the optimal control can be found by the following:

(i) remove all the basic noise sources, i.e. the $\boldsymbol{\xi}$,

(ii) find the control expression for the resulting deterministic system,

(iii) in the control expression found in (ii) replace the state vector components, and the constants multiplying any additive time-varying factors, by their *a posteriori* means, or equivalently by their least squares estimates. The resulting expression is the control expression for the partially observable system.

All factors multiplying the state vector, such as the matrix $\mathbf{A}$, must be known exactly.

This is an extension of the Certainty Equivalence Principle (Simon 1956). For a follower servo, in the stationary condition, having a command signal corrupted by noise, an elegant direct proof of the rule has been given by Roberts (1961).

4.2. *Approaches to Defining Observability*

A definition of observability is required in the following context. Some controlled object is specified by the general form of its dynamic equations. There are constraints on the control, and observations can only be made at certain points. A fundamental question is then: do the observations yield enough information to plan a rational control strategy? If so, do the control constraints allow an effective strategy to be carried out? So far these questions have been considered only by Kalman (1960, 1961); he has given definitions for linear systems in which observability does not depend on the use to which observations are to be put, and controllability does not depend on the information available. Such an approach has obvious merits; however, it does not appear to be possible to keep such clear-cut distinctions when general non-linear systems or general classes of distributions are considered.

To see one way in which the control and observation problems could be interlinked consider the simple regulator

$$x(r+1) = x(r) + c + \theta(r) + \xi(r),$$

where $x(r)$ can be observed, but c is an unknown constant $\geqslant 1$. Suppose the observation scheme has yielded a cumulative *a posteriori* distribution for c:

$$\begin{aligned} F(c) &= 1 - c^{-3} \text{ for } c \geqslant 1 \\ &= 0 \text{ for } c < 1. \end{aligned}$$

Then a performance index involving $E\{x^2(r+1)\}$ depends on $E\{c^2\}$, which exists, but one involving $E\{x^4(r+1)\}$ depends on $E\{c^4\}$, which does not exist. Thus, this hypothetical observation scheme can lead to effective control for some performance indices but not others. It is for this type of problem that partial observability is considered in relation to particular requirements.

4.3. *Solutions for Linear Systems*

Having decided to view the observation and control problems as a whole it is possible to suggest some conditions under which the solutions of the general linear problem of § 3.3 might be satisfactory.

(i) Assuming the matrix $\mathbf{A}$ to be non-singular, the system must be controllable in the following modified Kalman sense. Remove all noise sources. Take the smallest coupled set of state components which contain those components appearing in the performance index, to give a sub-state vector $\mathbf{X}'$. Partition $\mathbf{A}$ and $\mathbf{D}$ so as to determine the Kalman controllability of $\mathbf{X}'$. This is illustrated by a three-dimensional system with constant coefficients having

$$\mathbf{A}=\begin{bmatrix} a_{11}, & 0, & 0, \\ 0, & a_{22}, & a_{23} \\ 0, & a_{32}, & a_{33} \end{bmatrix}.$$

It is given that only component x_3 appears in the performance index, so the partition is

$$\begin{bmatrix} x_1(r+1) \\ \hline \mathbf{X}'(r+1) \end{bmatrix} = \left[\begin{array}{c|cc} a_{11}, & 0, & 0 \\ \hline 0, & a_{22}, & a_{23} \\ 0, & a_{32}, & a_{33} \end{array}\right] \begin{bmatrix} x_1(r) \\ \hline \mathbf{X}'(r) \end{bmatrix} + \left[\begin{array}{ccc} d_{11}, & d_{12}, & d_{13} \\ \hline d_{21}, & d_{22}, & d_{23} \\ d_{31}, & d_{32}, & d_{33} \end{array}\right] \boldsymbol{\theta}(r).$$

Put the required partitioned parts of $\mathbf{A}$ and $\mathbf{D}$ equal to $\mathbf{A}'$, $\mathbf{D}'$, then

$$\mathbf{X}'(r+1)=\mathbf{A}'\mathbf{X}'(r+1)+\mathbf{D}'\boldsymbol{\theta}(r).$$

This sub-system is Kalman controllable if $[\mathbf{A}'^{-1}\mathbf{D}', \mathbf{A}'^{-2}\mathbf{D}']$ is of rank 2, which is equivalent to saying that $\mathbf{X}'$ can be reduced to zero in two time periods by suitable choice of $\boldsymbol{\theta}(0)$ and $\boldsymbol{\theta}(1)$.

(ii) The system is observable in the following sense: in eqn. (23) each component of $\mathbf{M}(r)$ is actually up-dated when an observation $\mathscr{S}(r)$ is made. This essentially means that $\mathbf{K}(r)$ does not have a zero row.

This criterion differs in certain critical respects from Kalman's; to see this consider the two-dimensional system with uncoupled state components having

$$\mathbf{A}=\begin{bmatrix} 1, & 0 \\ 0, & 1 \end{bmatrix}.$$

Only x_1 can be observed, hence $\mathbf{H}=[1, 0]$. Using eqn. (24) it is found that

$$\mathbf{K}(r)=\frac{1}{(u_{11}+q_{11})}\begin{bmatrix} u_{11}+q_{11} \\ u_{21}+q_{21} \end{bmatrix},$$

where u_{ij}, q_{ij} are the elements of $\mathbf{U}(r-1)$ and $\mathbf{Q}$, which are constant. Thus, in the general case, both $m_1(r)$ and $m_2(r)$ are up-dated from each observation. However, if at time $r=0$ it is given that $u_{21}=q_{21}=0$ then $m_2(r)$ will not be up-dated. With Kalman's definition the system is classified as non-observable. The physical reason for the possibility of observation is that x_1 and x_2 can be coupled through their noise sources.

Note carefully that even if the observability condition given above is not satisfied, the system may still be partially observable. If an initial $m_2(0)$ is given the estimation computation can still be carried out, the result, of course, being that m_2 remains constant; this mean value can still be used to compute the optimal control, although the performance index deteriorates rapidly.

(iii) A possible stricter condition on the observation process is to require that if initially a state component x_i has an 'ignorance' distribution with

u_{11} arbitrarily large, then after one observation the variance assumes a fixed finite value. Thereafter the variance should remain finite regardless of the realized values of the observation sequence. It will be seen that this happens in the examples of §§ 3.1 and 3.2.

The behaviour of the variance will be noted. In the example of § 3.1 the variance of the estimate tends to zero in spite of the inaccurate instrument. In the example of § 3.2 the variance tends to a constant, uniformly increasing or decreasing, depending on the starting value.

§ 5. Conclusions

It has been pointed out that it is possible to compute an optimal control for a controlled object whose state vector cannot be measured without error. This also applies when there are fewer observation points than state vector components, and when there are unknown parameters in the controlled object.

Several simple examples with linear dynamics and uncertainties described by Gaussian distributions have been presented to illustrate the implementation of the principles. Problems with non-linear dynamics and non-Gaussian distributions may be tackled by the same methods, but are substantially less tractable.

The use of these principles gives further insight into the fundamental theory of control. It is suggested that the property of partial observability must be judged by whether enough information can be obtained to compute optimal control for particular performance requirements.

Acknowledgments

The author would like to acknowledge many stimulating discussions with his colleague Mr. David Q. Mayne.

The author has been informed that some of the results of this paper have been obtained independently by Mr. J. Dawkins.

References

Bellman, R., 1961, *Adaptive Control Processes* (Princeton: University Press).

Kalman, R. E., 1960, *Symposium on Engineering Applications of Random Function Theory and Probability*, Purdue University, November (New York: Wiley). Also published as *Technical Report* 61–1 (Baltimore: R.I.A.S.), 1961, *Proceedings First International Congress on Automatic Control*, Moscow, 1960 (London: Butterworths).

Kalman, R. E., Lapidus, L., and Shapiro, E., 1959, *Proceedings Joint Symposium on Instrumentation and Computation in Process Development*, May (London: Institution of Chemical Engineers).

Merriam III, C. W., 1960, *Inform. Contr.*, **3**, 32.

Raiffa, H., and Schlaifer, R., 1961, *Applied Statistical Decision Theory* (Harvard: School of Business Administration).

Roberts, A. P., 1961, *Trans. Soc. Instrum. Tech.*, **13**, 213.

Simon, H. A., 1956, *Econometrica*, **24**, 74.

Wald, A., 1950, *Statistical Decision Functions* (New York: Wiley).

Part IV.

Sub-optimal nonlinear control

INTRODUCTION TO PART IV

Part IV treats the replacement of the complicated nonlinear control function, which is optimal, by a simpler sub-optimal control function. Paper 17 investigates the use of a control function which is simply a linear combination of state coordinates. Although Paper 17 considers primarily the deterministic case, it suggests that a linear control function will often be nearly optimal in the stochastic case. Paper 18 suggests a computational method of finding the best coefficients for the linear control function in the stochastic case.

Note also that Paper 10 discusses sub-optimal techniques, for deterministic systems, and these techniques may be used in the stochastic case also.

Paper 17

Reprinted from INTERNATIONAL JOURNAL OF CONTROL, Vol. 5, No. 3, pp. 197–243, March, 1967.

Paper 18

Reprinted from INTERNATIONAL JOURNAL OF CONTROL, Vol. 10, No. 1, pp. 77–98, July, 1969.

Linear Control of Non-Linear Systems†

By A. T. Fuller

Engineering Department, Cambridge University

[Received March 18, 1967]

Abstract

For some simple relay control systems it is shown that if the control signal input to the relay is made an appropriate linear combination of the state variables, nearly optimum performance can be achieved. The performance index used is integral-square-error; its value when the switching function is linear is calculated and compared with the value obtained when the switching function is the optimal (usually non-linear) one. E.g. for a second-order plant and a command signal which is a step of random magnitude having a rectangular probability distribution, linear switching can give a mean integral-square-error which exceeds the optimal value by less than 1%.

Part I. Preliminaries

§ 1. Introduction

It is well known that for simple relay or saturating control systems the optimal controller is represented by a switching surface in the phase space or state space of the system (see e.g. Fuller 1960 a, b, c). In other words the controller preceding the relay or saturator generates a control signal which is an 'instantaneous' function of the state coordinates. With the aid of contemporary optimal control theories, e.g. Pontryagin's maximum principle, this switching function has been calculated in many cases, and has usually been found to be non-linear and too complicated to be synthesized readily. Consequently designers have tended to seek sub-optimal control methods or to rely on traditional linear design methods. The purpose of the present paper is to show that a sub-optimal controller which generates simply a linear combination of the state coordinates can in some cases give very nearly optimal performance. In such cases the traditional use of a linear controller preceding the relay or saturator, a practice which has developed empirically during the nineteenth and twentieth centuries (see e.g. Fuller 1963 a for history), is justified.

It has often been suspected that the gains to be obtained from using the exactly optimal non-linear controller would be small. In 1957 Oldenburger reported‡ that he had used such controllers and had found the results 'disappointing'. Several writers, for example Oldenburger (1957, 1960), Flügge-Lotz and Lindberg (1959), Fuller (1959 a, 1963 a), Flügge-Lotz

---

† Communicated by the Author.

‡ In the discussion of his paper.

(1960), Flügge-Lotz and Ishikawa (1960), and Jackson (1960), have suggested that a linear switching function might be nearly optimal. However, the arguments given in support of this suggestion have been somewhat vague and qualitative. In the present paper the argument will be strengthened by a quantitative comparison of the optimal and sub-optimal performances of simple relay control systems.

§ 2. Performance Indices

In optimal non-linear theory it is usual to adopt settling time as the performance index which is to be minimized. Settling time is the time for the system to travel from a given starting point in the state space to a desired state (usually the origin of the state space). This performance index is adopted because the optimal controller is then relatively easy to calculate.

If we use a sub-optimal controller which gives a linear switching surface, we find that usually the settling time is infinite. This is because the final part of the trajectory is in the switching surface, and corresponds to 'chattering' or 'sliding' motion, which decays exponentially†. Thus if we apply naïvely the settling time performance index we reach the conclusion that linear switching functions are not even approximately optimal. Admittedly, provided the starting point in the state space is fixed, the coefficients in the switching function can be chosen so that no sliding motion takes place (Flügge-Lotz and Lindberg 1959, Flügge-Lotz and Ishikawa 1960). The settling time is then finite, and indeed often optimal. However, if the controller coefficients are fixed in this way on the basis of a given starting point, then for only a slightly changed starting point the settling time will usually be infinite. Thus if we adopt settling time as the performance index it is difficult to convince ourselves of the adequacy of linear switching functions.

On reflection, however, we see that it is the choice of performance index, not the linear controller, which is at fault. Settling time is too idealized an index—it penalizes all non-zero errors equally, and only when the state point is exactly at the origin is there no further contribution to the performance index. In actual practice, however, small errors are tolerable, and the occurrence of sliding motion in the vicinity of the origin is not an important drawback. Thus settling time is an inappropriate index for judging the performance of these sub-optimal control systems.

The performance index we shall adopt is integral-square-error. This index is well known in the context of linear control theory. Other performance indices will be discussed in § 18. The relation between performance indices such as settling time and integral-square-error has been discussed by Fuller (1959 b). The exactly optimal non-linear controller which minimizes integral-square-error is known only in simple

† This phenomenon is well known. See e.g. Flügge-Lotz (1953). It will be discussed further in § 6.

cases. For this reason the investigation below is restricted to plants consisting of one, two and three integrators respectively. Although such plants are over-idealized, the results to be obtained are sufficiently emphatic to suggest that they can be extrapolated to more general systems.

§ 3. The System

The control system to be considered is shown in fig. 1. The plant has a given transfer function $g(s)$ between its input u and its output x. The relay has input z and output u which satisfy†

$$u = \operatorname{sgn} z \quad (z \neq 0) \tag{1}$$

The controller output is a function of the state coordinates of the system.

Fig. 1

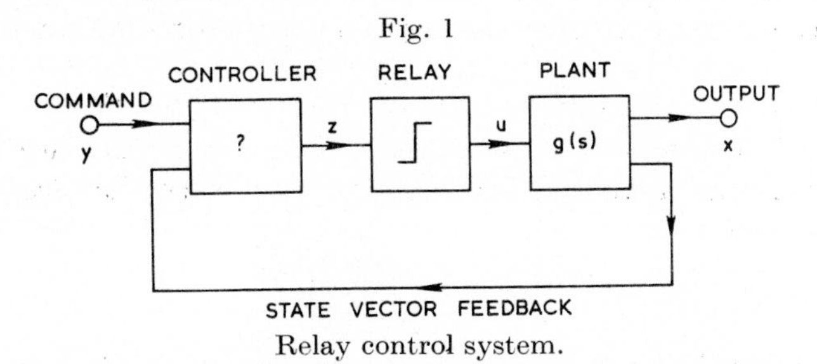

Relay control system.

The command signal is y, and the system error e is defined as

$$e = x - y \tag{2}$$

The performance index w is

$$w = \int_0^\infty e^2(t)\,dt \tag{3}$$

where t is time.

We shall often have in mind the calculation of step response, i.e. the case when the command signal is a step applied at $t = 0$ and the initial values of the plant state coordinates are zero. However, the calculations are essentially the same if instead we consider the command signal to be zero and the initial value of the plant output to be non-zero (see e.g. Fuller 1960 b). We shall adopt this procedure since it enables the notation to be simplified, but we shall still call the system output a step response, when appropriate.

Thus for the purposes of calculation we shall take

$$y = 0 \tag{4}$$

and then the performance index is

$$w = \int_0^\infty x^2(t)\,dt \tag{5}$$

The problem is to find which function z of the state coordinates of the plant minimizes w, for given initial values of the state coordinates, first when z is unrestricted, secondly when z is restricted to be linear.

† The interpretation of sgn z when z is zero will be discussed in §§ 6 and 9.

§ 4. First-order Plant

When the plant transfer function is $1/as$, i.e. when the plant consists of a single integrator:

$$\frac{dx}{dt} = \frac{u}{a} \quad (a > 0) \tag{6}$$

the problem is trivial. Integral-square-error is minimized by a switching function

$$z(x) = -x \tag{7}$$

(Fuller, 1960 d), i.e. the linear switching function (7) is optimal not only in the class of linear switching functions but also in the class of general switching functions. Thus for the plant $1/as$ there is no increase in performance index incurred by the use of a linear switching function.

§ 5. Optimal Control of a Second-order Plant

Most of the remainder of the paper will be concerned with the case when the plant has a transfer function $1/as^2$, i.e. when the plant consists of two integrators:

$$\frac{d^2x}{dt^2} = \frac{u}{a} \quad (a > 0) \tag{8}$$

In this section the known results for the optimal controller and minimal performance index will be stated and discussed. The plant state coordinates are $x_1(t)$ and $x_2(t)$ where

$$x_1 = x \tag{9}$$

$$x_2 = \frac{dx}{dt} \tag{10}$$

The optimal non-linear switching function is

$$z(x_1, x_2) = -x_1 - H_0 a x_2 |x_2| \tag{11}$$

where

$$H_0 = \sqrt{\left(\frac{\sqrt{33}-1}{24}\right)} = 0{\cdot}444623560 \tag{12}$$

Thus the optimal switching line is

$$x_1 + H_0 a x_2 |x_2| = 0 \tag{13}$$

For a starting point C on the x_1-axis:

$$x_1(0) = x_1^{C} \tag{14}$$

$$x_2(0) = 0 \tag{15}$$

i.e. for what is effectively a step response, the minimum value of integral-square-error (obtained with the non-linear switching function (11)) is

$$w_{\text{opt}} = G_0 a^{1/2} |x_1^{C}|^{5/2} \tag{16}$$

where

$$G_0 = \tfrac{1}{20}\sqrt{(222 + 2\sqrt{33})} = 0{\cdot}764017548 \tag{17}$$

More generally, for a starting point B

$$\left.\begin{aligned} x_1(0) &= x_1^{\mathrm{B}} \\ x_2(0) &= x_2^{\mathrm{B}} \end{aligned}\right\} \qquad \begin{aligned}(18)\\(19)\end{aligned}$$

let us represent the minimal value of the performance index as $v(x_1^{\mathrm{B}}, x_2^{\mathrm{B}})$. For notational economy let us agree also to drop the superscripts in the arguments of v, so that for a starting point (x_1, x_2)

$$w_{\mathrm{opt}} = v(x_1, x_2) \qquad (20)$$

Then for a starting point to the right of the switching curve (13), i.e. where the switching function $z(x_1, x_2)$ given by (11) is negative,

$$v(x_1, x_2) = a x_1^2 x_2 + \tfrac{2}{3} a^2 x_1 x_2^3 + \tfrac{2}{15} a^3 x_2^5 + G_0 a^{1/2} (x_1 + \tfrac{1}{2} a x_2^2)^{5/2} \qquad (21)$$

and for a starting point to the left of the switching curve (13), i.e. where the switching function $z(x_1, x_2)$ given by (11) is positive,

$$v(x_1, x_2) = -a x_1^2 x_2 + \tfrac{2}{3} a^2 x_1 x_2^3 - \tfrac{2}{15} a^3 x_2^5 + G_0 a^{1/2} (-x_1 + \tfrac{1}{2} a x_2^2)^{5/2} \qquad (22)$$

This problem of minimizing integral-square-error for a relay control system with plant $1/as^2$ has been studied in several papers. Originally results were obtained by the use of the theory of dimensions (Fuller 1959 a, 1960 d). The results were shown to satisfy Pontryagin's maximum principle by Pearson (1961), using digital computation, by Brennan and Roberts (1962), using analogue computation, and by Fuller (1963 b), using algebraic techniques. The results were shown to satisfy Bellman's equation by Wonham (1963) and Fuller (1964 a). Methods of deriving, not merely verifying, the results by means of Bellman's equation were given by Wonham (1963) and Fuller (1966). Similarly a method of deriving, not merely verifying, the results by means of Pontryagin's maximum principle was given by Grensted and Fuller (1965). In addition the results have been shown to correspond to an absolute, not merely a local, optimum (Fuller 1964 b); and the theory of dimensions, of which applications are made in some of the above papers, has been rigorized (Fuller 1966).

§ 6. Sliding Motion

Phase plane studies of second-order relay control systems with linear controllers go back to Léauté (1885, 1891), Houkowsky (1896), Synge (1933), Nikolski (1934) and Teodorchik (1938). The case with the control signal a linear combination of both phase coordinates was investigated in Germany by Houkowsky (1896) and Bilharz (1941, 1942). Their approach was followed up by a group consisting of Flügge-Lotz, Klotter, Meissinger, Hodapp and Scholz (Flügge-Lotz *et. al.* 1943–5, Flügge-Lotz 1947, Flügge-Lotz and Klotter 1948). Other work was done by MacColl (1945) and Weiss (1946) in America. An exposition and continuation of the early work of Flügge-Lotz and her colleagues is given in her book (1953).

Consider the system of fig. 1 (the plant transfer function being $1/as^2$), with a linear switching function:

$$z(x_1, x_2) = -(k_1 x_1 + k_2 x_2) \tag{23}$$

where k_1 and k_2 are constants. As follows from the geometrical treatment of Flügge-Lotz (1953), for stability k_1 and k_2 must be positive†. Since the presence of the relay implies that only the sign of z affects the system performance, we may take k_1 to be unity without loss of generality.

Fig. 2

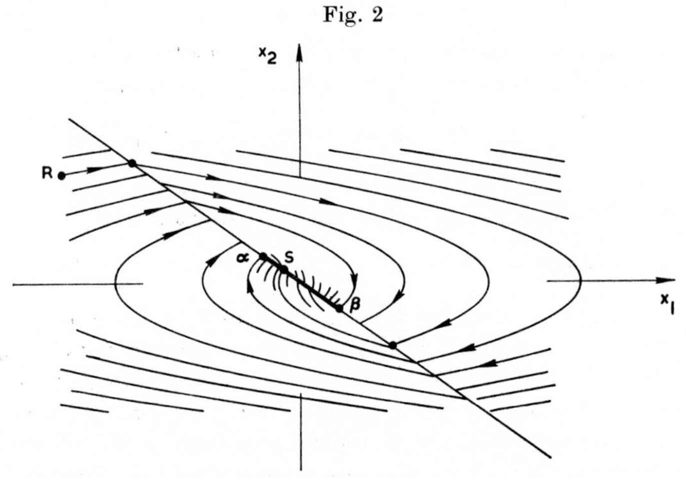

Phase portrait of a second-order relay control system with linear switching.

Dropping the subscript from k_2 we can therefore write the linear switching function as

$$z = -(x_1 + kx_2) \tag{24}$$

The phase portrait of the system with controller (24) is shown in fig. 2. The straight line through α and β is the switching line: above it the relay output is $u = -1$, and below it the relay output is $u = +1$. On each side of the switching line the trajectories consist of parabolae (see § 8). Inspection of the portrait shows that the phase point tends to spiral in towards the origin (a typical trajectory is represented by RS). After a number of switches and a number of overshoots (these numbers depend on the starting point) the phase point reaches the segment of the switching line marked $\alpha\beta$ in fig. 2. This segment is characterized by the property that, on both sides of it, the direction of the trajectories is towards the switching line. What happens when the phase point arrives at a point, such as S, on this segment? The phase portrait indicates that the phase point cannot move into either of the two regions bounded by the switching line. As is well known (e.g. Flügge-Lotz 1953, André and Seibert 1956), in practice the phase point oscillates about the switch line at a high

† See also § 8.

frequency dependent on parasitic delays in the switching, and also tends to move along the switching line (towards the origin in the present case). Mathematically the motion is no longer well defined by the differential equations so far adopted for the system†. We therefore adopt, as a mathematical definition of the motion for starting points on $\alpha\beta$, motion along the switching line‡. This motion along the switching line is variously called sliding, gliding, chattering, or after-endpoint motion||. Of these terms, it is preferable to adopt sliding (Weissenberger 1966).

On the switching line we have

$$x_1 + kx_2 = 0 \tag{25}$$

i.e. in view of (9) and (10),

$$k\frac{dx_1}{dt} + x_1 = 0 \tag{26}$$

Thus during the motion along the switching line, $x_1(t)$ is differentiable and satisfies differential eqn. (26). Since the only solution of this equation, starting at $x_1 = x_1^S$ at time $t = t_s$, is

$$x_1 = x_1^S \exp\left(-\frac{t - t_s}{k}\right) \tag{27}$$

the sliding motion is exponential, with time constant determined by the slope of the switching line.¶

Our immediate problem is to find the value of k that results in minimum integral-square-error, for a given starting point in the phase plane. Too low a value of k will cause large initial overshoots, and too high a value will cause the final sliding motion to be slow.

§ 7. Results for Linear Switching

In this section the main results to be obtained for the plant $1/as^2$ with linear switching will be summarized. The detailed derivation of these results will be given in subsequent sections.

Consider a fixed starting point $(x_1, 0)$ on the x_1-axis in the (x_1, x_2) phase plane, i.e. consider step response. The value of the coefficient k (i.e. minus the reciprocal of the slope of the switching line) that minimizes integral-square-error is

$$k^* \simeq 0{\cdot}460960545 a^{1/2} |x_1|^{1/2} \tag{28}$$

---

† There exists no solution of eqns. (1), (8), (9), (10), (24) such that $x_1(t)$ and $x_2(t)$ are continuous, $|u(t)| = 1$ and $u(t)$ is piece-wise constant, for a starting point on $\alpha\beta$.

‡ The motion along the switching line can also be derived by considering it as the motion in the unsaturated region of a saturating control system, in the limiting case when the saturation characteristic approaches a relay characteristic.

|| According to Tsypkin (1955), sliding motion was already discussed in Nikolski's (1934) paper. Andronov and Bautin (1944) treated a third-order system with sliding motion.

¶ This remark is due to Golling, apparently in a German war-time report around 1943. See Flügge-Lotz *et. al.* (1943) and Flügge-Lotz and Klotter 1948).

and yields as the minimum value of the integral-square-error w obtainable with linear switching

$$w^* \simeq 0{\cdot}764345704a^{1/2}|x_1|^{5/2} \tag{29}$$

Value (29) is 1·00042951 times the integral-square-error obtained with the optimal non-linear switching line (13), i.e. (29) is about 0·043% worse than the optimal value.

Consider a fixed starting point $(0, x_2)$ on the x_2-axis in the (x_1, x_2) phase plane (i.e. consider ramp response). The best value of k is now

$$k^* \simeq 0{\cdot}325948328a|x_2| \tag{30}$$

and yields as the performance index with linear switching

$$w^* = 0{\cdot}268451841a^3|x_2|^5 \tag{31}$$

Value (31) is 1·00021614 times the integral-square-error obtained with the optimal non-linear switching line (13), i.e. (31) is about 0·022% worse than the optimal value.

For other starting points, various values are obtained for the fractional discrepancy between the performance indices obtained with the best linear switching function and the optimal non-linear switching function. In the most favourable case this fraction is 0·020%, being realized for points on the locus

$$x_1 + H_1 a x_2 |x_2| = 0 \tag{32}$$

where

$$H_1 = 0{\cdot}236605249 \tag{33}$$

In the least favourable case, the fraction is 2·7%, being realized for points on the locus

$$x_1 + H_2 a x_2 |x_2| = 0 \tag{34}$$

where

$$H_2 = 0{\cdot}354654565 \tag{35}$$

The above results hold when for each starting point the best value of k is chosen; thus in general different starting points have different values of k. The question arises: with a fixed value of k and a variety of starting points, is the performance appreciably worse than optimal? This question is more realistic than that for which the above results apply; i.e. in practice we are interested in the sensitivity of the system to a change of initial conditions. (Indeed this was why we rejected settling time as the performance index in § 2.) The following result illustrates that when the performance index is integral-square-error and with a fixed straight switch line, nearly optimal performance can still be obtained when the initial conditions are allowed to vary.

Consider an ensemble of starting points distributed uniformly along a segment of the x_1-axis. The segment is centred on the origin and has length $2c$. Consider each member of the ensemble to have the same k. The value of k

$$k^* = 0{\cdot}407428021a^{1/2}c^{1/2} \tag{36}$$

yields a minimum value of the expected integral-square-error, $\bar{w}$, namely

$$\bar{w}^* = 0{\cdot}220078713a^{1/2}c^{5/2} \tag{37}$$

(37) is only 1·00819084 times greater than the expected integral-square-error obtained with the optimal non-linear switching line (13); i.e. is only about 0·82% worse.

In parts II and III the calculations leading to the above results will be given. In part IV, more general problems will be discussed.

Part II. Analysis of the System with a given Switching Line

§ 8. Derivation of the Phase Portrait

Let us derive the phase portrait discussed in § 6. To begin with we revert to the general† linear switching function (23), so that the system is

$$\left.\begin{aligned} \frac{dx_1}{dt} &= x_2 \qquad (38) \\ \frac{dx_2}{dt} &= \frac{u}{a} = -\frac{1}{a}\operatorname{sgn}(k_1 x_1 + k_2 x_2) \qquad (39) \end{aligned}\right\}$$

8.1. *Trajectories*

On one side of the switching line the control is

$$u = +1 \tag{40}$$

and there the system satisfies

$$\left.\begin{aligned} \frac{dx_1}{dt} &= x_2 \qquad (41) \\ \frac{dx_2}{dt} &= \frac{1}{a} \qquad (42) \end{aligned}\right\}$$

Formal division‡ of (41) by (42) yields

$$\frac{dx_1}{dx_2} = ax_2 \tag{43}$$

as the differential equation of the trajectory. Integration of (43) yields as the equation of the trajectory

$$x_1 = \tfrac{1}{2}ax_2^2 + x_1^{\mathrm{C}} \tag{44}$$

where x_1^{C} is an arbitrary constant. (44) is called a *P*-arc since the control (40) is positive.

Similarly, on the other side of the switching line the control is

$$u = -1 \tag{45}$$

† We can confine attention to switching lines which pass through the origin, since otherwise the origin is not a point of equilibrium.

‡ The results can be verified by integrating (41) and (42) directly, then eliminating t.

and the equation of a trajectory there is

$$x_1 = -\tfrac{1}{2}ax_2^2 + x_1^C \tag{46}$$

where x_1^C is an arbitrary constant. (46) is called an N-arc since the control is negative.

8.2. *Stability*

For the present system a set of necessary and sufficient conditions for (in-the-small) asymptotic stability† of the origin is

$$\left.\begin{aligned} k_1 &> 0 \qquad (47)\\ k_2 &> 0 \qquad (48)\end{aligned}\right\}$$

These conditions are as given for general relay systems by Tsypkin (1955). Tsypkin's derivation of his stability criteria, while moderately convincing, lacked rigour. A more rigorous derivation of Tsypkin's criteria has been given by Anosov (1959). An earlier rigorization was given by Boltyanski and Pontryagin (1956); however, only an abstract of their paper seems to have been published.

It is usually required to have (in-the-small) asymptotic stability of the origin, for the system to be practical, and also for the performance index to be finite. Thus we restrict attention to the case specified by (47) and (48). System (38, 39) then reduces to‡ (see § 6):

$$\left.\begin{aligned} \frac{dx_1}{dt} &= x_2 \qquad (49)\\ \frac{dx_2}{dt} &= \frac{u}{a} = -\frac{1}{a}\operatorname{sgn}(x_1 + kx_2) \qquad (50)\end{aligned}\right\}$$

In § 9 it will be shown that (47) and (48) are also necessary and sufficient conditions for in-the-large asymptotic stability of the origin.

8.3. *The Phase Portrait*

Points to the left of the switching line (25) satisfy

$$x_1 + kx_2 < 0 \tag{51}$$

so that from (50)

$$u = +1 \tag{52}$$

Hence the trajectories to the left of the switching line are the P-arcs given by (44). When x_1^C varies, (44) represents a family of parabolae, any one of which is obtained from any other by translation along the x_1-axis. Similarly, to the right of the switching line the trajectories are N-arcs and constitute a family of parabolae (46). The complete phase portrait is as shown in fig. 2.

---

† See e.g. Kalman and Bertram (1960) and Lasalle and Lefschetz (1961) for discussion of various types of stability.

‡ The interpretation on the switching line of the sgn function in (50) will be specified in § 9.

§ 9. Properties of the Phase Portrait

9.1. *The Sliding-régime Segment of the Switching Line*

Let us find the segment of the switching line on which sliding motion takes place. The condition for sliding is that the directions of trajectories on both sides of the switching line are towards the switching line (§ 6). Consider, to begin with, the part of the switching line below the x_1-axis. It is obvious from fig. 2 that the direction of the N-arcs on the right of the switching line is towards the switching line. On a P-arc just to the left of the switching line the horizontal component of the motion is to the left, since $\dot{x}_1 = x_2 < 0$. Therefore the motion here is towards or away from the switching line, according as the slope of the P-arc is algebraically less or greater than that of the switching line.

From (43) the slope of the P-arc is $1/ax_2$, and from (25) the slope of the switching line is $-1/k$. Hence, on the switching line, the condition for sliding is

$$\frac{1}{ax_2} \leqslant -\frac{1}{k} \quad (x_2 < 0) \tag{53}$$

i.e.

$$|x_2| \leqslant \frac{k}{a} \tag{54}$$

for $x_2 < 0$; and by symmetry (54) holds also for $x_2 > 0$.

From (25) and (54) the extreme points α and β of the sliding-régime segment of the switching line are $[-k^2/a, k/a]$ and $[k^2/a, -k/a]$ respectively. Sliding motion occurs everywhere on the segment $\alpha\beta$, and nowhere else†.

9.2. *Definitions*

At this stage, to avoid ambiguities, it is helpful to adopt a few definitions:

A general trajectory consists of an alternating sequence of P-arcs and N-arcs followed by a sliding trajectory. We call the junction points of the P-arcs and the N-arcs, and the initial point of the sliding trajectory, *transition points* or simply *transitions*. We call the junction points of the P-arcs and the N-arcs *switch points*, or simply *switches*. Thus we do not count as a switch the transition to a sliding régime.

Therefore the sliding-régime segment of the switching line‡ consists of transition points which are not switches. All other points of the switching line are switches.

If a trajectory to the origin has n switches we call it an *n-switch trajectory*. If a trajectory starts at a switch point of the switching line, this initial point is counted in evaluating n.

---

† At the origin of the phase plane, the phase point remains stationary. For convenience we count this state as sliding 'motion'.

‡ Strictly we ought to call this line a transition line—not a switching line. However, the latter term is well established.

9.3. *System Specification*

We can now specify the system equations more precisely than hitherto. For points off the switching line, i.e. for

$$x_1 + kx_2 \neq 0 \tag{55}$$

the system is described by (49) and (50). For points on the sliding segment, i.e. for

$$x_1 + kx_2 = 0, \quad |x_2| \leqslant \frac{k}{a} \tag{56}$$

the system is described by

$$\left.\begin{aligned} \frac{dx_1}{dt} &= x_2 \qquad (57) \\ \frac{dx_2}{dt} &= -\frac{1}{k}x_2 \qquad (58) \end{aligned}\right\}$$

((58) is obtained by differentiation of (26).)

For points on the switching line but not on the sliding segment, i.e. for

$$x_1 + kx_2 = 0, \quad |x_2| > \frac{k}{a} \tag{59}$$

it is immaterial whether we specify dx_2/dt as $+1/a$ or $-1/a$. However, to place in evidence our above convention that a trajectory starting at a point on the switching line satisfying (59) has an initial switch, we shall describe the system in region (59) by†

$$\left.\begin{aligned} \frac{dx_1}{dt} &= x_2 \qquad (60) \\ \frac{dx_2}{dt} &= \frac{1}{a}\operatorname{sgn} x_2 \qquad (61) \end{aligned}\right\}$$

9.4. *Relation between Successive Transitions*

For a trajectory from a given starting point let us find the relation between a pair of successive transition points. Consider a P-arc starting at a point A (not on the sliding segment) on the switching line (fig. 3). Its points of intersection A and E with the switching line are easily calculated: solving (25) and (44) we find that the ordinate of A is

$$x_2^{\mathrm{A}} = -\frac{k}{a} - \left(\frac{k^2}{a^2} - \frac{2x_1^{\mathrm{C}}}{a}\right)^{1/2} \tag{62}$$

and the ordinate of E is

$$x_2^{\mathrm{E}} = -\frac{k}{a} + \left(\frac{k^2}{a^2} - \frac{2x_1^{\mathrm{C}}}{a}\right)^{1/2} \tag{63}$$

† Strictly, (61) is a one-sided (backward) derivative, whereas (58) is a one-sided (forward) derivative.

Addition of (62) and (63) yields

$$x_2^{\mathrm{E}} = -x_2^{\mathrm{A}} - \frac{2k}{a} \tag{64}$$

(64) shows that to locate E, we find the point A′ which is on the switching line and symmetrically opposite the point A, then move from A′ a fixed distance in the direction of the origin, i.e. a distance such that the x_2 coordinate decreases algebraically by $2k/a$.

Fig. 3

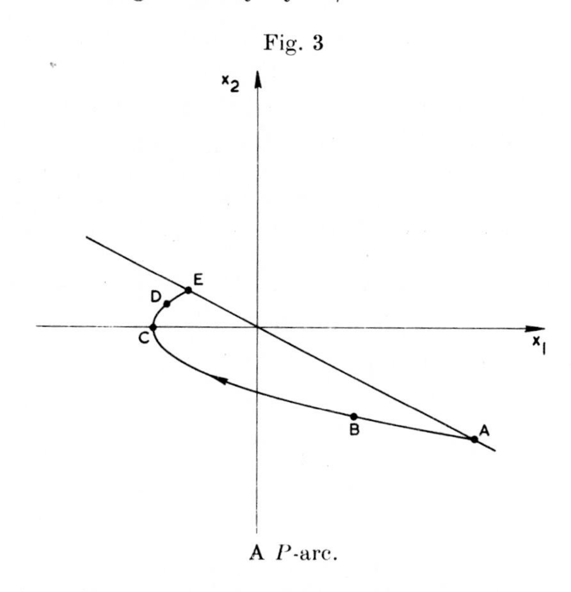

A P-arc.

If E is on the other side of the origin from A, it follows that the second intersection (E) is closer to the origin than the first intersection (A). By symmetry the same is true for N-arcs. Thus for a complete trajectory from a given starting point, any transition point occurs a constant distance closer to the origin than its predecessor; at least until two successive transition points occur on the same side of the origin.

If E is on the same side of the origin as A, we can show as follows that E is in the sliding motion segment. E is now below the x_1-axis, i.e.

$$x_2^{\mathrm{E}} < 0 \tag{65}$$

Also, since A is not in the sliding motion segment, (53) does not hold, i.e.

$$x_2^{\mathrm{A}} < -\frac{k}{a} \tag{66}$$

From (64) and (66)

$$x_2^{\mathrm{E}} > -\frac{k}{a} \tag{67}$$

From (65) and (67)

$$|x_2^{\mathrm{E}}| < \frac{k}{a} \tag{68}$$

Comparison of (54) and (68) shows that E is in the sliding motion segment.

We conclude that for a general complete trajectory, the transition points are successively closer to the origin by a constant distance, except possibly for the final transition point. The latter point corresponds to the onset of sliding motion, during which the phase point moves to the origin.

It follows that conditions (47) and (48) are sufficient (and thus necessary and sufficient) for in-the-large asymptotic stability of the origin.

§ 10. Dependence of the Number of Switches on the Starting Point

Each point in the phase plane is the starting point of a trajectory which goes to the origin and has a certain number of switches. We shall divide the phase plane into regions of starting points which correspond to given numbers of switches. In principle it is simple to find these regions, by tracing trajectories backwards from the sliding-régime segment.

10.1. *Number of Switches for Starting Points on the Switching Line*

Let us show that the switching line is divided into equal segments, corresponding to starting points of 0-switch, 1-switch, ... trajectories (see fig. 4).

Fig. 4

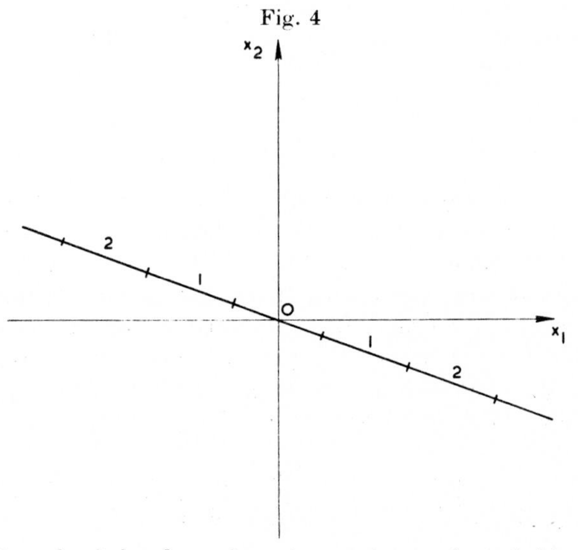

Number of switches for trajectories starting on the switching line.

All starting points on the sliding-régime segment of the switching line satisfy, from (54),

$$-\frac{k}{a} \leqslant x_2 \leqslant \frac{k}{a} \tag{69}$$

and yield 0-switch trajectories. The P-arcs traced backwards from segment (69) meet the switching line in a segment which corresponds to

1-switch trajectories. Using (64) to calculate the extremes of the latter segment from the extremes of segment (69), we find that the segment yielding 1-switch trajectories satisfies

$$-3\frac{k}{a} \leqslant x_2 < -\frac{k}{a} \tag{70}$$

By symmetry, or by tracing backwards N-arcs from the sliding-régime segment, we find that starting points on the switching line satisfying

$$\frac{k}{a} < x_2 \leqslant 3\frac{k}{a} \tag{71}$$

also yield 1-switch trajectories.

More generally, on tracing backwards from an n-switch segment of the switching line ($n > 0$), we arrive at an $(n+1)$-switch segment on the other side of the origin. From (64), any point on the $(n+1)$-switch segment is obtained from a corresponding point on the n-switch segment by a reflection in the origin† and a constant translation. Therefore the $(n+1)$-switch segment has the same length as the n-switch segment. By induction all segments have the same length. Also the segments are contiguous and are arranged in order of n; their common boundary points being found by tracing backwards trajectories from the boundary points $|x_2| = k/a$ at the junction of the 0 and 1-switch segments (69), (70) and (71). The result may be stated as follows: trajectories with starting points on the switching line and satisfying

$$(2n-1)\frac{k}{a} < |x_2| \leqslant (2n+1)\frac{k}{a} \tag{72}$$

are n-switch trajectories ($n = 0, 1, 2, \ldots$).

10.2. *Number of Switches for a General Starting Point*

The switching line and the trajectories through the boundary points of its segments divide the phase plane into contiguous regions (fig. 5). These regions correspond to starting points of 0-switch, 1-switch, etc., trajectories. The algebraic specification of these regions may be obtained as follows.

Let us trace backwards a P-arc from a point E on the (upper if $n > 0$) n-switch segment of the switching line to the point A on the switching line (fig. 3). The points on this arc and excluding its end points satisfy, from (44),

$$x_1 - \tfrac{1}{2}ax_2{}^2 = x_1{}^{\mathrm{E}} - \tfrac{1}{2}a(x_2{}^{\mathrm{E}})^2 \tag{73}$$

and, from (51),

$$x_1 + kx_2 < 0 \tag{74}$$

† A reflection in the origin means a reflection in the x_1-axis followed by a reflection in the x_2-axis (see e.g. Klein 1908).

while the point E satisfies, from (25),

$$x_1^{\mathrm{E}}+kx_2^{\mathrm{E}} = 0 \tag{75}$$

and, from (69) or (72),

$$(2n-1)\frac{k}{a} < x_2^{\mathrm{E}} \leqslant (2n+1)\frac{k}{a} \quad (n = 0, 1, \ldots) \tag{76}$$

Fig. 5

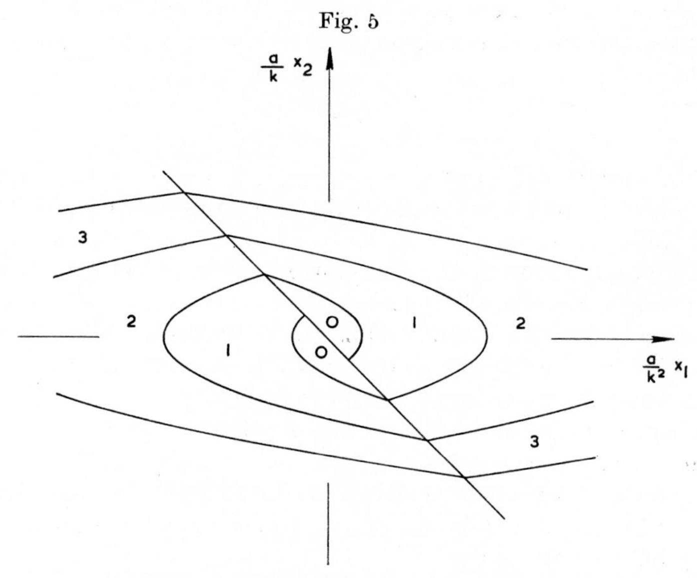

Number of switches for general starting points.

The system of equations and inequalities (73), (74), (75) and (76) specifies a region of starting points of n-switch trajectories. Let us simplify this system.

Elimination of x_1^{E} from (73) and (75) yields

$$x_1-\tfrac{1}{2}ax_2^2 = -kx_2^{\mathrm{E}}-\tfrac{1}{2}a(x_2^{\mathrm{E}})^2 \tag{77}$$

i.e.

$$\left(\frac{a}{k}x_2^{\mathrm{E}}+1\right)^2 = \frac{a}{k^2}(ax_2^2-2x_1)+1 \tag{78}$$

Also, (76) may be written

$$2n < \frac{a}{k}x_2^{\mathrm{E}}+1 \leqslant 2n+2 \tag{79}$$

Since the three members of the inequality chain (79) are non-negative, it may be squared:

$$4n^2 < \left(\frac{a}{k}x_2^{\mathrm{E}}+1\right)^2 \leqslant 4(n+1)^2 \tag{80}$$

From (78) and (80)

$$4n^2 < \frac{a}{k^2}(ax_2^2-2x_1)+1 \leqslant 4(n+1)^2 \quad (n = 0, 1, \ldots) \tag{81}$$

Inequalities (74) and (81) specify the region to the left of the switching line which corresponds to starting points of n-switch trajectories.

Similarly, for starting points to the right of the switching line; i.e. for

$$x_1 + kx_2 > 0 \tag{82}$$

we find

$$4n^2 < \frac{a}{k^2}(ax_2^2 + 2x_1) + 1 \leqslant 4(n+1)^2 \tag{83}$$

The simultaneous pair of inequalities (74, 81) and the simultaneous pair of inequalities (82, 83) are both included in the following inequality

$$4n^2 < \frac{a}{k^2}[ax_2^2 + 2x_1 \operatorname{sgn}(x_1 + kx_2)] + 1 \leqslant 4(n+1)^2 \tag{84}$$

which therefore specifies the regions of starting points of n-switch trajectories ($n = 0, 1, \ldots$).

Inequalities (84) are not well defined for points on the switching line; here inequalities (72) should be used instead.

Note that for starting points on the x_1-axis (step response), (84) shows that the ranges of x_1 giving n-switch trajectories satisfy†

$$2n^2 - \tfrac{1}{2} < \frac{a}{k^2}|x_1| \leqslant 2(n+1)^2 - \tfrac{1}{2} \quad (n = 0, 1, \ldots) \tag{85}$$

§ 11. Dimensionless Variables and Symmetry Relations

11.1. *Dimensions*

It is occasionally convenient to work with the dimensionless variables X_1 and X_2 instead of x_1 and x_2, where

$$X_1 = \frac{a}{k^2}x_1 \tag{86}$$

$$X_2 = \frac{a}{k}x_2 \tag{87}$$

To check that X_1 and X_2 are dimensionless, let x_1 have the dimension of signal S, and write this statement in the notation (Fuller 1966)

$$D(x_1) = S \tag{88}$$

If the dimension of time is represented by T,

$$D(t) = T \tag{89}$$

we have from (49)

$$D(x_2) = ST^{-1} \tag{90}$$

from (50) and (90)

$$D(a) = S^{-1}T^2 \tag{91}$$

and from (25) and (90)

$$D(k) = T \tag{92}$$

† For $n = 0$ the left member of (85) is negative; and can then be replaced by 0, since the central member of (85) is non-negative.

From (86)–(92)

$$D(X_1) = S^0 T^0 \quad (93)$$

$$D(X_2) = S^0 T^0 \quad (94)$$

i.e. X_1 and X_2 are dimensionless.

If the axes of the phase plane are taken as X_1 and X_2, a single portrait is obtained which applies for all positive† values of a and k. See e.g. fig. 5.

In terms of the dimensionless variables, the inequalities (84), specifying the regions of starting points of n-switch trajectories, are

$$4n^2 < X_2^2 + 2X_1 \operatorname{sgn}(X_1 + X_2) + 1 \leqslant 4(n+1)^2 \quad (n = 0, 1, \ldots) \quad (95)$$

Similarly we shall sometimes work with the dimensionless variable W instead of the performance index w, where

$$W = \frac{a^2}{k^5} w \quad (96)$$

For a given trajectory let us define q as the ordinate of the first point to lie on the switching line. If this point is labelled F

$$q = x_2^{\mathrm{F}} \quad (97)$$

A parameter we shall often work with is the related dimensionless variable Q, where

$$Q = \frac{a}{k} q \quad (98)$$

Q can also be expressed in terms of the coordinates of F as

$$Q = -\frac{a(x_2^{\mathrm{F}})^2}{x_1^{\mathrm{F}}} \quad (99)$$

which follows from (75), (97) and (98).

11.2. *Symmetry*

Since the system equations are unchanged if we write $-x_1$ for x_1, and $-x_2$ for x_2, the integral-square-error for a starting point (x_1, x_2) is the same as for a starting point $(-x_1, -x_2)$, i.e.

$$w(-x_1, -x_2) = w(x_1, x_2) \quad (100)$$

This relation will be used occasionally, as will its dimensionless version

$$W(-X_1, -X_2) = W(X_1, X_2) \quad (101)$$

§ 12. Integrals with Respect to State Variables

In evaluating performance indices along various trajectories, the calculations are simplified if we work with phase variable integrals rather than time integrals. The technique is similar to a well-known method

† Care is needed in applying dimensional methods when the parameters change sign (see Fuller 1966).

for evaluating time on the phase plane. Thus suppose we wish to evaluate

$$w_{IJ} = \int_{t_I}^{t_J} f_0(x_1, x_2)\, dt \tag{102}$$

for a trajectory satisfying

$$\left.\begin{aligned} \frac{dx_1}{dt} &= f_1(x_1, x_2) \qquad (103) \\ \frac{dx_2}{dt} &= f_2(x_1, x_2) \qquad (104) \end{aligned}\right\}$$

with starting point I

$$x_1(t_I) = x_1^{I}, \quad x_2(t_I) = x_2^{I} \tag{105}$$

and finishing point J

$$x_1(t_J) = x_1^{J}, \quad x_2(t_J) = x_2^{J} \tag{106}$$

t will be some function of x_2

$$t = t(x_2) \tag{107}$$

Formal substitution of (107) in (102) yields

$$w_{IJ} = \int_{x_2^{I}}^{x_2^{J}} f_0(x_1, x_2) \frac{dt}{dx_2}\, dx_2 \tag{108}$$

From (104) and (108)

$$w_{IJ} = \int_{x_2^{I}}^{x_2^{J}} \frac{f_0(x_1, x_2)}{f_2(x_1, x_2)}\, dx_2 \tag{109}$$

Similarly

$$w_{IJ} = \int_{x_1^{I}}^{x_1^{J}} \frac{f_0(x_1, x_2)}{f_1(x_1, x_2)}\, dx_1 \tag{110}$$

The performance index in form (109) has been used by Zachary (1966), and in form (110) by Fuller (1960 d).

When the trajectory is known, x_1 is a known function of x_2; thus the integrand in (109) is a known function of x_2 and the integral is often easy to evaluate. In our application

$$f_0(x_1, x_2) = x_1^2 \tag{111}$$

and $f_2(x_1, x_2)$ is a simple function, often a constant; this technique is then expedient.

§ 13. Evaluation of Integral-square-error

For a fixed switching line and a starting point (x_1, x_2) the integral-square-error w is some function $w(x_1, x_2)$ which it is our aim to calculate. We begin with starting points on the switching line.

13.1. *Integral-square-error for Starting Points on the Sliding-régime Segment*

Consider a starting point A on the switching line, and on the sliding-régime segment, with

$$x_2(0) = x_2^{A} \tag{112}$$

Comparison of (58) and (104) yields

$$f_2(x_1, x_2) = -\frac{1}{k}x_2 \tag{113}$$

Substituting from (25), (111) and (113) in (109) we find

$$w = \int_{x_2^A}^{0} (-k^3 x_2)\, dx_2 \tag{114}$$

$$= \tfrac{1}{2}k^3(x_2^A)^2 \tag{115}$$

Putting q for x_2^A in view of (97), we write that for a starting point $(-kq, q)$ on the switching line and on the sliding-régime segment (56) the performance index is

$$w = \tfrac{1}{2}k^3 q^2 \tag{116}$$

13.2. *Cost of a P-arc from or to the x_1-axis*

Let us calculate the cost w_{CD} for a P-arc from a point C on the x_1-axis (fig. 3) to a point D with coordinates (x_1^D, x_2^D). From (44) the P-arc satisfies

$$x_1 = \tfrac{1}{2}ax_2^2 + x_1^D - \tfrac{1}{2}a(x_2^D)^2 \tag{117}$$

Also, from (42) and (104),

$$f_2(x_1, x_2) = \frac{1}{a} \tag{118}$$

From (109), (111), (117) and (118), the cost of the arc CD is

$$w_{CD} = \int_0^{x_2^D} [\tfrac{1}{2}ax_2^2 + x_1^D - \tfrac{1}{2}a(x_2^D)^2]^2\, dx_2 \tag{119}$$

$$= a(x_1^D)^2 x_2^D - \tfrac{2}{3}a^2 x_1^D (x_2^D)^3 + \tfrac{2}{15}a^3(x_2^D)^5 \tag{120}$$

Similarly the cost w_{BC} for a P-arc from a point B to the point C on the x_1-axis (fig. 3) is

$$w_{BC} = -a(x_1^B)^2 x_2^B + \tfrac{2}{3}a^2 x_1^B (x_2^B)^3 - \tfrac{2}{15}a^3(x_2^B)^5 \tag{121}$$

Let us define the cost of any trajectory traced backwards as minus the cost of the trajectory traced forwards:

$$w_{IJ} = -w_{JI} \tag{122}$$

Then (120) and (121) imply that the cost of a P-arc with one end at C on the x_1-axis and the other end at a general point J (which may precede or follow C) is

$$w_{JC} = -a(x_1^J)^2 x_2^J + \tfrac{2}{3}a^2 x_1^J (x_2^J)^3 - \tfrac{2}{15}a^3(x_2^J)^5 \tag{123}$$

13.3. *Cost of a General P-arc*

Let us calculate the cost w_{BD} of the P-arc which joins two points B and D. Suppose the P-arc, extrapolated if necessary, cuts the x_1-axis at point C, and suppose B, C, D are in any order along the P-arc (not necessarily as in fig. 3). We have

$$w_{BD} = w_{BC} + w_{CD} \tag{124}$$

From (122), (123) and (124)

$$\begin{aligned} w_{BD} = & -a(x_1{}^B)^2 x_2{}^B + \tfrac{2}{3}a^2 x_1{}^B (x_2{}^B)^3 - \tfrac{2}{15}a^3 (x_2{}^B)^5 \\ & + a(x_1{}^D)^2 x_2{}^D - \tfrac{2}{3}a^2 x_1{}^D (x_2{}^D)^3 + \tfrac{2}{15}a^3 (x_2{}^D)^5 \end{aligned} \tag{125}$$

13.4. *Cost of a P-arc between two Successive Transition Points*

Let us calculate the cost w_{AE} for a P-arc from a starting point A on the switching line to a point E on the switching line (fig. 3). Note that A must be below the x_1-axis, to be the starting point of a P-arc, but E can be above or below the x_1-axis.

Our aim is to express w_{AE} in terms of the ordinates of points A and E. To simplify the notation let us write these ordinates as

$$q = x_2{}^A \tag{126}$$

and

$$r = x_2{}^E \tag{127}$$

(126) agrees with the definition of q in (97), since A is the first point of the arc to lie on the switching line. Since A and E are on the switching line

$$x_1{}^A = -kq \tag{128}$$

and

$$x_1{}^E = -kr \tag{129}$$

Writing A for B and E for D in (125), and using (126)–(129), we find

$$\begin{aligned} w_{AE} = & -ak^2 q^3 - \tfrac{2}{3}a^2 kq^4 - \tfrac{2}{15}a^3 q^5 \\ & + ak^2 r^3 + \tfrac{2}{3}a^2 kr^4 + \tfrac{2}{15}a^3 r^5 \end{aligned} \tag{130}$$

13.5. *Integral-square-error for Starting Points on the Switching Line*

Let us denote by $w_n(x_2{}^A)$ the integral-square-error for the complete trajectory to the origin, starting from a point A on an n-switch segment of the switching line. Suppose A is on the lower half of the switching line:

$$x_2{}^A < 0 \tag{131}$$

Suppose $n > 0$, the first arc being a P-arc which cuts the switching line at A and E (fig. 3). Then

$$w_n(x_2{}^A) = w_{AE} + w_{n-1}(x_2{}^E) \tag{132}$$

which expresses the fact that the total cost is the cost of the first arc plus the cost of the subsequent trajectory.

(132) can be used to calculate w_1, since w_0 is known, being given by (115). However, we cannot immediately use (132) to calculate $w_2, w_3, \ldots$ iteratively. The difficulty is that in (132) the argument of w_n is negative (see (131)) whereas the argument of w_{n-1} is positive, if $n > 1$ (this follows

from (64) and (72)). To avoid this difficulty we use symmetry and write, from (100),

$$w_{n-1}(x_2{}^{\mathrm{E}}) = w_{n-1}(-x_2{}^{\mathrm{E}}) \tag{133}$$

so that (132) can be replaced by

$$w_n(x_2{}^{\mathrm{A}}) = w_{\mathrm{AE}} + w_{n-1}(-x_2{}^{\mathrm{E}}) \tag{134}$$

In (134) w_n and w_{n-1} have arguments with the same sign if† $n > 1$, consequently the direct solution of (134) can be carried out without involving a contradiction of inequality (131) at any stage of the iteration.

From (126), (127), (130) and (134)

$$w_n(q) = ak^2(r^3-q^3)+\tfrac{2}{3}a^2k(r^4-q^4)+\tfrac{2}{15}a^3(r^5-q^5)+w_{n-1}(-r) \tag{135}$$

If, for brevity, we define b as

$$b = \frac{k}{a} \tag{136}$$

then q and r are related by

$$q+r = -2b \tag{137}$$

in view of (64), (126) and (127). Eliminating r and k from (135), (136) and (137), we obtain a recurrence relation or difference equation for w_n:

$$w_n(q) = -\tfrac{2}{15}a^3(2q^5+10bq^4+15b^2q^3+5b^3q^2+10b^4q+12b^5)+w_{n-1}(q+2b) \tag{138}$$

The initial condition for this difference equation is, from (116) and (136),

$$w_0(q) = \tfrac{1}{2}a^3b^3q^2 \tag{139}$$

In terms of the dimensionless variables W_n and Q where (see (96) and (98))

$$W_n = a^{-3}b^{-5}w_n = a^2k^{-5}w_n \tag{140}$$

$$Q = b^{-1}q = ak^{-1}x_2{}^{\mathrm{A}} \tag{141}$$

(138) and (139) are

$$W_n(Q) = W_{n-1}(Q+2)-\tfrac{2}{15}(2Q^5+10Q^4+15Q^3+5Q^2+10Q+12) \tag{142}$$

and

$$W_0(Q) = \tfrac{1}{2}Q^2 \tag{143}$$

The solution of (142) and (143) is straightforward (see Appendix 1) and yields

$$\begin{aligned} W_n(Q) = -\tfrac{1}{90}[&24nQ^5+120n^2Q^4+(320n^3-140n)Q^3 \\ &+(480n^4-420n^2-45)Q^2+(384n^5-560n^3+116n)Q \\ &+128n^6-280n^4+116n^2] \end{aligned} \tag{144}$$

which holds for negative values of Q, in view of (131) and (141). The corresponding result for $Q > 0$ is obtained by symmetry:

$$W_n(Q) = W_n(-Q) \tag{145}$$

---

† If $n = 1$, the sign of the argument of w_0 is immaterial. Thus (134) can be solved for all $n \geqslant 1$.

from (101). Thus, in general, for a starting point with ordinate x_2 $(= a^{-1}kQ)$ on an n-switch segment of the switching line, the integral-square-error w_n $(= a^{-2}k^5 W_n)$ is determined by

$$W_n(Q) = \tfrac{1}{90}[24n|Q|^5 - 120n^2Q^4 + (320n^3 - 140n)|Q|^3$$
$$- (480n^4 - 420n^2 - 45)Q^2 + (384n^5 - 560n^3 + 116n)|Q|$$
$$- 128n^6 + 280n^4 - 116n^2] \quad (2n-1 < |Q| \leqslant 2n+1) \qquad (146)$$

in which the range specified for Q follows from (72).

In particular, for $n = 1$, (146) yields

$$W_1(Q) = \tfrac{1}{30}[8|Q|^5 - 40Q^4 + 60|Q|^3 - 5Q^2 - 20|Q| + 12] \quad (1 < |Q| \leqslant 3) \qquad (147)$$

and, for $n = 2$, (146) yields

$$W_2(Q) = \tfrac{1}{30}[16|Q|^5 - 160Q^4 + 760|Q|^3$$
$$- 1985Q^2 + 2680|Q| - 1392] \quad (3 < |Q| \leqslant 5) \qquad (148)$$

Results (147) and (148) will be used in § 15.

13.6. *Integral-square-error for Starting Points on the x_1-axis*

To find the integral-square-error for a starting point C on the negative x_1-axis we shall find the cost of the P-arc CE to the switching line, and add the cost of the trajectory for a starting point E on the switching line:

$$w(x_1^{\mathrm{C}}, 0) = w_{\mathrm{CE}} + w_n(x_2^{\mathrm{E}}) \qquad (149)$$

(120) with E written for D is

$$w_{\mathrm{CE}} = a(x_1^{\mathrm{E}})^2 x_2^{\mathrm{E}} - \tfrac{2}{3}a^2 x_1^{\mathrm{E}}(x_2^{\mathrm{E}})^3 + \tfrac{2}{15}a^3(x_2^{\mathrm{E}})^5 \qquad (150)$$

Since E is the first transition point and is above the x_1-axis we can write

$$x_2^{\mathrm{E}} = q > 0 \qquad (151)$$

(see (97)); and also

$$x_1^{\mathrm{E}} = -kq \qquad (152)$$

From (150), (151) and (152)

$$w_{\mathrm{CE}} = ak^2q^3 + \tfrac{2}{3}a^2kq^4 + \tfrac{2}{15}a^3q^5 \qquad (153)$$

or, in terms of the dimensionless variables (96) and (98),

$$W_{\mathrm{CE}} = Q^3 + \tfrac{2}{3}Q^4 + \tfrac{2}{15}Q^5 \quad (Q > 0) \qquad (154)$$

Addition of (146) and (154) shows that the total cost of the trajectory from C is determined by

$$W(X_1^{\mathrm{C}}, 0) = \phi_n(Q) \quad (X_1^{\mathrm{C}} < 0,\ Q > 0) \qquad (155)$$

where the function $\phi_n(\rho)$ is defined as

$$\phi_n(\rho) = \tfrac{1}{90}[(24n+12)\rho^5 - (120n^2 - 60)\rho^4 + (320n^3 - 140n + 90)\rho^3$$
$$- (480n^4 - 420n^2 - 45)\rho^2 + (384n^5 - 560n^3 + 116n)\rho$$
$$- 128n^6 + 280n^4 - 116n^2] \qquad (156)$$

and X_1^C is the dimensionless variable (see (86))

$$X_1^C = \frac{a}{k^2} x_1^C \tag{157}$$

(155) holds for starting points on the negative x_1-axis. The results for starting points on the positive x_1-axis (thus for $Q < 0$) is obtained from the symmetry eqn. (101). Thus in general

$$W(X_1^C, 0) = \phi_n(|Q|) \tag{158}$$

(156) and (158) express W in terms of Q and n, which turn out to be suitable parameters for subsequent manipulations. For interest, however, let us also express W directly in terms of x_1^C, the abscissa of the starting point, or rather in terms of the dimensionless variable X_1^C given by (157).

From (63), (98), (151) and (157)

$$Q = -1 + (1 - 2X_1^C)^{1/2} \quad (X_1^C < 0) \tag{159}$$

Dropping the superscript C, and using symmetry for $X_1 > 0$, we generalize (159) to

$$|Q| = -1 + (1 + 2|X_1|)^{1/2} \tag{160}$$

From (156), (158) and (160)

$$\begin{aligned} W(X_1, 0) = &-\tfrac{1}{45}[(240n^2 + 240n) X_1^2 \\ &+ (480n^4 + 960n^3 + 540n^2 + 60n - 15)|X_1| \\ &+ 64n^6 + 192n^5 + 340n^4 + 360n^3 + 118n^2 - 30n - 9] \\ &+ \tfrac{1}{45}[(48n + 24) X_1^2 + (320n^3 + 480n^2 + 148n - 6)|X_1| \\ &+ 192n^5 + 480n^4 + 360n^3 + 60n^2 - 30n - 9](1 + 2|X_1|)^{1/2} \end{aligned} \tag{161}$$

which determines the integral-square-error w $(= a^{-2} k^5 W)$ for an n-switch trajectory with starting point on the x_1-axis with abscissa x_1 $(= a^{-1} k^2 X_1)$, i.e. for a starting point in the range (see (85)):

$$4n^2 < 1 + 2|X_1| \leqslant 4(n+1)^2 \tag{162}$$

(161) implies that w is a somewhat involved function of the parameter k. We shall find the parameter Q more convenient to work with.

13.7. *Integral-square-error for General Starting Points*

To find the integral-square-error for a starting point B on the left of the switching line we shall find the cost of the P-arc BE to the switching line, and add the cost of the trajectory for a starting point E on the switching line:

$$w(x_1^B, x_2^B) = w_{BE} + w_n(x_2^E) \tag{163}$$

(125) with E written for D is

$$\begin{aligned} w_{BE} = &-a(x_1^B)^2 x_2^B + \tfrac{2}{3}a^2 x_1^B (x_2^B)^3 - \tfrac{2}{15}a^3 (x_2^B)^5 \\ &+ a(x_1^E)^2 x_2^E - \tfrac{2}{3}a^2 x_1^E (x_2^E)^3 + \tfrac{2}{15}a^3 (x_2^E)^5 \end{aligned} \tag{164}$$

Since E is the first transition point we can write

$$x_2^{\mathrm{E}} = q \tag{165}$$

(see (97)); also, from (75),

$$x_1^{\mathrm{E}} = -kq \tag{166}$$

Dropping the superscript B, and substituting from (165) and (166) in (164), we obtain

$$\begin{aligned} w_{\mathrm{BE}} = & -ax_1^2 x_2 + \tfrac{2}{3}a^2 x_1 x_2^3 - \tfrac{2}{15}a^3 x_2^5 \\ & + ak^2 q^3 + \tfrac{2}{3}a^2 kq^4 + \tfrac{2}{15}a^3 q^5 \end{aligned} \tag{167}$$

or, in terms of the dimensionless variables (86), (87), (96) and (98),

$$\begin{aligned} W_{\mathrm{BE}} = & -X_1^2 X_2 + \tfrac{2}{3}X_1 X_2^3 - \tfrac{2}{15}X_2^5 \\ & + Q^3 + \tfrac{2}{3}Q^4 + \tfrac{2}{15}Q^5 \end{aligned} \tag{168}$$

The range of Q in (168) may be determined as follows. E is a point of the switching line which is approached by a P-arc. Thus E can be any point inside the sliding-régime segment given by (69) and any point of the switching line above the x_1-axis:

$$-\frac{k}{a} < x_2^{\mathrm{E}} < \infty \tag{169}$$

From (98), (165) and (169) the range of Q is

$$-1 < Q < \infty \tag{170}$$

The cost of the trajectory from E is given by (146). For $Q > 0$ (146) is

$$\begin{aligned} W_n(Q) = \tfrac{1}{90}[& 24nQ^5 - 120n^2 Q^4 + (320n^3 - 140n)Q^3 - (480n^4 - 420n^2 - 45)Q^2 \\ & + (384n^5 - 560n^3 + 116n)Q - 128n^6 + 280n^4 - 116n^2] \end{aligned} \tag{171}$$

For $-1 < Q \leqslant 0$, the inequality appended to (146) shows that $n = 0$, so that (146) is then

$$W_0(Q) = \tfrac{1}{2}Q^2 \tag{172}$$

(172) is the same as (171) when $n = 0$; hence (171) is the correct expression for $W_n(Q)$, not merely for $Q > 0$, but for all Q in range (170).

Addition of (168) and (171) shows that the total cost of the trajectory from a point (x_1, x_2) on the left of the switching line is determined by

$$W(X_1, X_2) = -X_1^2 X_2 + \tfrac{2}{3}X_1 X_2^3 - \tfrac{2}{15}X_2^5 + \phi_n(Q) \quad (X_1 + X_2 < 0) \tag{173}$$

where ϕ_n is the function defined by (156).

By symmetry (see (101)), for a starting point to the right of the switching line, the total cost is determined by

$$W(X_1, X_2) = X_1^2 X_2 + \tfrac{2}{3}X_1 X_2^3 + \tfrac{2}{15}X_2^5 + \phi_n(-Q) \quad (X_1 + X_2 > 0) \tag{174}$$

In (173) Q is a function of X_1 and X_2. This function can be determined by substituting for x_1^{C} from (44) in (63) and then using (86), (87), (98) and (165). The result is

$$Q = -1+(1-2X_1+X_2^2)^{1/2} \quad (X_1+X_2<0) \tag{175}$$

By symmetry, in (174) Q is the following function

$$Q = 1-(1+2X_1+X_2^2)^{1/2} \quad (X_1+X_2>0) \tag{176}$$

Q could be eliminated from (173) and (175) and thus w could be expressed in terms of the parameter k. For our purposes, however, it will be more convenient to retain Q as an explicit parameter, and to eliminate instead k. To find k in terms of x_1, x_2 and Q, note that from (44)

$$x_1-\tfrac{1}{2}ax_2^2 = x_1^{\mathrm{E}}-\tfrac{1}{2}a(x_2^{\mathrm{E}})^2 \tag{177}$$

for (x_1, x_2) to the left of the switching line. From (98), (165), (166) and (177)

$$x_1-\tfrac{1}{2}ax_2^2 = -\frac{k^2}{a}Q-\frac{1}{2}\frac{k^2}{a}Q^2 \tag{178}$$

i.e.

$$k = a^{1/2}\left(\frac{-x_1+\frac{1}{2}ax_2^2}{Q+\frac{1}{2}Q^2}\right)^{1/2} \tag{179}$$

From (86), (87), (96), (156), (173) and (179)

$$w(x_1, x_2) = -ax_1^2x_2+\tfrac{2}{3}a^2x_1x_2^3-\tfrac{2}{15}a^3x_2^5$$
$$+\theta_n(Q)\,a^{1/2}(-x_1+\tfrac{1}{2}ax_2^2)^{5/2} \quad (x_1+kx_2<0) \tag{180}$$

where the function $\theta_n(\rho)$ is defined as

$$\theta_n(\rho) = \frac{\phi_n(\rho)}{(\frac{1}{2}\rho^2+\rho)^{5/2}} \tag{181}$$

where $\phi_n(\rho)$ is the function defined in (156).

By symmetry, for starting points (x_1, x_2) to the right of the switching line

$$w(x_1, x_2) = ax_1^2x_2+\tfrac{2}{3}a^2x_1x_2^3+\tfrac{2}{15}a^3x_2^5$$
$$+\theta_n(-Q)\,a^{1/2}(x_1+\tfrac{1}{2}ax_2^2)^{5/2} \quad (x_1+kx_2>0) \tag{182}$$

Equations (180) and (182) express the integral-square-error for linear switching, and a trajectory which, being n-switch, starts in region (84). They closely resemble eqns. (21) and (22) which express the integral-square-error for optimal non-linear switching†. (Note, however, that in (180) and (182) Q is a function of x_1 and x_2, being given by (175) and (176).)

Part III. Selection of the Best Straight Switching Line

§ 14. *Best Linear Switching for a Starting Point on the x_1-axis*

In this section we shall find the straight switching line which minimizes integral-square-error for a given starting point $(x_1, 0)$ on the x_1-axis (step response).

---

† Note that in (180) the term $(-x_1+\frac{1}{2}ax_2^2)^{5/2}$ is imaginary for points (x_1, x_2) to the right of the P-arc through the origin. However, $\theta_n(Q)$ is then also imaginary, so that w is real, as it should be. A similar remark applies to (182).

The algebra turns out to be simpler if we work with Q as the minimizing parameter rather than k. For the starting point $(x_1, 0)$, we have $Q > 0$ if $x_1 < 0$ and $Q < 0$ if $x_1 > 0$. Thus from (180) and (182) the integral-square-error is

$$w(x_1, 0) = \theta_n(|Q|)\, a^{1/2} |x_1|^{5/2} \tag{183}$$

where θ_n is given by (181). With x_1 fixed, Q and n vary when k varies. Thus (183) shows that we have to find the minimum of $\theta_n(|Q|)$ when Q and n (which is a function of Q) vary. By symmetry we may restrict attention to non-negative values of Q.

14.1. 0-*switch Trajectories*

When Q is in the range

$$0 \leqslant Q \leqslant 1 \tag{184}$$

it follows from (69) and (98) that

$$n = 0 \tag{185}$$

When $n = 0$, (181) yields

$$\theta_0(Q) = \frac{4Q^5 + 20Q^4 + 30Q^3 + 15Q^2}{30(\tfrac{1}{2}Q^2 + Q)^{5/2}} \tag{186}$$

Differentiation of (186) yields

$$\frac{d\theta_0(Q)}{dQ} = -\frac{Q^3 + Q^2}{4(\tfrac{1}{2}Q^2 + Q)^{7/2}} < 0 \quad (Q > 0) \tag{187}$$

Hence θ_0 decreases when Q increases in the range (173). Hence there is no internal minimum of Q in the range corresponding to a 0-switch trajectory.

14.2. 1-*switch Trajectories*

When Q is in the range

$$1 < Q \leqslant 3 \tag{188}$$

it follows from (72) and (98) that

$$n = 1 \tag{189}$$

When $n = 1$, (181) yields

$$\theta_1(Q) = \frac{12Q^5 - 20Q^4 + 90Q^3 - 5Q^2 - 20Q + 12}{30(\tfrac{1}{2}Q^2 + Q)^{5/2}} \tag{190}$$

Differentiation of (190) yields

$$\frac{d\theta_1(Q)}{dQ} = \frac{16Q^5 - 48Q^4 + 21Q^3 + 17Q^2 - 12}{12(\tfrac{1}{2}Q^2 + Q)^{7/2}} \tag{191}$$

From (191), it is easy to verify that $d\theta_1/dQ$ is negative at $Q = 1$ and positive at $Q = 3$. Hence there is a minimum of θ_1 for a value of Q in the range (188), and this value of Q is a root of

$$16Q^5 - 48Q^4 + 21Q^3 + 17Q^2 - 12 = 0 \tag{192}$$

There is only one real root of (192) in range (188); this may be shown as follows†. Let us define P as

$$P = Q - 1 \tag{193}$$

Then (192) is

$$16P^5 + 32P^4 - 11P^3 - 48P^2 - 15P - 6 = 0 \tag{194}$$

Descartes' rule (see e.g. Burnside and Panton 1904) states that the number of positive roots of a polynomial equation is not greater than the number of sign changes in the coefficients. Hence there is not more than one positive root of (194). Hence, from (193), there is not more than one root of (192) which is greater than unity.

We conclude that the unique root of (192) in range (188) yields a minimum value of θ_1. By computation this root Q^* is

$$Q^* \simeq 2{\cdot}22683215 \tag{195}$$

From (190) and (195) the minimum value of θ_1 is

$$\theta^* \simeq 0{\cdot}764345704 \tag{196}$$

14.3. *Multi-switch Trajectories*

When $n > 1$ it can be shown (Appendix 2) that $\theta_n(Q)$ increases with Q. Hence there is no minimizing value of Q for n-switch trajectories with $n > 1$.

Thus, in general, as Q increases from 0 to ∞ (and n increases correspondingly from 0 to ∞) $\theta_n(Q)$ first decreases to a minimum value θ^*, and thereafter increases. Also from (179), k decreases as Q increases. Therefore as the switching line is rotated clockwise from the horizontal to the vertical, the integral-square-error (for a starting point on the x_1-axis) first decreases to a minimum value, and thereafter increases.

14.4. *Results*

From (179) and (195), the value of k (minus the reciprocal of the slope of the switching line) that minimizes integral-square-error for a starting point $(x_1, 0)$ is

$$k^* = 0{\cdot}460960545 a^{1/2} |x_1|^{1/2} \tag{197}$$

The trajectory is then a 1-switch trajectory.

From (183) and (196) the minimum integral-square-error for a starting point $(x_1, 0)$ and linear switching is

$$w^* = G_1 a^{1/2} |x_1|^{5/2} \tag{198}$$

where

$$G_1 = 0{\cdot}764345704 \tag{199}$$

† Alternatively one can apply the theorem of Fourier and Budan. See Burnside and Panton (1904) for a statement of this theorem, and Grensted and Fuller (1965) for an application.

From (16), (17), (198) and (199) the ratio of the integral-square-errors with the best linear switching and with the best non-linear switching is

$$\frac{G_1}{G_0} = 1{\cdot}00042951 \tag{200}$$

Thus we have verified the results asserted in (28) and (29).

§ 15. Best Linear Switching for a General Starting Point

In this section we shall find the straight switching line which minimizes integral-square-error for an arbitrary fixed starting point (x_1, x_2).

Consider first a starting point C on the negative x_1-axis, and suppose that k has been selected to be optimum for this starting point. The P-arc traced forwards cuts the switching line at E (fig. 3). Then for any starting point D on the arc CE, the same k is optimum†, since the trajectory from D is the continuation of the trajectory from C.

Similarly, let the P-arc through C traced backwards cut the switching line at A (fig. 3). Consider a given starting point B on the arc AC and excluding the point A. Assume that the optimal straight switching line for starting point B is to the right of B. Then the optimal straight switching line for starting point B is the same as that for starting point C, since the trajectory from C is the continuation of the trajectory from B.

It remains to verify the assumption made in the last paragraph. Suppose the assumption is not valid. Then on tracing the P-arc backwards from C we shall arrive at a point B (before A is reached) such that the optimal straight switching line for starting point B passes through B. But the original straight switching line (through A and E) gives a trajectory from B with a smaller integral-square-error, since it gives a trajectory from C with a smaller integral-square-error, while the arc BC is the same for both switching lines. Hence the straight switching line through B is not optimal. Hence the above assumption that all points B are to the left of their respective optimal straight switching lines is valid.

Therefore for all starting points on the arc AE (except possibly the point A) the optimal straight switching line is the straight line through A and E.

These remarks can be verified by inspection of (180). Thus for fixed (x_1, x_2), (180) indicates that $w(x_1, x_2)$ is minimized by minimizing $\theta_n(Q)$. The minimization of $\theta_n(Q)$ with respect to Q and n is the same as carried out in § 14 for a starting point on the x_1-axis, and yields the same k.

Note, however, that when varying Q (and thus k) in (180) we must avoid violating the inequality appended in (180). Thus our argument does not apply to points preceding A or following E.

---

† Bellman's principle of optimality (Bellman 1957) applies for points on the trajectory from C up to the point E. Once E is reached, however, the switching line is determined, and the principle of optimality cannot be applied to subsequent points.

o

When the point C varies along the negative x_1-axis, the arc ACE (excluding the point A) sweeps out a region of the phase plane which, together with its reflection in the origin†, we call region γ. The remaining region we call region δ, and is shown shaded in fig. 6.

Fig. 6

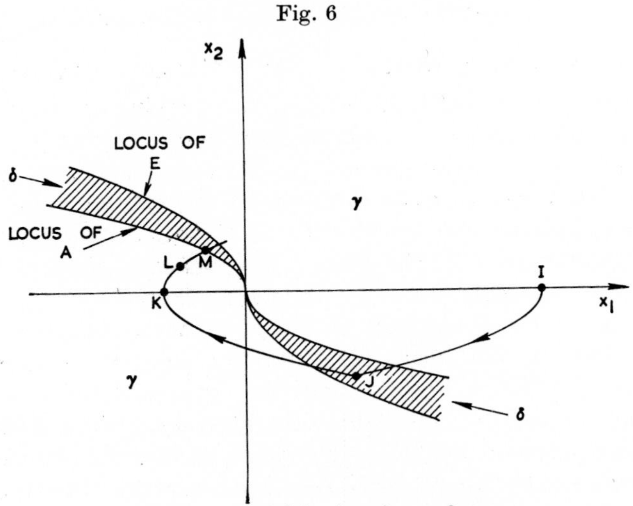

Regions γ and δ in the phase plane.

15.1. *Determination of the Regions γ and δ*

Let us determine the regions γ and δ. We begin by finding the locus of the point E.

Since $x_2{}^{\mathrm{E}}$ is the ordinate q of the first switch point of a trajectory starting from C

$$x_2{}^{\mathrm{E}} = \frac{k}{a}Q \tag{201}$$

(see (98) and (165)). The value of Q in (201) is that found optimum for starting points on the x_1-axis, namely

$$Q = Q^* \tag{202}$$

where Q^* is given by (195). From (75), (201) and (202)

$$x_1{}^{\mathrm{E}} + \frac{a}{Q^*}(x_2{}^{\mathrm{E}})^2 = 0 \quad (x_2{}^{\mathrm{E}} > 0) \tag{203}$$

Using symmetry to obtain a similar relation for $x_2{}^{\mathrm{E}} < 0$, and dropping the superscript, we find that the locus of the first switch point E (when the starting point C varies along the whole x_1-axis) is

$$x_1 + H_{\mathrm{E}}\, a x_2 |x_2| = 0 \tag{204}$$

where

$$H_{\mathrm{E}} = \frac{1}{Q^*} = 0{\cdot}449068421 \tag{205}$$

† Defined in a footnote to § 10, p. 211.

Next let us find the locus of the point A. Since A is on the switching line

$$x_1^{\mathrm{A}} + kx_2^{\mathrm{A}} = 0 \quad (x_2^{\mathrm{A}} < 0) \tag{206}$$

Eliminating k, x_2^{E} and Q from (64), (201), (202) and (206), we find

$$x_1^{\mathrm{A}} - \frac{a}{Q^* + 2}(x_2^{\mathrm{A}})^2 = 0 \quad (x_2^{\mathrm{A}} < 0) \tag{207}$$

Using symmetry to obtain a similar relation for $x_2^{\mathrm{A}} > 0$, and dropping the superscript, we find that the locus of A (when C varies along the whole x_1-axis) is

$$x_1 + H_{\mathrm{A}}\, ax_2 |x_2| = 0 \tag{208}$$

where

$$H_{\mathrm{A}} = \frac{1}{Q^* + 2} = 0{\cdot}236583797 \tag{209}$$

Now consider a general locus

$$x_1 + Hax_2|x_2| = 0 \tag{210}$$

where H is a parameter. When H varies from $-\infty$ to $+\infty$, locus (210) sweeps over the entire phase plane. The shaded region δ in fig. 6 corresponds to parameter values in the range

$$H_{\mathrm{A}} \leqslant H < H_{\mathrm{E}} \tag{211}$$

From (210) and (211) the region δ consists of the points (x_1, x_2) satisfying

$$H_{\mathrm{A}} \leqslant -\frac{x_1}{ax_2|x_2|} < H_{\mathrm{E}} \tag{212}$$

where H_{A} and H_{E} are given by (209) and (205).

The remaining points, namely the points satisfying

$$-\infty < -\frac{x_1}{ax_2|x_2|} < H_{\mathrm{A}} \tag{213}$$

and also the points satisfying

$$H_{\mathrm{E}} \leqslant -\frac{x_1}{ax_2|x_2|} < \infty \tag{214}$$

constitute region γ.

15.2. *Starting Points in Region γ*

From the above, the best value of Q for a starting point (x_1, x_2) in region γ (given by (213) and (214)) and to the left of region δ is Q^* as given by (195). From (179) the best value of k is then

$$k^* = \frac{a^{1/2}(-x_1 + \frac{1}{2}ax_2^2)^{1/2}}{(Q^* + \frac{1}{2}Q^{*2})^{1/2}} \tag{215}$$

i.e.

$$k^* = 0{\cdot}460960545a^{1/2}(-x_1 + \tfrac{1}{2}ax_2^2)^{1/2} \tag{216}$$

The trajectory is then a 1-switch trajectory. From (180) and (196), the minimum integral-square-error with this starting point and linear switching is

$$w^*(x_1, x_2) = -ax_1^2 x_2 + \tfrac{2}{3}a^2 x_1 x_2^3 - \tfrac{2}{15}a^3 x_2^5 + G_1 a^{1/2}(-x_1 + \tfrac{1}{2}ax_2^2)^{5/2} \tag{217}$$

where G_1 is the constant given by (199).

Similarly, for a starting point in region γ and to the right of region δ, the best value of k is

$$k^* = 0{\cdot}460960545a^{1/2}(x_1 + \tfrac{1}{2}ax_2^2)^{1/2} \tag{218}$$

the trajectory is a 1-switch trajectory, and the minimum integral-square-error with this starting point and linear switching is

$$w^*(x_1, x_2) = ax_1^2 x_2 + \tfrac{2}{3}a^2 x_1 x_2^3 + \tfrac{2}{15}a^3 x_2^5 + G_1 a^{1/2}(x_1 + \tfrac{1}{2}ax_2^2)^{5/2} \tag{219}$$

where G_1 is given by (199).

The optimal non-linear switching curve (13) is in region δ (this follows from (13), (205), (209) and (212)). Hence for a starting point B (fig. 3) an initial part of the trajectory, as far as E, is the same for both the non-linear switching and the linear switching. Hence the ratio of the integral-square-errors with the best linear switching and with the best non-linear switching, for the starting point B, is

$$\frac{w_{\mathrm{BC}} + w^*(x_1^{\mathrm{C}}, 0)}{w_{\mathrm{BC}} + v(x_1^{\mathrm{C}}, 0)} \tag{220}$$

This ratio changes monotonically with w_{BC}, the cost of the P-arc BC. Thus the ratio increases as the point B describes the P-arc from A to E. We conclude that the ratio is, in region γ, minimum for points at the locus of A, i.e. locus (208), and maximum for points on the locus of E, i.e. locus (204). Thus, in seeking the maximum and minimum values of the ratio, we may confine attention to starting points in region δ or on its boundaries.

15.3. *Starting Points on the x_2-axis*

Starting points on the x_2-axis (ramp response) are in region γ. Thus (216) and (218) apply, and yield for a starting point $(0, x_2)$

$$k^* = 0{\cdot}325948328a|x_2| \tag{221}$$

Similarly, from (217) and (219)

$$w^*(0, x_2) = G_2 a^3 |x_2|^5 \tag{222}$$

where

$$G_2 = \frac{2}{15} + \frac{\surd(2)\, G_1}{8} = 0{\cdot}268451841 \tag{223}$$

From (17), (21), (22), (222) and (223) the ratio of the integral-square-errors with the best linear switching and with the best non-linear switching is

$$\frac{G_2}{\frac{2}{15}+[\sqrt{(2)}\,G_0/8]} = 1{\cdot}00021614 \tag{224}$$

Thus we have verified the results asserted in (30) and (31).

15.4. *Starting Points in Region δ*

For a given starting point J in region δ, the best straight switching line is the straight line through J. This assertion may be proved as follows.

J must be in one (or both) of the two following regions: (i) the region to the left of and on the switching line; (ii) the region to the right of and on the switching line.

Suppose J is in region (i). Let the P-arc from J cut the x_1-axis at K (fig. 6). Then rotation of the switching line clockwise makes the integral-square-error of the trajectory from K decrease (see § 14), since the rotation makes the first switch point L move towards the best first switch point M (fig. 6) for starting point K. Hence the rotation makes the integral-square-error of the trajectory from J to the origin decrease. The rotation can be continued until the switching line meets J; at this stage the minimum integral-square-error is reached, subject to the restriction that J is in region (i).

On the other hand, suppose J is in region (ii). Let the N-arc traced backwards from J cut the x_1-axis at I (fig. 6). Then rotation of the switching line anti-clockwise makes the integral-square-error of the trajectory from I to the origin decrease (see § 14). Hence the rotation makes the integral-square-error of the trajectory from J decrease. The rotation can be continued until the switching line meets J; at this stage the minimum integral-square-error is reached, subject to the restriction that J is in region (ii).

We conclude that the best straight switching line is such that J is on the common boundary of regions (i) and (ii), i.e. the switching line passes through J.

Hence for a starting point (x_1, x_2) in region δ, the best value of k is

$$k^* = -\frac{x_1}{x_2} \tag{225}$$

Since the first switch point is the starting point, (99) shows that the best value of Q is

$$Q = Q^* = -\frac{a x_2^{\,2}}{x_1} \tag{226}$$

For a starting point in the part of region δ (see (212)) defined by

$$\frac{1}{3} \leqslant -\frac{x_1}{a x_2 \,|\, x_2 \,|} < H_{\mathrm{E}} \tag{227}$$

it follows from the inequality appended in (146) and from (226) that the best trajectory is a 1-switch trajectory. Then (96) and (147) show that the minimum integral-square-error with this starting point and linear switching is

$$w^*(x_1, x_2) = \frac{(k^*)^5}{30a^2}[8|Q^*|^5 - 40|Q^*|^4 + 60|Q^*|^3 - 5|Q^*|^2 - 20|Q^*| + 12] \tag{228}$$

$$= \frac{1}{30a^2|x_2|^5}[8a^5 x_2^{10} - 40a^4|x_1|x_2^8 + 60a^3 x_1^2 x_2^6 - 5a^2|x_1|^3 x_2^4 - 20ax_1^4 x_2^2 + 12|x_1|^5] \tag{229}$$

For a starting point in the remaining part of region δ (see (212)), defined by

$$H_A \leqslant \frac{x_1}{ax_2|x_2|} < \frac{1}{3} \tag{230}$$

it follows from the inequality appended in (146), (209) and (226) that the best trajectory is a 2-switch trajectory. Then (96) and (148) show that the minimum integral-square-error with this starting point and linear switching is

$$w^*(x_1, x_2) = \frac{(k^*)^5}{30a^2}[16|Q^*|^5 - 160|Q^*|^4 + 760|Q^*|^3 - 1985|Q^*|^2 + 2680|Q^*| - 1392] \tag{231}$$

$$= \frac{1}{30a^2|x_2|^5}[16a^5 x_2^{10} - 160a^4|x_1|x_2^8 + 760a^3 x_1^2 x_2^6 - 1985a^2|x_1|^3 x_2^4 + 2680ax_1^4 x_2^2 - 1392|x_1|^5] \tag{232}$$

For a starting point in region δ the ratio of the integral-square-errors with the best linear switching and with the best non-linear switching can be computed from formulae (21), (22), (229) and (232). This ratio can be written as a function of the single parameter H defined by (210). Computation shows that the ratio is minimum when H takes the value H_1 given by

$$H_1 = 0{\cdot}236605249 \tag{233}$$

and maximum when H takes the value H_2 given by

$$H_2 = 0{\cdot}354654565 \tag{234}$$

The minimum and maximum values of the ratio are found to be 1·00019878 and 1·02726623 respectively.

Thus we have verified the results asserted in (32)–(35).

§ 16. Best Linear Switching for a Random Starting Point

In this section we shall seek the straight switching line which minimizes the expected integral-square-error for an ensemble of starting points. The motivation for this investigation was given in § 7. We shall confine attention to the case when the ensemble of starting points is distributed uniformly along a segment of the x_1-axis of length $2c$ and centred on the origin. Thus if the probability density of the starting points is $p(x_1)$

$$p(x_1) = \begin{cases} \dfrac{1}{2c} & (|x_1| < c) \\ 0 & (|x_1| > c) \end{cases} \tag{235}$$

In general, for a given k, some members of the ensemble have 0-switch trajectories, some have 1-switch trajectories, and so on. When k is sufficiently large, all members of the ensemble have 0-switch trajectories (see (85)). Such a value of k is not optimum, since the integral-square-error of each member of the ensemble can be decreased by decreasing k (see § 14). Thus we need consider only values of k such that some of the trajectories are 1-switch. To abbreviate the calculations, moreover, we shall not consider values of k so small that a trajectory has more than one switch. Thus we shall restrict attention to the range

$$\frac{2}{15} \leqslant \frac{k^2}{ac} < \frac{2}{3} \tag{236}$$

obtained by substituting $n = 1$ and $x_1 = c$ in (85).

It will turn out that the range of k given by (236) is sufficient to yield at least a local minimum of expected integral-square-error. It is reasonable to assume that this local minimum is also the absolute minimum. However, we need not make this assumption; we can content ourselves with comparing the locally minimal linear switching with the absolutely optimal non-linear switching.

In the deterministic case we found it convenient to work with the parameter Q instead of k. Q was the ordinate of the first transition point, expressed as a dimensionless variable. Similarly in the present case we shall work with the parameter L which is the largest of the values of Q yielded by all members of the ensemble. Thus if the trajectory from the most negative starting point $(-c, 0)$ has a first transition point with ordinate l

$$L = \frac{a}{k} l \tag{237}$$

Alternatively L can be expressed as

$$L = -1 + \left(1 + \frac{2ac}{k^2}\right)^{1/2} \tag{238}$$

(compare (157) and (159)).

16.1. *Expected Cost of the 0-switch Trajectories*

For a fixed k in range (236), let us calculate the contribution e_0 of the members of the ensemble with 0-switch trajectories to the expected integral-square-error. We have

$$e_0 = \int_{-3/2\,(k^2/a)}^{+3/2\,(k^2/a)} w(x_1, 0)\, p(x_1)\, dx_1 \tag{239}$$

where the limits of integration are obtained from (85). From (235), (236) and (239), and from symmetry

$$e_0 = \frac{1}{c}\int_{-3/2\,(k^2/a)}^{0} w(x_1, 0)\, dx_1 \tag{240}$$

With the dimensionless variables X_1 and W defined by (86) and (96), and with C and E_0 similarly defined by

$$C = \frac{a}{k^2}c \tag{241}$$

and

$$E_0 = \frac{a^2}{k^5}e_0 \tag{242}$$

(240) may be written

$$E_0 = \frac{1}{C}\int_{-3/2}^{0} W(X_1, 0)\, dX_1 \tag{243}$$

Also, from (86) and (178) (with $x_2 = 0$)

$$X_1 = -Q - \tfrac{1}{2}Q^2 \tag{244}$$

Substitution of (244) in (243) and use† of (155) yields

$$E_0 = \frac{1}{C}\int_{0}^{1} (Q+1)\,\phi_0(Q)\, dQ \tag{245}$$

where $\phi_0(Q)$ is given by (156) with $n = 0$:

$$\phi_0(Q) = \tfrac{1}{30}(4Q^5 + 20Q^4 + 30Q^3 + 15Q^2) \tag{246}$$

Substituting (246) in (245) and evaluating the integral we find

$$E_0 = \frac{863}{840}\cdot\frac{1}{C} \tag{247}$$

16.2. *Expected Cost of the* 1-*switch Trajectories*

The contribution e_1 of the members of the ensemble with 1-switch trajectories to the expected integral-square-error is

$$e_1 = \left[\int_{-c}^{-3/2\,(k^2/a)} + \int_{3/2\,(k^2/a)}^{c}\right] w(x_1, 0)\, p(x_1)\, dx_1 \tag{248}$$

in which the ranges of integration are obtained by subtracting the range of integration of (239) from the range of the distribution. From (235), (248) and symmetry

$$e_1 = \frac{1}{c}\int_{-c}^{-3/2\,(k^2/a)} w(x_1, 0)\, dx_1 \tag{249}$$

Introducing the dimensionless variable

$$E_1 = \frac{a^2}{k^5}e_1 \tag{250}$$

† The superscript in (155) is now dropped.

we write (249) as

$$E_1 = \frac{1}{C}\int_{-C}^{-3/2} W(X_1, 0)\, dX_1 \tag{251}$$

where X_1, W and C are defined by (86), (96) and (241).

Also, from (238) and (241)

$$L = -1 + (1+2C)^{1/2} \tag{252}$$

Substitution of (244) in (251), and use of (155) and (252), yields

$$E_1 = \frac{1}{C}\int_{1}^{L} (Q+1)\,\phi_1(Q)\, dQ \tag{253}$$

where $\phi_1(Q)$ is given by (156) with $n = 1$:

$$\phi_1(Q) = \tfrac{1}{30}(12Q^5 - 20Q^4 + 90Q^3 - 5Q^2 - 20Q + 12) \tag{254}$$

Substituting (254) in (253) and evaluating the integral we find

$$E_1 = (\tfrac{2}{35}L^7 - \tfrac{2}{45}L^6 + \tfrac{7}{15}L^5 + \tfrac{17}{24}L^4 - \tfrac{5}{18}L^3 - \tfrac{2}{15}L^2 + \tfrac{2}{5}L - \tfrac{593}{504})/C \tag{255}$$

16.3. *Total Expected Cost*

Let us define the expected integral-square-error for the ensemble of starting points (235) as $\bar{w}$. Thus

$$\bar{w} = \int_{-\infty}^{\infty} w(x_1, 0)\, p(x_1)\, dx_1 \tag{256}$$

The corresponding dimensionless variable is $\overline{W}$ where, from (96),

$$\overline{W} = \frac{a^2}{k^5}\bar{w} \tag{257}$$

Thus

$$\overline{W} = E_0 + E_1 \tag{258}$$

From (247), (255) and (258)

$$\overline{W} = \frac{1}{C}\xi(L) \tag{259}$$

where $\xi(\rho)$ is the function defined by

$$\xi(\rho) = \tfrac{1}{2520}(144\rho^7 - 112\rho^6 + 1176\rho^5 + 1785\rho^4 - 700\rho^3 - 336\rho^2 + 1008\rho - 376) \tag{260}$$

Also, putting $x_1 = -c$, $x_2 = 0$ and $Q = L$ in (179) we obtain

$$k = \left(\frac{ac}{L + \frac{1}{2}L^2}\right)^{1/2} \tag{261}$$

From (241), (257), (259) and (261)

$$\bar{w} = a^{1/2}\, c^{5/2}\, \zeta(L) \tag{262}$$

where $\zeta(\rho)$ is the function defined by

$$\zeta(\rho) = \frac{\xi(\rho)}{(\frac{1}{2}\rho^2 + \rho)^{7/2}} \tag{263}$$

where $\xi(\rho)$ is given by (260).

16.4. *Minimizing Value of k*

(236), (238) and (262) show that to find a minimum of $\bar{w}$ we have to find a minimum of $\zeta(L)$ when L varies in the range (compare (188))

$$1 < L \leqslant 3 \tag{264}$$

From (260) and (263)

$$\frac{d\zeta(L)}{dL} = \frac{160L^7 - 416L^6 - 261L^5 + 655L^4 + 340L^3 - 720L^2 - 344L + 376}{720(\frac{1}{2}L^2 + L)^{9/2}} \tag{265}$$

From (265) $[d\zeta(L)/dL]$ is negative at $L = 1$ and positive at $L = 3$. Hence there is a minimum of $\zeta(L)$ for a value of L in the range (264), and this value of L is a root of

$$160L^7 - 416L^6 - 261L^5 + 655L^4 + 340L^3 - 720L^2 - 344L + 376 = 0 \tag{266}$$

As in § 14, it can be shown that there is only one real root of (266) in range (264). To do this we define K as

$$K = L - 1 \tag{267}$$

so that (266) is

$$160K^7 + 704K^6 + 603K^5 - 1290K^4 - 2370K^3 - 1260K^2 - 825K - 210 = 0 \tag{268}$$

By Descartes' rule (§ 14) (268) has not more than one positive root; hence (266) has not more than one root which is greater than unity.

We conclude that the unique root of (266) in range (264) yields a minimum value of $\zeta(L)$. By computation this root L^* is

$$L^* \simeq 2{\cdot}61225242 \tag{269}$$

From (260), (263) and (269) the minimum of $\zeta(L)$ is

$$\zeta^* \simeq 0{\cdot}220078713 \tag{270}$$

From (261) and (269) the minimizing value of k is

$$k^* \simeq 0{\cdot}407428021a^{1/2}c^{1/2} \tag{271}$$

From (262) and (270) the minimum expected integral-square-error with linear switching is

$$\bar{w}^* \simeq 0{\cdot}220078713a^{1/2}c^{5/2} \tag{272}$$

16.5. *Comparison with Optimal Non-linear Switching*

Let us calculate the expected integral-square-error $\bar{v}$ for the ensemble of starting points (235) when the optimal non-linear switching curve (13) is used. (Since this switching curve is optimal for each member of the ensemble separately, it is optimal for the ensemble as a whole.) We have

$$\bar{v} = \int_{-\infty}^{\infty} v(x_1, 0)\, p(x_1)\, dx_1 \tag{273}$$

From (17), (21), (22), (235) and (273)

$$\bar{v} = \frac{G_0 a^{1/2}}{2c} \int_{-c}^{+c} |x_1|^{5/2} dx_1 \tag{274}$$

$$= \tfrac{2}{7} G_0 a^{1/2} c^{5/2} \tag{275}$$

$$= \tfrac{1}{70}(222 + (33)^{1/2})^{1/2} a^{1/2} c^{5/2} \tag{276}$$

$$\simeq 0{\cdot}218290728 a^{1/2} c^{5/2} \tag{277}$$

From (272) and (277) the ratio of the expected integral-square-errors with the best linear switching and the optimal non-linear switching is

$$\frac{\bar{w}^*}{\bar{v}} \simeq 1{\cdot}00819084 \tag{278}$$

i.e. the linear switching is only about 0·8% worse than the non-linear switching. Thus we have verified the results asserted in (36) and (37).

Part IV. Complement

§ 17. Plants of Higher Order

It would be useful to repeat the previous investigation for a plant with transfer function $1/as^3$, i.e. when the plant consists of three integrators†:

$$\frac{d^3 x}{dt^3} = \frac{u}{a} \tag{279}$$

In this case the linear switching surface is a plane in three dimensions. Results for the system with optimal non-linear switching have been given by Grensted and Fuller (1965).

One computational result from the latter paper is that the minimum settling time controller gives, for a step response, an integral-square-error which is about 1·2% greater than that yielded by the minimum integral-square-error controller. Also the linear switching controller can be chosen to give the same step response as the minimum settling time controller (see § 2). We conclude that linear switching gives a step response (for a fixed step) with an integral-square-error which is less (probably much less) than 1·2% greater than that given by the optimal non-linear switching. This preliminary result suggests that our justification of linear controllers for first and second-order systems will extend to higher-order systems.

† Some early investigations of third-order plants with linear switching were as follows. Andronov and Bautin (1944, 1955) treated the case when the plant consists of two integrators and a lag. Flügge-Lotz (1953) treated the case when the plant consists of one integrator and a second-order element with complex characteristic roots. This analysis was complicated by the choice of a discontinuous transformation of the state coordinates. In a subsequent treatment (Flügge-Lotz and Ishikawa 1960) the analysis was simplified by the choice of a continuous transformation of state coordinates.

However, this justification of linear controllers for higher-order systems should not be interpreted too literally. Thus with an appropriate linear switching surface and plant (279), the system will be asymptotically stable in-the-small, but cannot be asymptotically stable in-the-large. This assertion is suggested by an approximate analysis using describing function techniques† and can be verified by exact piecewise linear calculations as discussed in Appendix 3. The correct design procedure seems to be as follows. The controller should be linear in the central region of the state space where the state point is usually to be found. Outside this central region the controller should be non-linear, if necessary, to secure in-the-large stability. Since the excursions of the state point outside the central region are usually rare, they do not contribute appreciably to the performance index. Hence the controller non-linearities can be designed quite crudely. Thus it will usually be sufficient to make the controller output a linear combination of functions of single state variables:

$$z = f_1(x_1) + f_2(x_2) + \ldots + f_r(x_r) \tag{280}$$

(see e.g. Persson (1963) and Frederick and Franklin (1966)); and to make the $f_i(x_i)$, outside the central region, piecewise linear functions selected on the basis of describing function considerations.

§ 18. Other Performance Indices

In some applications it may be that integral-square-error is a performance index which penalizes large errors too much and small errors too little. If so, we should use a performance index which is intermediate between settling time and integral-square-error. Such a performance index is integral-modulus-error (Fuller 1959 b). It would be possible and profitable to repeat the present investigation with the performance index changed to integral-modulus-error.

Occasionally it may be desirable to choose a performance index which penalizes other variables, as well as the system error. Wonham and Johnson (1964) and Johnson and Wonham (1965) have investigated the optimal saturating control of linear plants when the performance index is the integral of a quadratic form of the state variables (and possibly also of the control variable u). It turns out that the optimal non-linear controller is precisely linear in a central region of the state space. Thus for such performance indices we expect the best linear controller to be even more nearly optimal than for integral-square-error.

---

† See e.g. Grensted (1962) for an account of these techniques. Note that Woodside (1965) studied system (279) with linear switching and stochastic disturbances but did not mention instability. However, in his thesis (Woodside 1964) he reported that he had encountered instability, and his analogue computer results were obtained by monitoring the system and rejecting those runs which turned out to be unstable.

§ 19. More General Systems

For simplicity in the present investigation attention has been given to relay control systems. If the relay is replaced by a saturating element, then the best linear controller must be at least as good as in the relay system, since the relay system is a limiting case of the saturating system.

When the plant is subject to stochastic disturbances of white noise type, and the performance index is mean-square-error, computational investigations suggest that the optimal switching surface is closely linear in a central region of the state space, at least in simple cases (Woodside 1964, 1965, Buhr 1966). In such cases we expect a switching surface which is exactly linear in a central region to be nearly optimal.

When the measurement of the state variables is contaminated by noise, it appears that the optimal saturating control is not bang-bang, and tends to be more nearly linear as the noise increases (Rosenbrock 1963). Here we expect a control signal which (in a central region) is a linear combination of estimated state variables to be nearly optimal.

§ 20. Conclusions

For some simple relay control systems we have found that if the optimal non-linear switching surface is replaced by a linear switching surface, the loss in performance can be small. In some cases the loss is so remarkably small that it would be difficult to measure on an analogue computer.

It is not unexpected that linear controllers give roughly optimal performance. In many situations the ordinary linearization procedures (e.g. describing function methods) turn out to be approximately valid for the analysis of non-linear control systems. If an optimal non-linear controller can be replaced by a linear controller for the purposes of such analysis, then the linear controller itself will (usually) yield nearly optimal performance. This argument might suggest that the linear controller would yield a performance index within say 10% or 20% of the optimum. However, the present paper indicates that the actual loss due to a linear controller is much smaller.

The main use of contemporary optimal non-linear theory using state space methods may be to justify established empirical and quasi-linear design methods, rather than to initiate new control systems. The present paper supports this point of view.

APPENDIX 1

Cost of Trajectories Starting on the Switching Line

In this Appendix the solution will be derived of eqns. (142) and (143), namely the equations

$$W_n(Q) = W_{n-1}(Q+2) - \tfrac{2}{15}(2Q^5 + 10Q^4 + 15Q^3 + 5Q^2 + 10Q + 12) \quad (281)$$

and

$$W_0(Q) = \tfrac{1}{2}Q^2 \tag{282}$$

Inspection of (281) shows that if $W_{n-1}(Q)$ is a polynomial in Q of not greater than fifth degree, then so is $W_n(Q)$. From (282) $W_0(Q)$ is such a polynomial. Hence by induction, $W_n(Q)$ is such a polynomial for all $n \geqslant 0$:

$$W_n(Q) = A_n Q^5 + B_n Q^4 + C_n Q^3 + D_n Q^2 + E_n Q + F_n \tag{283}$$

To find $A_n, B_n, \ldots$ we substitute (283) in (281) and equate coefficients of powers of Q:

$$\left.\begin{aligned} A_n &= A_{n-1} && -\tfrac{4}{15} \\ B_n &= 10A_{n-1} + B_{n-1} && -\tfrac{4}{3} \\ C_n &= 40A_{n-1} + 8B_{n-1} + C_{n-1} && -2 \\ D_n &= 80A_{n-1} + 24B_{n-1} + 6C_{n-1} + D_{n-1} && -\tfrac{2}{3} \\ E_n &= 80A_{n-1} + 32B_{n-1} + 12C_{n-1} + 4D_{n-1} + E_{n-1} && -\tfrac{4}{3} \\ F_n &= 32A_{n-1} + 16B_{n-1} + 8C_{n-1} + 4D_{n-1} + 2E_{n-1} + F_{n-1} && -\tfrac{8}{5} \end{aligned}\right\} \tag{284}$$

Also, from (282) and (283)

$$\left.\begin{aligned} A_0 = B_0 = C_0 = E_0 = F_0 &= 0 \\ D_0 &= \tfrac{1}{2} \end{aligned}\right\} \tag{285}$$

The first of eqns. (284) is

$$A_n - A_{n-1} + \tfrac{4}{15} = 0 \tag{286}$$

which is a simple recurrence relation, i.e. a linear difference equation for A_n. Its general solution, obtained by standard methods (see e.g. Ferrar 1947), is

$$A_n = -\tfrac{4}{15}n + A_0 \tag{287}$$

From (285) and (287)

$$A_n = -\tfrac{4}{15}n \tag{288}$$

Substituting from (288) in the second of eqns. (284) we obtain a linear difference equation for B_n:

$$B_n - B_{n-1} + \tfrac{8}{3}n - \tfrac{4}{3} = 0 \tag{289}$$

which may be readily solved.

Solving the rest of equations (284) successively in a similar manner, and using (285) we find

$$\left.\begin{aligned} A_n &= -\tfrac{4}{15}n \\ B_n &= -\tfrac{4}{3}n^2 \\ C_n &= -\tfrac{2}{9}(16n^3 - 7n) \\ D_n &= -\tfrac{1}{6}(32n^4 - 28n^2 - 3) \\ E_n &= -\tfrac{2}{45}(96n^5 - 140n^3 + 29n) \\ F_n &= -\tfrac{2}{45}(32n^6 - 70n^4 + 29n^2) \end{aligned}\right\} \tag{290}$$

From (283) and (290) the expression for $W_n(Q)$ is as given in (144).

APPENDIX 2

Multi-switch Trajectories Starting on the x_1-axis

In this Appendix it is shown that the integral-square-error for a starting point on the x_1-axis increases with $|Q|$ if $n > 1$. By symmetry we can confine attention to starting points $(x_1, 0)$ on the negative x_1-axis, so that

$$x_1 < 0 \tag{291}$$

$$Q > 0 \tag{292}$$

and

$$n > 1 \tag{293}$$

From (180) the integral-square-error is

$$w(x_1, 0) = \theta_n(Q)\, a^{1/2}(-x_1)^{5/2} \tag{294}$$

From (156) and (181)

$$\begin{aligned}\theta_n(Q) = \tfrac{1}{90}[\tfrac{1}{2}Q^2+Q]^{-5/2}[&(24n+12)Q^5-(120n^2-60)Q^4\\ &+(320n^3-140n+90)Q^3-(480n^4-420n^2-45)Q^2\\ &+(384n^5-560n^3+116n)Q-128n^6+280n^4-116n^2]\end{aligned} \tag{295}$$

Differentiation of (295) yields

$$\frac{d\theta_n(Q)}{dQ} = \frac{\psi(n,Q)}{180(\frac{1}{2}Q^2+Q)^{7/2}} \tag{296}$$

where

$$\begin{aligned}\psi(n,Q) = {}&120(n^2+n)Q^5-40(16n^3+9n^2-7n)Q^4\\ &+5(288n^4+64n^3-252n^2-28n-9)Q^3\\ &-(1536n^5-480n^4-2240n^3+420n^2+464n+45)Q^2\\ &+4(160n^6-288n^5-350n^4+420n^3+145n^2-87n)Q\\ &+20(32n^6-70n^4+29n^2)\end{aligned} \tag{297}$$

We have to show that $\psi(n,Q)$ as given by (297) is positive for integer values of $n > 1$ and Q varying in the range (see (146) and (292))

$$2n-1 < Q \leqslant 2n+1 \tag{298}$$

It is convenient to make the following change of variables

$$P = Q-2n+1 \tag{299}$$

and

$$m = n-1 \tag{300}$$

Thus P represents the difference between Q and the lower boundary of its range (298), and m represents the difference between n and the boundary (293) of its range. Then (297) becomes

$$\begin{aligned}\psi(m+1, P+2m+1) \\ &= 120(m^2+3m+2)P^5+80(7m^3+24m^2+23m+6)P^4 \\ &\quad +5(224m^4+896m^3+1108m^2+412m-33)P^3 \\ &\quad +2(672m^5+3360m^4+5740m^3+3600m^2+143m-360)P^2 \\ &\quad +(896m^6+5376m^5+11480m^4+10080m^3+2564m^2-696m-225)P \\ &\quad +2(128m^7+896m^6+2296m^5+2520m^4+732m^3-716m^2-501m-45)\end{aligned} \tag{301}$$

where, from (293) and (298)–(300),

$$P > 0 \tag{302}$$

and

$$m > 0 \tag{303}$$

(301) shows that ψ is a polynomial in P, with coefficients which are themselves polynomials in m. If we can show that these coefficients are all positive, then in view of (302) ψ will be proved positive, and our objective will be gained.

From (303) the coefficients of P^5 and P^4 in (301) are clearly positive. Let us consider the coefficient of P^3, namely†

$$5(224m^4+896m^3+1108m^2+412m-33) \tag{304}$$

and let us temporarily consider m as a continuous variable, in (304). Then (304) has a zero between $m = 0$ and $m = 1$ since (304) is respectively negative and positive at these values. However, by Descartes' rule (§ 14) (304) has not more than one positive real zero. Hence (304) has no real zero for $m \geqslant 1$. Hence (304) is positive for all continuous real values of $m \geqslant 1$. In particular (304) is positive for all positive integer values of m.

In the same way the remaining coefficients of the powers of P in (301) are seen to be positive for integer $m > 0$. Hence ψ is positive for P and m satisfying (302) and (303), and thus for Q and n satisfying (298) and (293). In view of (296) we conclude that $\theta_n(Q)$ increases with Q for all $n > 1$.

APPENDIX 3

Triple-integrator Plant

In this Appendix some results on stability are stated (without proof) for the case when the plant consists of three integrators. This case was discussed in § 17. (The more general case when the plant consists of two integrators and one lag has been treated by Andronov and Bautin 1944, 1955.)

---

† (304) can be seen to be positive for $m \geqslant 1$ by mere inspection. The alternative proof given above is intended to show how to deal with the other coefficients in (301).

The system is

$$\left.\begin{aligned} \frac{dx_1}{dt} &= x_2 && (305)\\ \frac{dx_2}{dt} &= x_3 && (306)\\ \frac{dx_3}{dt} &= -\frac{1}{a}\operatorname{sgn}(k_1 x_1 + k_2 x_2 + k_3 x_3) && (307) \end{aligned}\right\}$$

A set of necessary and sufficient conditions for (in-the-small) asymptotic stability of the origin is

$$k_1 > 0, \quad k_2 > 0, \quad k_3 > 0 \tag{308}$$

When conditions (308) are satisfied, the following results are valid.

The sliding region is the strip of the switching plane

$$k_1 x_1 + k_2 x_2 + k_3 x_3 = 0 \tag{309}$$

between the planes

$$k_1 x_2 + k_2 x_3 = k_3/a \tag{310}$$

and

$$k_1 x_2 + k_2 x_3 = -k_3/a \tag{311}$$

(Planes (310) and (311) are parallel to themselves and to the x_1-axis.)

An approximate analysis by means of the describing function method suggests that the system has a limit cycle which is unstable. Exact analysis confirms the existence of a limit cycle. If

$$k_2^2 \geqslant \tfrac{1}{3} k_1 k_3 \tag{312}$$

the limit cycle consists of a P-arc and an N-arc, its period is

$$4\left(3\frac{k_3}{k_1}\right)^{1/2} \tag{313}$$

and switches occur at the two points

$$\left.\begin{aligned} x_1 &= \pm\frac{1}{a}\left(3\frac{k_3^3}{k_1^3}\right)^{1/2} && (314)\\ x_2 &= 0 && (315)\\ x_3 &= \mp\frac{1}{a}\left(3\frac{k_3}{k_1}\right)^{1/2} && (316) \end{aligned}\right\}$$

If

$$k_2^2 < \tfrac{1}{3} k_1 k_3 \tag{317}$$

the limit cycle is more complicated†. It consists of a P-arc from a starting point on one edge of the sliding strip to the interior of the sliding strip, a sliding trajectory to the other edge of the sliding strip, an N-arc to the interior of the sliding strip, and a sliding trajectory back to the

† When (317) holds, points (314–316) are in the sliding motion strip, and thus can no longer be switch points of a limit cycle consisting of a P-arc and an N-arc.

starting point. (When (317) is satisfied the sliding motion in the neighbourhood of the origin is very under-damped. Thus in practice the coefficients would be chosen to satisfy (312) rather than (317).)

We conclude that the system does not have in-the-large asymptotic stability of the origin.

References

André, J., and Seibert, P., 1956, *Arch. Math.*, **7**, 148.

Andronov, A. A., and Bautin, N. N., 1944, *Dokl. Akad. Nauk. SSSR*, **43**, 197; 1955, *Izv. Akad. Nauk. SSSR, OTN*, No. 3, 3, No. 6, 54.

Anosov, D. V., 1959, *Avtomat. Telemech.*, **20**, 135.

Bellman, R., 1957, *Dynamic Programming* (Princeton: Univ. Press).

Bilharz, H., 1941, *Luftfahrtforschung*, **18**, 317; 1942, *Z. angew. Math. Mech.*, **22**, 206.

Boltyanski, V. G., and Pontryagin, L. S., 1956, *Trudi Tret. Vseso. Mat. Sez.* (*Trans. Third All-union Math. Conf.*) (Izd. AN SSSR), **1**, 217.

Brennan, P. J., and Roberts, A. P., 1962, *J. Electron. Contr.*, **12**, 345.

Buhr, R. J. A., 1966, Ph.D. Thesis, Cambridge.

Burnside, W. S., and Panton, A. W., 1904, *Theory of Equations*, Fifth edition (Dublin: University Press).

Ferrar, W. L., 1947, *Higher Algebra* (Oxford: Univ. Press).

Flügge-Lotz, I., 1947, *Z. angew. Math. Mech.*, **25/27**, 97; 1953, *Discontinuous Automatic Control* (Princeton); 1960, *Proc. I.F.A.C. Congress, Moscow* (London: Butterworths, 1961), **1**, 390.

Flügge-Lotz, I., Hodapp, H., Klotter, K., Meissinger, H., and Scholz, K., 1943–5, *Zentrale für wissenschaftliches Berichtswesen der Luftfahrtforschung des Generalluftzeugmeisters* (ZWB, Berlin), Untersuchungen und Mitteilungen, Nos. 1326–1329.

Flügge-Lotz, I., and Ishikawa, T., 1960, *NASA tech. Note*, D–428.

Flügge-Lotz, I., and Klotter, K., 1948, *Z. angew. Math. Mech.*, **28**, 317.

Flügge-Lotz, I., and Lindberg, H. E., 1959, *NASA tech. Note*, D–107.

Frederick, D. K., and Franklin, G. F., 1966, Preprints, *Joint Aut. Contr. Conf.* (U.S.A.), p. 594.

Fuller, A. T., 1959 a, Ph.D. Thesis, Cambridge; 1959 b, *J. Electron. Contr.*, **7**, 456; 1960 a, *Ibid.*, **8**, 381; 1960 b, *Ibid.*, **8**, 465; 1960 c, *Ibid.*, **9**, 65; 1960 d, *Proc. I.F.A.C. Congress, Moscow* (London: Butterworths, 1961), **1**, 510; 1963 a, *Automatica*, **1**, 289; 1963 b, *J. Electron. Contr.*, **15**, 63; 1964 a, *Ibid.*, **17**, 283; 1964 b, *Ibid.*, **17**, 301; 1966, *Int. J. Contr.*, **3**, 359.

Grensted, P. E. W., 1962, *Progress in Control Engineering*, edited by R. H. MacMillan, **1**, 105.

Grensted, P. E. W., and Fuller, A. T., 1965, *Int. J. Contr.*, **2**, 33.

Houkowsky, A., 1896, *Z. Ver. dt. Ing.*, **40**, 839, 871.

Jackson, R., 1960, Discussion in *Proc. I.F.A.C. Congress, Moscow* (London: Butterworths, 1961), **1**, 518.

Johnson, C. D., and Wonham, W. M., 1965, *J. bas. Engng*, **87**, 81.

Kalman, R. E., and Bertram, J. E., 1960, *J. bas. Engng*, **82**, 371.

Klein, F., 1908, *Elementary Mathematics from an Advanced Standpoint. Geometry* (translation from the German) (New York: Dover Publications).

Lasalle, J. P., and Lefschetz, S., 1961, *Stability by Liapunov's Direct Method* (New York: Academic Press).

Léauté, H., 1885, *J. Éc. polytech.*, **55**, 1; 1891, *Ibid.*, **61**, 1.

MacColl, L. A., 1945, *Servomechanisms* (New York: Van Nostrand).

Nikolski, G. N., 1934, *Trudi Tsentral. Lab. Provod. Svyazi.*, No. 1, 34.

Oldenburger, R., 1957, *Trans. Am. Soc. mech. Engrs*, **79**, 527; 1960, Discussion in *Proc. I.F.A.C. Congress, Moscow* (London: Butterworths, 1961), **1**, 528.

PEARSON, J. D., 1961, *J. Electron. Contr.*, **10**, 323.
PERSSON, E. V., 1963, *Proc. I.F.A.C. Congress* (*Basle*). *Theory* (London: Butterworths, 1965), 210.
ROSENBROCK, H. H., 1963, *Automatica*, **1**, 263.
SYNGE, J. L., 1933, *Am. math. Mon.*, **40**, 202.
TEODORCHIK, K. F., 1938, *Zh. tekh. Fiz.*, **8**, 960.
TSYPKIN, YA. Z., 1955, *Theory of Relay Systems of Automatic Control* (in Russian; Moscow). (German translation; Munich: Oldenbourg, 1958.)
WEISS, H. K., 1946, *J. aeronaut. Sci.*, **13**, 364.
WEISSENBERGER, S., 1966, *J. bas. Engng*, **83**, 419.
WONHAM, W. M., 1963, *J. Electron. Contr.*, **15**, 59.
WONHAM, W. M., and JOHNSON, C. D., 1964, *J. bas. Engng*, **86**, 107.
WOODSIDE, C. M., 1964, Ph.D. Thesis, Cambridge; 1965, *Int. J. Contr.*, **2**, 285, 409.
ZACHARY, D. H., 1966, *Int. J. Contr.*, **4**, 357.

A computational approach to optimal control of stochastic saturating systems†

By W. M. Wonham‡ and W. F. Cashman
NASA/Electronics Research Center, 575 Technology Square, Cambridge, Mass.

[Received August 6,1968]

Abstract

A systematic procedure is given for digital computation of a sub-optimal non-linear feedback control $u=q(x)$, $|\phi \leqslant 1$, for a system $\dot{x}=Ax-Bu+C\dot{w}$, where $\dot{w}$ is Gaussian white noise. The performance criterion is expected quadratic cost relative to the stationary probability distribution of x. A control of form $\phi(x)=\text{sat}\ (f^1x)$ or sgn (f^1x) is obtained, by a combination of dynamic programming and statistical linearization; computation time is a few seconds. Applications are given to design of an eighth-order attitude control and of a second-order tracking system. For the latter system, the elliptic differential equation of dynamic programming is solved by finite differences to yield the true optimal control, and results are compared with those for the sub-optimal control. It is shown finally how the procedure can be modified if the observations of the state include additive noise.

1. Introduction

This paper reports on a computational method of determining a 'good' sub-optimal feedback control for the following class of stochastic systems. A time-invariant linear system is subject to additive white noise. Amplitude bounds are imposed on the components of the control vector. The cost functional is the expectation, under stochastically stationary conditions, of a quadratic function of state and control variables. For this problem the optimal control is, of course, non-linear. An exact approach by dynamic programming would require the solution of Bellman's functional equation: in this case a quasi-linear partial differential equation of elliptic type. Solution of Bellman's equation by finite difference methods is troublesome for systems of dynamic order 2 and is apparently beyond feasible computation for systems of order greater than 3. To find a *sub-optimal* control we approximate the solution by a quadratic form, from which is determined a non-linear feedback control satisfying the amplitude constraints. Statistical linearization is used to achieve consistency of the non-linear control with the quadratic approximation. The procedure yields a set of finite (not differential) equations, the solution of which is readily computed by a version of Howard's policy improvement algorithm. In this way the difficulty of 'exactly' solving Bellman's equation is bypassed. Although general estimates are not available, evidence suggests that the resulting design is not far from optimal in terms of performance. Our viewpoint is that Bellman's equation is a natural point of departure for approximations; and we are guided by the heuristic principle

† Communicated by Dr. A. T. Fuller.
‡ On leave from Division of Applied Mathematics, Brown University, Providence, R.I.

that an apparently crude approximation to the optimal control may, if properly chosen, yield near-optimal results. The present approach was first suggested by Wonham (1966 a).

The paper is organized as follows. After a statement of the problem in § 2, we digress in § 3 to a description of statistical linearization from a state-space viewpoint. The design algorithm is given in § 4 and in § 5 numerical results are presented. The latter include design of a two-axis attitude control, with two non-linearities, for a system of order 8. The results of analogue simulation are presented as numerical checks and in § 6 some practical details are included of the digital and analogue computations. In § 7 we discuss briefly a finite-difference 'exact' solution of Bellman's equation for a second-order system and compare the result with the algorithmic solution. Finally in § 8 it is shown how the same method can be used when noise is added to observations of the system state. In the Appendix we prove the existence of a solution of the algorithm equations.

2. Dynamic programme statement of the problem

The type of system considered is described by a formal stochastic differential equation:

$$\dot{x}(t) = Ax(t) - Bu(t) + C\dot{w}(t) \tag{2.1 a}$$

or, in the notation of Itô:

$$dx(t) = Ax(t)dt - Bu(t)dt + Cdw(t). \tag{2.1 b}$$

Here and below all vectors and matrices have real-valued components; A, B, C are constant matrices of dimension respectively $n \times n$, $n \times m$, $n \times d$; x is the state vector, u the control vector and $\{w(t)\}$ the standard d-dimensional Wiener (Brownian motion) process: that is, the components of w are independent Wiener processes with

$$\mathscr{E}\{(w_i(t_2) - w_i(t_1))(w_j(t_2) - w_j(t_1))\} = \delta_{ij}|t_2 - t_1|$$

$(i, j = 1, \ldots, d)$. (In the following, $\mathscr{E}$ denotes expectation.)

If, in (2.1), the control is given by a feedback law:

$$u(t) = \phi[x(t)],$$

and if ϕ is suitably regular (see, e.g., Dynkin 1965), then (2.1) determines a Markov process X_ϕ of diffusion type (Dynkin 1965). We shall be interested only in those ϕ which (i) satisfy the amplitude constraint:

$$|\varphi_i(x)| \leq 1 \quad (i = 1, \ldots, m;\ \text{all } x), \tag{2.2}$$

and (ii) are such that X_ϕ has a unique stationary probability distribution p_ϕ with respect to which the function $|x|^2$ has finite expectation. Denote this class of *admissible* controls by Φ. The index of performance for X_ϕ will be the stationary expectation of a quadratic form:

$$L(x, u) = x'Mx + u'Nu. \tag{2.3}$$

namely:

$$\mathscr{E}\{L(x, \phi)\} \equiv \int L(x, \varphi(x)) p_\phi(dx). \tag{2.4}$$

In (2.3), M and N are positive semi-definite constant matrices.

The optimization problem is to find a control $\phi^\circ \in \Phi$ which is optimal in the sense that

$$\mathscr{E}_{\phi^\circ}\{L(x, \varphi^\circ)\} = \min_{\varphi \in \Phi} \mathscr{E}_{\phi}\{L(x, \varphi)\}.$$

Our approach to the problem by dynamic programming will be somewhat heuristic and non-rigorous. Thus it will be tacitly assumed that an admissible control exists (Φ is non-empty) and that Bellman's equation, below, has a solution. A rigorous treatment along these lines of a similar (but somewhat easier) problem can be found in Wonham (1967 a).

Let $V(x)$ be a real-valued smooth function and let $\mathscr{L}_u$ be the elliptic operator given by:

$$\mathscr{L}_u V(x) \equiv \tfrac{1}{2} \operatorname{tr}[C' V_{xx}(x) C] + (Ax - Bu)' V_x(x), \tag{2.5}$$

where tr denotes trace and $V_x(V_{xx})$ the vector (matrix) of first (second) partial derivatives of V. The operator $\mathscr{L}_\phi$, obtained by substituting $u = \phi(x)$ in (2.5), is called the *differential generator* (sometimes *backward operator*) of the process X_ϕ (see Dynkin (1965)).

Now suppose there exist a function $V(x)$, a control $\phi^\circ \in \Phi$, and a constant λ with the properties:

(i) $V(x)$ and $\mathscr{L}_\phi V(x)$ are *finitely integrable relative to p_ϕ*, for all $\phi \in \Phi$.

(ii) $\mathscr{L}_{\phi^\circ} V(x) + L(x, \phi^\circ(x)) = \lambda$ *(all x)*.

(iii) $\lambda \le \mathscr{L}_u V(x) + L(x, u)$ *for all x, and all u such that $|u_i| \le 1, i = 1, \ldots, m$.*

Under these conditions ϕ° is optimal.

Combining (ii) and (iii) we obtain the appropriate version of Bellman's equation:

$$\min_{|u| \le 1} \{\mathscr{L}_u V(x) + L(x, u)\} = \lambda, \tag{2.6}$$

where $|u| = \max\{|u_i| : i = 1, \ldots, m\}$.

To verify formally the italicized statement, observe first that

$$\mathscr{E}_\phi\{\mathscr{L}_\phi V(x)\} = 0 \quad (\phi \in \Phi). \tag{2.7}$$

Indeed, if $p_\phi(t, x, dy)$ is the transition function of X_ϕ then, by definition of $p_\phi(dx)$:

$$p_\phi(\mathscr{B}) = \int p_\phi(dx) p_\phi(t, x, \mathscr{B}),$$

on measurable x sets $\mathscr{B}$. Thus (dropping the subscript ϕ):

$$\begin{aligned}
\int p(dx) \mathscr{L} V(x) &= \int p(dx) \lim_{t \downarrow 0} \frac{\left\{\int p(t, x, dy) V(y) - V(x)\right\}}{t} \\
&= \lim_{t \downarrow 0} t^{-1} \left\{\int p(dx) \int p(t, x, dy) V(y) - \int p(dx) V(x)\right\} \\
&= \lim_{t \downarrow 0} t^{-1} \left\{\int p(dy) V(y) - \int p(dx) V(x)\right\} \\
&= 0.
\end{aligned}$$

Combining this result with properties (ii) and (iii) one obtains:

$$\begin{aligned}
\mathscr{E}_{\phi^\circ}\{L(x, \phi^\circ)\} &= \lambda \\
&\le \mathscr{E}_\phi\{L(x, \phi)\} \quad (\phi \in \Phi)
\end{aligned}$$

This inequality states that ϕ° is optimal.

With L, $\mathscr{L}_u$ given by (2.3), (2.5), and on the assumption that N^{-1} exists, Bellman's eqn. (2.6) can be written:

$$\tfrac{1}{2}\operatorname{tr}(C'V_{xx}C)+(Ax-Bu)'V_x+x'Mx+u'Nu=\lambda \tag{2.8 a}$$

with

$$\begin{aligned} u&=\phi^\circ(x)\\ &=\operatorname{sat}[\tfrac{1}{2}N^{-1}B'V_x(x)]. \end{aligned} \tag{2.8 b}$$

In (2.7 *b*), sat$(\cdot)$ is an m vector with components:

$$[\operatorname{sat}(y)]_i=\begin{cases} y_i, & |y_i|\le 1\\ \operatorname{sign} y_i, & |y_i|\ge 1 \end{cases}$$

$(i=1,\ldots,m)$.

Equation (2.8) is a quasi-linear elliptic equation defined on the whole n space†. Both $V(x)$ and λ are initially unknown; λ is in fact determined by the implicit integrability condition (i) stated above. Observe that V is determined only up to an additive constant; since the control law ϕ° is a function of the gradient V_x this lack of determinacy is of no particular importance and will be ignored.

In practice a difficulty in solving (2.8) is that the p_ϕ are unknown and the integrability condition (i) is therefore not explicitly available. In 'exact' numerical solution of (2.8) the integrability condition can be replaced by a reflecting barrier condition at boundaries placed suitably far from the origin: this approach is described in § 7. However, the integrability condition plays no explicit role in the approximate method of § 4.

Analytical solution of (2.8) is generally not feasible even if $C=0$ and the problem is nonstochastic: for a discussion of that case see, for instance, Johnson and Wonham (1965).

A useful iterative method for solving functional equations like (2.6) is quasi-linearization and successive approximation (Howard 1960). This procedure is the basis of the algorithm described in § 4. One chooses $\phi^{(1)}\in\Phi$ arbitrarily and solves the linear equation:

$$\mathscr{L}^{(1)}V^{(1)}(x)+L(x,\phi^{(1)}(x))=\lambda^{(1)}$$

for $V^{(1)}$ and $\lambda^{(1)}$ (here $\mathscr{L}^{(1)}\equiv\mathscr{L}_{\phi^{(1)}}$). The minimizing operation of (2.6) applied to $V^{(1)}$ then determines $\phi^{(2)}$, and so on. For $r=1,2,\ldots,$

$$\begin{aligned} \lambda^{(r)}&=\mathscr{L}^{(r)}V^{(r)}(x)+L(x,\phi^{(r)}(x))\\ &\ge\mathscr{L}^{(r+1)}V^{(r)}(x)+L(x,\phi^{(r+1)}(x)). \end{aligned} \tag{2.9}$$

From (2.7) and (2.9) it is readily seen that $\lambda^{(r)}\ge\lambda^{(r+1)}(r=1,2,\ldots)$ and thus at each stage of approximation the control law is improved. Since $\lambda^{(r)}=\mathscr{E}_{\phi^{(r)}}\{L(x,\phi^{(r)})\}\ge 0$, the sequence $\{\lambda^{(r)}\}$ converges, but additional considerations are necessary to guarantee that $\{\phi^{(r)}\}$ converges and, if so, that the limit is actually an optimal control (see, for instance, Wonham 1967 a).

† We shall call (2.8) elliptic even in cases where the operator is formally degenerate: i.e. rank $(CC')<n$.

3. Statistical linearization

As introduced by Booton (1954) and Kazakov (1956), statistical linearization is a method of determining the (approximately) Gaussian stationary probability distribution of the output of a feedback system with linear dynamics and a single zero-memory non-linearity, when the system input is stationary and Gaussian. The method consists in replacing the non-linear element by a linear element which has, on the average, an 'equivalent' gain. The assumptions that underlie this approach are that the output process is actually stationary and is 'nearly' Gaussian. Both assumptions are difficult to justify *a priori* but are satisfied for, roughly speaking, stable systems in which the linear part has a low-pass transfer characteristic. Analogue simulation of numerous examples (e.g. Smith (1966)) has verified that the method is capable of reasonable accuracy—predicting output variance usually within about 15% of its true value and often within 5%.

In this section we describe a state-space version of this technique which can be used when the system has several non-linearities and which is well adapted to programming for digital computer. It is assumed that the system can be written in the form of (2.1 *b*) with $u=\phi(x)$:

$$dx=Axdt-B\phi(x)dt+Cdw. \tag{3.1}$$

A further assumption is that the process X_ϕ determined by (3.1) is stationary and (approximately) Gaussian, with expectation μ and positive definite covariance matrix Q:

$$\begin{aligned} p_\phi(dx) &\doteqdot g(x;\mu,Q)dx \\ &=(2\pi)^{-n/2}(\det Q)^{-1/2}\exp\{-\tfrac{1}{2}(x-\mu)'Q^{-1}(x-\mu)\}dx. \end{aligned} \tag{3.2}$$

To determine μ and Q, (3.1) is replaced by the equation of a *linear model*:

$$dy=Aydt-B(Ky+h)dt+Cdw, \tag{3.3}$$

where K is an $m\times n$ matrix and h is an m vector.

The row vectors $k_i'(i=1,\ldots,m)$ of K are the 'equivalent' gains corresponding to the m scalar components $\phi_i(x)$ of the vector-valued non-linear element $\phi(x)$. The gain vectors k_i and scalar bias levels h_i are determined by the conditions:

$$\int_{R^n}[\phi_i(x)-k_i'x-h_i]^2 g(x;\mu,Q)dx=\text{minimum}$$

for fixed μ and Q. The result, using (3.2), is:

$$k_i=Q^{-1}\int_{R^n}(x-\mu)\phi_i(x)g(x;\mu,Q)dx, \tag{3.4 a}$$

$$h_i=\int_{R^n}\phi_i(x)g(x;\mu,Q)dx-k_i'\mu \tag{3.4 b}$$

$(i=1,\ldots,m)$. Next, μ and Q are computed for the 'equivalent' $y(t)$ process of (3.3). By straightforward linear analysis:

$$(A-BK)Q+Q(A-BK)'+CC'=0 \tag{3.5 a}$$

and

$$\mu=(A-BK)^{-1}Bh. \tag{3.5 b}$$

P

For the equations to have meaning it is necessary, of course, that the $y(t)$ process be stationary and Q^{-1} exist. This is true if and only if the matrix $A-BK$ is stable (i.e. all its eigenvalues have negative real parts) and the pair $(A-BK, C)$ is controllable. The computational problem is to solve (3.4) and (3.5) for K, h, Q, μ. In the authors' experience a straightforward successive approximation scheme is effective: choose $K^{(0)}$ so that $A-BK^{(0)}$ is stable, let $h^{(0)}=0$, and construct an approximating sequence according to the rule:

$$\left.\begin{matrix} K^{(r)} \\ h^{(r)} \end{matrix}\right\} \to (3\cdot5) \to \left\{\begin{matrix} Q^{(r)} \\ \mu^{(r)} \end{matrix}\right\} \to (3\cdot4) \to \left\{\begin{matrix} K^{(r+1)} \\ h^{(r+1)} \end{matrix}\right. \tag{3.6}$$

$(r=0, 1, \ldots)$.

In the following special case the equations simplify in an important way. For $i=1, \ldots, m$, assume that feedback into the non-linear elements is linear:

$$\phi_i(x) = \hat{\phi}_i(f_i'x)$$

and that $\hat{\phi}_i$ is an odd function of its argument:

$$\hat{\phi}_i(-y) = -\hat{\phi}_i(y).$$

Then $h=0$, $\mu=0$ and $k_i=\gamma_i f_i$ for scalars γ_i. A computation from (3.4 *a*) shows that

$$\gamma_i = q_i^{-1} \int_{-\infty}^{\infty} \xi\hat{\phi}(\xi) g(\xi, q_i)\, d\xi \tag{3.7}$$

$(i=1, \ldots, m)$, where

$$q_i = f_i'Qf_i \tag{3.8}$$

and

$$g(\xi, q) = (2\pi q)^{-1/2} \exp(-\xi^2/2q).$$

On writing $\Gamma = \operatorname{diag}(\gamma_1, \ldots, \gamma_m)$, $F=[f_1, \ldots, f_m]$, and $K=\Gamma F'$, (3.5 *a*) becomes:

$$(A-B\Gamma F')Q + Q(A-B\Gamma F')' + CC' = 0. \tag{3.9}$$

The iteration scheme (3.6) can now be replaced by the simpler scheme:

$$\Gamma \to (3.9) \to Q \to (3.8) \to \left\{\begin{matrix} q_1 \\ \vdots \\ q_m \end{matrix}\right\} \to (3.7) \to \Gamma. \tag{3.10}$$

Observe that the n-dimensional integrals (3.4 *a*) have been replaced by 1-dimensional integrals (3.7).

Of special interest later are the relay characteristic:

$$\phi_i(x) = \operatorname{sgn}(f_i'x),$$

for which

$$k_i = (2/\pi)^{1/2} (f_i'Qf_i)^{-1/2} f_i, \tag{3.11 a}$$

and the saturation characteristic:

$$\phi_i(x) = \operatorname{sat}(f_i'x),$$

for which

$$k_i = \operatorname{erf}\{(f_i'Qf_i)^{-1/2}\} f_i \tag{3.11 b}$$

$(i=1, \ldots, m)$. In (3.11 *b*):

$$\operatorname{erf}(y) \equiv (2/\pi)^{1/2} \int_0^y \exp(-t^2/2)\, dt.$$

The foregoing discussion raises several theoretical questions which are the subject of current research. These include:

(i) Under what conditions does the non-linear Itô eqn. (3.1) determine a stationary diffusion process with $\mathscr{E}\{|x|^2\} < \infty$?

(ii) Under what conditions do (3.4) and (3.5) have a unique solution for K, h, Q, μ, with $A-BK$ stable and Q positive definite?

(iii) Under what conditions do the iterative processes (3.6) or (3.10) converge?

(iv) Assuming that $\mathscr{E}\{|x|^2\} < \infty$ and the parameters Q, μ exist, what is the accuracy with which Q represents the covariance matrix of x?

A Lyapunov function approach to (i) is described by Wonham (1966b); of course, it suffers from the usual limitations of this technique. Question (ii) can be answered affirmatively for the simplest case (3.7)–(3.9) and $m=1$, subject to natural stability conditions on ϕ and A, B; this result will be reported elsewhere. Little is known about (iii), although convergence under broad conditions is made plausible by inspection of first-order examples and by computational experience with higher-order problems. Question (iv) is almost entirely open, although a rule of thumb for estimating accuracy has been proposed (Smith 1966) and recent results of Kolovskii (1966) and Holtzman (1968) are applicable in certain cases.

In the authors' opinion statistical linearization is not of much current importance as a practical method of computing stationary averages, because of the computational demands of multi-dimensional integration and because more accurate results can be obtained directly by digital or analogue simulation. However, the method can offer real advantages when combined with dynamic programming, as will now be shown.

4. Algorithm for approximate solution of Bellman's equation

For simplicity of notation assume first that the control u is scalar-valued ($m=1$) and write $B=b$, $N=\nu$. We consider the two cases:

$$L(x,u)=x'Mx+\nu u^2, \quad (\nu=0 \text{ or } 1).$$

Performing the minimization in (2.6) (cf. (2.8 b)) one obtains:

$$\phi(x)=\begin{cases}\operatorname{sgn}[b'V_x(x)] & (\nu=0),\\ \operatorname{sat}[\tfrac{1}{2}b'V_x(x)] & (\nu=1).\end{cases} \tag{4.1}$$

To solve (2.6) approximately we make the intuitively reasonable assumption that V can be crudely represented by a quadratic form:

$$V(x)\doteqdot x'Px, \tag{4.2}$$

where P is a positive semi-definite matrix to be determined. Then (4.1) yields:

$$\phi^\circ(x)=\begin{cases}\operatorname{sgn}(b'Px) & (\nu=0),\\ \operatorname{sat}(b'Px) & (\nu=1).\end{cases} \tag{4.3}$$

To be consistent with (4.2), the control u in (2.8 a) must be linear. To linearize the non-linear controls (4.3) we write $\phi^\circ(x)\doteqdot k'x$ and determine k by (3.11) (with $f=Pb$):

$$k=\begin{cases}(2/\pi)^{1/2}(b'PQPb)^{-1/2}Pb & (\nu=0),\\ \operatorname{erf}\{(b'PQPb)^{-1/2}\}Pb & (\nu=1).\end{cases} \tag{4.4}$$

The (approximate) covariance Q is determined by (3.5 a) with $B=b$ and $K=k'$:

$$(A-bk')Q+Q(A-bk')'+CC'=0. \tag{4.5}$$

Finally, P is determined by substitution of $u=k'x$ and $V=x'Px$ in (2.8 a); this yields:

$$(A-bk')'P+P(A-bk')+M+\nu kk'=0 \tag{4.6}$$

and

$$\lambda=\mathrm{tr}\,(C'PC). \tag{4.7}$$

The computation problem is to solve (4.4), (4.5) and (4.6) for k, Q, P. With these parameters determined, the resulting sub-optimal control is given by (4.3) and an estimate of the corresponding expected cost by (4.7).

It will be noted that the sub-optimal control (4.3) is a non-linear function of a single linear combination of the state variables. As such it is much simpler in form than the true optimal control, which would be obtained by exact solution of (2.6): this fact is inferred from the rather complicated structure of the true optimal control even in the deterministic case ($C=0$), as illustrated by examples in Johnson and Wonham (1965). Equations (4.4)–(4.6) determine an approximation to that control which is optimal in the sub-class $\hat{\Phi}$, of admissible controls Φ, defined by imposing the additional constraint:

$$\phi(x)=\hat{\phi}(f'x) \quad \text{for some vector } f.$$

The control found need not be optimal even relative to $\hat{\Phi}$, because of the error inherent in the approximation (4.2). Disregarding the latter error, however, it is plausible that restriction of the control to $\hat{\Phi}$ results in very little loss of performance in problems of the present type: this conjecture is borne out by the example of § 7 and the results of Fuller (1967).

To solve (4.4)–(4.6) we use a version of the method of successive approximation outlined at the end of § 2. Choose $k^{(0)}$ so that $A-bk^{(0)\prime}$ is stable and solve (4.5), (4.6) for Q, P. Then $k^{(1)}$ is determined from (4.4), and so on. For $r=0, 1, \ldots$:

$$k^{(r)}\rightarrow\begin{Bmatrix}(4.5)\\(4.6)\end{Bmatrix}\rightarrow\begin{Bmatrix}Q^{(r)}\\P^{(r)}\end{Bmatrix}\rightarrow(4.4)\rightarrow k^{(r+1)}. \tag{4.8}$$

Convergence of this scheme in a neighbourhood of the solution has been proved, on the assumption that the noise coefficient C is sufficiently small; convergence is geometrically fast. The proof, omitted here, is based on straightforward but tedious estimation of successive differences in (4.8). In applications convergence has been found satisfactory for suitable A (see Appendix) and for values of C in the range of practical interest; some details are given in § 6. As might be expected from § 2, convergence of the sequence $\lambda^{(r)}=\mathrm{tr}\,\{C'P^{(r)}C\}$ is in many cases nearly monotone.

A more basic question is whether a solution of (4.4)–(4.6) exists. It is shown in the Appendix that a solution does exist under broad conditions on the parameters (A, b, C, M). The question of uniqueness can be settled in specific cases (see the Remark at the end of Appendix).

It is straightforward to extend (4.4)–(4.6) to the case of vector-valued control. For simplicity we restrict attention to the case where N in (2.3) is positive definite. If $k_i'(i=1,\ldots,m)$ are the row vectors of K then, repeating (3.11 *b*):

$$k_i = \operatorname{erf}\{(f_i'Qf_i)^{-1/2}\}f_i. \tag{4.9}$$

From (2.8 *b*) and (4.2), f_i is the ith column vector of the $n \times m$ matrix:

$$F = PBN^{-1}. \tag{4.10}$$

The additional equations are:

$$(A-BK)Q + Q(A-BK)' + CC' = 0 \qquad \text{(3.5 } a\text{) bis}$$

and

$$(A-BK)'P + P(A-BK) + M + K'NK = 0. \tag{4.11}$$

Finally the computation algorithm is:

$$K^{(r)} \rightarrow \begin{Bmatrix} (3.5\,a) \\ (4.11) \end{Bmatrix} \rightarrow \begin{Bmatrix} Q^{(r)} \\ P^{(r)} \end{Bmatrix} \rightarrow \begin{Bmatrix} (4.9) \\ (4.10) \end{Bmatrix} \rightarrow K^{(r+1)} \tag{4.12}$$

$(r=0,1,\ldots)$.

Experience with one example (§ 5) suggests that (4.12) may be an effective scheme in practice.

5. Numerical results

Example 1 $(n=2)$

$$dx_1 = x_2 dt + c\,dw,$$
$$dx_2 = -u\,dt \quad |u| \le 1,$$
$$\mathscr{E}\{x_1^2\} = \min.$$

The system represents a second-order servo tracking a Brownian motion (x_1 = position error). The control (4.3) is of form:

$$u = \operatorname{sgn}(f^{(1)}x_1 + f^{(2)}x_2). \tag{5.1}$$

It is possible to calculate analytically the exact and estimated values of $\mathscr{E}\{x_1^2\}$, for both the sub-optimal control (4.3) and for the strictly optimal control in the class (5.1) (cf. Barrett 1960, Merklinger 1963). Write $u = \operatorname{sgn}(x_1 + \alpha x_2)$, $\alpha > 0$. The Fokker–Planck equation for the stationary probability density $p(x_1, x_2)$ is:

$$\frac{c^2}{2}\frac{\partial^2 p}{\partial x_1^2} - x_2 \frac{\partial p}{\partial x_1} + \frac{\partial}{\partial x_2}[\operatorname{sgn}(x_1 + \alpha x_2)p] = 0.$$

This yields:

$$p(x_1, x_2) = \text{const.} \exp\left[\frac{-2\alpha}{c^2}|x_1 + \alpha x_2| - \frac{\alpha}{c^2}x_2^2\right]$$

and

$$\mathscr{E}\{x_1^2\} = \int\int x_1^2 p(x_1, x_2)\,dx_1\,dx_2$$
$$= \frac{c^2}{2}\left(\frac{c^2}{\alpha^2} + \alpha\right). \tag{5.2}$$

From (5.2):

$$\alpha_{\text{opt}} = (2c^2)^{1/3} = 1{\cdot}26c^{2/3},$$
$$\mathscr{E}\{x_1^2\}_{\min} = 3(2^{-5/3})c^{8/3}$$
$$= 0{\cdot}94c^{8/3}. \tag{5.3}$$

For the sub-optimal control (4.3), there results $k^{(1)}=(4/\pi c^2)^{2/3}$, $k^{(2)}=+\sqrt{k^{(1)}}$, or $\varphi(x)=\operatorname{sgn}(x_1+\alpha^0 x_2)$, where

$$\alpha^0=(\pi c^2/4)^{1/3}=0{\cdot}92c^{2/3}. \tag{5.4}$$

From (5.2):

$$\mathscr{E}\{x_1^2\}_{\alpha=\alpha^0}=2^{-5/3}(4+\pi)\pi^{-2/3}c^{8/3}$$
$$=1{\cdot}05c^{8/3}. \tag{5.5}$$

Comparison of (5.3) and (5.5) shows that the sub-optimal control (4.3) yields m.s. error about 12% greater than the best achievable in the restricted class of controls (5.1).

The estimate (4.7), of m.s. error using control (4.3), turns out to be:

$$\mathscr{E}\{x_1^2\}_{\alpha=\alpha^0}\doteqdot(\pi/4)^{1/3}c^{8/3}$$
$$=0{\cdot}92c^{8/3},$$

or about 12% below the true value (5.5).

Example 2 ($n=2$)

$$dx_1=x_2dt+cdw,$$
$$dx_2=-udt, \qquad |u|\leq 1.$$
$$\mathscr{E}\{x_1^2+u^2\}=\min.$$

System dynamics are as in Ex. 1. Results are given in table 1. For comparison, data were obtained for a 'standard' control:

$$u^s=\operatorname{sat}(f^{*\prime}x)=\operatorname{sat}(x_1+\sqrt{2}x_2). \tag{5.6}$$

Here $u^*=f^{*\prime}x$ would be optimal if the constraint $|u|\leq 1$ were absent, and so (5.6) is very nearly optimal for small c.

The 'exact' solution of Bellman's equation for this example is outlined in § 7 and the results given in the last column of table 1. At moderate noise levels ($c\sim 2{\cdot}0$) the sub-optimal control (4.3) yields quadratic cost about 10% greater than the minimum achievable in the class of all feedback controls u with $|u|\leq 1$.

Table 1. Results for Ex. 2

Noise level	Sub-optimal control				Standard control		Optimal control
	Equations (4.3)–(4.7)			Exact	Est.	Exact	
c	f_1	f_2	λ	λ	λ	λ	λ
0·5	1·00	1·41	0·354	0·354	0·355	0·355	0·354
0·75	1·00	1·42	0·796	0·795	0·796	0·797	0·795
1·0	1·01	1·43	1·42	1·45	1·42	1·44	1·43
1·5	1·09	1·60	3·33	3·43	3·37	3·61	3·28
2·0	1·29	2·07	6·43	6·62	6·90	7·74	5·98
5·0	3·68	10·0	69·3	75·2	135·0	175·0	†

Est.: By statistical linearization.
Exact: By Fokker–Planck equation.

† Excessive running time required.

Example 3 ($n=8$)

$$\begin{aligned}
\dot{x}_1 &= x_2, \\
\dot{x}_2 &= -0{\cdot}7x_4 - 0{\cdot}004x_6 + c\dot{w}_1, \\
\dot{x}_3 &= x_4, \\
\dot{x}_4 &= 0{\cdot}7x_2 - 0{\cdot}004x_8 + c\dot{w}_2, \\
\dot{x}_5 &= x_6, \\
\dot{x}_6 &= -0{\cdot}001x_5 - 1{\cdot}001x_6 - u_1, \\
\dot{x}_7 &= x_8, \\
\dot{x}_8 &= -0{\cdot}001x_7 - 1{\cdot}001x_8 - u_2, \\
u_1 &= \mathrm{sat}\,(f_1'x), \quad u_2 = \mathrm{sat}\,(f_2'x), \\
&\mathscr{E}\{x_1^2 + x_3^2 + u_1^2 + u_2^2\} = \min.
\end{aligned}$$

This system is a two-axis description of a spinning space station with one reaction wheel torquer on each axis. The variables x_1, x_3 are respectively pitch and yaw angle; x_2 to x_8 are state variables of the torquers.

The 'standard' controls, chosen on the same basis as in Ex. 2, are:

$$\begin{aligned}
u_1^{\mathrm{S}} = \mathrm{sat}\,(&2{\cdot}71x_1 + 1{\cdot}46x_2 + 0{\cdot}503x_3 - 1{\cdot}24x_4 \\
&+ 0{\cdot}16x_5 - 0{\cdot}0075x_6 - 0{\cdot}048x_7 - 0{\cdot}0028x_8), \\
u_2^{\mathrm{S}} = \mathrm{sat}\,(&-0{\cdot}256x_1 + 0{\cdot}794x_2 + 2{\cdot}39x_3 + 1{\cdot}1x_4 \\
&+ 0{\cdot}0295x_5 + 0{\cdot}0035x_6 + 0{\cdot}001x_7 - 0{\cdot}0075x_8).
\end{aligned}$$

The results (table 2) indicate a reduction in quadratic cost (compared to the standard) of about 20% at moderate levels of noise ($c \sim 0{\cdot}2$).

Discussion

In all cases, performance estimates computed by statistical linearization are close (within 5%–10%) to the performance measured for the same control by analogue simulation. In general, estimation by linearization is accurate if the non-linear element is followed by two or more stages of low-pass filtering. Furthermore, the optimized control computed by the algorithm of § 4 is uniformly superior to the standard control. For controls of type $u = \mathrm{sat}\,(Fx)$, the performance improvement achieved is negligible if the noise level is low, since the system operates nearly always in the linear mode, but may become significant (10%–20%) if noise level is moderate to large, i.e.

$$\mathscr{E}\{|Fx|^2\} \gtrsim 1.$$

The tabulated results include variation of F_{opt} with noise level. This variation may be quite marked: in Exs. 1 and 2, note the counterclockwise rotation, in the phase plane, of the characteristic line:

$$f'x = 0,$$

as noise level increases. In general, the orientation of the characteristic hyperplanes

$$f_i'x = 0 \quad (i = 1, \ldots, m)$$

is much more sensitive to noise level than the corresponding percentage improvements of performance might lead one to expect.

Table 2. Results for Ex. 3

Noise level	Sub-optimal control										Standard	
	Equations (4.3)–(4.7)									Simul.	Est.	Simul.
c	f_{11} f_{12}	f_{21} f_{22}	f_{31} f_{32}	f_{41} f_{42}	f_{51} f_{52}	f_{61} f_{62}	f_{71} f_{72}	f_{81} f_{82}	λ	λ	λ	λ
0·01	2·71 −0·256	1·46 0·794	0·50 2·39	−1·24 1·1	0·016 0·0295	−0·0075 0·0035	−0·048 0·001	−0·0028 −0·0075	0·0022	0·0022	0·0022	0·0022
0·02	As above								0·0089	0·0089	0·0089	0·0089
0·05	As above								0·051	0·051	0·051	0·051
0·1	2·9 −0·29	1·8 0·841	0·55 2·58	−1·31 1·3	0·018 0·035	−0·008 0·004	−0·051 0·0015	−0·003 −0·008	0·21	0·22	0·23	0·24
0·2	3·4 −0·31	2·5 0·89	0·61 3·1	−1·35 1·95	0·021 0·039	−0·0085 0·0044	−0·054 0·0021	−0·0033 −0·0084	0·84	0·89	1·05	1·12
0·5	4·2 −0·33	3·5 0·95	0·68 3·9	−1·41 2·97	0·024 0·042	−0·009 0·0049	−0·06 0·0025	−0·0037 −0·0089	5·2	5·7	7·5	8·3
1·0	8·3 −0·7	7·1 1·8	1·2 7·7	−1·9 6·5	0·037 0·056	−0·011 0·006	−0·09 0·0035	−0·007 −0·013	21·2	23·1	30·9	33·7

6. Details of the digital optimization programme and analogue simulations

6.1. *Digital programme*

As seen from (4.12) the computational procedure requires solving two linear matrix equations at each stage of iteration. In this section we outline a technique, developed by Potter (1966) for matrix quadratic equations, which was found effective in solving the linear equations for systems of dynamic order up to 8†.

Let $\hat{A}$ be an $n \times n$ matrix with characteristic polynomial:

$$\sum_{i=0}^{n} \alpha_i \lambda^{n-i} \quad (\alpha_0 = 1).$$

The coefficients, α_i $(i=1,\ldots,n)$ are given by Faddeeva (1959):

$$\alpha_i = -i^{-1}\operatorname{tr}(\hat{A}D_i),$$

where

$$D_1 = I, \quad D_{i+1} = \hat{A}D_i + \alpha_i I \quad (i=1,\ldots,n-1).$$

Now consider (4.11), written as:

$$\hat{A}'P + P\hat{A} + \hat{M} = 0, \tag{6.1}$$

where $\hat{A} = A - BK$, $\hat{M} = M + K'NK$. In standard form, (6.1) becomes:

$$[P, I]H[I, -P]' = 0,$$

where

$$H = \begin{bmatrix} \hat{A} & 0 \\ \hat{M} & -\hat{A}' \end{bmatrix}$$

Having computed the coefficients α_i, form the partitioning:

$$\sum_{i=0}^{n} \alpha_i H^{n-i} = \begin{bmatrix} S & T \\ U & V \end{bmatrix},$$

where S, T, U, V are $n \times n$ blocks. Then with $\hat{A}$ stable, it can be shown that (Potter 1966)

$$P = V^{-1}U.$$

With this method only a single $n \times n$ matrix must be inverted. Equation (3.5 *a*) for Q is treated similarly.

To start the iteration (4.12) it is necessary to find a matrix $K^{(0)}$ such that $A - BK^{(0)}$ is stable (for the possibility of doing so, see Wonham (1967 b)). We shall assume that (A, B) is actually controllable and use a method of Bass (1961). Then $(A + \beta I, B)$ is controllable for every scalar β. Let $\beta > |A|$ ($|\cdot|$ denotes Euclidean norm).‡ Then $-(A + \beta I)$ is stable, and the equation

$$-(A + \beta I)R - R(A + BI)' + 2BB' = 0$$

has a unique positive definite solution R. Setting $K^{(0)} = B'R^{-1}$ it follows that $A - BK^{(0)}$ is stable.

---

† For $n = 8$, the number of linear equations at each stage of computation is $2 \times (n/2)(n+1) = 72$.

‡ The inequality holds if $\beta > \sqrt{n} \max_i \sum_{j=1}^{n} |a_{ij}|$.

In case BB' is large it may be useful to re-scale $K^{(0)}$ according to:

$$A - BK^{(0)} = A - (rB)(r^{-1}K^{(0)}),$$

where r is chosen so that $|rB| \simeq 1$.

An alternative and useful choice of $K^{(0)}$ is the solution for the standard linear regulator problem with $C=0$ and u not constrained. Although this $K^{(0)}$ requires more computation than the method just outlined, some advantage may be gained if system order is high, because an 'arbitrary' choice of stabilizing $K^{(0)}$ is likely to differ greatly from the final solution.

Computation was performed on a UNIVAC 1108; running times were of the order of 1 to 2 sec. On the average about ten iterations were required for convergence, with stopping rule:

$$|P^{(r)}B - P^{(r-1)}B| \lesssim 0{\cdot}1 |P^{(r)}B|.$$

Here the 10% factor is not unreasonable in view of the relative insensitivity of performance λ to $P^{(r)}B$, near the solution.

6.2. *Analogue simulations*

To eliminate random fluctuations in computation of stationary performance, a standard sample of 'white' noise was recorded with Gaussian amplitude distribution and spectral density flat up to 200 radians/sec. Run length was 100 sec, real time, or about 12 sec after scaling. An Ampex SP–300 tape recorder was used because of its good low-frequency characteristics.

7. Solution of Bellman's equation by finite differences

In this section we describe briefly a method for solving Bellman's eqn. (2.8) by finite differences. Again, the approach is heuristic. Suppose a control $\phi \in \Phi$ is selected and consider the process X_ϕ. If p_ϕ is the stationary distribution, there exists for every $\epsilon > 0$ a number ρ such that

$$\int_{|x| \le \rho} p_\phi(dx) > 1 - \epsilon.$$

By the ergodic property, the state $x_\phi(\cdot)$ spends at least the fraction of time $1-\epsilon$ in the cube:

$$\mathscr{D} = \{x : |x| \le \rho\}.$$

Consider the modified process $\tilde{X}_\phi$ in $\mathscr{D}$, determined by the same Itô equation as X_ϕ in the interior of $\mathscr{D}$, and by a reflecting barrier condition at the boundary:

$$\partial\mathscr{D} = \{x : |x| = \rho\}.$$

If $\tilde{p}_\phi$ is the stationary distribution of $\tilde{X}_\phi$ (which exists, since $\mathscr{D}$ is compact) it follows that averages

$$\int_{R^n} \tilde{p}_\phi(dx)L(x) = \int_{\mathscr{D}} \tilde{p}_\phi(dx)L(x)$$

are good approximations to averages

$$\int_{R^n} p_\phi(dx)L(x)$$

if ρ is taken suitably large. On this basis we confine attention to $\tilde{X}_\phi$.

Write $\tilde{p}_\phi(dx) = \tilde{p}_\phi(x)dx$ and assume the density $\tilde{p}_\phi(x)$ is smooth. In the interior of $\mathscr{D}$, $\tilde{p}_\phi$ satisfies (dropping subscript ϕ) the Fokker–Planck equation:

$$\tfrac{1}{2}\,\mathrm{tr}\left[\frac{\partial^2}{\partial x^2}\{CC'\tilde{p}(x)\}\right] - \frac{\partial'}{\partial x}[\{Ax - B\phi(x)\}\tilde{p}(x)] = 0, \tag{7.1}$$

where $\partial^2/\partial x^2$ is the matrix $[\partial^2/\partial x_i \partial x_j]$ and $\partial/\partial x$ the vector $[\partial/\partial x_i]$. Analytically the reflection condition is that the normal component of probability flow should vanish at the boundary; i.e.

$$n(x)'\left\{-\frac{\partial'}{\partial x}[CC'\tilde{p}(x)] + [Ax - B\phi(x)]\tilde{p}(x)\right\} = 0 \quad (x \in \partial\mathscr{D}), \tag{7.2}$$

where $n(x)$ denotes the outward normal at x. Noting as in § 2 that

$$\int_{\mathscr{D}} \tilde{p}(dx)\mathscr{L}\tilde{V}(x) = 0 \tag{7.3}$$

for arbitrary smooth functions $\tilde{V}$ we infer, from (7.1), (7.2) and the divergence theorem, that (7.3) holds provided

$$n(x)'CC'\tilde{V}_x(x) = 0 \quad (x \in \partial\mathscr{D}). \tag{7.4}$$

The adjoint condition (7.4) states that the derivative of $\tilde{V}$ must vanish in the co-normal direction at the boundary.

The control problem for the family $\tilde{X} = \{\tilde{x}_\phi : \phi \in \Phi\}$ is solved by using approximation in policy space as described in § 2, where at the rth stage the function $\tilde{V}^{(r)}(x)$ is determined from:

$$\mathscr{L}^{(r)}\tilde{V}^{(r)}(x) + L[x, \phi^{(r)}(x)] = \lambda^{(r)} \quad (|x| < \rho)$$

and the boundary condition (7.4). Let $\rho_0 > 0$ be given. If $\rho > \rho_0$ is chosen sufficiently large, the optimal control $\tilde{\phi}(x)$ for $\tilde{X}$ would be expected to represent closely the control $\phi^0(x)$ in the region $|x| < \rho_0$.

To illustrate the method in R^2, write $x_1 = x$, $x_2 = y$ and consider the problem (Example 2, § 5):

$$dx = y\,dt + c\,dw,$$

$$dy = -u\,dt, \quad |u| \leq 1.$$

$$\mathscr{E}\{x^2 + u^2\} = \min.$$

Equation (2.8) becomes

$$\tfrac{1}{2}c^2\frac{\partial^2 V}{\partial x^2} + y\frac{\partial V}{\partial x} - u\frac{\partial V}{\partial y} + x^2 + u^2 = \lambda \tag{7.5a}$$

$$u = \mathrm{sat}\left(\tfrac{1}{2}\frac{\partial V}{\partial y}\right)$$

in the region $|x|, |y| < \rho$; and (7.4) yields:

$$\partial V/\partial x = 0, \quad |x| = \rho, \quad |y| \leq \rho. \tag{7.5b}$$

We normalize $V(x, y)$ by setting:

$$V(0, 0) = 0. \tag{7.5c}$$

Equation (7.5) presents a quasi-linear elliptic boundary problem of Neumann type, in which both V and λ are to be determined. Using approximation in policy space, the problem was solved by finite differences, with $\rho \sim 4\sqrt{(\mathscr{E}\{|x|^2\})}$ and a uniform mesh of squares of side 0·07 unit. Running time was about 30 min on a UNIVAC 1108.

Values of λ for $c = 1, 1{\cdot}5, 2, 5$ are given in table 1; if $c < 1$ the effect of saturation is not significant. Further details of computation and results for the geometry of the optimal control law will be reported in a separate publication.

8. Optimal control with noisy measurements of state

It is easy to extend the method of § 4 to the case where observations of the system state are mixed with additive white noise. We merely install a Wiener–Kalman filter to obtain the conditional expectation $\hat{x}(t)$ of the dynamic state $x(t)$, given the observations, and then reformulate the dynamic programme in the state-space of values of $\hat{x}$. The problem is thus reducible to the one already considered.

To be precise, adjoin to (2.1 *a*) the equation of observation:

$$\dot{y}(t) = Hx(t) + D\dot{w}(t), \tag{8.1 a}$$

or in Itô's notation:

$$dy(t) = Hx(t)dt + Ddw(t). \tag{8.1 b}$$

Here H and D are constant matrices and $w(t)$ is the same Wiener process which appears in (2.1 *b*)†. Since the control $u(t)$ can be stored as it evolves (i.e. at time t we know $u(s)$ for $s \le t$) it is clear that the conditional distribution of $x(t)$, given the observations for $s \le t$, is Gaussian: let the mean be $\hat{x}(t)$ and covariance matrix be $R(t)$. Since we are interested only in stationary control we consider the limiting covariance:

$$R = \lim R(t) \quad (t \to \infty), \tag{8.2}$$

and this is given by (Kalman 1963):

$$AR + RA' + CC' - (CD' + RH')(DD')^{-1}(CD' + RH')' = 0. \tag{8.3}$$

The discussion so far is formal: but it can be shown (Wonham 1968 b) that the limit (8.2) exists and is the unique positive definite solution of (8.3) provided:

(i) DD' is non-singular,

(ii) the pair (H, A) is detectable (see Wonham 1968 b),

(iii) the pair $[A - CD'(DD')^{-1}H, S]$ is controllable, where S is any matrix such that

$$CC' - CD'(DD')^{-1}DC' \equiv SS'.$$

In particular the condition holds if dynamic noise and observation noise are independent ($CD' = 0$) and (A, C) is controllable.

With R determined by (8.3) it is known (Kalman 1963) that $\hat{x}(t)$ is given by:

$$d\hat{x} = (A - EH)\hat{x}dt - Budt + Edy, \tag{8.4}$$

where

$$E = (CD' + RH')(DD')^{-1}. \tag{8.5}$$

---

† If dynamic noise ($\dot{w}_1$, say) and observation noise ($\dot{w}_2$) are independent, write $w' = (w_1', w_2')$, $C = [C_1, 0]$, $D = [0, D_2]$.

Because $\hat{x}(t)$ is a sufficient statistic for optimal control (cf. the reasoning by Wonham (1968 a)) we assume that

$$u(t) = \phi[\hat{x}(t)]. \tag{8.6}$$

Under this condition it can be shown (cf. Wonham 1968 a) that $\{\hat{x}(t)\}$ is a diffusion process specified by the Itô equation:

$$d\hat{x} = A\hat{x}dt - B\phi(\hat{x})dt + (CD' + RH')(DD')^{-1/2}d\hat{w}. \tag{8.7}$$

In (8.6), $\hat{w}$ is a standard Wiener process formally given by:

$$d\hat{w} = (DD')^{-1/2}H(x - \hat{x})dt + (DD')^{-1/2}Ddw.$$

Now

$$\mathscr{E}\{x(t)'Mx(t)|y(s), u(s), -\infty < s \leq t\} = \hat{x}(t)'M\hat{x}(t) + \text{tr}\,(MR). \tag{8.8}$$

Since R is independent of the choice of control law ϕ, it follows from (8.8) that only the quadratic term is a candidate for optimization.

The foregoing discussion shows that the problem of minimizing

$$\mathscr{E}\{x'Mx + u'Nu\}$$

is formally identical with the problem of § 2 provided we replace (x, w) by $(\hat{x}, \hat{w})$, (2.1 *b*) by (8.7) and the matrix C by $(CD' + RH')(DD')^{-1/2}$. R is computed from (8.3) and the optimal control law is now $u = \phi^0(\hat{x})$, where $\hat{x}$ is determined by (8.4), i.e. by a Wiener–Kalman filter. The final estimate of cost will be, noting the additional term of (8.8):

$$\mathscr{E}\{x(t)'Mx(t) + u(t)'Nu(t)\} = \lambda + \text{tr}\,(MR).$$

9. Conclusions

The algorithm presented for sub-optimal controls is straightforward to implement by computer, and yields a design which improves performance over that obtained by design on a noise-free basis. The degree of improvement over that of a nominal standard design is usually small; but this relative insensitivity to disturbances is inherent in the class of systems considered, and should be characteristic of a good 'deterministic' design in general. Perhaps more important is the fact that the algorithm represents a rational and straightforward design procedure, of modest computational size, which can be used for problems of relatively high dynamic order.

As always with 'sub-optimal' design, it is desirable to estimate how far performance falls short of the optimal. Unfortunately, for the present class of problems a good estimate of optimal cost is not known; and it seems unlikely that such an estimate could be computed at a level of effort comparable with that of the sub-optimal algorithm. Our explicit comparisons (Exs. 1, 2) suggest that the sub-optimal quadratic cost typically lies within 10% of the optimal cost: but such a conclusion is necessarily tentative. The authors suspect that a significant further improvement in design could be obtained only by an order-of-magnitude jump in computational effort.

Further research is in progress to improve efficiency of computation in high-order cases, and to verify that the design obtained is stochastically stable in the appropriate sense.

ACKNOWLEDGMENT

Part of this research was conducted while the first author (W.M.W.) was pursuing a NRC Postdoctoral Resident Research Associateship supported by the National Aeronautics and Space Administration.

Appendix

Existence of solution of (4.4)–(4.6)

It will be shown that eqns. (4.4)–(4.6), in the case $\nu=1$, have a solution (P, Q, k) under the following conditions:

(i) The pair (A, b) is stabilizable (i.e. there exists a vector k such that $A-bk'$ is stable: see Wonham (1967 b)).

(ii) The pair $(\sqrt{M}, A)$ is observable.

(iii) The matrix CC' is positive definite.

(iv *a*) The matrix A is stable, or

(iv *b*) In a suitable coordinate system A has the block form:

$$A=\begin{bmatrix} A_1 & 0 \\ 0 & A_2 \end{bmatrix}$$

where A_2 is stable and A_1 is a 2×2 matrix with minimal polynomial $\alpha^2=0$. The physical system here represented is a pure inertia (transfer function $1/s^2$) coupled to auxiliary dynamics which are inherently stable.

The two conditions (iv) are those of greatest interest in current applications: the conclusion also holds if $A_1=0$ is a scalar. Condition (iii) can sometimes be relaxed to controllability of the pair (A, C): it is in fact enough to ensure that Q be positive definite.

Re-stating eqns. (4.4)–(4.6) we have:

$$(A-bk')'P+P(A-bk')+M+kk'=0, \tag{A 1}$$

$$(A-bk')Q+Q(A-bk')'+CC'=0, \tag{A 2}$$

$$k=\gamma Pb, \tag{A 3}$$

where

$$\gamma=\operatorname{erf}\{(b'PQPb)^{-1/2}\}. \tag{A 4}$$

Observe that $0<\gamma<1$. Substitution of (A 3) in (A 1) yields:

$$A'P+PA-\gamma(2-\gamma)Pbb'P+M=0. \tag{A 5}$$

Under conditions (i), (ii) it is well known (cf. Wonham 1968 b, Theorem 4.1) that (A 5) has for every $\gamma\in(0,2)$ a unique positive definite solution $P(\gamma)$ and that, for the corresponding value $k(\gamma)$, the matrix $A-bk(\gamma)'$ is stable. The latter fact together with condition (iii) implies that (A 2) has a unique solution

$Q(\gamma)$ which is positive definite. Since $P(\gamma)b \neq 0$, the right side of (A 4) is a number in (0, 1); thus (A 4) can be written:

$$\gamma = T(\gamma), \tag{A 6}$$

where

$$T(\gamma) = \mathrm{erf}\{(b'P(\gamma)Q(\gamma)P(\gamma)b)^{-1/2}\}$$

is a mapping of the interval (0, 1) into itself.

It is enough to show that T has a fixed point. Clearly T is continuous in (0, 1] and $T(1) < 1$. If A is stable, $P(0)$ and $Q(0)$ are defined and positive definite, T is defined and continuous at $\gamma = 0$, and $T(0) > 0$. This proves the result for case (iv *a*).

In case (iv *b*) it will be shown that

$$T(\gamma) \sim \text{const.} \quad \gamma^{7/8} \quad (\gamma \downarrow 0) \tag{A 7}$$

and the result will follow. We use the following notation:

$$M = \begin{bmatrix} M_1 & M_3' \\ M_3 & M_2 \end{bmatrix}, \quad M_1 = \begin{bmatrix} m_{11} & m_{13} \\ m_{13} & m_{12} \end{bmatrix}, \quad M_3 = [m_{31}, m_{32}] \tag{A 8}$$

where $\dim(M_1) = 2 \times 2$, $\dim(M_3) = (n-2) \times 2$, $\dim(M_2) = (n-2) \times (n-2)$, $\dim(m_{31}) = \dim(m_{32}) = (n-2) \times 1$. A similar decomposition with corresponding notation is made of the matrices P, Q, and of $CC' \equiv G$. We also write:

$$b = \begin{bmatrix} b_1 \\ b_2 \end{bmatrix}, \quad k = \begin{bmatrix} k_1 \\ k_2 \end{bmatrix}, \quad k_1 = (k_{11}, k_{12})' \tag{A 9}$$

where b_1 is a 2 vector and b_2 an $(n-2)$ vector, and similarly for k.

Since (A, b) is stabilizable, the pair (A_1, b_1) is controllable (see Wonham 1967 b), and it is no loss of generality to assume that

$$A_1 = \begin{bmatrix} 0 & 1 \\ 0 & 0 \end{bmatrix}, \quad b_1 = \begin{bmatrix} 0 \\ 1 \end{bmatrix}. \tag{A 10}$$

Furthermore, on writing $M_1 = \tilde{M}_1'\tilde{M}_1$ (where $\tilde{M}_1$ is of dimension, say, $\tilde{m}_1 \times 2$) and considering (ii), it is easily seen that $m_{11} > 0$.

Using these definitions to re-write (A 5) there results:

$$A_1'P_1 + P_1A_1 + M_1 - \gamma(2-\gamma)(P_1b_1 + P_3'b_2)(P_1b_1 + P_3'b_2)' = 0, \tag{A 11 a}$$

$$A_2'P_2 + P_2A_2 + M_2 - \gamma(2-\gamma)(P_3b_1 + P_2b_2)(P_3b_1 + P_2b_2)' = 0, \tag{A 11 b}$$

$$A_2'P_3 + P_3A_1 + M_3 - \gamma(2-\gamma)(P_3b_1 + P_2b_2)(P_1b_1 + P_3'b_2)' = 0. \tag{A 11 c}$$

Writing out (A 11 *a*) in detail gives:

$$\left.\begin{aligned} &m_{11} - \gamma(2-\gamma)(p_{13} + p_{31}'b_2)^2 = 0, \\ &2p_{13} + m_{12} - \gamma(2-\gamma)(p_{12} + p_{32}'b_2)^2 = 0, \\ &p_{11} + m_{13} - \gamma(2-\gamma)(p_{13} + p_{31}'b_2)(p_{12} + p_{32}'b_2) = 0. \end{aligned}\right\} \tag{A 12}$$

Assume temporarily that

$$P_3(\gamma) = 0(1). \tag{A 13}$$

Here and below, order relations hold in the limit $\gamma \downarrow 0$, and if V is a vector or matrix, $V = 0(\cdot)$ means $|V| = 0(\cdot)$. The solution of (A 12) is then:

$$\left.\begin{aligned} p_{11} &= 2^{1/4} m_{11}^{3/4} \gamma^{-1/4} + 0(1), \\ p_{12} &= 2^{-1/4} m_{11}^{1/4} \gamma^{-3/4} + 0(1), \\ p_{13} &= 2^{-1/2} m_{11}^{1/2} \gamma^{-1/2} + 0(1). \end{aligned}\right\} \tag{A 14}$$

Now (A 11 c) can be written:

$$\begin{aligned} A_2'[p_{31}, p_{32}] + [p_{31}, p_{32}]A_1 + [m_{31}, m_{32}] \\ -\gamma(2-\gamma)(p_{32} + P_2 b_2)\{[p_{13}, p_{12}] + b_2' P_3\} = 0. \end{aligned} \tag{A 15}$$

Assuming temporarily that

$$P_2(\gamma) = 0(1), \tag{A 16}$$

we obtain from (A 14), (A 15):

$$\begin{aligned} A_2' p_{31} + m_{31} + 0(\gamma^{1/2}) &= 0, \\ A_2' p_{32} + p_{31} + m_{32} + 0(\gamma^{1/4}) &= 0, \end{aligned}$$

with the solution:

$$\left.\begin{aligned} p_{31} &= -(A_2')^{-1} m_{31} + 0(\gamma^{1/2}), \\ p_{32} &= (A_2')^{-2} m_{31} - (A_2')^{-1} m_{32} + 0(\gamma^{1/4}). \end{aligned}\right\} \tag{A 17}$$

By (A 16) and (A 17), (A 11 b) yields:

$$A_2' P_2 + P_2 A_2 + M_2 + 0(\gamma) = 0,$$

whence

$$P_2(\gamma) = P_2(0) + 0(\gamma), \tag{A 18}$$

where $P_2(0)$ is positive definite. For brevity we shall write $P_2(0) = P_2$. Since the temporary assumptions (A 13) and (A 16) are consistent with the results (A 17), (A 18), the former relations are now justified. [More rigorously set $p_{11} = \gamma^{-1/4} \hat{p}_{11}$, etc., and observe that the equations determine the variables ($\wedge$) as analytic functions of $\gamma^{1/4}$ in a neighbourhood of 0.]

From (A 3), (A 9) and the expressions for the P_i there follows:

$$\left.\begin{aligned} k_1' &= [2^{-1/2} m_{11}^{1/2} \gamma^{1/2}, 2^{-1/4} m_{11}^{1/4} \gamma^{1/4}] + 0(\gamma), \\ k_2 &= \gamma(P_3(0) b_1 + P_2 b_2) + 0(\gamma^{5/4}). \end{aligned}\right\} \tag{A 19}$$

Equation (A 2) for Q is treated in similar fashion. One obtains:

$$(A_1 - b_1 k_1')Q_1 + Q_1(A_1 - b_1 k_1')' - b_1 k_2' Q_3 - Q_3' k_2 b_1' + G_1 = 0, \tag{A 20 a}$$

$$(A_2 - b_2 k_2')Q_2 + Q_2(A_2 - b_2 k_2')' - b_2 k_1' Q_3' - Q_3 k_1 b_2' + G_2 = 0, \tag{A 20 b}$$

$$Q_3(A_1 - b_1 k_1')' - Q_2 k_2 b_1' - b_2 k_1' Q_1 + (A_2 - b_2 k_2')Q_3 + G_3 = 0 \tag{A 20 c}$$

In detail, (A 20 a) becomes:

$$\left.\begin{aligned} 2q_{13} + g_{11} &= 0, \\ -2(k_{11} q_{13} + k_{12} q_{12}) + g_{12} - 2k_2' q_{32} &= 0, \\ q_{12} - k_{11} q_{11} - k_{12} q_{12} + g_{13} - k_2' q_{31} &= 0. \end{aligned}\right\} \tag{A 21}$$

Assume temporarily that

$$q_{31}=0(\gamma^{-1/4}),\ q_{32}=0(1). \tag{A 22}$$

Then

$$k_2'q_{31}=0(\gamma^{3/4}),\ k_2'q_{32}=0(\gamma).$$

Since $CC'>0$, there follows $G_1>0$, hence $g_{12}>0$.

Then (A 21) gives:

$$\left.\begin{aligned} q_{11}&=2^{-1/4}m_{11}{}^{-3/4}g_{12}\gamma^{-3/4}+0(\gamma^{1/4}),\\ q_{12}&=2^{-3/4}m_{11}{}^{-1/4}g_{12}\gamma^{-1/4}+0(\gamma^{1/4}),\\ q_{13}&=-\tfrac{1}{2}g_{11}.\end{aligned}\right\} \tag{A 23}$$

Writing out (A 20 c) in detail we get:

$$q_{32}-(k_{11}q_{11}+k_{12}q_{13})b_2+[A_2+0(\gamma)]q_{31}+g_{31}=0,$$

$$-k_{11}q_{31}-k_{12}q_{32}-Q_2k_2-(k_{11}q_{13}+k_{12}q_{12})b_2+[A_2+0(\gamma)]q_{32}+g_{32}=0.$$

By (A 19) and (A 23) this yields:

$$[A_2+0(\gamma)]q_{31}+q_{32}=2^{-3/4}m_{11}{}^{-1/4}g_{12}\gamma^{-1/4}b_2-g_{31}+0(\gamma^{1/4}),$$

$$[2^{-1/2}m_{11}{}^{1/2}\gamma^{1/2}+0(\gamma)]q_{31}+[A_2+0(\gamma^{1/4})]q_{32}$$

$$=-Q_2k_2-2^{-1}g_{12}+g_{32}+0(\gamma^{1/2}).$$

Assume temporarily that

$$Q_2=0(1). \tag{A 24}$$

Then

$$\left.\begin{aligned} q_{31}&=2^{-3/4}m_{11}{}^{-1/4}g_{12}\gamma^{-1/4}A_2{}^{-1}b_2+0(1),\\ q_{32}&=A_2{}^{-1}(g_{32}-2^{-1}g_{12})+0(\gamma^{1/4}).\end{aligned}\right\} \tag{A 25}$$

Thus:

$$Q_3k_1=k_{11}q_{31}+k_{12}q_{32}$$

$$=0(\gamma^{1/4}),$$

and (A 20 b) becomes:

$$A_2Q_2+Q_2A_2'+G_2+0(\gamma^{1/4})=0,$$

with solution:

$$Q_2=Q_2(0)+0(\gamma^{1/4}). \tag{A 26}$$

Clearly the assumptions (A 22) and (A 24) are consistent with the results (A 25) and (A 26).

For brevity write $Q_2(0)=Q_2$. Collecting results we find:

$$\begin{aligned} b'PQPb&=(P_1b_1+P_3'b_2)'Q_1(P_1b_1+P_3'b_2)\\ &\quad+2(P_3b_1+P_2b_2)'Q_3(P_1b_1+P_3'b_2)\\ &\quad+(P_3b_1+P_2b_2)'Q_2(P_3b_1+P_2b_2)\\ &=2^{-1/4}m_{11}{}^{1/4}g_{12}\gamma^{-7/4}+0(\gamma^{-3/2}).\end{aligned}$$

Thus:

$$T(\gamma)=\mathrm{erf}\{2^{1/8}m_{11}{}^{-1/8}g_{12}{}^{-1/2}\gamma^{7/8}+0(\gamma^{9/8})\},$$

and so, as $\gamma\downarrow 0$:

$$T(\gamma)=2^{5/8}\pi^{-1/2}m_{11}{}^{-1/8}g_{12}{}^{-1/2}\gamma^{7/8}+0(\gamma^{9/8}),$$

the required result.

Remark

Uniqueness of the solution γ (in the open interval $0<\gamma<1$) has not been proved, but would follow from the conjecture that $T(\gamma)$ is a concave function of γ for $\gamma \in [0, 1]$ (where $T(0)$ is defined by continuity). Of course, in any numeral instance the graph of $T(\gamma)$ could be computed and, for practical purposes, uniqueness verified by inspection. It is emphasized that this conjecture applies only to cases where A is stable, or has at most one unstable block component of type $[0]$ (1×1) or $\begin{bmatrix} 0 & 1 \\ 0 & 0 \end{bmatrix}$ (2×2). If A has one or more eigenvalues α with $\operatorname{Re}\alpha>0$, in general a solution need not exist for sufficiently large noise levels ($|C|$) or, if there is a solution for some C, it need not be unique. This fact is confirmed by inspection of first-order systems.

References

Barrett, J. F., 1960, *Proc. IFAC Automation and Remote Control* (London: Butterworths, 1961), **2,** 724.
Bass, R. W., 1961, Lecture Notes on Control Synthesis and Optimization, presented at NASA/LaRC, August.
Booton, R. C., Jr., 1954, *I.R.E. Trans. Circuit Theory,* **1,** 9.
Dynkin, E. B., 1965, *Markov Processes* (New York: Academic Press).
Faddeeva, V. N., 1959, *Computational Methods of Linear Algebra* (New York: Dover Publications).
Fuller, A. T., 1967, *Int. J. Control,* **5,** 197.
Holtzman, J. M., 1968, *SIAM Jl Control,* **6,** 235.
Howard, R. A., 1960, *Dynamic Programming and Markov Processes* (New York: Wiley).
Johnson, C. D., and Wonham, W. M., 1965, *J. bas. Engng,* March, 81.
Kalman, R. E., 1963, *Proc. First Symposium on Engineering Applications of Random Function Theory and Probability* (New York: Wiley), p. 270.
Kazakov, I. E., 1956, *Avtomatika Telemekh.,* **17**.
Kolovskii, M. Z., 1966, *Avtomatika Telemehk.,* (10), 43.
Merklinger, K. J., 1963, *Proc. Second IFAC Congress, Theory* (London: Butterworths, 1964), p. 81.
Potter, J. E., 1966, *SIAM Jl appl. Math.,* **14,** 496.
Smith, H. W., 1966, M.I.T. Res. Monograph No. 34 (Cambridge, Mass.: M.I.T. Press).
Wonham, W. M., 1966 a, Third IFAC Congress, London (to be published); 1966 b, *J. diff. Eqns,* **2,** 365; 1967 a, *SIAM Jl Control,* **5,** 486; 1967 b, *I.E.E.E. Trans. autom. Control,* **12,** 660; 1968 a, *SIAM Jl Control,* **6,** 312; 1968 b, *Ibid.,* **6,** 681.